See also the companion textbook *Semantics and the Syntax of Algebra* by the author.

Style in Technical Math

Afshin Azari-Vala

To Keon

Contents

III   Applications and Base-Ten Notations   197

5   Communication Needs of Applications   199

6   Working with Base-Ten Notations   203

Preface

This textbook and its companion, *Semantics and the Syntax of Algebra*, present a novel approach to the teaching of the fundamentals of mathematics that is coherent, accessible and immediately applicable.[1] The main objective is to bring about fluency[2] in the use of mathematical tools in working with everyday life problems as well as a wide range of problems in pure and applied sciences.

To make sure that the presentation is coherent, we have imposed an overarching structure that connects the many seemingly disparate concepts, tools and techniques in mathematics into a single narrative. The underlying theme in this narrative is that mathematical notation and algorithms emerge naturally as a result of our increasingly more complex problem-solving needs. We hope that the reader will find this view sufficiently stimulating to spark her or his interest in finding out more and thus keep reading.

To make the subject matter accessible, we have adopted a concrete to abstract approach in our coverage of the material. Each section presents a selection of word problems whose solutions require the use of similar techniques. We start with the familiar and wade into the unfamiliar in stages to ensure continuity.

To make the tools and techniques applicable, we promote algorithms that are meaningful and efficient. Where possible, alternative approaches to the art of problem-solving are presented and circumstances under which the use of one or some other alternative is in order are listed. Particular attention is given to algorithms that are used by those who exhibit fluency in the use of math.

This brings us to the notion of style. By *style* we mean the particular selection of tools that one makes when solving a given problem. Style mat-

---

[1] The novelty is in the use of semantics to teach the subject matter at the level of fundamentals. Such use of semantics has already been made in the teaching of other subject matters notably in teaching the fundamentals of logic in the publication *Leblanc, Hugues & Wisdom, William. Deductive Logic. 2nd Edition. Boston: Allyn and Bacon Inc., 1976. ISBN 0-205-05496-X.*

[2] By *fluency* we mean the ability to naturally select and seamlessly use tools that are meaningful and efficient and to do so with confidence.

ters. Good style ensures that what one does is meaningful, i.e., that the activity is in line with the way we naturally reason, and that the activity is efficient enough to keep the conversation going. It is this combination of meaningfulness and efficiency that makes a mathematical tool truly powerful.

The textbook caters to the needs of a wide variety of readers. It is recommended for adult learners who wish to strengthen their understanding of the fundamentals of math particularly those who intend to move on to the study of pure and applied sciences. It is written for parents who wish to have an understanding of the organization of the subject matter and its application to everyday life problems and problems in the sciences in order to help them teach the subject to their kids and to do so with confidence. It is suitable for high school students in their senior years and college and university students in their junior years. It provides a good reading for those active in the pure and applied sciences including scientists and engineers who wish to have a highly organized understanding of the subject matter at the level of fundamentals. And it is written for those active in the field of linguistics as it offers deep insight into the syntactic evolution of mathematical language to deal with our daily needs and the needs of the sciences.

The author has taught the fundamentals of mathematics to adult learners for over 25 years and, in addition to briefly pursuing studies in music, the sciences (physics and biology), and dentistry, holds bachelor's and master's degrees in aerospace engineering, with unfinished studies at the level of PhD. The author has taught courses in mathematics, physics, chemistry and computer programming at various institutions and colleges and currently teaches mathematics at George Brown College in Toronto, Canada. The present textbook and its companion, *Semantics and the Syntax of Algebra*, advance the point of view of the author on the manner in which mathematical knowledge at the level of fundamentals should be organized and taught.

The author wishes to thank his current and former students for the many engaging discussions and questions that helped immensely in shaping this textbook. Words of gratitude are also due to current and former Deans, Ian Wigglesworth and Georgia Quartaro, and Chairs, Alex Irwin, Gerry Conrad, Susan Toews, and Tony Priolo, for their support, as well as the many professors who used this textbook or earlier versions of it in their courses and provided the author with valuable feedback. This includes professors Marie Jaffe, John Waters, Sorina Zota, Michael Matisko, Bartek Roszak, Nader Afrand, Negica Popovic, Rebecca Pali, Mazdak Nik-Bakht, Steven Konvalinka, Jeff McManus, Natalie Drumonde and Elisa Romeo.

The author is indebted to Prof. Jeff McManus in particular for the few intriguing weekly discussions at the cafe across the street. It was reflection on your objection to the association made in the textbook between direct proportion and multiplication during these discussions that led the author to the formulation of the standard and conservation forms of the models for the many problem types[3] which greatly improved the structure of the presentation on alternative models for the various problem types.

An additional note of thanks to Dr. Bartek Roszak for your recommendation on the resequencing of the topics covered in the chapters on working with the formal and base-ten notations to bring the presentation in these chapters in line with the main theme of the textbooks.[4] This recommendation has not been incorporated in the textbook for two reasons: First, the recommendation was given at a time when most of the work on the textbook was complete and such resequencing of the topics was deemed to be quite time consuming. More important, there are some inherent problems with this approach[5] that one needs to resolve before it can be deployed. This recommendation will be taken into consideration in possible future editions of this work.

Special thanks are due to my lovely wife, Nooshin Mohtashami-Maali, for the love, patience and understanding over the past twenty years that I spent working on these pages as well as my son, Keon, to whom this work is dedicated, for livening up the red and the blue with green.

---

[3] The terminology is borrowed from aerodynamics.

[4] The current coverage in these two chapters presents the manner in which one works with (i.e., adds, subtracts, etc.) natural numbers, followed by the manner in which one works with whole numbers, followed by the manner in which one works with integers, etc. The recommendation by Dr. Roszak would resequence this presentation by discussing the manner in which one adds natural numbers, whole numbers, integers, etc., followed by the manner in which one subtracts natural numbers, whole numbers, integers, etc. Such a sequence falls in line with the main theme of the textbook that mathematical tools and algorithms evolve in response to the increasing level of complexity of the problems that we solve.

[5] As an example of such problems, addition of rational numbers requires a knowledge of multiplication of whole numbers, a skill which, in the modified sequence, would have to be slated for coverage later.

Introduction

In Part I of this textbook we focus on problem-solving needs that lead to the classification of the various types of numbers for use in dealing with everyday life problems and problems that arise in the study of pure and applied sciences. In Chapter 1 we use the concept of *closure* to show how the various types of numbers come into existence while Chapter 2 presents two notations, referred to in this textbook as the *formal notation* and *base-ten notations*, that were invented for use in conveying ideas in theoretical and applied contexts.

In Part II we put the spotlight on the formal notation and the manner in which its features meet the demands of theoretical discourse while Part III presents a similar treatment of base-ten notations and their suitability in working with applied problems. In each part we discuss practical needs that justify the use of the relevant notation (Chapter 3 for the formal notation and Chapter 5 for base-ten notations) and move on to a detailed discussion of the manner in which one works with the numbers written in that notation (Chapter 4 for the formal notation and Chapter 6 for base-ten notations). Various techniques for working with the two notations are discussed and their strengths and weaknesses are assessed. Throughout we promote algorithms that are meaningful and efficient, a combination of features that help one become fluent in communicating theoretical and applied ideas.

In Part IV we shift our focus to the study of measurement systems. Chapter 7 introduces the key concepts of *quantity* and its *value* while Chapter 8 is devoted entirely to the study of the *International System of Units, SI*: the default measurement system used in the sciences. Chapter 9 presents three alternative schemes for working with measured and exact values.

In Part V we turn our attention to the study of algebraic systems. Chapter 10 provides an overview of equations, their structure, genesis, analysis and schemes used to solve them with emphasis on algorithms that are based on natural semantics. The treatment covers solution techniques for solving select equations in one unknown (Chapter 11), tools for establishing relationships between quantities in equations in two unknowns (Chapter 12) and an extension of these tools for relating quantities in and working with formulas (Chapter 13).

Appendices are used to discuss related material that would have taken too much space as footnote. Some expand on the material covered in the main body of the text or present alternative algorithms for working with mathematical structures while others provide proofs whose inclusion in the main body of the text would have interfered with the flow of the material.

Numerous problems are included in the exercises at the end of each section and subsection. The level of the problems ranges from simple to challenging. Answers and detailed solutions to the problems in the exercise sets are given in the companion solution manual.[6]

Ideally the textbook should be covered in tandem with the companion textbook *Semantics and the Syntax of Algebra* by the author in the same course with equal times allocated to the coverage of each textbook. If both texts are used simultaneously, high school curricula and upgrading programs may opt to cover Chapters 1 to 10 in one term and the rest in another while the entire textbook may be covered in a single term at college and university level programs. If the companion text is not covered, the times recommended above may be cut in half.

---

[6]In order to bring about fluency in solving mathematical problems, one would have to have access to detailed solutions to the problems that are posed in the exercises in the textbook and not just the final answer. For this reason, we have not provided the answers to the problems in the exercises at the end of the textbook and since appending full solutions to the problems would require the addition of many pages, it was decided to place the solutions in a separate document.

Part I
Number Systems

In this part of the book we introduce the scheme used to formally classify numbers and present alternative notations for the representation of these numbers. In introducing the formal classification scheme we point to the connection between the problem-solving need for closure and the introduction of new numbers as well as the connection between increasing levels of complexity of the problems that we solve and the order in which the new numbers are introduced into the system. Following this, we introduce two classes of notations that are commonly used to represent numbers with: The formal notation, which is based directly on the formal classification scheme, and base-ten notations. We discuss the strengths and weaknesses of both notations and present circumstances under which the use of one or the other notation is deemed appropriate.

Chapter 1
The Formal Classification Scheme

The **formal classification scheme** is the default scheme used to classify numbers for use in math and the sciences. The formal classification scheme begins with the simplest of numbers, i.e., the numbers 1, 2, 3, 4, and so on, and continually introduces new numbers into the system to meet the demands of our problem-solving needs as we move from the simplest of the problems to increasingly more complex problems. The relation between the order in which the new numbers are introduced into the system and increasing level of complexity of the types of problems that we solve is important as it provides an underlying theme for the classification of numbers in the formal classification scheme. An understanding of this theme helps us see that the formal classification scheme is not arbitrary, but coherent, logical and necessary. It also enables us to invent notation to formally represent the new numbers as they are introduced into the system.

1.1 Natural Numbers

The formal classification scheme begins with the set of **natural numbers**, also called **counting numbers**, i.e., the numbers 1, 2, 3, 4, and so on. Presumably, the terminology relates to the fact that we arrive at these numbers through the *natural* process of counting and that we use these numbers to *count* with. We use the symbol $\mathbb{N}$ to represent the set of natural numbers[1]

---

[1] In some texts the letter N is used to represent the set of natural numbers.

and write[2]

$$\mathbb{N} = \{\, 1, 2, 3, 4, \ldots \,\}$$

The primary use of natural numbers is in the counting and ordering of entities.

Numbers are, of course, operated on: When we solve problems, we add numbers, subtract them, multiply them and divide them. We raise them to an exponent, take their roots and so on. We would, of course, want to know whether the application of a given operation to our existing numbers would always generate another existing number. If so, we say that our existing set of numbers is **closed** under that operation.

Knowledge of closure of our existing set of numbers under a given operation addresses a practical concern: It tells us that, when solving problems that require that we apply that operation to our existing numbers, the result will be a number and can, therefore, be written down, interpreted and worked with.

Closure Under Addition

The set of natural numbers is closed under the operation of addition. This means that the sum of two natural numbers is a natural number.

The reason it is important to know that $\mathbb{N}$ is closed under addition is that it assures us that, when solving a problem that involves natural numbers, if we need to add these numbers, the result will be a number[3] and can, therefore, be written down, interpreted and worked with.

Closure Under Subtraction

We started with the set of natural numbers and tested the set for closure under addition. The reason we chose addition as the first operation to test closure against is that the operation is associated with the simplest type of

---

[2]In what follows the curly brackets signify that we are talking about a collection or a set. The **elements** of the set are listed within the curly brackets with the word **list** referring to a comma separated sequence of the names of the elements in the set or symbols that represent the elements in the set. The ellipsis, i.e., the '...', at the end of the list should be read as *and so on* indicating that the pattern in the list will continue without end. Putting all this together, we read the statement above as *the set of natural numbers is the set that contains the numbers 1, 2, 3, 4, and so on.*

[3]A natural number of course as these are the only numbers that we have formally recognized as such so far.

problems that we, as humans, solve.[4]

Having tested $\mathbb{N}$ for closure under addition, we move on to test the set for closure under subtraction, the operation associated with the next simplest type of problems that we, as humans, solve.[5]

The set of natural numbers is *not* closed under the operation of subtraction. While the subtraction of some natural numbers results in a value that is a natural number (e.g., $7 - 4$), some subtractions result in values that are not. Notable among the latter is subtraction of equal quantities (e.g., $6 - 6$). There are, of course, problems for which the subtraction $6 - 6$ makes sense: If one has \$6 and owes \$6, what can one say about one's net worth? There are no symbols in the set of natural numbers that can be used to represent the results of such subtractions.[6]

1.2   Whole Numbers

The question here is whether it is reasonable to call the result of the subtraction of equal values a *number*. If so, we can come up with a name (zero) and a symbol (0) for it. And when we come to consider what counts as a number, we note that it is such properties of theirs as the fact that they can be ordered, added, subtracted, etc. But a bit of reflection shows that the result of subtraction of equal quantities can be placed in order relative to the numbers in $\mathbb{N}$, that it can be added to the numbers in $\mathbb{N}$ and itself in meaningful ways,[7] that it can be subtracted from the numbers in $\mathbb{N}$ and itself in meaningful ways,[8] and so on. We will not get into the technicalities of what counts as a number in this textbook but to mention that such considerations show that there is no reason why we cannot consider the result of subtraction of equal quantities to be a number. In fact, regardless of whether one considers such a result to be a number or not, it is a concept that can be used along with the numbers in $\mathbb{N}$ to communicate the solution of certain practical problems.

The problem of lack of closure of the set of natural numbers, $\mathbb{N}$, with

---

[4]These are **direct superposition** problems which are problems in which the last step in the solution process requires that we add the values of certain quantities to arrive at the result. For more on direct superposition problems please see the companion textbook *Semantics and the Syntax of Algebra* by the author.

[5]These are **inverse superposition** problems which are problems in which the last step in the solution process requires that we subtract the values of certain quantities to arrive at the result. For more on inverse superposition problems please see the companion textbook *Semantics and the Syntax of Algebra* by the author.

[6]0 has not yet been recognized as a number.

[7]As in *I have \$4 in my left pocket and nothing in my right pocket. How much money do I have in total?*

[8]As in *I had \$7 and lost none of it. How much money do I have left?*

respect to subtraction is partially addressed by an extension of the set to include 0 and we refer to the resulting set as the set of **whole numbers**.[9] We use the symbol $\mathbb{W}$ to represent the set of whole numbers[10] and write

$$\mathbb{W} = \{\, 0, 1, 2, 3, \ldots \,\}$$

Note that every natural number is a whole number as well so that the set of whole numbers represents an extension of the set of natural numbers.

Closure with respect to addition is preserved in going from the set of natural numbers to the set of whole numbers. However, the set of whole numbers is *not* closed under subtraction. While the set of whole numbers is able to represent the result of subtraction of equal quantities, it is incapable of representing the results of other subtractions such as the subtraction $4 - 7$. And as with the case of subtraction of equal quantities, there are problems for which the subtraction $4 - 7$ makes sense: If one has \$4 and owes \$7, what can one say about one's net worth? There are no symbols in the set of whole numbers that can be used to represent the results of such subtractions.

1.3   Integers

Once again we are faced with the question as to whether it is reasonable to recognize the results of such subtractions as $4 - 7$ as numbers. And once again, when we consider what counts as a number, we note that it is such properties of theirs as the fact that they can be ordered, added, subtracted, etc. But a bit of consideration shows that the results of such subtractions as $4 - 7$ can be placed in order relative to our existing numbers and each other, that they can be added to our existing numbers and each other in meaningful ways,[11] that they can be subtracted from our existing numbers and each other in meaningful ways,[12] and so on. And if these entities behave like numbers in such ways, we can refer to them as numbers by analogy.

The set of **integers** extends the set of whole numbers by adding the negatives of the natural numbers to the set of whole numbers. The terminology arises from the fact that the entities in the set of integers are integral in

---

[9]One might be tempted to object to the claim that 0 is new by pointing out that the symbol is already used in the representation of some natural numbers (e.g., 10, 204, etc.). However, note that the use of 0 as a place-holder for use in the place-value scheme is different from its use as a number in its own right. If, as an example, we used the symbol $\sqcap$ to represent *ten* with, then we could write $4 + 6 = \sqcap$ without any need to use the symbol 0.

[10]In some texts the letter W is used to represent the set of whole numbers.

[11]As in *I owe \$7 and I have \$3. What is my net worth?*

[12]As in *The temperature was 2 °C below zero and it dropped by 5 °C. What is the temperature now?*

nature. We use the symbol $\mathbb{I}$ to represent the set of integers[13] and write

$$\mathbb{I} = \{ \ldots, -3, -2, -1, 0, 1, 2, 3, \ldots \}$$

Note that every whole number is an integer so that the set of integers represents an extension of the set of whole numbers.

The set of integers is closed under both addition and subtraction: The sum of any two integers is an integer and the difference between any two integers is an integer. This knowledge assures us that, when solving a problem that involves integers, if we need to add or subtract these numbers, the result will be a number[14] and can, therefore, be written down, interpreted and worked with.

Closure Under Multiplication

We now move on to test the next operation in the hierarchy against closure. This is multiplication, corresponding to the next simplest type of problems beyond superposition that we, as humans, solve.[15]

The set of integers is closed under the operation of multiplication.[16] This knowledge assures us that, when solving a problem that involves integers, if we need to multiply these numbers, the result will be a number[17] and can, therefore, be written down, interpreted and worked with.

Closure Under Division

The next operation in the hierarchy to test against closure is division, corresponding to the next simplest type of problems that we, as humans, solve.[18]

---

[13] In some texts the letter I is used to represent the set of integers.

It should also be noted that the alternative symbol $\mathbb{Z}$ or the corresponding letter Z which derive from the German word for *integer* are most commonly used in mathematics textbooks. In this textbook we have adopted the use of the less common but more natural choice, $\mathbb{I}$, to refer to the set of integers.

[14] An integer of course as these are the only numbers that we have formally recognized as such so far.

[15] These are **direct proportion** problems which are problems in which the last step in the solution process requires that we multiply the values of certain quantities to arrive at the result. For more on direct proportion problems please see the companion textbook *Semantics and the Syntax of Algebra* by the author.

[16] The set of natural numbers and the set of whole numbers are also closed under multiplication.

[17] An integer of course as these are the only numbers that we have formally recognized as such so far.

[18] These are **inverse proportion** problems which are problems in which the last step in the solution process requires that we divide the values of certain quantities to arrive at

The set of integers is *not* closed under the operation of division.[19] While the division of certain integers results in a value that is an integer (e.g., $-8 \div 2$, $12 \div 4$, etc.), division of certain other integers does not (e.g., $-7 \div 2$, $3 \div 8$, etc.). There are, of course, problems for which divisions such as $3 \div 8$ make sense: If 3 pizzas should be equally divided among 8 people, how much pizza should each person get? There are no symbols in the set of integers that can be used to represent the results of such divisions.

1.4   Rational Numbers

We are back to the question as to whether it is reasonable to refer to the results of such divisions as $3 \div 8$ as numbers. We remind ourselves that the reason numbers are numbers is that they can be ordered, added, subtracted, etc. But a little consideration shows that the results of such divisions as $3 \div 8$ can be ordered in relation to our existing numbers and each other, that they can be added to our existing numbers and each other in meaningful ways,[20] that they can be subtracted from our existing numbers and each other in meaningful ways,[21] and so on. Since the results of such divisions behave like numbers, we refer to them as numbers by analogy and add them to the set of numbers that we have built so far.

To represent the results of such divisions we use the fraction notation. As an example, we can write $\frac{3}{8}$ to refer to the result of the division of 3 things into 8 equal parts. Such notation can be used to represent the result of the division of all integers except division by 0 which is inherently problematic. The issues surrounding division by 0 are discussed at length in Appendix A and give rise to two special cases.[22] These are the undefined case

$$\frac{n}{0}, \qquad n \neq 0$$

and the indeterminate case

$$\frac{0}{0}$$

both of which are excluded from our formulation of the extended set that can represent the results of division of integers. The extended set is called the set

---

the result. For more on inverse proportion problems please see the companion textbook *Semantics and the Syntax of Algebra* by the author.

[19]The set of natural numbers and the set of whole numbers also fail to exhibit closure under division.

[20]As in *There are 2 pizzas here and $\frac{2}{3}$ of a pizza there. How much pizza do we have in total?*

[21]As in *I bought 2 pizzas and ate $\frac{2}{3}$ of a pizza on the way home. How much pizza do I have left?*

[22]We advise the reader to read this appendix before continuing further.

of **rational numbers** in reference to the word *ratio*.[23] For us, then, the set of rational numbers is the set whose members have the form of a fraction $\frac{p}{q}$, where p and q are integers and q is not 0.[24] Using the symbol $\mathbb{Q}$ to refer to the set of rational numbers[25] we can use mathematical notation to write the statement above as[26]

$$\mathbb{Q} = \left\{ \frac{p}{q} \mid p \wedge q \in \mathbb{I} \wedge q \neq 0 \right\}$$

Note that every integer is a rational number[27] so that the set of rational numbers represents an extension of the set of integers.

The set of rational numbers is closed under addition, subtraction, multiplication and, except for division by 0, under division. As such, the set of rational numbers has the notation necessary to represent the results of performing the basic arithmetic operations of addition, subtraction, multiplication and division on the numbers that we have described so far with the exception of division by 0.

Closure Under Other Operations

When we come to consider mathematical operations that are more complex than addition, subtraction, multiplication and division, or when we arrive at certain values following special investigations, we run, once again, into

---

[23] A **ratio** is a comparison of the values of two similar quantities by division.

[24] The requirement that p and q must be integers is somewhat obvious in light of the fact that integers are the only numbers that we have recognized as such so far and, therefore, they are the only numbers that can be divided to generate the new numbers. The requirement that q should not be 0 excludes the two special cases that arise when one attempts to divide an integer by 0.

[25] The more natural choice, $\mathbb{R}$, is used to refer to the set of real numbers which will be introduced shortly. The symbol $\mathbb{Q}$ is taken from the letter Q which stands for the word *quotient* which in turn refers to the result of division.

In some texts the letter Q is used to represent the set of rational numbers.

[26] In the statement that follows, the curly brackets indicate that we are talking about a set. The expression $\frac{p}{q}$ says that the elements of the set have the form of a fraction. The symbol | is read as *such that* and indicates that there will be limitations on what p and q can be. The symbol $\wedge$ stands for *and* and the symbol $\in$ stands for *is an element of.* The expression $p \wedge q \in \mathbb{I}$, therefore, says that p and q are elements of $\mathbb{I}$, i.e., p and q are *integers.* Finally, we impose the additional condition that the denominator, q, cannot be 0, excluding expressions that are undefined or indeterminate. Putting all this together, we read the statement above as *the set of rational numbers is the set whose elements have the form of a fraction, $\frac{p}{q}$, such that the numerator and denominator of the fraction are integers and the denominator is not 0.*

[27] An integer can be placed over 1 to take the form of a rational number. As an example, 4 can be written as $\frac{4}{1}$ (read *four ones*), 0 can be written as $\frac{0}{1}$ and -6 can be written as $\frac{-6}{1}$.

notational shortcomings. As an example, the set of rational numbers is *not* closed under the operation of exponentiation: While some rational numbers raised to an exponent that is a rational number evaluate to rational numbers (e.g., $\left(\frac{1}{4}\right)^2$ is $\frac{1}{16}$), others (e.g., $2^{\frac{1}{2}}$, which is usually written as $\sqrt{2}$) cannot be expressed as rational numbers.[28] There are, of course, problems for which such operations as taking the square root of 2 make sense: As shown in Appendix B, if one has a right-angle triangle with the short sides each 1 unit long, the length of the hypotenuse can be computed by finding the square root of 2, i.e., $\sqrt{2}$, which, as we just noted, is not expressible as a rational number (the largest set of numbers that we have constructed so far). But it is not unreasonable at all to expect that we should be able to associate *some* number to that length.[29]

1.5   Real Numbers

One last time we face the question as to whether it is reasonable to refer to the results of these more advanced operations and special investigations as numbers. One last time we remind ourselves that the reason numbers are numbers is that they can be ordered, added, subtracted, etc. However, it can be shown that the results of these more advanced operations and special investigations can be ordered in relation to our existing numbers and each other, that they can be added to our existing numbers and each other in meaningful ways, that they can be subtracted from our existing numbers and each other in meaningful ways, and so on. We, therefore, refer to them as numbers by analogy and add them to our existing set of numbers.

As for notation, to represent those new numbers that arise from the application of a more advanced operation, we use the notation used to represent the operation itself in their expression as in $\sqrt{2}$ where the radical symbol is used to represent the square root of 2. For those numbers that arise from special investigations, we invent new symbols, e.g., π.

The set of **real numbers** extends the set of rational numbers by adding the numbers that arise as a result of the application of the more advanced

---

[28] For a proof that $\sqrt{2}$ cannot be represented as a rational number please see Appendix B.

[29] In addition to certain roots such as $\sqrt{2}$, there are values arising from studies involving other operations as well as special investigations in other areas of mathematics that are shown not to be expressible as rationals. Examples of the former are $\log_2 3$ in the study of logarithms, $\sin 1°$ in the study of trigonometry, and the like, and examples of the latter are the number pi, π, arising from the study of the circle in geometry, Euler's Number, e, arising from the study of derivatives in calculus, and the like.

operations[30] and special investigations.[31] We use the symbol $\mathbb{R}$ to represent the set of real numbers.[32]

Note that every rational number is also a real number so that the set of real numbers represents an extension of the set of rational numbers.

The set of real numbers is closed under addition, subtraction, multiplication, division[33] and other operations[34]. By including *all* other operations[35] and special investigations, the set of real numbers already includes any numbers that might arise as a result of the introduction of new operations and special investigations.[36]

In summary, the formal classification scheme begins with the set of natural numbers, $\mathbb{N}$, and extends this set in stages to impose as much closure as possible under the many operations that we use when we solve problems. As for the order in which closure is imposed under the many operations, the order of difficulty of the types of problems that we solve is used as a guide. This order begins with direct superposition problems, followed by inverse superposition problems, followed by direct proportion problems, followed by inverse proportion problems, followed by all other types of problems beyond direct superposition, inverse superposition, direct proportion, and inverse proportion with these higher order problem types lumped together in one group. The operations associated with this order are addition which is used in the last step in the solution of direct superposition problems, subtraction which

---

[30] With the exception of an even root of a negative number which provides yet another extension from the set of real numbers, $\mathbb{R}$, to the set of complex numbers, $\mathbb{C}$.

[31] The description of real numbers given here is sufficient for our purposes but the reader should note that it is not rigorous. A rigorous definition of the reals is too advanced for an introductory textbook on applied math and will not be covered in this textbook. The interested reader is referred to such topics as Dedekind Cuts, Cauchy Sequences, and others for rigorous definitions.

[32] In some texts the letter R is used to refer to the set of real numbers.

[33] Except for division by 0.

[34] Except for an even root of a negative number. In addition, depending on the operation, there may be special exclusions as in the exclusion of division by 0 in the case of closure of the set of rational numbers with respect to division.

[35] We note once again that this does not include an even root of a negative number.

[36] These numbers are referred to as **irrational numbers** with the word *irrational* literally meaning *not rational*. The symbol $\overline{\mathbb{Q}}$ is used to represent the set of irrational numbers. The line over the symbol Q is read as *not* so that $\overline{\mathbb{Q}}$ is read as *not rational*.

In some texts the symbol $\overline{\text{Q}}$ is used to refer to the set of irrational numbers.

Much has been said about the wonder of irrational numbers. The irrationals are the same as the set $\mathbb{R} - \mathbb{Q}$, i.e., the reals minus the rationals. In what way is this so special? We never speak of the wonder of the set of *non-integers* which we could define as the set $\overline{\mathbb{I}}$ or $\mathbb{R} - \mathbb{I}$ (or $\mathbb{Q} - \mathbb{I}$, if you like) or the wonder of the set of *non-wholes* which we could define as the set $\overline{\mathbb{W}}$ or $\mathbb{R} - \mathbb{W}$ (or $\mathbb{I} - \mathbb{W}$) or the wonder of the set of *non-naturals* which we could define as the set $\overline{\mathbb{N}}$ or $\mathbb{R} - \mathbb{N}$ (or $\mathbb{W} - \mathbb{N}$). Why is it that the set of irrational numbers is given such prominence but not the set of non-integers or non-wholes or non-naturals as defined above? If there is any wonder, it is in the recognition of 0 as a number from which came our first extension.

is used in the last step in the solution of inverse superposition problems, multiplication which is used in the last step in the solution of direct proportion problems, division which is used in the last step in the solution of inverse proportion problems, and all other operations (exponentiation, taking roots, logarithms, etc.) which are used in the last step of higher order types of problems with these higher order operations lumped together in one group. Starting with $\mathbb{N}$ and imposing closure operation by operation following the order above, we arrived at the sets $\mathbb{W}$, $\mathbb{I}$, $\mathbb{Q}$, and $\mathbb{R}$. Each set extends the one prior to it in the sense that the new set includes the numbers in the set prior to it. This relationship is illustrated in Figure 1.5.1.

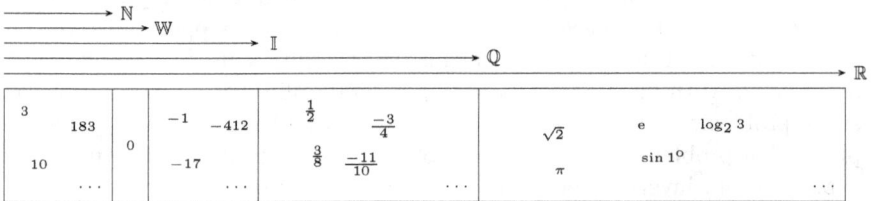

Figure 1.5.1: Relationship between the sets $\mathbb{N}$, $\mathbb{W}$, $\mathbb{I}$, $\mathbb{Q}$, and $\mathbb{R}$

1.6   Further Extensions

The set of real numbers has been extended to include numbers that represent an even root of a negative number. The largest such set is the set of **complex numbers**. A different extension of the set of real numbers includes notions for dealing with the arithmetic of the infinitely large and the infinitely small. The largest such set is the set of **surreal numbers**. Such sets are needed in more advanced applications of mathematics and will not be discussed further in this text.

1.7   Meta Notes

Notwithstanding the practical needs that justify the introduction of such entities as 0, -6, $\frac{3}{8}$, $\sqrt{2}$ and π into mathematics, some may still feel uneasy with the classification of them as *numbers*. After all, when we use the word *number*, are we not simply referring to $1, 2, 3, 4, \ldots$?

Obviously, if by the word *number* one means 1, 2, 3, 4, ..., then there is no doubt that such entities as 0, −6, $\frac{3}{8}$, $\sqrt{2}$, π and the like are *not* numbers. However, as we have shown, it is difficult and somewhat foolish to discard these entities if it is possible for us to use them in mathematics to help us solve such practical problems as *I have \$2 and I owe \$8. What is my net worth?* or the problem *Divide 3 pizzas equally among 8 people. How much pizza should each person get?*. Even more important, such entities as 0, −6, $\frac{3}{8}$, $\sqrt{2}$, π, etc. behave very much like the numbers 1, 2, 3, 4, ..., in that we can do to them what we do to 1, 2, 3, 4, ..., e.g., order them, perform mathematical operations on them, and the like. As such, they enter mathematics and can be mixed with the numbers 1, 2, 3, 4, ... in our calculations. It is, therefore, not unreasonable at all for us to refer to such entities as numbers *by analogy*.

Generalization of the meaning of established terminology used to refer to common entities, to represent new entities that behave in a manner that is similar to the behaviour of these common entities is quite prevalent in mathematics: We speak of *linear* equations in contexts that have nothing to do with geometry or we speak of 6-*dimensional space* knowing very well that physical space has 3 dimensions. All such references are made by analogy: In the case of *linear equations* we use the adverb *linear* in analogy to the behaviour of lines in the plane and in the case of 6-*dimensional space* we use the word *space* because, although the problem at hand involves 6 independent unknowns, it behaves in a manner that is similar to those problems in physical space that have 3 independent unknowns of type length.

To be sure, in ordinary discourse the word *number* is used to refer to the natural numbers, i.e., 1, 2, 3, 4, ... However, in mathematics and in the sciences, we often find that we make conclusions that hold true for, not just natural numbers, but other numbers as well. In these disciplines, therefore, the word *number* is used in its more general sense.

Exercise Set 1

1. In each case list all the sets, $\mathbb{N}$, $\mathbb{W}$, $\mathbb{I}$, $\mathbb{Q}$, $\mathbb{R}$ that the given number belongs to or specify whether the given expression is undefined or indeterminate.

a. 2400 f. 0 k. $\frac{-8}{-3}$ p. 304 u. $\frac{15}{5}$

b. −36 g. $\frac{72}{0}$ l. π q. −29 v. $\frac{4}{-5}$

c. $\frac{0}{0}$ h. 1 m. $\frac{12}{3}$ r. $\frac{\sqrt{5}}{0}$ w. $-\sqrt{16}$

d. $\frac{3}{4}$ i. −2000 n. $\sqrt{9}$ s. $\frac{-1}{-6}$ x. −1

e. $\sqrt{5}$ j. $\frac{0}{9}$ o. $\frac{2}{3}$ t. $\sqrt{2}$ y. $\sqrt{0}$

2. Each symbol below represents a set of numbers. Name that set.

   a. $\mathbb{I}$ b. $\mathbb{Q}$ c. $\mathbb{R}$ d. $\mathbb{N}$ e. $\overline{\mathbb{Q}}$ f. $\mathbb{W}$

3. a. Describe the set of rational numbers.
   b. Why do we exclude expressions that involve division by 0 from the set of rational numbers?
   c. Explain the use of the adverb *rational* in *rational numbers*.

4. a. What is an irrational number?
   b. Is there a number that is both rational and irrational?

5. True or false?

   a. There is a whole number that is not a natural number.
   b. All integers are rational.
   c. All rational numbers are real.
   d. All rational numbers are integers.
   e. There is an irrational number that is not real.
   f. All natural numbers are rational.
   g. Every whole number is an integer.

6. Which of the sets $\mathbb{N}$, $\mathbb{W}$, $\mathbb{I}$, $\mathbb{Q}$ and $\mathbb{R}$ are closed under

   a. addition   c. multiplication
   b. subtraction   d. division

7. What does it mean when we say that *the set of integers is closed under multiplication?*

8. The following are examples of closure of a set under an operation. In each case explain what practical information we get from the fact.

   a. The set of natural numbers is closed under addition.
   b. The set of integers is closed under subtraction.

9. a. What does it mean when we say that the expression $\frac{0}{0}$ is indeterminate?
   b. What does it mean when we say that the expression $\frac{-2}{0}$ is undefined?

Chapter 2
Notation

In the study of math and its applications two classes of notations for the representation of numbers will be of use to us. These are the *formal notation* and *base-ten notations*. In this chapter we will present an overview of these notations and discuss their strengths, weaknesses and intended uses.

2.1 The Formal Notation

We have already seen examples of **the formal notation** in our coverage of the formal classification scheme in Chapter 1. A summary of the notations used to formally represent the numbers in our selected sets is given below:

- 1, 2, 3, 4, ... to represent the naturals.
- 0, 1, 2, 3, ... to represent the wholes.
- ..., -2, -1, 0, 1, 2, ... to represent the integers.
- $\frac{p}{q} \mid p \wedge q \in \mathbb{I} \wedge q \neq 0$ to represent the rationals.
- Extension of use of symbols for operations under which we seek maximum closure (e.g., $\sqrt{}$, log, etc.) or extension of the symbol system (e.g., π, e, etc.) to represent the reals.

Let us take a closer look at the themes that are used by the formal notation to represent the numbers in the selected sets: The use of symbols, the place value scheme and the use of the notation for the operation under which we seek maximum closure.

Natural Numbers

The set of natural numbers uses symbols to represent the numbers one through nine. These symbols are 1, 2, 3, ..., 9. Obviously it is not practical to go on forever inventing new symbols to represent the rest of the naturals (ten, eleven, twelve, etc.). In this sense, the use of the place value scheme to reuse the digits 1, 2, 3, ..., 9 to represent natural numbers that are greater than nine is quite smart.[1]

Whole Numbers

The symbolism used to represent the set of natural numbers is extended by the set of whole numbers to include a new symbol, 0, for the representation of zero.

Beyond the set of whole numbers, to the extent that is possible, the formal notation adopts the notation for the relevant operation under which we seek maximum closure, to represent integers, the rationals and the reals. Extension of the symbolism used in the representation of natural numbers and whole numbers is adopted in cases where this scheme fails.

Integers

To illustrate matters, let us consider the act of subtracting 7 from 4, i.e., $4 - 7$. Obviously the notations afforded to us by the set of natural numbers and the set of whole numbers are not able to express the result of such a subtraction. However, we could "reduce" such subtractions.[2] The reduction process requires that we subtract the smaller number from both operands. This is illustrated below.

It is easy to see that the process will always lead to a 0 as the first operand.[3] Here are a few more examples:

$$15 - 17 = 0 - 2$$
$$9 - 20 = 0 - 11$$
$$1 - 7 = 0 - 6$$

and so on. This allows us to simplify notation by dropping 0 as the first operand. In the case of $0 - 3$, this yields -3 as the answer.

The fact that we can drop the first operand is quite convenient[4] as it

---

[1] While we have not acknowledged 0 as a symbol yet, we do use it to represent certain naturals using the place value scheme. We could circumvent this apparent conflict by replacing the place value scheme with some other until we introduce 0. Following the introduction of 0, we could then introduce the place value scheme and impose the notation on the naturals retroactively.

[2] We call this *reduction* as the process is reminiscent of the reduction of fractions.

[3] The reason we wish to reduce subtractions is that it makes it easier for us to assess the size of the difference using 0 as reference.

[4] A convenience that will not be available to us when we reduce division as we will soon see.

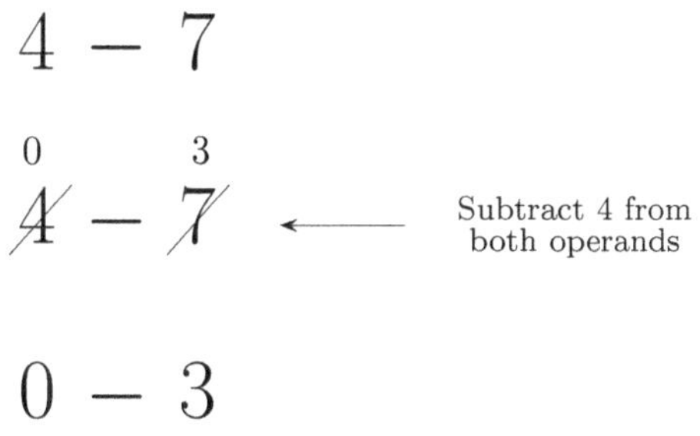

$$4 - 7$$

$$\overset{0}{\cancel{4}} - \overset{3}{\cancel{7}} \quad \longleftarrow \quad \text{Subtract 4 from both operands}$$

$$0 - 3$$

Figure 2.1.1: Reduction in the sense of subtraction

allows us to make it appear as if -3 is the value that we *arrive* at. In other words, while -3 is the simplified form of the more elaborate $0 - 3$, we assign to it the duty of representing the number that we arrive at as a result of the subtraction $0 - 3$ (itself generated through the reduction of $4 - 7$).

The negative sign, then, has its root in the symbol used for the operation of subtraction. The negative sign is used to refer to the number that one arrives at when one performs subtractions where a larger number is subtracted from a smaller number. It arises from dropping 0 as the first argument of reduced subtractions.

The understanding that the negative sign is derived from the subtraction sign has many important practical applications[5] and figures in the subtraction algorithm for integers, rearranging algebraic equations using semantic tools and the like. We will remind the reader of this equivalence later in the text when we come to discuss such topics in detail.

Note that it is quite fitting to represent our new numbers using the symbol for the operation of subtraction as it is closure with respect to subtraction that we wish to impose by introducing the new numbers. Figure 2.1.2 demonstrates the interpretations that are associated with a subtraction, its associated reduced subtraction and notation used to communicate the *result* of the subtraction.[6]

---

[5]It also explains why the symbol used to represent the negative sign is identical to the symbol used to represent subtraction.

[6]The same can be said about the notation used to represent positive integers: $+4$ derives from $0 + 4$ which represents a kind of "reduction" for related additions such as $1 + 3$. This also illustrates why the symbol used to represent the positive sign is identical to the symbol used to represent addition.

$$4 - 7 \qquad 0 - 3 \qquad -3$$

a subtraction   the associated   the result
four minus seven   reduced subtraction   *negative three*
                                      zero minus three

Figure 2.1.2: Interpretations of a subtraction, its associated reduced subtraction, and the result of that subtraction

Rational Numbers

Consider the act of dividing 4 things into 6 equal parts, i.e., $4 \div 6$, or, in our preferred notation, $\frac{4}{6}$. We can show that the result of dividing 4 things into 6 parts is the same as the result of dividing 2 things into 3 parts, i.e., $\frac{2}{3}$.[7] This is shown below.

$$\frac{4}{6}$$

$$\frac{\overset{2}{\cancel{4}}}{\underset{3}{\cancel{6}}} \quad \longleftarrow \quad \text{Divide both operands by 2}$$

$$\frac{2}{3}$$

Figure 2.1.3: Reduction in the sense of division

Unlike reduction of subtractions which always yielded 0 as its first operand, reduction of divisions can generate any number for its operands. Here are

---

[7]The reason we care to know this is that it allows us to work with smaller numbers. This enhances understanding.

some examples:

$$\frac{6}{15} = \frac{2}{5}$$
$$\frac{21}{28} = \frac{3}{4}$$
$$\frac{2}{6} = \frac{1}{3}$$

Therefore, we can not drop the first (or the second) operand in the case of reduction of divisions as we did with the reduction of subtractions. Unlike reduction of subtractions where we used the opportunity to drop an operand to generate a new notation to refer to the *result* of reduced subtractions, here the two notations overlap although our reading of them makes it clear which of the two we have in mind. We will therefore have to train ourselves to see the notation $\frac{2}{3}$ as both a reduced division (read *two divided by three*) *and* the result of that division (read *two thirds*[8]). This is illustrated in Figure 2.1.4 below.

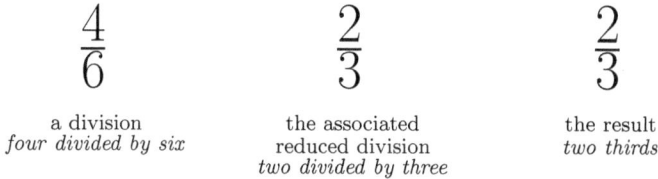

$$\frac{4}{6} \qquad\qquad \frac{2}{3} \qquad\qquad \frac{2}{3}$$

a division   the associated   the result
four divided by six   reduced division   *two thirds*
                    two divided by three

Figure 2.1.4: Interpretations of a division, its associated reduced division, and the result of that division

The "fraction line", then, has its root in the symbol used for the operation of division. Like integers which used the subtraction symbol to provide us with the means to express the result of certain subtractions, rationals use the division symbol to provide us with the same capability for certain divisions.

The ability to switch back and forth between the two interpretations of the horizontal line as the symbol for a division (of the top by the bottom) and a fraction (of 1) representing the size of the result of that division is an important skill.[9]

---

[8] This is short for *two thirds of one*, i.e., the size of the result is compared to the size of 1.

[9] As an example, it is important to be able to see $\frac{4}{6}$ both as a division (of four things into six parts, i.e., 4 divided by 6) or a fraction (i.e., four sixths of 1: divide 1 into 6 parts and take 4 parts). In fact, if we were to divide 4 things into 6 parts, we would arrive at

Note once again how fitting it is to use the symbol for division to represent our new numbers as it is closure with respect to division that we seek to enforce by the introduction of the new numbers.

Real Numbers

Beyond division we do not extend our number sets by imposing closure operation by operation. Rather, we impose maximum closure with respect to *any and all other operations* beyond addition, subtraction, multiplication and division.[10] To do so, we extend the set of rational numbers by adding new numbers to the set and we refer to the new set as the set of real numbers. The new numbers added to the set of rational numbers are called irrational numbers. For those irrationals that result from the application of an operation (e.g., taking roots, finding logarithms, etc.), we use the symbol for that operation (e.g., $\sqrt{}$, log, etc.) to represent them with. As before, we view such notations in two different ways: as directives instructing us what to do and as single entities referring to the value that one arrives at following that directive. The multiple views of the notation are illustrated in Figure 2.1.5 below.

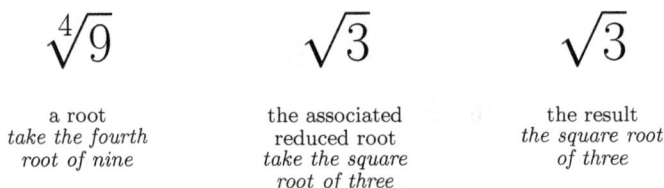

$$\sqrt[4]{9} \qquad\qquad \sqrt{3} \qquad\qquad \sqrt{3}$$

a root
take the fourth
root of nine

the associated
reduced root
take the square
root of three

the result
the square root
of three

Figure 2.1.5: Interpretations of a root, its associated reduced root, and the result of taking that root

For those irrationals which arise through special investigations (e.g., divide the circumference of a circle by its diameter) we invent new symbols (e.g., π).

There are still those irrationals whose size is communicated by the use of multiple operations and symbols. An example of this is the irrational number $4 + \sqrt{2}$. As before, the notation may be seen as a detailed account

---

the same size for a part as what we would get if we divided 1 thing into 6 parts and took 4 parts. These alternate views of $\frac{4}{6}$ are discussed at length in Appendix C and the reader is advised to read this appendix before continuing further.

[10]We remind the reader that this does not include an even root of a negative number.

of what needs to be done (i.e., take the square root of 2 and add the result to 4) or it may be seen as the overall size of the number that one arrives at following the directives. As a second example, the notation 2π may be seen as a directive (i.e., divide the circumference of a circle by its diameter and multiply the result by 2) or the totality that one arrives at following that directive. The need to use such mixed notations arises frequently in our dealings with irrational numbers as the mixed notations used to refer to them can not be simplified further.[11]

Strengths, Weaknesses, and Applications of the Formal Notation

One of the main strengths of the formal notation is in its ability to represent real numbers exactly. Other notations are either unable to do so or are more cumbersome to use. As an example, the decimal equivalent of $\sqrt{2}$ is $1.414\,213\,56\ldots$. This sequence of digits never ends and no part of it ever repeats. How does one work with such a representation?[12]

The formal notation is also more descriptive. As an example, consider the fraction $\frac{1}{4}$. The immediate comprehension of the size of $\frac{1}{4}$ is difficult to match using other notations. That this is the case should become apparent when one notes that, when solving a problem using other notations to represent $\frac{1}{4}$ with, we naturally revert back to the use of $\frac{1}{4}$ in the solution process. As an example, if one seeks to find 25% of the value of a quantity, one hardly uses the numbers 25 and 100 which are those that figure in the making of 25%. Instead, one makes an automatic association between 25% and $\frac{1}{4}$ and divides the value of the quantity by 4. Moving away from the more familiar values, it becomes even more difficult to make such mental associations: 0.625 maps onto $\frac{5}{8}$, $0.\overline{142\,857}$ maps onto $\frac{1}{7}$, etc.[13] In all cases the fraction notation is much more descriptive. As a second example, consider the expression 2π

---

[11]At times we do use mixed notations to work with rationals, integers, whole numbers, and natural numbers. An example of this is the common mixed number notation in expressing the sizes of such numbers as $2\frac{3}{4}$ or negative fractions as in $-\frac{1}{2}$. Strictly speaking, however, such mixed expressions are not needed: The mixed number $2\frac{3}{4}$, which is short for the longer $2 + \frac{3}{4}$, may be written as the rational number $\frac{11}{4}$ and $-\frac{1}{2}$ may be written as $\frac{-1}{2}$ or $\frac{1}{-2}$. In practice, however, we prefer the former presentations when quoting the results of our calculations as they are more descriptive: Comprehension of the size of $2\frac{3}{4}$ is immediate and $-\frac{1}{2}$ makes it clear that we are talking about a "loss" of one half. Note that the place value scheme is also a mixed notation: The representation 2478 is short for the longer $2000 + 400 + 70 + 8$ and refers to the totality that one arrives at if one adds 2000, 400, 70, and 8.

[12]Rounding this infinite sequence of digits is out of the question if one insists on being exact.

[13]The line over the digits in $0.\overline{142\,857}$ indicates a repeating pattern and, as such, is short for the longer $0.142\,857\,142\,857\,142\,857\ldots$.

and its decimal equivalent representation of $6.283\,185\ldots$. With the latter representation we may not be able to even notice that π plays a significant role in the problem under investigation.

A major weakness of the formal notation is that it is a difficult notation to work with as it requires the use of many algorithms for working with each of the operations, depending on the type of the operands. As an example, the use of the formal notation requires that we learn many addition algorithms: one for adding integers, another for adding rationals, a third for adding roots, etc. Certain other notations, such as the decimal notation, have an advantage over the formal notation in this respect as they use a more uniform notation to represent the various types of numbers in $\mathbb{R}$. This allows for fewer algorithms for working with each of the operations. As an example, we can use a single algorithm to add natural numbers, whole numbers, integers, rationals, and the reals if we use the decimal notation to represent our numbers with.

The formal notation should be used when one engages in activities that require that we represent our numbers exactly and/or situations in which we need to be as descriptive as possible. Under such constraints we have no option but to work with the formal notation and the operational algorithms that come with it no matter how numerous or complex these algorithms may be. Communication of theoretical ideas and discussions, especially in the sciences, provides one such area for the use of the formal notation.

2.2   Base-Ten Notations

Base-ten notations represent an attempt to impose as much uniformity on the representation of the various numbers in $\mathbb{R}$ as possible through the extension of the place value scheme used in the representation of natural numbers, whole numbers, and integers. Full uniformity through the extension of the place value scheme is out of the question as the place value scheme deals with the representation of the *size* of a number. This means that the extension of the scheme is inherently unable to represent negative numbers. For this reason base-ten notations adopt the use of the subtraction symbol to represent the negatives following the lead of the formal notation.

An example of a base-ten notation is the **decimal notation**. The notation extends the place value scheme for the representation of the sizes of the various numbers in $\mathbb{R}$ by appending a point, called the **decimal point**, to the right end of an integer followed by place values that are sub-powers of 10. Examples of numbers written using the decimal notation are 3.5, -0.012, and 147.0. In this scheme, the number 2.56 is understood to represent a total of 2 units, 5 tenths, and 6 hundredths.

Further structure can be imposed on the decimal notation by writing each of our numbers as the product of a decimal number whose magnitude is between 1 and 10 (inclusive on 1 but not 10), multiplied by a power of 10 with an integer exponent to place the decimal point.[14] Such notation is referred to as **scientific notation** and is useful in the representation of numbers that are very large with a string of right-end 0s or very small with a string of left-end 0s.[15] Examples of such numbers are

$$602\,000\,000\,000\,000\,000\,000\,000\,000$$

which represents the number of particles in 1 mole of a substance or

$$0.000\,000\,000\,000\,000\,000\,000\,000\,000\,000\,911$$

which represents the mass of an electron in grams. We can use scientific notation to write the former as 6.02×10^{23} and the latter as 9.1×10^{-28}. In both cases the exponent of 10 sets the location of the decimal point in the associated decimal representations of the numbers. In the case of 6.02×10^{23} the associated decimal representation of the number would have the decimal point 23 digits to the right of its location in the representation that uses scientific notation and in the case of 9.1×10^{-28} the associated decimal representation of the number would have the decimal point 28 digits to the left of its location in the representation that uses scientific notation.

Strengths, Weaknesses, and Applications of Base-Ten Notations

One major advantage of base-ten notations is that they place more uniformity on the representation of the various types of numbers in $\mathbb{R}$. This naturally leads to the employment of fewer algorithms to process the various operations compared to the multitude of algorithms that are needed to process the operations using the formal notation. As an example, we can use a single algorithm to add real numbers if we use the decimal notation. However the use of the formal notation requires that we learn many addition algorithms (one to add integers, another to add rationals, a third to add roots, etc.).

However, as we noted earlier, base-ten notations are not practical in situations where one needs to be exact or descriptive. As an example, the value of π in decimal notation is $3.141\,592\,65\ldots$, a sequence of digits that never ends and no part of which ever repeats. It is not quite clear how one can add such numbers or subtract them or perform any other operations on them. Furthermore, the representation 0.125 can never match the level of clarity that is afforded to us using its formal representation of $\frac{1}{8}$.

---

[14]This scheme works for all numbers other than 0. In this more structured representation of our numbers we write 0 simply as 0.

[15]Such values are routinely encountered in basic applications of scientific theories.

The strengths and weaknesses of base-ten notations imply that they should be used in situations where being exact or descriptive is not of concern. Applications of theories, especially science theories, provide one such area for the use of base-ten notations. Applications of science theories require that we make measurements and feed the measured values into science formulas to process them. Measured values, however, are rarely exact and the processing of measured values is subject to rounding rules. Furthermore, there is hardly any significance to the precise height of an individual or the mass of a specific object. The approximate nature of measured and processed values and their natural lack of significance allow us to reject the use of the formal notation in favour of base-ten notations which are much more pleasant to work with (compare, round, add, subtract, etc.). This is one reason why the International System of Units, SI, which is the default measurement system used in the sciences, recommends the use of base-ten notations in expressing the numerical values of quantities.

Exercise Set 2

1. a. List the strengths and weaknesses of the formal notation.
   b. List the strengths and weaknesses of base-ten notations.

2. Name two examples of base-ten notations.

3. a. When is it proper to use the formal notation in the sciences?
   b. When is it proper to use base-ten notations in the sciences?

Part II
Theory and the Formal Notation

In this part of the book we will discuss the communication needs of theoretical discourse and show how the strengths of the formal notation can be utilized to meet the demands of such needs. Following this we will present a detailed account of the manner in which one can work with the formal notation by adopting styles that are both meaningful and efficient.

Chapter 3
Communication Needs of Theory

Theoretical statements are about relationships between the values of quantities.[1] Such relationships are often expressed as algebraic equations. Examples of such equations are Newton's formula, $F = ma$, the weight formula, $W = mg$, the kinetic energy formula, $E = \frac{1}{2}mv^2$, and so on. Each of these formulas expresses a relationship between the values of the various quantities involved in that formula. The formula $F = ma$, as an example, relates the values of the quantities force, mass and acceleration.[2] It tells us that the value of the force experienced by an entity is equal to the product of the values of the mass of that entity and its acceleration. The formula $W = mg$ expresses a relationship between the values of the quantities weight, mass and acceleration due to gravity near the surface of the Earth. It tells us that the value of the weight of an entity near the surface of the Earth is equal to the product of the values of the mass of that entity and the value of its acceleration due to gravity near the surface of the Earth. The formula $E = \frac{1}{2}mv^2$ expresses a relationship between the values of the quantities kinetic energy, mass and speed. It tells us that the value of the kinetic energy of an entity is equal to half the product of the value of its mass and the square of the value of its speed.[3]

Theoretical relationships between quantities are assumed to be exact.

---

[1] A *quantity* is a property of an entity that can be measured objectively. Examples of quantities are mass, time, speed, force, pressure and momentum, but not beauty, kindness or diligence. See the chapter on measurement for a more in-depth discussion on quantities and related topics.

[2] Implicit in the form of every formula is the units that are used to measure the values of the various quantities involved in that formula. This topic is discussed at length in the chapter on measurement. For now we simply note that the formulas given assume that the International System of Units, SI, is used in the measurement of the values of the various quantities involved.

[3] There is more to formulas than how the value of the quantity on the left side of the formula can be computed using the values of the quantities on its right side. For more on the semantics of formulas please see the companion textbook *Semantics and the Syntax of Algebra* by the author.

When we write $E = \frac{1}{2}mv^2$, we do not intend to imply that the value of the kinetic energy of the entity under consideration is *approximately* equal to half of the product of the value of the mass of that entity and the square of the value of its speed but that it is *exactly* so. And as theoretical discussions aim to expose relationships between the values of the quantities involved, it is expected that the language used in their expression is as descriptive as possible. This dual need for exactness and descriptiveness explains why science formulas employ the use of the formal notation in their expressions. This includes the use of the formal notation in the expression of numerical values, the use of symbolism to refer to exact values of constants and the use of symbols to represent operations. Let us look at a few theoretical relationships between the values of various quantities to demonstrate the manner in which the formal notation is used in their expressions.

Consider the kinetic energy formula, $E = \frac{1}{2}mv^2$. Note the use of the rational number, $\frac{1}{2}$, in the expression of this theoretical formula as opposed to the use of base-ten notations such as 0.5 or the percent notation 50%. No physics textbook worthy of its name would write the kinetic energy formula as $E = 0.5mv^2$ or $E = 50\%mv^2$. Why?[4] One might, of course, argue that the use of such other notations in the expression of the kinetic energy formula does not terribly affect descriptiveness and the expressions 0.5 and 50% are as exact as their formal counterpart $\frac{1}{2}$. However, note that, while this may be the case for the specific value, $\frac{1}{2}$, it is not so for many other values. As an example, the exact representation of $\frac{2}{3}$ as a decimal is $0.\overline{6}$ and its formulation as a percentage becomes $66.\overline{6}\%$ and we have given examples of decimal equivalencies for other fractions, such as $\frac{3}{8}$ or $\frac{1}{7}$ that are even worse.[5] We do *not*, of course, wish to allow the use of a certain notation under certain conditions but not others as this will require that we keep in mind circumstances under which such alternatives would and would not be acceptable. In addition, the use of dual notations becomes a problem when formulas that use alternative notations are combined with those that use the formal notation.

The need to be exact in the communication of theories also explains why we resort to the use of symbolic constants in the writing of such formulas as $F = \frac{Gm_1m_2}{r^2}$. In this formula the symbol G stands for the *exact* value of the universal gravitational constant. We can, and do, measure this value in the application of the theories; however, any such measured values are inherently approximate. The measured value of G is known to be approximately $6.7 \times$

---

[4]Before the advent of digital media as we know them today, editors would correct such errors. With the advent of self-publishing tools it has become more difficult to prevent an author from publishing textbooks that use nonstandard notations. However, such deviations only point to a lack in understanding matters that relate to the preference for the use of the formal notation in the expression of theoretical discussions.

[5]Rounding such representations is out of the question as they make the formulas approximate.

10^{-11} N·m^2/kg^2. A more accurate value is 6.674×10^{-11} N·m^2/kg^2 and still more accurate values have been measured using more sensitive instruments. However, replacement of G in the formula with any of these measured values will make the formula approximate. The symbol G, then, is used to represent the *exact* value of the universal gravitational constant; a value that we may never be able to *measure* exactly.

When numerical values are used in their formal form in a formula, it is implied that the values that they represent are exact and are based on theoretical derivations, not approximate measurements. In the kinetic energy formula $E = \frac{1}{2}mv^2$, the fraction $\frac{1}{2}$ (as well as its numerator, 1, and its denominator, 2) and the exponent 2 are all exact. These numerical values are not measured. They arise as a result of theoretical considerations that relate the value of the kinetic energy of an entity to the values of its mass and speed.

Theoretical relationships are of course useful because they are general statements that can be used to solve specific problems. The statement $E = \frac{1}{2}mv^2$ is quite general. As such, it can be used to solve any specific problem in which the values of kinetic energy, mass and speed are of interest.[6] An example of a problem that uses the theoretical formula $E = \frac{1}{2}mv^2$ in its solution is the problem *An object, moving at a speed of* 27.7 m/s *has a kinetic energy of* 1700 J. *Calculate its mass.* The solution to this problem can be communicated as follows:

$$E = \frac{1}{2}mv^2 \qquad \text{Theory}$$

$$m = \frac{2}{v^2}E$$

$$m = \frac{2}{27.7^2} \times 1700 \qquad \text{Application}$$

$$m = \frac{2}{767.29} \times 1700$$

$$m = 4.43 \text{ kg}$$

Note the clean separation of the theoretical argument and the application of the theory to solve the specific problem under consideration. So long as formal notation is used, the discussion is theoretical. Appearance of base-ten notations signals the shift from theory to applications of that theory.

---

[6]Formulas that result from observed relationships between quantities are considered to be correct within the range of experimental verification. Such relationships may or may not hold when applied outside the range within which they have been verified.

Exercise Set 3

1. What notation is suitable for use in the communication of theoretical ideas? Why?
2. Explain the use of symbols in the communication of the values of constants in science formulas.
3. What does the appearance of a numerical value in its formal form in a science formula imply?
4. How can you tell whether a discussion is theoretical or applied?

Chapter 4
Working with the Formal Notation

In this chapter we visit each of the sets $\mathbb{W}$, $\mathbb{I}$, $\mathbb{Q}$ and $\mathbb{R}$ in turn and for each set we will discuss the structure of the numbers in that set, the formal reading of the numbers in that set, the ordering of the numbers in that set, algorithms for performing arithmetic operations on the numbers in that set, the evaluation of expressions involving the numbers in that set, and the graphing of the numbers in that set.[1] Our focus will be on the development of tools that are meaningful and efficient as it is this combination of features that makes a mathematical tool practical.

4.1 Whole Numbers

In this section we will review aspects of whole numbers that are of interest to us. These include structure, formal reading, order, algorithms for carrying out arithmetic operations, evaluation of expressions, and graphing on the real line.

4.1.1 Structure

The symbols in Figure 4.1.1 are called **digits**.

$$0 \quad 1 \quad 2 \quad 3 \quad 4 \quad 5 \quad 6 \quad 7 \quad 8 \quad 9$$

Figure 4.1.1: Digits

Using the place value scheme, we write a whole number as a finite sequence of one or more digits, e.g., 362, 4200 and 0. In this scheme, the placement

---

[1]$\mathbb{N}$ will be covered under $\mathbb{W}$.

of a digit relative to the rightmost digit is important because it tells us what the true value of that digit is. As an example, by placing a 4 in the second place from the right (as in 249, 1240, 42, etc.) we intend to imply that we have 4 *tens* or 40 whereas placing a 4 in the third place from the right (as in 7400, 25 491, 402, etc.) implies that we have 4 *hundreds* or 400. In fact, the notation 3742 may be seen as a shorthand for the longer

$$3000 + 700 + 40 + 2$$

This is 3742 written in **expanded form**. As its expanded form clearly shows, 3742 is made up of 3 *thousands*, 7 *hundreds*, 4 *tens* and 2 *ones*. In this interpretation of 3742, the words *thousands*, *hundreds*, *tens* and *ones* set the true values of the digits due to their placement relative to the rightmost digit and are, for this reason, called **place values**. The more common place values are shown in Figure 4.1.2.

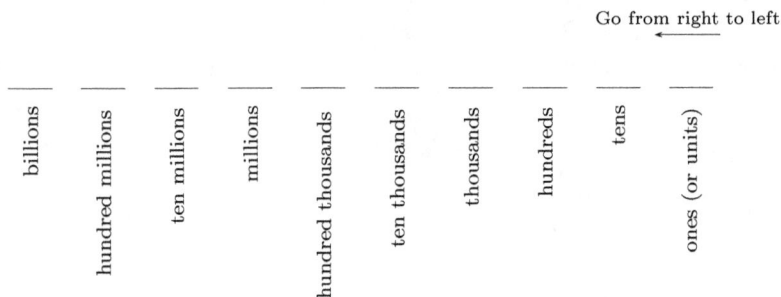

Go from right to left

billions	hundred millions	ten millions	millions	hundred thousands	ten thousands	thousands	hundreds	tens	ones (or units)

Figure 4.1.2: The more common place values

Relative Sizes of Place Values

The reading of the place values relates their sizes to that of the rightmost place value, i.e., *ones*. Therefore, the place value *tens* is 10 times larger than the place value *ones*, the place value *hundreds* is 100 times larger than the place value *ones*, and so on. However, in addition to relating the sizes of the many place values to that of *ones* (for which the naming of the place values provides the needed clue), it is possible to relate the sizes of *any* two place values relative to each other using the following structural feature of the place value scheme: Each place value is set to be 10 times larger than the one on its right and 10 times smaller than the one on its left. As an example, a *hundred* is 10 times larger than a *ten* and 10 times smaller than a *thousand*. The size relationship between adjacent place values can be extended to relate the sizes of place values that are farther apart. As an example each place value is 100 times larger than the place value that is two places on its right and 1000 times smaller than the place value that is three places on its left. This means that a *ten thousand* is 100 times larger than a *hundred* and 1000 times smaller than a *ten million*.

This observation leads to the following visual tool for determining the relative sizes of two place values: Place a 1 in the larger place value and place 0s on the right side of this 1 until you arrive at the smaller place value. The number generated by the sequence of the 1 and 0s shows how many times the larger of the two place values is larger than the other place value, or, equivalently, how many times the smaller of the two place values is smaller than the other place value. As an example, suppose we wish to determine the relative sizes of the place values *thousands* and *tens*. We begin by identifying the two place values as shown below.

$$\underline{\quad}\ \ \underline{\quad}\ \ \underline{\quad}\ \ \underline{\quad}\ \ \underline{\quad}\ \ \underline{\quad}\ \ \underset{\text{thousands}}{\underline{\quad}}\ \ \underline{\quad}\ \ \underset{\text{tens}}{\underline{\quad}}\ \ \underline{\quad}$$

We now place a 1 in the larger place value, i.e., *thousands*, followed by 0s until we get to the smaller place value, i.e., *tens*. This is shown below.

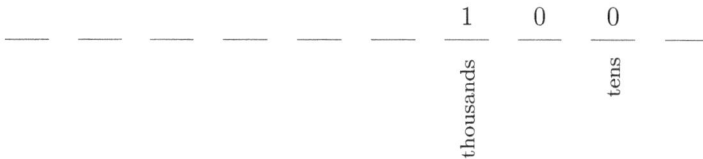

$$\underline{\quad}\ \ \underline{\quad}\ \ \underline{\quad}\ \ \underline{\quad}\ \ \underline{\quad}\ \ \underline{\quad}\ \ \underset{\text{thousands}}{\underline{1}}\ \ \underline{0}\ \ \underset{\text{tens}}{\underline{0}}\ \ \underline{\quad}$$

We can now say that the place value *thousands* is 100 times larger than the place value *tens*.[2]

---

[2]An alternative algorithm for the determination of the relative sizes of two place values requires that we divide the larger place value by the smaller place value to arrive at an

Triads

There is a cyclic scheme behind the naming of the place values that makes it easier to remember them. This cyclic scheme works as follows: Starting from the right end of a whole number and moving left, we divide the digits into groups of three.[3] Each group of digits generated in this manner is called a **triad**[4] and each triad is given a name. The names of the first four triads are shown in Figure 4.1.3.

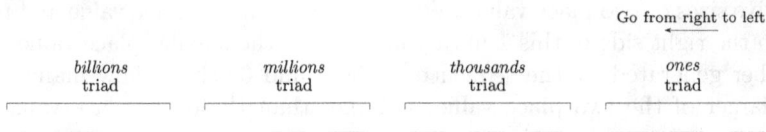

Go from right to left

| *billions* triad | *millions* triad | *thousands* triad | *ones* triad |

Figure 4.1.3: Names of the first four triads

Triad names appear in place value names as shown in Figure 4.1.4.[5] The cyclic naming of the place values within triads (e.g., *thousands, ten thousands, hundred thousands,* followed by *millions, ten millions, hundred millions,* followed by *billions, ten billions, hundred billions,* etc.) makes it easier to remember them.

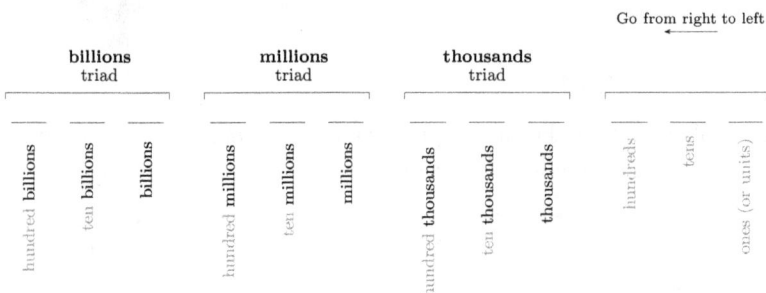

Go from right to left

billions triad   **millions** triad   **thousands** triad

hundred billions | ten billions | billions | hundred millions | ten millions | millions | hundred thousands | ten thousands | thousands | hundreds | tens | ones (or units)

Figure 4.1.4: Triad names as guides for place value names

---

answer. In the example just presented, we could divide 1000 by 10 which would quickly lead to the removal of a 0 in 1000 to yield 100. This means that the place value *thousands* is 100 times larger than the place value *tens*. The graphical tool above follows the same logic: In this diagram, we stopped at the *tens* place and did not write a 0 in the *ones* place as this is the 0 in 1000 that cancels with the 0 in 10.

[3] The leftmost group may end up with fewer than three digits.

[4] Also called a **period**.

[5] With the exception of the first triad: The place values within the first triad are labelled as *ones, tens,* and *hundreds* as opposed to *ones, ten ones,* and *hundred ones*.

The grouping of the digits in a whole number into triads and the cyclic naming convention for the naming of place values within a triad provide us with an effective tool to name place values quickly. As an example, suppose we are seeking the place value of the location pointed to by the arrow in the diagram below.

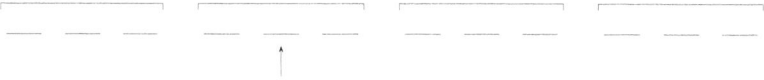

We first note the name of the triad within which the arrow is located. In the diagram above, this is the *millions* triad. The relative placement of the arrow within this triad tells us that it points to the place value *ten millions*.

To facilitate the task of naming the triad itself, we follow convention and, when writing a whole number with more than four digits, separate the triads with a bit of empty space,[6] e.g., we write 17 204 192 as opposed to 17204192. This convention helps us distinguish between the triads and tell which triad we are dealing with. The spacing of the triads as described above has the added advantage of making it easier to tell the overall size of a whole number at a glance: Comprehension of the size of 18 102 271 is immediate but that of 18102271 is not.

Exercise Set 4.1.1

1. In each case determine the name of the triad pointed to by the arrow.

(d)   (c)   (a)   (b)

---

[6]The use of the comma to break a sequence of digits into triads is strongly discouraged as the character disrupts the sequence and interferes with its use as separator in lists. In addition, in many countries the comma is used in place of the decimal point. Where a comma is used to denote a decimal point, apostrophes are used to break numbers into triads.

2. Consider the diagram below. In each case determine the place value of the location pointed to by the arrow.

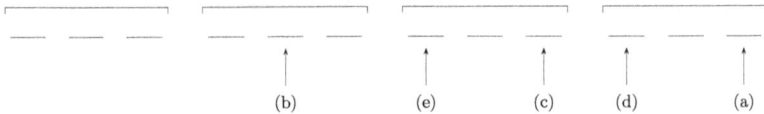

(b)   (e)   (c)   (d)   (a)

3. What is the place value of 7 in 271?
4. What is the place value of 1 in 1042?
5. What is the place value of digit 0 in 1019?
6. What is the place value of digit 8 in 42 008?
7. What is the place value of digit 3 in 32 020?
8. What is the place value of 2 in 12 318 184?
9. What is the place value of digit 4 in 418 166?

10. Write each of the following whole numbers in expanded form and interpret the result.

a. 45   c. 4027   e. 23 420

b. 9371   d. 400   f. 9 205 069

11. How does the place value of a given digit compare to the place value of the digit on its right?
12. How does the place value of a given digit compare to the place value of the third digit on its right?
13. How does the place value of a given digit compare to the place value of the second digit on its left?
14. How many times is a *hundred* larger than a *ten*?
15. How many times is a *ten thousand* larger than a *hundred*?
16. How many times is a *million* larger than a *thousand*?
17. How many times is a *ten* smaller than a *hundred thousand*?
18. How many times is a *hundred* smaller than a *million*?
19. How many times is a *ten million* larger than a *hundred thousand*?
20. How many *tens* fit in a *thousand*?
21. How many *ones* fit in a *ten thousand*?
22. How many *thousands* fit in a *million*?
23. How many *ten thousands* fit in a *hundred thousand*?
24. How many *hundreds* fit in a *hundred*?

25. Describe an efficient algorithm for determining the place value of a digit in a given whole number.

4.1.2   Formal Reading

In this section we will present the formal manner in which whole numbers are written using words. Cases where the given number contains one, two, or three digits are considered first. This is followed by a note on the manner in which numbers involving more than three digits are handled.

Single-Digit Numbers

The names of single-digit numbers as cardinals and ordinals are given in Table 4.1.1 below.[7]

0 zero, zeroth	5 five, fifth
1 one, first	6 six, sixth
2 two, second	7 seven, seventh
3 three, third	8 eight, eighth
4 four, fourth	9 nine, ninth

Table 4.1.1: Formal reading of single-digit numbers
as cardinals and ordinals

Double-Digit Numbers

The numbers 10, 20, 30, 40, 50, 60, 70, 80 and 90, as well as 11, 12, 13, 14, 15, 16, 17, 18 and 19 have special names. These are listed in Table 4.1.2 below.

To explain how to write other double-digit numbers in words, we will use 67 as an example. Break 67 into 60 and 7. Write 60 as a cardinal, followed by a hyphen, followed by 7 as a cardinal or an ordinal as the case may be. See Figure 4.1.5 for an illustration.

---

[7]**Cardinal numbers** are numbers that are used in counting. **Ordinal numbers** are numbers that describe the numerical position in a sequence.

10 ten, tenth	11 eleven, eleventh
20 twenty, twentieth	12 twelve, twelfth
30 thirty, thirtieth	13 thirteen, thirteenth
40 forty, fortieth	14 fourteen, fourteenth
50 fifty, fiftieth	15 fifteen, fifteenth
60 sixty, sixtieth	16 sixteen, sixteenth
70 seventy, seventieth	17 seventeen, seventeenth
80 eighty, eightieth	18 eighteen, eighteenth
90 ninety, ninetieth	19 nineteen, nineteenth

Table 4.1.2: Formal reading of special double-digit numbers as cardinals and ordinals

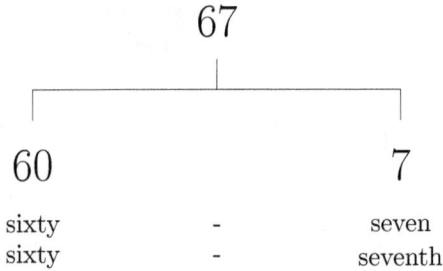

67

60		7
sixty	-	seven
sixty	-	seventh

Figure 4.1.5: Writing 67 as a cardinal (sixty-seven) and an ordinal (sixty-seventh)

Examples

Write each of the following as a cardinal and an ordinal.

1. 35

2. 82

Answers

1. thirty-five
   thirty-fifth

2. eighty-two
   eighty-second

Triple-Digit Numbers

The numbers 100, 200, 300, 400, 500, 600, 700, 800 and 900 have special names. To write each of these, write the digit on the left as a single-digit number, followed by the word 'hundred' or 'hundredth' depending on whether you want to write the number as a cardinal or an ordinal.

To explain how to write other triple-digit numbers in words, we will use 724 as an example. Break 724 into 700 and 24. Write 700 as a cardinal, followed by 24 as a cardinal or an ordinal as the case may be. See Figure 4.1.6 for an illustration.[8]

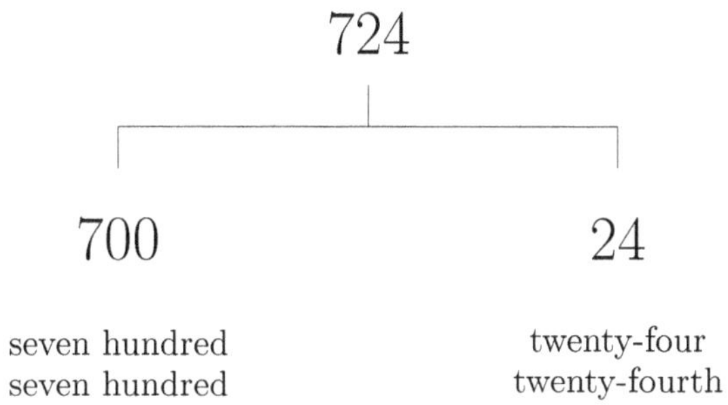

$$724$$

$$700 \qquad 24$$

seven hundred   twenty-four
seven hundred   twenty-fourth

Figure 4.1.6: Writing 724 as a cardinal (seven hundred twenty-four) and an ordinal (seven hundred twenty-fourth)

Examples

Write each of the following whole numbers as a cardinal and an ordinal.

1. 276   2. 150   3. 107   4. 214   5. 500

---

[8]The use of *and* is strictly forbidden as, officially, the word *and* is used to separate the whole part of a number from its fractional part as in 3.7 or $3\frac{7}{10}$ both of which are read as *three and seven tenths*. See the sections on the formal readings of rational numbers and numbers written using the decimal notation for guidelines on the proper use of the word *and*.

Answers

   1. two hundred seventy-six; two hundred seventy-sixth
   2. one hundred fifty; one hundred fiftieth
   3. one hundred seven; one hundred seventh
   4. two hundred fourteen; two hundred fourteenth
   5. five hundred; five hundredth

Numbers with More Than Three Digits

To write numbers with more than three digits as *cardinals*, break the number up into triads. Starting from left and going right, write the group of digits in each triad as a triple-digit, double-digit or single-digit number, followed by the name of the triad[9] without the plural 's'.[10] See Figure 4.1.7 for how we write 23 420 076 in words.

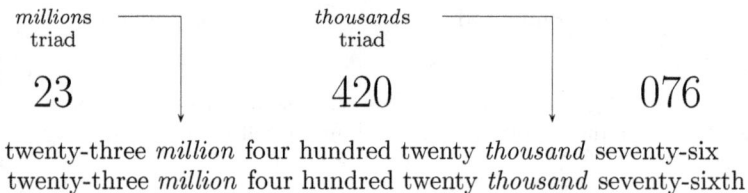

millions ———┐   *thousands* ———┐

triad   triad

23   420   076

twenty-three *million* four hundred twenty *thousand* seventy-six
twenty-three *million* four hundred twenty *thousand* seventy-sixth

Figure 4.1.7: Writing 23 420 076 as a cardinal (top row) and
an ordinal (bottom row)

If a triad consists entirely of zeros, then do not write any words down for that triad. As examples, we read 42 000 210 as *forty-two million two hundred ten* and we read 37 000 as *thirty-seven thousand*.

To write numbers with more than three digits as *ordinals*, write the group of digits in the rightmost triad as an ordinal. If this triad consists entirely of zeros, then write the name of the rightmost nonzero triad as an ordinal. To do so remove the plural 's' in the triad name and replace it with 'th'.

---

[9] Other than the *ones* triad.
[10] As before, the use of the word *and* is strictly forbidden.

Examples

Write each of the following whole numbers as a cardinal and an ordinal.

1. 5974
2. 12 205
3. 31 240 210
4. 2 000 015
5. 120 000
6. 10 000 000

Answers

1. five thousand nine hundred seventy-four
   five thousand nine hundred seventy-fourth

2. twelve thousand two hundred five
   twelve thousand two hundred fifth

3. thirty-one million two hundred forty thousand two hundred ten
   thirty-one million two hundred forty thousand two hundred tenth

4. two million fifteen
   two million fifteenth

5. one hundred twenty thousand
   one hundred twenty thousandth

6. ten million
   ten millionth

The names of the triads provide us with a handy tool to go from the English wording of the name of a number to its mathematical representation using digits. Since the names of triads are unique, the appearance of a triad name signals the end of that particular triad.[11] In addition, if a triad name is missing, it means that it consists entirely of zeros.[12]

As an example, to write *fifteen thousand twenty-seven* using digits, we start on the left and read until we arrive at a triad name, i.e., we read *fifteen* and the triad name *thousand*. This tells us that we have 15 in the *thousands* triad. We write 15 followed by a bit of empty space. Following this we have *twenty-seven* implying that we have 27 in the ones triad. Since 27 has two digits, we attach a 0 on its left and write 15 027.

---

[11]Triad names can never appear in the reading of the single-, double- or triple-digit numbers within a triad as these can run, at most, into the hundreds, not thousands, millions, billions, etc.

[12]This last convention does not apply to the ones triad as, by convention, we do not use the name of the ones triad in our reading of a whole number. The ones triad consists of zeros if the reading ends with the name of any other triad. We, therefore, write *seventy-eight thousand* as 78 000.

Examples

Write the following numbers using digits.

1. two thousand five hundred seventy-four
2. eighty-six thousand eleven
3. forty-two thousand six hundred twenty
4. five hundred sixty-seven thousand four hundred six
5. sixty-two million three hundred
6. fourteen thousand

Answers

1. 2574   3. 42 620   5. 62 000 300
2. 86 011   4. 567 406   6. 14 000

Exercise Set 4.1.2

1. Write each of the following whole numbers as a cardinal and an ordinal.

a. 25	d. 4000	g. 32 410	j. 230 000 196
b. 3	e. 7100	h. 2419	k. 438 000
c. 14	f. 4289	i. 36	l. 324 020

2. Write the following using digits.

   a. forty-six   b. eighteen   c. three hundred fifty

   d. eight thousand five hundred twenty-seven
   e. forty-five thousand
   f. nine thousand four hundred
   g. seventy-three thousand five hundred twenty-six
   h. two hundred fifteen
   i. two million five hundred thousand seventy
   j. fifty-five million one

4.1.3   Order

Formally, we use the phrases *is equal to*, *is greater than* and *is less than* to indicate that a number is equal to, larger than or smaller than another. We say, as an example, that 8 *is less than* 20 or that 15 *is equal to* 15. At times we also use the phrases *is greater than or equal to* and *is less than or equal to* to compare the sizes of two numbers. Table 4.1.3 lists the key phrases above along with their corresponding symbols using mathematical notation. It is important that you train yourself to recognize these symbols at a glance.[13]

English Phrase	Mathematical Notation
is equal to	$=$
is greater than	$>$
is less than	$<$
is greater than or equal to	$\geq$
is less than or equal to	$\leq$

Table 4.1.3: Key English phrases used to communicate order with corresponding symbols written using mathematical notation

Using mathematical notation we write $20 = 20$ to say that 20 *is equal to* 20 and we write $20 > 8$ to say that 20 *is greater than* 8. We can also write $20 < 30 < 40$ to say that 20 *is less than* 30 *and* 30 *is less than* 40.

A slash through each of these algebraic symbols negates the meaning of the symbol. As an example, the symbol $\not>$ is used in place of the English phrase *is not greater than* and the symbol $\neq$ is used in place of the phrase *is not equal to*. We can therefore write $7 \neq 4$ to say that 7 *is not equal to* 4.[14]

Two other phrases that are used frequently when ordering numbers are the phrases *increasing order* or *ascending order*, and *decreasing order* or *descending order*. To list a given set of numbers in *increasing order* or *ascending order* we must list the numbers so that, as we go through the list from left to right, the numbers *increase* in size. This means that the smallest number in the given set should be listed first, followed by the next smallest number in the given set, and onwards until all the given numbers in the given set have been listed. To list a given set of numbers in *decreasing order* or *descending order*, we must list the numbers so that, as we go through the list from left to right, the numbers *decrease* in size. This means that the largest number

---

[13] As you read from left to right, if you come across the open side of an inequality symbol, the inequality symbol relates to the *greater than* phrase. If, on the other hand, you come across the closed end of an inequality symbol, the inequality symbol relates to the *less than* phrase.

[14] Note that the symbol for negation is the slash (i.e., /) and not the backslash (i.e., \).

in the given set should be listed first, followed by the next largest number in the given set, and onwards until all the numbers in the given set have been listed. As an example of the use of these phrases, to list the numbers 24, 15, 43 and 6 in increasing order, we must list them from smallest to largest as

6, 15, 24, 43

Exercise Set 4.1.3

1. Write each of the following statements using mathematical notation.

   a. 36 is less than 74.
   b. 15 is greater than 2.
   c. 350 is greater than 200 and 200 is greater than 178.
   d. 241 is less than 320, 320 is less than 324 and 324 is less than 418.

2. Arrange 1242, 24, 620 and 4 in increasing order.
3. Arrange 4762, 5817, 4521 and 5416 in ascending order.
4. Arrange 54, 241, 560, 38 and 40 in decreasing order.
5. Arrange 3109, 420, 2410, 3729 and 197 in descending order.

4.1.4   Addition, Subtraction, Multiplication and Division

This section provides a review of the more challenging aspects of arithmetic operations on whole numbers. Notation and terminology are covered as needed.

Addition and Subtraction

We use the symbol $+$ to indicate that we want to add two numbers. To indicate that we want to add 47 and 15 we write $47 + 15$. Possible readings of this statement are *forty-seven plus fifteen, add forty-seven and fifteen, forty-seven raised by fifteen*, etc.

Formally we use the phrase *sum of* to refer to the result of an addition. As an example, viewed as a single total, the expression $47 + 15$ can be read as *the sum of forty-seven and fifteen*. And to say *the sum of forty-seven and fifteen is sixty-two*, we write

$$47 + 15 = 62$$

To add whole numbers we list them vertically making sure that corresponding place values line up. This can be done by lining up the rightmost digits (the digits in the ones place) which automatically lines up the rest of the place values. We then start with the ones, and add the digits in that column before moving on to the tens, hundreds, etc., carrying digits over as needed.[15]

Examples

1. Write each of the following phrases and statements using mathematical notation.

   a. the sum of 8 and 4
   b. Add 18 and 42.
   c. the sum of 126, 52 and 410
   d. The sum of 20, 18 and 32 is 70.

2. Evaluate the following expressions.

   a. $24 + 15$ c. $578 + 219 + 42$
   b. $900 + 25 + 160$ d. $9170 + 728 + 946 + 3840$

3. Calculate the sum of 236, 314 and 4120.
4. Add 314, 24 and 620.
5. What is the sum of 3149, 311 and 3580?

Answers

1. a. $8 + 4$ c. $126 + 52 + 410$
   b. $18 + 42$ d. $20 + 18 + 32 = 70$

2. a. 39   b. 1085   c. 839   d. 14 684

3. 4670   4. 958   5. 7040

We use the symbol $-$ to indicate that we want to subtract one number from another. To indicate that we want to subtract 14 from 29 we write $29 - 14$. Possible readings of this statement are *twenty-nine minus fourteen*, *subtract fourteen from twenty-nine*, *twenty-nine less fourteen*, *fourteen less than twenty-nine*, etc.

---

[15]The act of carrying is based on the structural feature of the place value scheme that relates the relative sizes of the place values. Since a given place value is 10 times smaller than the one on its left, each time the sum of the digits in a given column adds up to 10, the 10 is removed from that column and in its place a 1 is added to the column on its left.

Formally we use the phrase *difference between* to refer to the result of the subtraction of two numbers. Viewed as the result of a subtraction, we can read $29 - 14$ as *the difference between twenty-nine and fourteen*. To say *the difference between twenty-nine and fourteen is fifteen*, we write

$$29 - 14 = 15$$

Examples

Write each of the following phrases and statements using mathematical notation.

1. 42 less than 56   2. 18 less 12   3. 28 minus 10

4. The difference between 410 and 325 is 85.

Answers

1. $56 - 42$ 3. $28 - 10$
2. $18 - 12$ 4. $410 - 325 = 85$

To subtract two whole numbers we list them vertically making sure that corresponding place values line up. This can be done by lining up the right-most digits (the digits in the ones place) which automatically lines up the rest of the place values. We then start with the ones, and subtract the digits in that column before moving on to the tens, hundreds, etc., borrowing digits as needed.[16,17]

---

[16] The act of borrowing is based on the structural feature of the place value scheme that relates the relative sizes of the place values. Since a given place value is 10 times larger than the one on its right, we can take a 1 from one column and add 10 to the column on its right.

[17] We assume that the reader is familiar with the subtraction algorithm and continue with examples of subtraction. For a comment on the subtraction algorithm please see Appendix D.

Examples

1. Evaluate the following expressions.

   a. $138 - 26$ c. $8436 - 2618$ e. $5001 - 206$
   b. $4510 - 2300$ d. $2400 - 398$ f. $60\,000 - 54\,107$

2. Calculate the difference between 812 and 347.
3. Subtract 120 from 437.

Answers

1. a. 112   c. 5818   e. 4795
   b. 2210   d. 2002   f. 5893

2. 465   3. 317

Algorithm for Evaluation of Addition and Subtraction of Two Whole Numbers[18]

1. List the numbers vertically making sure that corresponding place values line up. This can be done by lining up the digits in the *ones* place which automatically lines up the rest of the place values.
2. Starting with the ones column and moving column by column to the left, add or subtract the digits in each column as instructed. Carry or borrow as needed.

Multiplication and Division

Formally we place the values that are to be multiplied within parentheses and place these parentheses side by side.[19] As an example, to imply that we wish to multiply 16 by 5 we write $(16)(5)$ and to imply that we wish to multiply 7, 12 and 8 we write $(7)(12)(8)$.

---

[18] As we assume that the reader already has the background to add and subtract whole numbers, the algorithm above does not provide detail on the mechanics of carrying and borrowing.

[19] In what follows, the term **parenthesis** (plural *parentheses*), also called **round brackets**, is used to refer to the symbols '(' and ')' while the phrase **square bracket** is used to refer to the symbols '[' and ']'. In this context, the term **bracket** is used to refer to either a parenthesis or a square bracket. **Curly brackets**, i.e., the symbols '{' and '}' should *never* be used in place of parentheses or square brackets as their use is limited to the representation of sets.

Occasionally, when working with more complex expressions of which multiplication is a part, we find that the use of parentheses to represent multiplication may not be quite as readily readable. The following conventions address this deficiency in the readability of the formal notation used to represent multiplication.

The first convention aims to simplify notation by requiring that we drop the brackets that enclose the first factor[20] so long as this factor contains a single whole number. Following this convention, to indicate that we want to multiply 16 by 5 we write $16\,(5)$ and to indicate that we wish to multiply 7, 12 and 8, we write $7\,(12)\,(8)$.

The second convention aims to improve the readability of expressions that involve **nested brackets**,[21] first through a reduction in their use by permitting the use of the $\times$ symbol in place of innermost parentheses (so long as the numbers being multiplied are whole numbers), and then through the incorporation of visual cues that help differentiate between the remaining parentheses. To explain how this convention is put in place, let us start with the expression for the calculation of the mass of 4 molecules of $Fe\,(NO_3)_2$, i.e.,[22]

$$4\,(55.85\,+\,2\,(14.01\,+\,3\,(16)))$$

We begin by removing the innermost parentheses and replacing them with $\times$:[23]

$$4\,(55.85\,+\,2\,(14.01\,+\,3\times16))$$

This reduces the number of parentheses from three to two. To help differentiate between the remaining parentheses, we allow the use of square brackets,

---

[20]Factors are entities that are being multiplied. As an example, $3\times7\times2$ has three factors: 3, 7 and 2.

[21]These are expressions that involve brackets that contain expressions that involve brackets.

[22]The notation $Fe(NO_3)_2$ implies that the molecule is made up of one atom of iron, Fe, and 2 groups of NO_3 with each group of NO_3 made up of one atom of nitrogen, N, and 3 atoms of oxygen, O. As such, the mass of a molecule of $Fe(NO_3)_2$ is equal to the sum of the mass of an atom of Fe and the mass of 2 groups of NO_3. The expression that evaluates to the mass of one group of NO_3 is $14.01+3\,(16)$ with 14.01 and 16 representing the values of the masses of N and O atoms in atomic mass unit (amu), a tiny unit of mass suitable for working with the masses of atoms and molecules. The mass of 2 groups of NO_3, i.e., $(NO_3)_2$, can now be written as $2\,(14.01+3\,(16))$. We can write an expression for the mass of a molecule of $Fe(NO_3)_2$ by adding the mass of an atom of Fe, i.e., 55.85 amu, to the expression above to get $55.85+2\,(14.01+3\,(16))$. The mass of 4 molecules of $Fe(NO_3)_2$ can now be written as $4\,(55.85+2\,(14.01+3\,(16)))$.

[23]The use of the *multiplication dot* is not recommended as it is easy to confuse the symbol with the one used to represent the decimal point even though the multiplication dot is centered along the font height (e.g., $4\cdot5$ which represents *the product of four and five*) while the decimal point appears on the baseline (e.g., 4.5 which represents *four and five tenths*).

i.e., [ ], in addition to parentheses with parentheses used at the lowest level and a switch made between the use of parentheses and square brackets as we move from the innermost brackets to the outermost brackets:

$$4\left[55.85 + 2\left(14.01 + 3 \times 16\right)\right]$$

And finally we use larger font sizes for enclosing brackets:

$$4\left[55.85 + 2\left(14.01 + 3 \times 16\right)\right]$$

The reader should compare the readability of this expression to that of the original expression:

$$(4)\left(55.85 + 2\left(14.01 + 3\left(16\right)\right)\right)$$

Possible readings of a statement such as 16×5 are *sixteen times five*, *multiply sixteen by five*, *sixteen multiplied by five*, etc.

Formally we use the phrase *product of* to refer to the result of multiplication. The formal reading of 16×5 is *the product of sixteen and five*. To say *the product of sixteen and five is eighty*, we write

$$16\left(5\right) = 80$$

or

$$16 \times 5 = 80$$

Examples

Write each of the following phrases and statements using mathematical notation.

1. the product of 16 and 5
2. 8 times 4 is 32.
3. Multiply 152 by 710.
4. The product of 10 and 24 is 240.

Answers

1. 16×5
2. $8 \times 4 = 32$

3. 152×710
4. $10 \times 24 = 240$

The multiplication algorithm can be used to evaluate the product of two numbers. We assume that the reader is familiar with the multiplication algorithm and present examples.[24]

Examples

1. Evaluate the following expressions.

   a. 43×24 b. $162\,(79)$ c. $9008\,(437)$

2. What is the product of 32 and 7?

3. Multiply 12 by 58.

Answers

1. a. 1032 b. $12\,798$ c. $3\,936\,496$

2. 224 3. 696

Multiplication by 10, 100, 1000, etc.

To multiply a nonzero whole number by 10, 100, 1000 and the like, add one, two, three or more 0s to the right end of the number, depending on whether you are multiplying by 10, 100, 1000, etc.

Examples

Evaluate each of the following.

1. 39×1000 2. $3491\,(10\,000)$ 3. $9200\,(10)$

---

[24]For a quick review of the multiplication algorithm please see Appendix D.

Answers

1. 39 000   2. 34 910 000   3. 92 000

The addition of the 0s reassigns the place values. In $37 \times 100 = 3700$, the addition of the two 0s in the answer reassign the place values of the digits 3 and 7 by making each 100 times larger. Note that digit 7 which was in the *ones* in 37 has its place value reassigned to *hundreds* in 3700, i.e., its value increased 100 times. Similarly, digit 3 which was in the *tens* in 37 has its place value reassigned to *thousands* in 3700, having its value increase 100 times. Since, in going from 37 to 3700, the value represented by each of the digits 3 and 7 increases by a factor of 100, the sum total represented by 37 increases by a factor of 100 as well.[25]

The official notation for division is the horizontal line. The use of other notations such as $\div$, : and / are not permitted as they tend to hinder our ability to work with expressions efficiently.[26] However, as these other symbols are still used by others, especially when working with basic math problems, we still need to know how to work with them.

To indicate that we want to divide 27 by 3 we write $\frac{27}{3}$. Possible readings of this statement are *twenty-seven divided by three*, *divide twenty-seven by three*, and *three into twenty-seven*.

Formally we use the phrase *quotient of* to refer to the result of a division. The formal reading of $\frac{27}{3}$ is *the quotient of twenty-seven and three*. To say *the quotient of twenty-seven and three is nine*, we write

$$\frac{27}{3} = 9$$

Examples

Write each of the following phrases and statements using mathematical notation.

1. The quotient of 1326 and 51 is 26.
2. 280 divided by 8

---

[25] We can appeal to *the distributive property of multiplication over addition* to justify this:

$$\begin{aligned} 37\,(100) &= (30 + 7)\,(100) \\ &= 3000 + 700 \\ &= 3700 \end{aligned}$$

[26] See Appendix E for a discussion on the advantages of the horizontal line over other notations for the representation of division.

Answers

1. $\frac{1326}{51} = 26$ 2. $\frac{280}{8}$

The division algorithm can be used to evaluate the quotient of two numbers. We assume that the reader is familiar with the division algorithm and present examples.[27]

Examples

1. Evaluate the following expressions.

    a. $\frac{184}{8}$ b. $\frac{432}{16}$ c. $\frac{36\,060}{12}$

2. What is the quotient of 4992 and 32?
3. Find the quotient and the remainder in $\frac{2853}{14}$.

Answers

1. a. 23   b. 27   c. 3005

2. 156   3. 203 R11

Division by 10, 100, 1000, etc.

To divide a whole number by 10, 100, 1000 and the like, remove one, two, three or more 0s from the right end of the number, depending on whether you are dividing by 10, 100, 1000, etc.[28] Following this shortcut, $\frac{900}{10}$ evaluates to 90 and $\frac{42\,000}{100}$ evaluates to 420.

The removal of the 0s from the right end of a whole number reassigns the place values in a manner that is analogous to that of multiplication but in reverse. We invite the reader to provide a detailed account of how this reassignment of place values generates the expected answer to such divisions. Follow our account given in the case for multiplication by 10, 100, 1000, etc., as a guide.

---

[27] For a quick review of the division algorithm please see Appendix D.

[28] See the sections on the division of rational numbers and decimals on how to deal with cases where there are not enough 0s in the given whole number to remove.

Algorithm for Evaluation of Multiplication and Division of Two Whole Numbers

We refer the reader to the sample problems worked out in Appendix D for the steps in multiplying and dividing two whole numbers.

Exercise Set 4.1.4

1. Write each of the following phrases and statements using mathematical notation.

   a. the product of 37 and 88
   b. 32 divided by 4 equals 8.
   c. the sum of 182, 16 and 4120
   d. Multiply 96 by 79.
   e. Add 15 and 23.
   f. 4200 less 3190
   g. Subtract 210 from 416.
   h. Divide 26 by 13.
   i. The product of 12 and 11 is 132.
   j. The sum of 36 and 70 is 106.
   k. 410, 9100 and 3162 add up to 12 672.
   l. the difference between 182 and 46
   m. 52 less than 67
   n. The difference between 3146 and 2050 is 1096.
   o. 18 times 4 equals 72.
   p. the quotient of 210 and 15

2. Evaluate the following expressions.

   a. $18 + 340 + 211$
   b. $1462 + 429 + 5000$
   c. $32 + 16 + 4 + 10 + 24$
   d. $96 + 0 + 25 + 89$
   e. $1297 - 184$
   f. $1520 - 108$
   g. $72\,008 - 2109$
   h. $6000 - 2086$
   i. $45\,000 - 23\,520$
   j. $1006 - 8$
   k. 82×9

   l. $200\,(310)$
   m. $76\,(10000)$
   n. 340×10
   o. $268 \div 4$
   p. $360 \div 24$
   q. $\frac{3624}{12}$
   r. $\frac{41\,400}{23}$
   s. $494/19$
   t. $664/44$
   u. $257 \div 8$

v. $200\,000 \div 10\,000$ x. $\frac{36\,000}{100}$

w. $\frac{142}{11}$

3. What does 20 less 12 equal to?
4. What is the sum of 26, 42 and 74?
5. What does 41 times 3008 equal to?
6. What is the difference between 48 and 12?
7. What is the quotient of 780 and 65?
8. Find the sum of 320, 415 and 620.
9. Add 3110, 425, 420 and 1200.
10. What is the product of 21 and 68?
11. What does 2184 divided by 91 equal to?
12. Find the difference between 400 and 365.

4.1.5   Expressions

In this section we will present an important algorithm for rapid evaluation of expressions that involve mixed arithmetic operations. The algorithm that we promote is in line with the semantic needs of applications,[29] is efficient and provides a skill that will be indispensable later when we start working with algebraic expressions and equations. In later sections we will show how this algorithm can be extended to evaluate expressions that involve numbers in $\mathbb{I}$, $\mathbb{Q}$ and $\mathbb{R}$.

An **expression** is a meaningful sequence of numbers and arithmetic operations.[30] The following are examples of expressions.

$4 \times 5 + 3 \times 2$

$20 + 12 \times 2 \times 4 - 7 \times 2$

$15 \times 2 \times 12 \times 4$

$5 \times 3 + 4\,(5 + 2 \times 6)$

The following, however, are *not* expressions.

$4 + 3\times$

$9 \times 2 + 3 \times \div 7$

In the first case it is not quite clear what we need to multiply 3 by and in the second case the sequence $\times \div$ makes no sense whatsoever.[31]

---

[29]See the companion textbook *Semantics and the Syntax of Algebra* by the author.

[30]Order operators such as $=$, $>$ and $\not<$ *are* not allowed. Presence of such operators points to the presence of equations and inequalities. Such mathematical constructs are not considered expressions. They are statements about relationships *between* expressions.

[31]This is why in our formulation of what counts as an expression above we noted that the sequence of numbers and operations must be *meaningful*.

Expressions evaluate to a value. By this we mean the following: After the arithmetic operations are performed in a certain order, we arrive at a single value. In the sample of expressions above, the first expression evaluates to 26, the second expression evaluates to 102, the third expression evaluates to 1440 and the last expression evaluates to 83.

Expressions can be evaluated using a technique that we refer to as the **analysis-synthesis technique.**[32] The word **analysis** refers to the process of breaking a complex entity into its logical components and these components into *their* logical components and so on until we arrive at the smallest entities that go into the making of the complex entity being analyzed. The opposite of analysis is **synthesis**, a process that requires that we combine the smallest entities into more complex entities and these into yet more complex entities until we arrive at the complex entity being synthesized. In the context of evaluating expressions, the entity that we speak of is the expression that we wish to evaluate. We wish to learn how to analyze an expression into smaller and smaller components until we arrive at the smallest components,[33] and how to synthesize the smaller components back into larger and larger components until we arrive at the expression itself. The ability to analyze and synthesize expressions is key to the mastery of the use of the analysis-synthesis technique in evaluating expressions.[34]

Analysis

The process of analysis of an expression begins with the breakup of that expression into pieces using additions and subtractions *outside* brackets.[35] Each piece generated in this manner is called a **term**. Here is how the sample

---

[32] For a discussion on the advantages of the analysis-synthesis technique over BEDMAS please see Appendix F.

[33] The smallest components are the numbers that appear in the original expression.

[34] It also helps one model word problems as algebraic equations and make semantic sense of such equations. See the textbook *Semantics and the Syntax of Algebra* by the author on the semantics behind syntactic components of an algebraic equation.

[35] Note that the choice of additions and subtractions as the first operations used in the analysis of expressions is in line with the theme used earlier in the text in the classification of numbers: The operations are associated with the simplest types of problems that we, as humans, solve. These are superposition problems, i.e., problems in which the last step in the solution process requires that we add and/or subtract the values of certain quantities to arrive at a net value.

expressions given earlier analyze into terms.

$$\boxed{4 \times 5} \; + \; \boxed{3 \times 2}$$

$$\boxed{20} \; + \; \boxed{12 \times 2 \times 4} \; - \; \boxed{7 \times 2}$$

$$\boxed{15 \times 2 \times 12 \times 4}$$

$$\boxed{5 \times 3} \; + \; \boxed{4\,(5 \; + \; 2 \times 6)}$$

The first expression analyzes into two terms. The second expression analyzes into three terms. The third expression analyzes into one term. The fourth expression analyzes into two terms.[36]

The terms themselves analyze into factors. **Factors** are entities within terms that are multiplied.[37]

In the first example in the sample set above

$$\boxed{4 \times 5} \; + \; \boxed{3 \times 2}$$

the first term, 4×5, analyzes into two factors: 4 and 5. The second term, 3×2, also analyzes into two factors: 3 and 2.

In the second example in the sample set above

$$\boxed{20} \; + \; \boxed{12 \times 2 \times 4} \; - \; \boxed{7 \times 2}$$

the first term, 20, analyzes into a single factor: 20. The second term, $12 \times 2 \times 4$ analyzes into three factors: 12, 2, and 4. The last term analyzes into two factors: 7 and 2.

In the third example in the sample set above

$$\boxed{15 \times 2 \times 12 \times 4}$$

the single term, $15 \times 2 \times 12 \times 4$, analyzes into four factors: 15, 2, 12, and 4.

In the last example in the sample set above

$$\boxed{5 \times 3} \; + \; \boxed{4\,(5 \; + \; 2 \times 6)}$$

---

[36] In some ways the ability to analyze an algebraic expression into terms is as fundamental as the ability to analyze an English sentence into words. And just as the words in an English sentence have meanings, so do the terms in a mathematical expression. See the companion textbook *Semantics and the Syntax of Algebra* by the author for more on the semantics of terms in applications.

[37] In keeping with our earlier theme, priority is given to multiplication now as multiplication deals with the next simplest type of problems beyond superposition problems that we, as humans, solve. These are direct proportion problems in which the last step in the solution process requires that we multiply the values of certain quantities to arrive at the result.

the first term analyzes into two factors: 5 and 3. The second term analyzes into two factors as well: 4 and the expression within brackets.[38]

Factors can be analyzed further into *their* sub-components. These sub-components may involve any of the higher order operations (e.g., division, exponents, roots, logarithms, etc.)[39] or they may involve sub-expressions within brackets. The former should be analyzed according to their own syntax; a topic that we will take up later in the text when we discuss expressions involving rationals and the reals. The sub-expressions within brackets can be processed using the same analysis-synthesis algorithm that is used to process the full expression. As an example, analysis of the sub-expression within the second factor of the second term in the last example above, i.e., $5 + 2 \times 6$, generates two terms. The first term is 5 which contains a single factor. The second term is 2×6 which contains two factors: 2 and 6.

Synthesis

Once we have analyzed an expression into terms, the terms into factors, and the factors into their components, we move on to the synthesis stage. Here, starting with the smallest of the components arrived at following analysis, we retrace our steps during the analysis stage backwards, at each step combining the various pieces until we have evaluated the factors. We continue to retrace our steps by multiplying the factors to evaluate the terms, and then add and subtract the terms to evaluate the expression. We will illustrate the ideas using the expressions in the sample set given at the beginning of this subsection.

Example

Evaluate

$$4 \times 5 + 3 \times 2$$

---

[38]Brackets perform two roles in expressions. In some cases, as here, they imply multiplication; in fact, they are the formal notation used to represent multiplication with. In such cases brackets enclose a factor regardless of whether there is a single number inside the brackets or if it contains an expression. In other cases brackets are used to enclose an expression to which a higher order operation is applied. An example of this is $5 \log (2 \times 40 + 20)$. Here it is understood that the expression within brackets should be worked out and then the log of the result should be taken. The result of this last step is then multiplied by 5. This means the expression has two factors: 5 and $\log (2 \times 40 + 20)$.

[39]Some may object to the view that we promote here by pointing to the fact that for some problems higher order operations might have priority over additions, subtractions and multiplications and point to $\frac{3+1}{2\times 3}$ as an example of such expressions. We prefer to see this as a single term with a single factor followed by division as the main operation.

Solution

We begin by analyzing the given expression into terms. This yields the following:

$$4 \times 5 + 3 \times 2 = \boxed{4 \times 5} + \boxed{3 \times 2}$$

We now analyze each term into factors. The first term contains two factors: 4 and 5. The second term also contains two factors: 3 and 2. We have now arrived at the smallest components of the original expression.[40] This marks the end of the analysis stage.

Having analyzed the expression into terms and the terms into factors, we now proceed to the synthesis stage. We multiply the factors to evaluate the terms:

$$= \boxed{20} + \boxed{6}$$

We now add and subtract the terms to evaluate the expression:

$$= 26$$

The full solution without the intervening explanations and the adorning boxes would look like the following.

$$4 \times 5 + 3 \times 2 = 20 + 6$$
$$= 26$$

Example

Evaluate

$$20 + 12 \times 2 \times 4 - 7 \times 2$$

Solution

We start by analyzing the expression into terms.

$$20 + 12 \times 2 \times 4 - 7 \times 2 = \boxed{20} + \boxed{12 \times 2 \times 4} - \boxed{7 \times 2}$$

We now analyze each term into factors. The first term contains a single factor: 20. The second term contains three factors: 12, 2 and 4. The third term contains two factors: 7 and 2. We have now

---

[40]These are the numbers 4, 5, 3 and 2 in the original expression.

arrived at the smallest components of the original expression.[41] This marks the end of the analysis stage.

We now move on to the synthesis stage by multiplying the factors to evaluate the terms. This yields

$$= \boxed{20} + \boxed{96} - \boxed{14}$$

We now add and subtract the terms to evaluate the expression.

$$= 102$$

The full solution without the intervening explanations and adorning boxes would look like the following.

$$20 + 12 \times 2 \times 4 - 7 \times 2 = 20 + 96 - 14$$
$$= 102$$

The last stage in the evaluation of the expression above involves a move from $20 + 96 - 14$ to 102. This move can be done in different ways. To understand how, you must adopt the point of view that associates each addition and each subtraction with the term that follows it. Adopting this point of view, we see the expression $20 + 96 - 14$ as

$$20 + 96 - 14$$

Starting with 20, we can perform the following operations in any order that we please: Add 96 and subtract 14. In fact, we could rewrite the expression above as

$$20 - 14 + 96$$

They both evaluate to 102.

These alternative points of view justify the validity of two commonly used algorithms for the evaluation of a string of terms that are being added and subtracted. The first simply goes from left to right and performs the additions and subtractions in sequence. This algorithm works well on a calculator but becomes inefficient if the work needs to be done by hand, especially when there are multiple subtractions. The alternative approach is to add the terms that are added,[42] add the terms that are subtracted,[43] and then subtract the latter sum from the former. This alternative algorithm reduces the number of subtractions to one.

---

[41]These are the numbers 20, 12, 2, 4, 7 and 2 in the original expression.

[42]Those with a + on their left side.

[43]Those with a − on their left side.

Example

Evaluate

$$927 - 82 - 360 + 725 - 250$$

Solution

$$927 - 82 - 360 + 725 - 250 = 960$$

We can arrive at the value 960 following either algorithm discussed above. The first alternative would lead to the following intermediate steps:

$$927 - 82 - 360 + 725 - 250$$
$$= 845 - 360 + 725 - 250$$
$$= 485 + 725 - 250$$
$$= 1210 - 250$$
$$= 960$$

while the second alternative would lead to the following intermediate steps:

$$927 - 82 - 360 + 725 - 250$$
$$= 927 + 725 - (82 + 360 + 250)$$
$$= 1652 - 692$$
$$= 960$$

The logic behind the second approach is in line with the way we normally reason as in when we add our sources of income, add our expenses, and then subtract the latter from the former to find our balance.

The second alternative above also works well when the given sequence of the additions and subtractions generates intermediate values that are negative. As an example of how this works, consider

$$2 - 7 + 8$$

Here it is much easier to add 2 and 8 first and *then* subtract 7 as this avoids the need to use negative numbers. In some cases, of course, the use of negative numbers may not be avoidable no matter what paraphrase is used. An example of this is the expression $4 - 8 + 1$. Such cases will be discussed in the section on integers but note that even here it is easier to add 4 and 1 first and *then* subtract 8 from the result. This generates a negative number at the very end of the calculation. Performing the operations as they appear from left to right would generate a negative number right away requiring that we add 1 to it next. As this argument shows, the second alternative for the evaluation of a string of additions and subtractions would, at most, generate a single negative number and it does so at the end of the evaluation process. Other alternatives may introduce negative numbers earlier which means that we will have to continue with the evaluation process with negative numbers in the game.

Algorithm for Evaluation of Expressions Involving Chain Addition/Subtraction of Whole Numbers

If you have access to a calculator, then go from left to right and perform the operations as you see them[44] otherwise

1. Add the terms that are being added.[45] Use the *Algorithm for Evaluation of Addition and Subtraction of Two Whole Numbers.*
2. Add the terms that are being subtracted[46] Use the *Algorithm for Evaluation of Addition and Subtraction of Two Whole Numbers.*
3. Subtract the latter sum from the former sum. Use the *Algorithm for Evaluation of Addition and Subtraction of Two Whole Numbers.*

We now take up the third problem in the Sample set that we listed at the beginning of this subsection.

Example

Evaluate
$$15 \times 2 \times 12 \times 4$$

Solution

This expression analyzes into a single term.
$$15 \times 2 \times 12 \times 4 \;=\; \boxed{15 \times 2 \times 12 \times 4}$$

The term above contains four factors: 15, 2, 12, and 4. We have now arrived at the smallest components of the original expression.[47] This marks the end of the analysis stage.

---

[44] This approach also works well by hand if the numbers involved are small.
[45] Those with a + on their left side.
[46] Those with a − on their left side.
[47] These are the numbers 15, 2, 12 and 4 in the original expression.

We now move on to the synthesis stage by multiplying the factors to evaluate the term.

$$= 1440$$

The full solution without the intervening explanations and adorning boxes would look like the following.

$$15 \times 2 \times 12 \times 4 = 1440$$

The manner in which we deal with division within a term depends on the notation used to represent division with. Formally, we use the horizontal line to represent division with.

Example

Evaluate

$$8 \times 2 + 5 \times \frac{6}{2}$$

Solution

This expression has two terms.

The first term, 8×2, has two factors: 8 and 2. Since the factors have been worked out, we can multiply them to evaluate the first term to get 16.

The second term, $5 \times \frac{6}{2}$, also contains two factors: 5 and $\frac{6}{2}$. The first factor has been worked out but the second factor can be simplified. We get

$$8 \times 2 + 5 \times \frac{6}{2} = 16 + 5 \times 3$$

The use of the horizontal line to represent division makes it easy to visually place division within a factor. In the example above, the division $\frac{6}{2}$ is within the second factor in the second term.

Having worked out the factors in the second term, we can now multiply them to evaluate the second term and then add the terms to evaluate the expression.

$$8 \times 2 + 5 \times \frac{6}{2} = 16 + 5 \times 3$$
$$= 16 + 15$$
$$= 31$$

While the horizontal line is the preferred notation to represent division with, alternative notations for division, e.g., $\div$, $/$, etc., are frequently used by some who may not be as well-versed in the art of communicating ideas using mathematical notation, and, therefore, we still need to know how to work with such alternative notations. We will discuss constructs using alternative notations for division using examples.

Example

Evaluate

$$20 - 12 \times 2 \div 4 + 8 \div 2$$

Solution

This expression analyzes into three terms as shown below.

$$20 - 12 \times 2 \div 4 + 8 \div 2 \;=\; \boxed{20} \;-\; \boxed{12 \times 2 \div 4} \;+\; \boxed{8 \div 2}$$

The first term, 20, contains a single factor which is 20. However, when we try to break the second term, $12 \times 2 \div 4$, into factors, we run into a problem as factors are entities within terms that are multiplied, not those that are divided. It is not immediately clear how one can deal with the $\div$ symbol within the second term.[48]

Expressions that involve a chain of $\times$ and $\div$ operations can always be evaluated from left to right. This yields the following

$$20 - 12 \times 2 \div 4 + 8 \div 2 \;=\; \boxed{20} \;-\; \boxed{12 \times 2 \div 4} \;+\; \boxed{8 \div 2}$$
$$= 20 - 6 + 4$$

We can now add and subtract the terms to get

$$= 18$$

---

[48]The problem arises from the use of the $\div$ notation to represent division with. Had we used the horizontal line instead, the expression would have been written as $12 \times \frac{2}{4}$ which would easily allow us to break the term into two factors: 12 and $\frac{2}{4}$. The use of $\div$ for division disrupts analysis and this is another reason why we prefer the use of the horizontal line to denote division over the use of such other notations as the $\div$ symbol. Similarly, if we had used the horizontal line to refer to the division in the third term, we would have had $\frac{8}{2}$ which would represent a single factor.

The horizontal line has many advantages over the $\div$ notation in addition to easing the task of analysis. See Appendix E for more on the advantages of the horizontal line over alternative notations for division.

The full solution without the intervening explanations and the adorning boxes would look like the following.

$$20 - 12 \times 2 \div 4 + 8 \div 2 = 20 - 6 + 4$$
$$= 18$$

Evaluation of the second term in the expression above, i.e., $12 \times 2 \div 4$, can be done in various ways. One approach is to go from left to right and perform the multiplications and divisions in the order that we see them. This is the approach that we chose above when we evaluated the second term. However, there are alternatives. The expression

$$12 \times 2 \div 4$$

requires that we begin with 12, and then perform the following activities in any order that we please: multiply by 2 and divide by 4. Note that each $\times$ and each $\div$ is associated with the value that follows it. The proper manner to look at $12 \times 2 \div 4$ is shown below.

$$12 \times 2 \div 4$$

The expression above is identical to

$$12 \div 4 \times 2$$

Although in some cases it may not make much difference which interpretation is used, in many cases one approach may be a lot more advantageous. In the expression above it is more convenient to divide 12 by 4 first and *then* multiply the result by 2 which keeps intermediate values small compared to the alternative of multiplying 12 by 2 to arrive at 24, and then dividing 24 by 4 to get 6.

In other cases a change in the order can help us avoid the need to use rational numbers. Consider the expression

$$4 \div 5 \times 10$$

While one can divide 4 by 5 and then multiply the result by 10 to arrive at the correct answer, the division of 4 by 5 will require the use of rational numbers. It would be much easier to multiply 4 by 10 and *then* divide the result by 5. In some cases, of course, the use of rational numbers may be unavoidable no matter what paraphrase is used. An example of this is the expression $3 \div 2 \times 5$. Such cases will be discussed in the section on rational numbers but note that even here it is easier to multiply 3 and 5 first and *then* divide the result by 2. This generates a rational number at the very end of the calculation. Performing the operations as they appear from left to right would generate a rational number right away requiring that we multiply it by 5 next.

Example

Evaluate

$$15 \div 2 \times 12 \div 3$$

Solution

As shown below, the expression contains a single term. We evaluate the expression inside the term to get

$$15 \div 2 \times 12 \div 3 \;=\; \boxed{15 \div 2 \times 12 \div 3}$$
$$= 30$$

In working out the expression within the term, we started with 15. Next we divided 15 by 3 to get 5. Next we multiplied 5 by 12 to get 60. We then divided 60 by 2 to get 30. This sequence avoids the need to use fractions.

We could still improve on this sequence by noting that any entity that is being multiplied with (with a $\times$ on its left, including the first number which may be preceded by $1\times$) can be divided by any entity that is being divided by (with a $\div$ on its left) and the results of such divisions may then be multiplied together to evaluate the chain. Following this approach, we would divide 15 by 3 to get 5, divide 12 by 2 to get 6, and then multiply 5 by 6 to get 30. The advantage is that this sequence keeps intermediate values down while still bypassing the need to use fractions.

Algorithm for Evaluation of Expressions Involving Chain Multiplication/Division of Whole Numbers

If you have access to a calculator, then go from left to right and perform the operations as you see them[49] otherwise

1. Divide any number that is being multiplied by[50] by any number that is being divided by[51]. Use the *Algorithm for Evaluation of Multiplication and Division of Two Whole Numbers.*
2. Multiply the results of the divisions from the previous step. Use the *Algorithm for Evaluation of Multiplication and Division of Two Whole Numbers.*

---

[49]Set your calculator's mode to formal notation to prevent the calculator from using base-ten notations.

[50]A number with a $\times$ on its left or the leading number which may be preceded by $1\times$ or a number on top of a horizontal line.

[51]A number with a $\div$ on its left or a number under a horizontal line.

We now move on to the last problem in the sample set listed at the beginning of this subsection.

Example

Evaluate

$$5 \times 3 + 4(5 + 2 \times 6)$$

Solution

We start by analyzing the expression into terms. This yields the following.

$$5 \times 3 + 4(5 + 2 \times 6) = \boxed{5 \times 3} + \boxed{4(5 + 2 \times 6)}$$

Note that we have used the additions and subtractions *outside* brackets to identify the terms.

The first term contains two factors: 5 and 3. Analysis of the first term is complete as we have arrived at the smallest components of that term.[52]

The second term contains two factors: 4, which cannot be analyzed further, and the expression within brackets, i.e., $5 + 2 \times 6$. This expression can be analyzed further: The expression contains two terms. The first term contains a single factor: 5. The second term contains two factors: 2 and 6. We have now arrived at the smallest components of the expression within brackets.[53] We have also arrived at the smallest components of the expression within the second term.[54] The analysis stage is now complete.

We move on to the synthesis stage. The factors within the first term are numbers. We can therefore multiply them to evaluate the term: 15. As far as the second term is concerned, we focus on the evaluation of the second factor within it. The first term of this factor is a number: 5. The second term can be worked out through the multiplication of 2 and 6 to yield 12. We now have

---

[52] These are the numbers 5 and 3.
[53] These are the numbers 5, 2 and 6.
[54] These are the numbers 4, 5, 2 and 6.

$$= \boxed{15} + \boxed{4\,(5\ +\ 12)}$$

We finish the evaluation of the second factor within the second term to get

$$= \boxed{15} + \boxed{4 \times 17}$$

Now that the factors within the second term have been evaluated, we multiply them to evaluate the term itself. This yields

$$= \boxed{15} + \boxed{68}$$

Now that the terms have been evaluated, we add them to evaluate the expression itself. This yields

$$= 83$$

The full solution without the intervening explanations and adorning boxes would look like the following.

$$
\begin{aligned}
5 \times 3 + 4\,(5\ +\ 2 \times 6) &= 15 + 4\,(5\ +\ 12) \\
&= 15 + 4 \times 17 \\
&= 15 + 68 \\
&= 83
\end{aligned}
$$

Beyond addition, subtraction and multiplication we do not assign priority to any higher order operations.[55] Rather, we let notation be our guide in pinpointing the next main operation. As an example, in

$$4\sqrt{5\ +\ 2 \times 3}$$

which is short for $4 \times \sqrt{5 + 2 \times 3}$, there is a single term within which there are two factors. The first is 4 and the second is $\sqrt{5 + 2 \times 3}$. We see this second

---

[55] Division covers a gray area. In classifying numbers it is given prominence through association with a set, i.e., $\mathbb{Q}$, that ensures maximum closure with respect to division. In other cases, as in here, division is grouped together with the higher order operations. From this point of view, higher order operations may be seen as part of a term and a factor within that term. Analysis of

$$4 \times \frac{1}{2}\ +\ 3\sqrt{2}$$

leads to two terms. The first is $4 \times \frac{1}{2}$ which contains two factors: 4 and $\frac{1}{2}$. The second term, $3\sqrt{2}$, is short for $3 \times \sqrt{2}$ and analyzes into two factors: 3 and $\sqrt{2}$. Note that, unless they appear as part of an expression within brackets or as arguments of higher order operations, the higher order operations of division in $\frac{1}{2}$ and taking roots in $\sqrt{2}$ are seen to form a factor of their own within the term in which they reside.

factor as a single entity as the horizontal arm of the radical sign covers the full expression $5+2\times3$. One can therefore say that the main operation within this second factor is the root. Under the arm of the radical sign, the main operation is addition, which breaks the expression into two terms and then multiplication within the second term that breaks the second term into two factors.

The discussion above leads to the following algorithm, called the *analysis-synthesis algorithm*, for the evaluation of expressions that involve whole numbers.

The Analysis-Synthesis Algorithm for Evaluation of Expressions Involving Addition, Subtraction, Multiplication and Division of Whole Numbers

1. Analyze:

   a. Analyze the expression into terms using additions and subtractions *outside* brackets.
   b. Analyze each term into factors using multiplications.
   c. Analyze each factor further if needed.

2. Synthesize:

   a. Evaluate the factors.
   b. Multiply the factors to evaluate the terms.[56] Use the *Algorithm for Evaluation of Expressions Involving Chain Multiplication/Division of Whole Numbers*.
   c. Add and subtract the terms to evaluate the expression.[57] Use the *Algorithm for Evaluation of Expressions Involving Chain Addition/ Subtraction of Whole Numbers*.

---

[56]Within a given term, we recommend that you wait until all the factors have been evaluated before you carry out the chain multiplication to evaluate the term. This is because the factors involved in the chain multiplication within a term are rich in semantics. Partial computations of the chain make it difficult to extract meaning from the values of the factors and their relative sizes. See the companion textbook *Semantics and the Syntax of Algebra* by the author for more on the semantics of factors within a term.

[57]We recommend that you wait until all the terms in the given expression have been evaluated before you carry out the chain addition/subtraction to evaluate the expression. The reason for this is similar to the one given for chain multiplication above: The terms involved in the chain addition/subtraction within an expression are rich in semantics. Partial computations of the chain make it difficult to extract meaning from the values of the terms and their relative sizes. See the companion textbook *Semantics and the Syntax of Algebra* by the author for more on the semantics of terms within an expression.

Example

Evaluate

$$3\,(8 \times 3 \ - \ 1) \ + \ 2\,(4 \ + \ 3 \times 2)$$

Solution

$$3\,(8 \times 3 \ - \ 1) \ + \ 2\,(4 \ + \ 3 \times 2)$$
$$= \boxed{3\,(8 \times 3 \ - \ 1)} \ + \ \boxed{2\,(4 \ + \ 3 \times 2)}$$
$$= \boxed{3\,(24 \ - \ 1)} \ + \ \boxed{2\,(4 \ + \ 6)}$$
$$= 3 \times 23 \ + \ 2 \times 10$$
$$= 69 \ + \ 20$$
$$= 89$$

Here is an expression involving nested brackets:

Example

Evaluate

$$2\left[4 \times 5 \ + \ 5\,(8 \times 2 \ - \ 3 \times 4)\right]$$

Solution

$$2\left[4 \times 5 \ + \ 5\,(8 \times 2 \ - \ 3 \times 4)\right]$$
$$= \boxed{2\left[4 \times 5 \ + \ 5\,(8 \times 2 \ - \ 3 \times 4)\right]}$$
$$= 2\left[20 \ + \ 5\,(16 \ - \ 12)\right]$$
$$= 2\,(20 \ + \ 5 \times 4)$$
$$= 2\,(20 \ + \ 20)$$
$$= 2 \times 40$$
$$= 80$$

Exercise Set 4.1.5

1. Evaluate each of the following expressions.

   a. $7 \times 6 + 9 \times 2$
   b. $15 \times 2 + 12 \times 3$
   c. $9 \times 4 - 2 \times 7$
   d. $17 \times 4 - 18 \times 3$
   e. $4 \times 3 + 7 \times 3 - 4 \times 2$
   f. $12 \times 5 - 8 \times 4 + 14 \times 3$
   g. $9 + 4 \times 6$
   h. $12 - 4 \times 2$
   i. $14 \times 11 - 10$
   j. $27 + 5 \times 4 - 2$
   k. $19 \times 2 + 3 - 2 \times 8$

   l. $14 \times 3 \times 4 + 15 \times 3 \times 6$
   m. $9 \times 3 \times 2 - 5 \times 3$
   n. $3 \times 4 + 5 \times 4$
   o. $6 + 7 \times 2$
   p. $6 + 6 \times 2 - 1$
   q. $13\,(2) + 14\,(3)$
   r. $5\,(4)\,(2) + 9\,(3)\,(7)$
   s. $52 - 5\,(3)\,(2)$
   t. $16\,(2)\,(5) + 12$
   u. $17\,(3) + 2\,(3)$

2. Evaluate each of the following expressions.

   a. $2\,(3 + 2 \times 5) + 3 \times 2$
   b. $5 \times 6 + 4\,(12 - 2 \times 2)$
   c. $35 - 3\,(4 \times 3 - 3 \times 2) + 4\,(3)$
   d. $2\,(3 \times 5 + 2 \times 3) + 4\,(24 - 2 \times 7)$
   e. $(3 + 4 \times 2)\,(2 \times 3 - 4) + 3\,(9 - 4 \times 2)$
   f. $(3 + 7)\,(9 - 2)\,(5 + 6)$
   g. $3\,(6 \times 2 + 2 \times 3) + 3 \times 8$
   h. $3\,(2 + 3 \times 7)\,(4 \times 2 - 1) + 4\,(15 \times 3 + 4 \times 0)$
   i. $5 \times 6 + 2\Big[8 + 3\,(4 - 2 \times 1)\Big]$
   j. $3\Big[2\,(4 \times 5 - 2 \times 6) + 5\,(3 \times 2 + 1)\Big]$
   k. $8\,(3 + 2 \times 6) - 4\Big[15 - (4 \times 5 - 6)\Big]$
   l. $3\,(2 + 6) - 4 \times 5 + 8\Big[6 - 2\,(4 - 1)\Big]$
   m. $12\Big[30 - 4\,(5 + 2 \times 6) + 8\,(19 - 2 \times 5)\Big]$
   n. $\Big[26 - 3\,(15 - 2 \times 6)\Big]\Big[4\,(3 \times 2 - 2 \times 1) + 16\Big]$
   o. $2\Big[3\,(14 - 4 \times 3) - 4\,(5 \times 6 - 3 \times 10)\Big] + 3\,(4 \times 6 - 3 \times 2)$

3. Evaluate each of the following expressions.

   a. $4 + 5 \times 2 - 10 \div 5$
   b. $14 \div 2 \times 7 + 3$
   c. $84 \div 2 \times 3 \div 7$
   d. $3 + 2 \times 5 + 7 \div 7 \times 2$
   e. $3 + 6 \times 3 \times 2 - 1 + 4 \div 2$

   f. $4 \div 2 + 5 \times 3 \div 15 + 4$

   g. $2\left(3 \times 2 + 2 \times 4\right) + 3 \div 3$

   h. $100 + 20 \times 6 \div 2 - 100 \div 50 \times 2$

   i. $4 \times 3 + 10 \div 5 \times 2 - 4\left(3 \times 2 - 6\right)$

4.1.6   Graphing

Draw a straight, horizontal line and place an arrow head on its right end. Somewhere on this line place a tick and label it as 0. To the right of 0 place a tick and label it as 1. Mark ticks to the right of 1 at distances equal to the distance from 0 to 1. Label these ticks 2, 3, 4, and so on. Mark ticks to the left of 0 at distances equal to the distance between 0 and 1 but do not label them.[58]

The result is called the **real line** shown in Figure 4.1.8.

Figure 4.1.8: Whole numbers on the real line

The real line does not have to be horizontal. It may be vertical or stand at an angle.

By the **graph** of 4 we mean the point on the real line that corresponds to 4. To make the point stand out we place a dot on the point. The dot should be large enough to be seen clearly. We refer to 4 as the **coordinate** of the point. Figure 4.1.9 below shows the graph of 4.

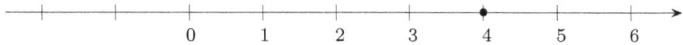

Figure 4.1.9: Graph of 4 on the real line

Step Size

Ticks do not have to mark every single whole number. As an example, in Figure 4.1.10 we have ticks at steps of 5.

---

[58]See the section on integers for how ticks on the left of 0 are labelled.

Figure 4.1.10: Real line with a step size of 5 units

Choose step sizes that are easy to use. Humans find it easy to count in steps of 1, 2, 5, 10, 20, etc. but not 4, 7, 19, etc.

If step size is larger than 1, then you might have to estimate the coordinate of a point on the line. As an example, one might estimate the coordinate of the point in Figure 4.1.11 to be approximately 45.

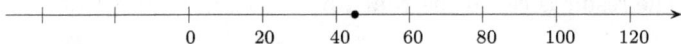

Figure 4.1.11: Estimating the coordinate of a point that does not fall on a tick

Exercise Set 4.1.6

1. In each case determine the coordinate of the point on the real line.

   a.

   b.

   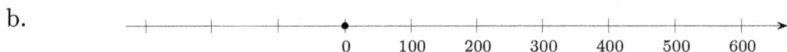

   c.

   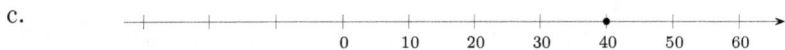

2. In each case estimate the coordinate of the point on the real line.

   a.

   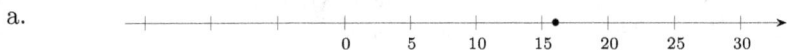

b.

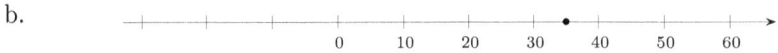

c.

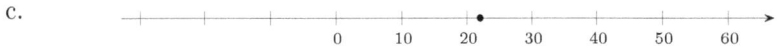

d.

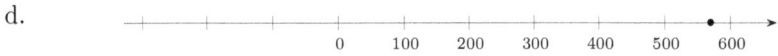

e.

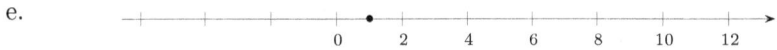

f.

3. Graph each of the following on the real line with a step size of 1.

    a. 6   b. 0   c. 4

4. Graph 20 on the real line with a step size of 5.
5. Graph 35 on the real line with a step size of 10.
6. Graph 12 on the real line with a step size of 20.
7. Graph 0 on the real line with a step size of 10.
8. Graph 50 on the real line with a step size of 50.

4.2 Integers

In this section we will discuss aspects of integers that are of interest to us. These include structure, formal reading, order, algorithms for carrying out arithmetic operations, evaluation of expressions, and graphing on the real line.

4.2.1 Structure

An integer is either negative, 0, or positive.[59]

---

[59] A **negative number** is a number that is less than 0. A **positive number** is a number that is greater than 0.

Integers that are less than 0 are negative. Such integers are written using the negative sign (which, for reasons that we discussed earlier in the text, is derived from the subtraction sign, i.e., $-$) followed by a *natural* number. As examples, -5, -22 and -1320 are all negative integers.[60]

Integers that are greater than 0 are positive. Such integers are written using an optional positive sign (which, for reasons similar to those presented earlier for the negative sign, is identical to the addition sign, i.e., $+$) followed by a *natural* number. As examples, $+5$, 22 and $+1320$ are all positive integers.[61]

0 is neither positive nor negative. This is due to our definitions of positive and negative numbers: A positive number is, by definition, greater than 0 (and so 0 cannot be positive as it is not greater than 0). Similarly, a negative number is, by definition, less than 0 (and so 0 cannot be negative as it is not less than 0).

Absolute Value

At times we are interested in, not whether a given number is negative or positive, but simply in how large that number is. As an example of this, consider a car, moving along a road, with a velocity of -125 km/h. The negative sign tells us about the direction of motion. If we assign positive to east, as an example, then the negative sign in the example above tells us that the car is travelling west. However, if the direction of motion is of no interest to us and we wish to talk about *how fast* the car is moving (as might be the case when one receives a ticket for speeding), then we speak of the *size* of the velocity of the car, i.e., 125 km/h for the car in the example above, a quantity that we call *speed*.

For us, then, it does make sense to talk about the *size* of a number, an attribute that we officially refer to as the **absolute value** of a number. To indicate that we are seeking the absolute value of a number, we place the number within vertical bars. As an example, $|-125|$ speaks of the absolute value of -125 which is equal to 125, i.e.,

$$|-125| = 125$$

---

[60]Note that, while numbers such as $-2\frac{3}{4}$, -5.6 and $-\sqrt{2}$ and the like are all negative numbers, they are not classified as integers as the numeric part is not a natural number. We will discuss these other negative numbers in later sections of the textbook.

[61]We follow the common approach and drop the $+$ sign (and, therefore, write 42 in place of $+42$) unless we wish to emphasize the fact that the number *is*, indeed, positive (in which case we write $+42$ in place of the common 42).

Note also that, while numbers such as $+2\frac{3}{4}$, $+5.6$ and $+\sqrt{2}$ and the like are positive numbers, they are not classified as integers as the numeric part is not a natural numbers. We will discuss these other positive numbers in later sections of the textbook.

We read this as *the absolute value of* −125 *is* 125 and by this we mean that the *size* of −125 is 125 units.

Note that we can also speak of the size of a positive number. We can, as an example, speak of the speed of a car with a velocity of +90 km/h with the + sign indicating that, according to our convention above, it is moving east. The speed would be 90 km/h and we write

$$|+90| = 90$$

We read this as *the absolute value of* +90 *is* 90. Since 90 can also be used as a shorthand for +90, we can also write

$$|90| = 90$$

In applications, we expect the size of 0 to be 0 and we define the absolute value of 0 accordingly:

$$|0| = 0$$

Examples

Evaluate each of the following expressions.

1. $|-154|$ 2. $|+32|$ 3. $|1400|$ 4. $|0|$

Answers

1. 154   2. 32   3. 1400   4. 0

Exercise Set 4.2.1

Evaluate each of the following expressions.

1. $|-2|$ 3. $|-4100|$ 5. $|32|$
2. $|+47|$ 4. $|0|$ 6. $|-1|$

4.2.2   Formal Reading

To write an integer formally, write the numeric part following the steps that we provided earlier on the reading of whole numbers preceded by the word *negative* if the given integer is negative and *positive* if the given integer is positive.[62]

If absolute values are sought, precede the reading above with the phrase *the absolute value of.*

Examples

Write each of the following in words using formal English.

1. −34   3. 18   5. |−17|
2. +15   4. 0   6. |25|

Answers

1. negative thirty-four   3. eighteen
2. positive fifteen   4. zero

    the absolute value of negative seventeen
6. the absolute value of twenty-five

Exercise Set 4.2.2

Write each of the following in words using formal English.

1. −2	5. −104	9. −15	13. 0
2. +6	6. |−7|	10. +205	14. −1
3. −198	7. |0|	11. 1200	15. −20 004
4. +782	8. |12|	12. |+410|	16. |1|

---

[62]Normally, if the positive sign is written, it implies that the author wishes to emphasize the fact that the number being discussed is positive and therefore the word *positive* should be used in its reading.

We should also mention that the use of the words *plus* and *minus* is strictly forbidden if we wish to remain formal. While the symbol used to denote a positive number is related to the symbol used to represent addition (which is why the two symbols are chosen to be the same) and although the symbol used to denote a negative number is related to the symbol used to represent subtraction (which is why the two symbols are chosen to be the same), we use the sign interpretation to point to the value that one arrives at as a result of the additions or subtractions that generate those values. And while the symbols used to represent the two interpretations are set to be the same (emphasizing the relationship between the two), the readings of the two differ (positive vs. plus or negative vs. minus), helping us differentiate which interpretation we have in mind. See Chapter 2 for more on the connection between the two interpretations of the + and − symbols.

4.2.3   Order

When we speak of *order*, we are *not* referring to order according to *size* (which would be a comparison of the absolute values of the integers) but the relative sequencing of the integers in the listing

$$\ldots, -3, -2, -1, 0, 1, 2, 3, \ldots$$

To order integers according to the listing above

- if the numbers being compared are both positive, follow the guidelines for ordering whole numbers.
- if one number is positive and the other negative, then the positive number is always greater than the negative number, irrespective of the relative sizes of the numbers. As an example $2 > -8$.
- If the numbers being compared are both negative, then the one with the larger size is *less* than the one with the smaller size. As an example $-7 < -2$.

If we wish to order integers according to size, then we compare their absolute values. We can, therefore, write

$$|-3| > |-1|$$

which tells us that the size of -3 (which is 3) is greater than the size of -1 (which is 1).

Exercise Set 4.2.3

1. In each case insert $>$ or $<$ between the pair of numbers to make a true statement.

   a. -20 -47 e. $|-25|$ $|1|$

   b. -6271 -5989 f. $|0|$ $|-20|$

   c. -240 15 g. $|34|$ $|12|$

   d. -6 0 h. $|18|$ $|-40|$

2. Arrange the following in descending order.

   a. −4, −8, −2, −16, 0   c. −746, −747, −744, −745
   b. 826, −950, 552, −700   d. −420, −310, −18, −84

3. Arrange the following in ascending order.

   a. 15, −18, 0, −3, 2   c. −425, 710, −1020, −95, 240
   b. −126, −324, −120, −150   d. −591, −593, −590, −592

4.2.4   Addition, Subtraction, Multiplication and Division

We begin this section with a discussion on addition and subtraction of integers. Following this, we will discuss the multiplication and division of integers.

Addition and Subtraction

Of the many scenarios that might arise, some are identical to cases that we have already covered in our study of whole numbers. An example of this is the addition of integers 4 and 6, i.e., $4 + 6$, and we already know how to simplify such expressions. Once we remove the familiar forms from the list of possibilities, we end up with four novel scenarios. These are illustrated using examples below.

Scenario 1:	$6 - 8$
Scenario 2:	$-5 + 9$
Scenario 3:	$-5 + 2$
Scenario 4:	$-2 - 7$

When working with such expressions, we can think of the terms that are being added (those with a + sign on their left) as terms that represent *gains* and those that are being subtracted (those with a − sign on their left) as *losses*. We will, therefore, think of the expression $6 - 8$ as *a gain of* 6 *followed by a loss of* 8 which we would expect to result in a net loss of 2. Similarly, we can think of the expression $-5 + 9$ as *a loss of* 5 *followed by a gain of* 9 which would result in a net *gain* of 4. Moving on to the other two expressions above, we think of $-5 + 2$ as *a loss of* 5 *followed by a gain of* 2, resulting in a net loss of 3, and we interpret $-2 - 7$ as *a loss of* 2 *followed by a loss of* 7 which would represent a net loss of 9.

As the four novel scenarios show, the sign of the answer in the first, second and third scenarios (in which there is mix of a gain and a loss) depends on whether the gain is larger than the loss (in which case the answer is positive) or whether the loss is larger than the gain (in which case the answer is negative). Having set the sign of the answer, the numerical value that follows can, in all three cases, be calculated by subtracting the smaller value from the larger value. The subtraction reflects the fact that a gain and a loss are opposites: One adds, the other subtracts.

As for the fourth scenario (in which one loss is followed by another), the losses add up to a net loss and, therefore, the sign of the answer will be negative. Having set the sign of the answer, the numerical value that follows can be calculated by adding the given numbers. This sum represents the size of the net loss.

The following algorithm summarizes the above.

Algorithm for Evaluation of Addition and Subtraction of Two Integers

1. If one term is being added and the other is being subtracted, then

   a. Determine the *sign* of the answer by comparing the relative sizes of the term that is being added and the term that is being subtracted. The answer will be positive if the term that is being added is larger in size than the one that is being subtracted and negative if the term that is being subtracted is larger in size than the one that is being added.

   b. Determine the *size* of the answer by subtracting the smaller value from the larger value. Use the *Algorithm for Evaluation of Addition and Subtraction of Two Whole Numbers*.

2. If both terms are being added or both terms are being subtracted, then

   a. Determine the *sign* of the answer by noting whether the terms are being added or subtracted. The answer will be positive if both terms are being added and negative if both terms are being subtracted.

   b. Determine the *size* of the answer by adding the two terms. Use the *Algorithm for Evaluation of Addition and Subtraction of Two Whole Numbers*.

Examples

Evaluate each of the following expressions.

1. $15 - 20$	3. $-12 + 8$	5. $82 - 150$
2. $-92 - 14$	4. $-14 + 17$	6. $-150 + 82$

Answers

1. -5	3. -4	5. -68
2. -106	4. 3	6. -68

Having studied the scenarios above, we can now explain how one deals with cases when a signed number is added to, or subtracted from, another. Here are the possibilities:[63],[64]

$$\ldots + (+5)$$
$$\ldots + (-5)$$
$$\ldots - (+5)$$
$$\ldots - (-5)$$

---

[63] The ellipsis at the beginning of each of these phrases implies that some work has been done and now we arrive at the specifics of what needs to be done next. And as usual, we associate each addition and subtraction to the term that follows it. As an example, the second line, i.e.,

$$\ldots + (-5)$$

should be viewed as

$$\ldots + \boxed{(-5)}$$

i.e., *starting with whatever we are given or whatever we have calculated, now we want to add negative five.* A similar comment applies to the other constructs in the cases above.

[64] The use of brackets around the signed number is required as their use facilitates analysis: It is easy to analyze the following expression into four terms (as the additions and subtractions stand out)

$$3\,(-2) + (-5) - (-2)\,(-5) - (-3)$$

but the following formulation is very difficult to decode:

$$3 \times -2 + -5 - -2 \times -5 - -3$$

The first case speaks of adding positive 5 and it is somewhat obvious that this is the same as adding 5 itself, i.e.,

$$\ldots + (+5) = \ldots + 5$$

The second case speaks of adding negative 5. We expect this to be the same as subtracting 5. This makes sense as *adding a loss* is the same as *subtracting the size of the loss*. We arrive at

$$\ldots + (-5) = \ldots - 5$$

The third case speaks of subtracting positive 5. This is obviously the same as subtracting 5 itself, i.e.,

$$\ldots - (+5) = \ldots - 5$$

The last case speaks of subtracting negative 5. The expected behaviour of this is to add 5. To see why, consider the scenario where you start with $20. We write this as

20

Suppose that you owe your grandma $5 and you pay back the $5 to your grandma. We write this as

$$\underbrace{\text{pay}}\ \underbrace{\text{the \$5 that you owe}}$$
$$20 - 5$$

If your kindly grandma forgives the $5 that you owe, we arrive at

$$\underbrace{\text{pay}}\ \underbrace{\text{the \$5 that you owe}}$$
$$20 - 5 - (-5)$$
$$\underbrace{\text{forgive}}\ \underbrace{\text{the owed}}\ \underbrace{\text{\$5}}$$

Of course we expect that the loan forgiveness means that your grandma returns the $5 that you gave back to her so that we can add it back to your net worth, i.e.,

$$20 - 5 - (-5) = 20 - 5 + 5$$
$$= 20$$

We arrive at

$$\ldots - (-5) = \ldots + 5$$

Examples

Evaluate each of the following expressions.

1. $38 + (+5)$ 3. $-38 + (+5)$ 5. $-20 + (+35)$
2. $38 - (+5)$ 4. $-38 - (-5)$ 6. $-20 + (-35)$

Solutions

1.
$$38 + (+5) = 38 + 5$$
$$= 43$$

2.
$$38 - (+5) = 38 - 5$$
$$= 33$$

3.
$$-38 + (+5) = -38 + 5$$
$$= -33$$

4.
$$-38 - (-5) = -38 + 5$$
$$= -33$$

5.
$$-20 + (+35) = -20 + 35$$
$$= 15$$

6.
$$-20 + (-35) = -20 - 35$$
$$= -55$$

Multiplication and Division

Here we will focus on the rules surrounding the multiplication of two nonzero signed numbers.[65] Since division can be viewed as inverse multiplication, the same rules apply to the division of signed numbers.

To understand how multiplication of signed numbers work, we present three scenarios:

Scenario 1: $4(-5)$

Scenario 2: $-4(5)$

Scenario 3: $-4(-5)$

These three scenarios cover all the possibilities where at least one of the operands is negative.

The first scenario evaluates to -20, i.e.,

$$4(-5) = -20$$

To explain why this makes sense, consider the case where we owe \$5 to each of 4 people. It should make sense that we should be able to multiply 4 by -5 to arrive at total amount owed and that we expect that the result will tell us that we owe \$20, i.e., that our net worth is -20.

The second scenario also evaluates to -20, i.e.,

$$-4(5) = -20$$

Here, we see the expression as a loss of 4, counted 5 times which adds up to a loss of 20, i.e., -20.

The third scenario evaluates to 20, i.e.,

$$-4(-5) = 20$$

We may view this as a forgiveness of 4 loans of \$5 each resulting in a gain of \$20.

The following algorithm summarizes these observations.

Algorithm for Evaluation of Multiplication and Division of Two Integers

1. Determine the *sign* of the answer: The product or quotient of two nonzero integers with the same signs is positive. The product or quotient of two nonzero integers with different signs is negative.

---

[65]Multiplication by 0 is trivial as it generates 0.

2. Determine the *size* of the answer by multiplying or dividing the given numbers. Use the *Algorithm for Evaluation of Multiplication and Division of Two Whole Numbers.*

Examples

Evaluate each of the following expressions.

1. $-2(-20)$ 3. $-14(-8)$ 5. $-150 \div 75$
2. $\frac{45}{-9}$ 4. $14 \div (-7)$ 6. $-32(14)$

Answers

1. 40   3. 112   5. -2
2. -5 4. -2 6. -448

Exercise Set 4.2.4

1. Evaluate the following expressions.

   a. $5 + 6$ i. $245 + 359$ q. $42 + 0$
   b. $5 - 6$ j. $245 - 359$ r. $42 - 0$
   c. $-5 + 6$ k. $-245 + 359$ s. $-42 + 0$
   d. $-5 - 6$ l. $-245 - 359$ t. $-42 - 0$
   e. $18 + 12$ m. $2450 + 325$ u. $0 + 15$
   f. $18 - 12$ n. $2450 - 325$ v. $0 - 15$
   g. $-18 + 12$ o. $-2450 + 325$
   h. $-18 - 12$ p. $-2450 - 325$

2. Evaluate the following expressions.

   a. $8 + (+6)$ e. $42 + (-12)$ i. $23 + (+34)$
   b. $20 - (-18)$ f. $31 - (+1)$ j. $627 - (+123)$
   c. $-3 + (+7)$ g. $-8 + (-140)$ k. $-195 + (-62)$
   d. $-21 - (-8)$ h. $-1 - (+10)$ l. $425 + (-510)$

3. Evaluate the following expressions.

a. 14×5
b. $14\,(-5)$
c. -14×5
d. $-14\,(-5)$
e. 59×0
f. -72×0
g. $32 \div 8$
h. $32 \div (-8)$

i. $-32 \div 8$
j. $-32 \div (-8)$
k. $0 \div 18$
l. $0 \div (-18)$
m. $-25 \div 0$
n. $+54 \div 0$
o. $0 \div 0$

p. $\frac{-15}{-3}$
q. $\frac{12}{-4}$
r. $\frac{-100}{-2}$
s. $\frac{32}{-16}$
t. $\frac{-18}{-9}$
u. $\frac{-3}{0}$

4.2.5   Expressions

We will begin with an algorithm for the evaluation of expressions that involve chain addition/subtraction of integers. Next we will present an algorithm for the evaluation of expressions that involve chain multiplication/division of integers. We will end this subsection with the extension of the analysis-synthesis algorithm, introduced in the section on whole numbers, for rapid evaluation of expressions that involve integers with a mix of the four basic arithmetic operations.

Once the terms have been evaluated and the signs simplified, expressions involving integers degenerate into chain addition/subtraction of whole numbers with the possibility of a leading negative number. The algorithm for the evaluation of such chains remains the same as the one given for whole numbers.

Algorithm for Evaluation of Expressions Involving Chain Addition/Subtraction of Integers

If you have access to a calculator, then go from left to right and perform the operations as you see them[66] otherwise

1. Add the terms that represent gains.[67] Use the *Algorithm for Evaluation of Addition and Subtraction of Two Whole Numbers*.
2. Add the terms that represent losses.[68] Use the *Algorithm for Evaluation of Addition and Subtraction of Two Whole Numbers*.
3. Subtract the sum of the losses from the sum of the gains. Use the *Algorithm for Evaluation of Addition and Subtraction of Two Integers*.

---

[66]This approach also works well by hand if the numbers involved are small.
[67]Those with a + on their left side.
[68]Those with a − on their left side.

Example

Evaluate

$$-25 + 18 - 72 - 12 + 19 - 31$$

Solution

$$-25 + 18 - 72 - 12 + 19 - 31 = -103$$

We can arrive at this result following either alternative given above. While either alternative generates -103 as the result, the intermediate steps that are taken in the evaluation of this expression are not the same. Following the first alternative we would have the following intermediate steps:

$$
\begin{aligned}
-25 + 18 - 72 &- 12 + 19 - 31 \\
&= -7 - 72 - 12 + 19 - 31 \\
&= -79 - 12 + 19 - 31 \\
&= -91 + 19 - 31 \\
&= -72 - 31 \\
&= -103
\end{aligned}
$$

while following the second approach we would have the following intermediate steps:

$$
\begin{aligned}
-25 + 18 - 72 &- 12 + 19 - 31 \\
&= 18 + 19 - (25 + 72 + 12 + 31) \\
&= 37 - 140 \\
&= -103
\end{aligned}
$$

The algorithm developed in the previous section on evaluating the products and quotients of nonzero integers can be generalized into a two-step algorithm for the evaluation of expressions that contain chain multiplication and/or division of integers.

Algorithm for Evaluation of Expressions Involving Chain Multiplication/Division of Integers

If you have access to a calculator, then go from left to right and perform the operations as you see them[69] otherwise

1. Determine the *sign* of the result: If there are an even number of negative signs in the chain, the result is positive otherwise the result is negative.[70] From this point on, ignore all signs.
2. Determine the *size* of the result: Multiply and/or divide the numbers as requested. Use the *Algorithm for Evaluation of Expressions Involving Chain Multiplication/Division of Whole Numbers*.

The logic behind the first step is that, according to the *Algorithm for Evaluation of Multiplications and Divisions of Two Integers* presented in the previous subsection, pairs of negative signs multiply or divide to generate positive results so that an even number of negative signs will eventually turn positive while an odd number of negative signs will eventually turn negative.

Note that other than the additional requirement to determine the sign of the result, the algorithm above is identical to the corresponding algorithm given for whole numbers.

Example

Evaluate

$$-5\,(-2)\,(4)\,(-3)$$

Solution

There are three negative signs. Since three is odd, the result will be negative. We write:

$$-5\,(-2)\,(4)\,(-3)\ =\ -$$

We now ignore all the negative signs and evaluate $5 \times 2 \times 4 \times 3$ which evaluates to 120. We write

$$-5\,(-2)\,(4)\,(-3)\ =\ -120$$

---

[69]Set your calculator's mode to formal notation to prevent the calculator from using base-ten notations.

[70]We recommend that you write the sign down if it happens to be negative otherwise chances are you might forget to do so at the end.

Examples

1. $-5\,(-6)\,(3) \;=\; 90$
2. $3\,(-4) \;=\; -12$
3. $-15 \div (-3)\,(2) \;=\; 10$
4. $\dfrac{-10\,(-6)\,(-4)}{3\,(-2)\,(5)} \;=\; 8$
5. $\dfrac{-4}{-2} \times \dfrac{-15}{3} \;=\; -10$
6. $-3\,(-2)\,(-4)\,(3)\,(2) \;=\; -144$

Step-by-step BEDMAS takes much longer and involves a lot more mental activity.[71] Here is the solution to the last example above using step-by-step BEDMAS.

$$
\begin{aligned}
-3\,(-2)\,(-4)\,(3)\,(2) &= 6\,(-4)\,(3)\,(2) \\
&= -24\,(3)\,(2) \\
&= -72\,(2) \\
&= -144
\end{aligned}
$$

The analysis-synthesis technique introduced in the section on whole numbers can be extended to deal with the evaluation of expressions that involve the addition, subtraction, multiplication and division of integers.

The Analysis-Synthesis Algorithm for Evaluation of Expressions Involving Addition, Subtraction, Multiplication and Division of Integers

1. Analyze:
   a. Analyze the expression into terms using additions and subtractions *outside* brackets.
   b. Analyze each term into factors using multiplications.
   c. Analyze each factor further if needed.

---

[71]One might woder whether *more mental activity* is not better than less given that the subject of study is math. While this may be the case at times, as we tackle more and more complex problems, the need to resort to simple logic to process an expression in the step-by-step manner of the BEDMAS variety tends to interfere with the larger logic that tries to keep track of the larger steps that need to be taken to solve the problem. Here and elsewhere, the tools that we promote provide shortcuts that facilitate the processing of calculations that would otherwise take too long. Interestingly, the more efficient techniques promoted in this textbook tend to be the ones that are semantically loaded as well.

2. Synthesize:

   a. Evaluate the factors.

   b. Multiply the factors to evaluate the terms. To do so

      i. Determine whether the term should be added or subtracted. Count the number of negative signs in the term *including any subtraction sign that may sit on the left side of the term.*[72] If there is an even number of negative signs in the chain, the term should be added otherwise it should be subtracted.[73]

     ii. Determine the size of the term: Ignore all the signs and multiply and/or divide the numbers involved. To do so, follow the *Algorithm for Evaluation of Expressions Involving Chain Multiplication/Division of Whole Numbers.*

   c. Add and subtract the terms to evaluate the expression. To do so, use the *Algorithm for Evaluation of Expressions Involving Chain Addition/Subtraction of Integers.*

Note that other than the additional requirement to determine whether a term should be added or subtracted, the algorithm above is identical to the corresponding algorithm given for whole numbers.

Example

Evaluate

$$-3\,(-4)\,(2)\ -\ 2\,(-5)\,(-1)\ -\ 4\,(-3)$$

Solution

$$-3\,(-4)\,(2)\ -\ 2\,(-5)\,(-1)\ -\ 4\,(-3)$$
$$=\ 24\ -\ 10\ +\ 12$$
$$=\ 26$$

---

[72]This combines the rules for setting the sign of a chain multiplication/division and the rules that set the sign of expressions that involve the subtraction of a signed number.

[73]We recommend that you write the sign down if it happens to be negative otherwise chances are you might forget to do so at the end.

Exercise Set 4.2.5

1. Evaluate each of the following expressions.

    a. $-9 \times 4 + 5(-6)(8)$

    b. $-6 \times 4 \times 2 - (-7)(2)$

    c. $-12 \div (-3) \div 2 + (-5)(-2) - 18 \div (-2)(-7)$

    d. $-2 \times 4 + 6(-8)$

    e. $3(-2)(-4) - (-5)(6)$

    f. $-4 \times 2 \times 3 - 5(-3)(-8)$

    g. $4(-7) + (-6)(-4)$

    h. $-5(-8) - 3(-2)$

    i. $4(-3) + 2(-1) - (-6)$

    j. $3\Big[2 + 8(-1)\Big] - 2(-3)$

    k. $5 \times 4 + (-6)\Big[3 - 2(-2)\Big]$

    l. $-2\Big[4(-3) - (-2)(5)\Big]$

    m. $-(2 + 3 \times 6)$

    n. $-\Big[-8 - (-3)(-2)\Big]$

    o. $-2(8 - 3) - (-4 + 2)$

4.2.6   Graphing

In the section on whole numbers we introduced the real line and described the manner in which whole numbers can be placed on the real line. Following the size of the unit, we can locate the negative integers on the left side of 0 in a manner that is similar to the process for locating whole numbers on the right side of 0. The result is shown in Figure 4.2.1 below.

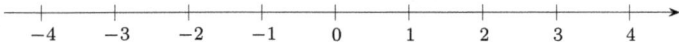

Figure 4.2.1: Integers on the real line

By the graph of -3 we mean the point on the real line that corresponds to -3. To make the point stand out we place a dot on the point. We refer to -3 as the coordinate of the point. Figure 4.2.2 below shows the graph of -3.

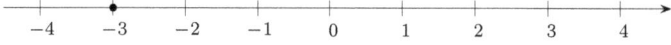

Figure 4.2.2: Graph of -3 on the real line

As before, ticks do not have to mark every single integer. As an example, in Figure 4.2.3 we have ticks at steps of 5.

Figure 4.2.3: Real line with a step size of 5 units

And as before, one should choose step sizes that are easy to use. Humans find it easy to count in steps of 1, 2, 5, 10, 20, etc. but not 4, 7, 19, etc.

If step size is larger than 1, then you might have to estimate the coordinate of a point on the line. As an example, one might estimate the coordinate of the point in Figure 4.2.4 to be approximately -28.

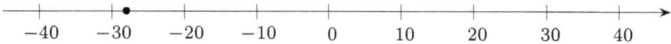

Figure 4.2.4: Estimating the coordinate of a point that does not fall on a tick

The reader should take note not to read this as -32. When reading the coordinate of a point on the line, it may help avoid this error by starting at 0 and then moving toward the point.

Exercise Set 4.2.6

1. In each case estimate the coordinate of the point on the real line.

a.

b.

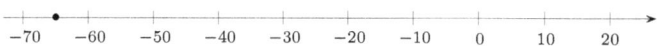

c.

d.

e.

f.

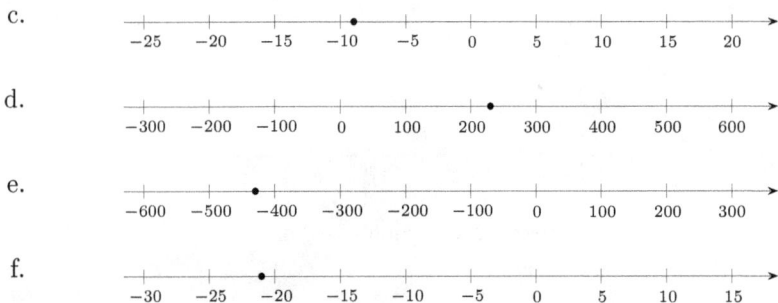

2. Graph -24 on a real line with a step size of 10.
3. Graph 24 on a real line with a step size of 10.
4. Graph -7 on a real line with a step size of 5.
5. Graph 7 on a real line with a step size of 5.
6. Graph -360 on a real line with a step size of 100.
7. Graph 360 on a real line with a step size of 100.

4.3   Rational Numbers

We begin this section with a discussion on background that is needed to work with rational numbers efficiently. Following this we will discuss aspects of the rational numbers that are of interest to us. These include structure, formal reading, order, algorithms for carrying out arithmetic operations, evaluation of expressions, and graphing on the real line. Initially we will focus on the positive rationals which, in this text, we refer to as **fractions**. Negative rationals are discussed whenever the opportunity to do so presents itself.

4.3.0   Fundamental Tools

Mastery of three basic tools is required if one wishes to develop the skill to work with rational numbers efficiently. These relate to a solid, working knowledge of the multiplication table, a knowledge of prime numbers and a knowledge of divisibility tests for 2, 3, 5 and 10.

The Multiplication Table

As the reader will soon note, the need to divide features prominently when working with rational numbers. This is not an altogether unexpected state of affairs as rational numbers deal with the representation of the results of division.

The ability to divide efficiently, in turn, rests on one's mastery of the multiplication table; not just in *multiplying* numbers as in '3 times 8' which should immediately bring 24 to mind, but also in dividing numbers as in knowing that, as an example, 8 *does* go into 24 and that it goes into 24, 3 times.

We assume that the reader does have a working knowledge of the multiplication table as described above.[74]

Prime Numbers

A knowledge of prime numbers greatly reduces the amount of work that one needs to do when reducing, ordering, adding, subtracting, multiplying and dividing rational numbers. Let us see how.

Consider the reduction of the rational number $\frac{6}{8}$. One should have little trouble noting that the rational number can be reduced to $\frac{3}{4}$ through the division of both the numerator and the denominator by 2.[75]

In the case of the rational number $\frac{6}{8}$ the numbers involved are small and quite familiar. We can tell at a glance that 2 goes into both 6 and 8. But what if the numbers were more complicated as in $\frac{52}{65}$? We invite the reader to spend a few minutes and try to reduce this rational number before reading further.[76]

One way to reduce $\frac{52}{65}$ is to use brute force: Start at 2 and work your way successively through 3, 4, 5, etc. as laid out below:

$$2 \quad 3 \quad 4 \quad 5 \quad 6 \quad 7 \quad 8 \quad 9 \quad 10 \quad 11 \quad 12 \quad 13 \quad 14 \quad \ldots$$

---

[74]Lack of a working knowledge of the multiplication table can result in the adoption of inefficient algorithms that are mentally heavy and can significantly increase the time that it takes to solve a problem. It will become a matter of time before the reader becomes frustrated with the amount of work that needs to be done to complete a set of problems. Access to a multiplication table or a calculator will not help as one would not know which numbers should be tried and will have to constantly look up the values on the table or punch the keys on the calculator to check over and over whether one number goes into another. The reader should also note that shortcuts in the working out of the results of multiplication in place of memorization are not helpful in division which requires the association of the product to factors that can generate it. *The only way to be functional with the multiplication table is to memorize it.*

[75]We will soon discuss reduction and other related topics in detail. However, we do assume that the reader has *some* familiarity with fractions and rely on that background here.

[76]You should do this. The activity will give you a feel for the amount of work that needs to be done if inefficient techniques are utilized.

We go on and on, trying to see whether we can divide both the numerator and the denominator of $\frac{52}{65}$ by 2, 3, 4, etc. until something works. For the current problem, the smallest number that goes into both the numerator and the denominator is 13.[77] If we were to adopt such an approach, we would succeed in reducing $\frac{52}{65}$ after 15 divisions.[78] Clearly this is unacceptable.

In an attempt to reduce the amount of work that needs to be done, we note the following: First, if we have tried to reduce a rational number using 2 and it did not work, then there is no need to try 4, 6, 8, 10, etc. (which are multiples of 2 other than 2 itself). This is because if a number (such as the numerator or denominator of a rational number) is not divisible by 2, then it will not be divisible by 4 either. Nor will it be divisible by 6, 8, 10, etc. Crossing out the numbers that do not need to be tested we arrive at the following:

$$2 \quad 3 \quad \cancel{4} \quad 5 \quad \cancel{6} \quad 7 \quad \cancel{8} \quad 9 \quad \cancel{10} \quad 11 \quad \cancel{12} \quad 13 \quad \cancel{14} \quad \ldots$$

Further, we can say that if we have tried to reduce a rational number using 3 and it did not work, then there is no need to try 6, 9, 12, 15, etc. (which are multiples of 3 other than 3 itself). The reason for this is similar to the argument that we provided above for 2. Crossing out these numbers as well, we arrive at the following:

$$2 \quad 3 \quad \cancel{4} \quad 5 \quad \cancel{6} \quad 7 \quad \cancel{8} \quad \cancel{9} \quad \cancel{10} \quad 11 \quad \cancel{12} \quad 13 \quad \cancel{14} \quad \ldots$$

Next we have 4 on the list. However, since 4 has already been crossed out, we move on to 5. As before, if we have tried to reduce a rational number using 5 and it did not work, then there is no need to try 10, 15, 20, etc. (which are multiples of 5 other than 5 itself). We cross out these numbers and then move down the list to 7 and its multiples and then 11 and its multiples and so on.

---

[77]Any number on the list that works should be tried repeatedly until it no longer works. One can stop if none of the numbers on the list work by the time one gets to half of the size of the smaller of the numerator and denominator.

[78]There will be one division for each of the numbers 3, 5, 6, 7, 8, 9, 10, 11 and 12 as the numerator, 52, is not divisible by any of them and therefore, once we have performed a division of 52 by any of these, we do not need to try the denominator, 65. In addition, there will be two divisions for each of 2 and 4 as they divide the numerator so they have to be tried for the denominator as well. Finally, there will be two divisions for 13 which works for both the numerator and the denominator. This adds up to a total of 15 divisions.

The numbers that are *not* crossed out are typed in bold in the list below.[79]

2   **3**   A̸   **5**   6̸   **7**   8̸   9̸   1̸0̸   **11**   1̸2̸   **13**   1̸4̸   ...

The reader might recognize these numbers as **prime numbers** and our interest in them is rooted in the fact that, by limiting ourselves to testing these numbers[80] when reducing rational numbers, we speed up the process significantly. In fact, if we limit ourselves to testing only the primes for the reduction of $\frac{52}{65}$, we will need to perform only 8 divisions, almost half as many as was needed with the brute force approach.

Divisibility Tests for 2, 3, 5 and 10

Returning for the moment to the reduction of $\frac{6}{8}$, we note that we did not need to perform long divisions to conclude that 6 and 8 are divisible by 2. We simply decided that this was the case by looking at the numbers. This is not so much because 6 and 8 are small; We can tell as quickly that a number such as 5736 is divisible by 2 but that another such as 6827 is not. How is it that we can say this so quickly?

The answer of course is that we already know that even numbers are divisible by 2 and that odd numbers are not. We also know that even numbers have 0, 2, 4, 6 or 8 on their right end whereas odd numbers have 1, 3, 5, 7 or 9 on their right end. Such knowledge allows us to quickly tell whether a given number is divisible by 2 or not by simply looking at the digit on its right end.

The ability to tell whether a number is divisible by 2 or not at a glance is a very useful tool as it frees us from carrying out a long division to find out whether this is so. We refer to a tool that allows us to determine quickly whether a given number is divisible by 2 or not as a *divisibility test for 2*.

The divisibility test for 2 given above is summarized below:

A Divisibility Test for 2
A given number is divisible by 2 if it is even (i.e., if its rightmost digit is 0, 2, 4, 6 or 8), otherwise it is not divisible by 2.

As for 2, there are divisibility tests for other numbers. The tests for primes are of special interest to us as they help us speed up the process of working with rational numbers significantly. And of the tests for the rest of

---

[79]This list never ends.

[80]Except when we already know that a certain number goes into both the numerator and the denominator in which case the number should be used right away, whether or not that number is prime.

the primes, two are very efficient: The divisibility tests for 3 and 5.[81] Here is a divisibility test for 3:

A Divisibility Test for 3
A given number is divisible by 3 if the sum of its digits is divisible by 3 otherwise it is not divisible by 3.[82]

As an example of this, the number 4251 is divisible by 3 as the sum of its digits, i.e., $4 + 2 + 5 + 1$, which is 12, is divisible by 3. However, the number 413 is not divisible by 3 as the sum of its digits, i.e., $4 + 1 + 3$, which is 8, is not divisible by 3.

Here is a divisibility test for 5:

A Divisibility Test for 5
A given number is divisible by 5 if its rightmost digit is either 0 or 5 otherwise it is not divisible by 5.

As an example of this, 270 and 185 are both divisible by 5 but 558 is not.

Other than primes, we find the divisibility test for 10 useful.

A Divisibility Test for 10
A given number is divisible by 10 if its rightmost digit is 0 otherwise it is not divisible by 10.

As an example of this, 450 and 300 are both divisible by 10 but 382 is not.

With the new knowledge that we have gained, we can now reduce the fraction $\frac{52}{65}$ after 4 divisions. While 2, 3, and 5 must be tried, we can quickly determine whether they work or not using the divisibility tests above. Actual division will have to be done only for 7 (one long division as it does not work for the numerator), 11 (one long division as it does not work for the numerator) and 13 (two long divisions as it works for both the numerator and the denominator).

The fact that we can reduce $\frac{52}{65}$ after 4 long divisions is significant. This is half as much work as the technique that relies on a knowledge of primes only and about a quarter of the work that is needed using brute force.[83]

While the relative savings computed above are particular to the fraction $\frac{52}{65}$, in practical situations it is generally the case that the savings are substan-

---

[81] The divisibility test for 7, the next prime on the list, is time-consuming and often it is faster to simply divide the given number by 7 to find out whether 7 goes into that number evenly. With few exceptions, as we move down the list of primes, the divisibility tests become more and more complex. However, since the majority of practical problems get resolved sooner rather than later, we will be content with a knowledge of the divisibility tests for 2, 3 and 5.

[82] For a proof of the divisibility test for 3 as given above, please see Appendix G.

[83] And, therefore, with such knowledge, it would take only 1 hour to do a homework that would otherwise take 4 hours.

tial when we understand that we need to limit ourselves to testing the primes and that, at least for 2, 3, 5 and 10, we do not have to resort to long division to find out whether they work or not. Add a solid, working knowledge of the multiplication table to the mix and we have the tools that we need to work with rational numbers efficiently and with confidence. Keep in mind also that primes and divisibility tests are important not just in the reduction of rational numbers, but also in the ordering, addition, subtraction, multiplication and division of these numbers as well.

Exercise Set 4.3.0

1. List the primes up to 20.[84]

2. In each case determine whether the given number is divisible by each of 2, 3, 5 and 10.

a. 18	d. 21	g. 511	j. 343
b. 25	e. 312	h. 810	k. 1000
c. 16	f. 420	i. 4522	l. 4

4.3.1   Structure

Whole numbers are used to refer to whole things. We talk about 4 cars, 7 people, 3 pies and the like. But what if we had the amount of pie shaded in Figure 4.3.1 below? 2 is not enough and 3 is too much.

Figure 4.3.1: Whole numbers cannot be used to refer to bits and pieces of a whole

Fractions were invented in an attempt to refer to a section of a whole.[85]

---

[84]We strongly recommend that you memorize this list. In fact, if you have the patience, it would be nice to memorize the list up to 100.

[85]This attempt worked but only partially. Today we know that fractions cannot represent the size of *all* possible sections of a whole. See our earlier coverage of irrational numbers.

Here's what a fraction looks like.

$$\frac{3}{4}$$

A fraction is written as two natural numbers aligned vertically and separated by a horizontal line.[86]

The number below the line is called the **denominator**. It tells us how many pieces of equal size a whole should be cut into. For example, the fraction $\frac{}{4}$ cuts a whole into 4 pieces of equal size.[87] This is shown in Figure 4.3.2.

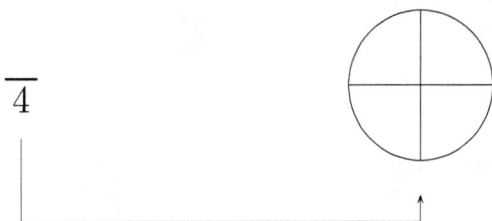

$$\frac{}{4}$$

Figure 4.3.2: A denominator of 4 requires that a whole should be cut into 4 pieces of equal size

The number above the line is called the **numerator**. This number tells us how many pieces we are interested in. As an example, the fraction $\frac{3}{4}$ tells us to cut a whole into 4 pieces and take 3 pieces. When drawing diagrams, we use shading to represent the numerator of a fraction. For $\frac{3}{4}$ we shade 3 pieces. A pie graph of $\frac{3}{4}$ is shown in Figure 4.3.3.

Fraction Types

The fraction $\frac{3}{4}$ consists of 3 pieces each of which is a quarter of a whole. This is shown in Figure 4.3.4(a). We can rearrange these pieces to form pies as shown in Figure 4.3.4(b).

---

[86]The use of forward slash, i.e., '/', to represent the fraction line is strictly forbidden. For a detailed discussion on the superiority of the horizontal line over other notations such as forward slash, please see Appendix E.

[87]It makes no difference what the top number is.

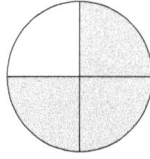

Figure 4.3.3: A pie graph for the fraction $\frac{3}{4}$

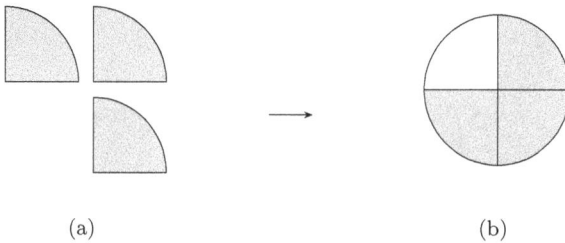

(a)   (b)

Figure 4.3.4: Three quarters is less than 1 whole

In the case of $\frac{3}{4}$ we end up with less than 1 whole pie. Fractions that add up to less than 1 whole pie are called **proper fractions**. $\frac{3}{4}$ is a proper fraction.

In some cases we may have enough pieces to make 1 whole pie. An example of this is $\frac{4}{4}$; See Figure 4.3.5.

And some fractions add up to *more* than 1 whole pie. An example of this is $\frac{5}{4}$; See Figure 4.3.6.

Fractions that add up to 1 whole pie or more than 1 whole pie are called **improper fractions**. $\frac{4}{4}$ is an improper fraction. So is $\frac{5}{4}$. In general

- In a proper fraction the numerator is less than the denominator.
- In an improper fraction either the numerator and the denominator are equal or the numerator is greater than the denominator.

Let us elaborate on these guidelines.

Consider a fraction with a denominator of 5, i.e., $\frac{}{5}$. This divides a whole into 5 equal parts so that we need at least 5 parts to make a whole pie. So long as we take 1, 2, 3, or 4 parts, i.e., so long as we have $\frac{1}{5}$, $\frac{2}{5}$, $\frac{3}{5}$, or $\frac{4}{5}$,

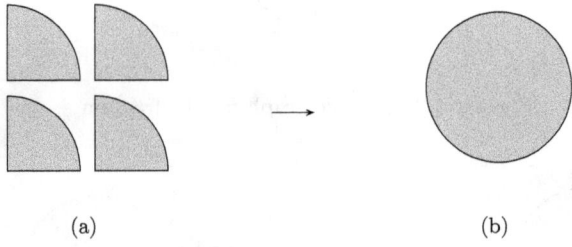

(a)   (b)

Figure 4.3.5: Four quarters make 1 whole

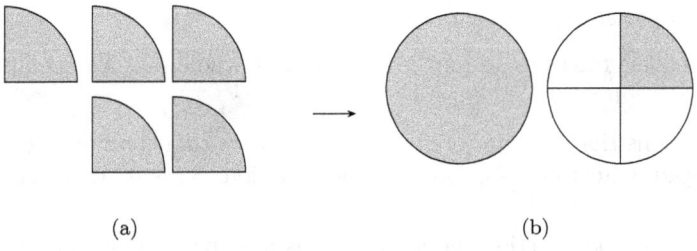

(a)   (b)

Figure 4.3.6: Five quarters is greater than 1 whole

we have less than 1 whole pie which implies that we have a proper fraction. With 5 pieces, i.e., $\frac{5}{5}$, we can make 1 whole pie which implies that $\frac{5}{5}$ is an improper fraction. If we have 6 pieces or more, i.e., $\frac{6}{5}$, $\frac{7}{5}$, etc., we will have more than 1 whole pie and so such fractions as $\frac{6}{5}$, $\frac{7}{5}$, etc. are also improper fractions.

Mixed Numbers

The graph of $\frac{7}{4}$ consists of 7 pieces each of which is a quarter. This is shown in Figure 4.3.7(a).

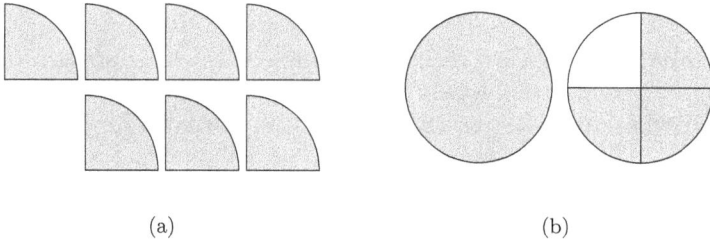

(a)   (b)

Figure 4.3.7: Equality of $\frac{7}{4}$ and $1\frac{3}{4}$ in size

We can rearrange these pieces to form pies. This is shown in Figure 4.3.7(b). In total we get 1 whole pie and an extra $\frac{3}{4}$. We write this as

$$1 + \frac{3}{4}$$

A more compact notation is[88]

$$1\frac{3}{4}$$

This is an example of a **mixed number**. The mix consists of 1, which is called the *whole* part of the mixed number, and $\frac{3}{4}$ is referred to as the *fractional* part of the mixed number. *Note that the fractional part of a mixed number must be proper.*[89]

---

[88]Negative mixed numbers such as $-1\frac{3}{4}$ are interpreted as the negative of the totality that $1\frac{3}{4}$ represents, i.e., as either $-\left(1 + \frac{3}{4}\right)$ or $-1 - \frac{3}{4}$ but not $-1 + \frac{3}{4}$.

[89]For the reason why, please see Appendix H.

We just saw that the improper fraction $\frac{7}{4}$ and the mixed number $1\frac{3}{4}$ have the same overall size (See Figure 4.3.7 above). This means we can write[90]

$$\frac{7}{4} = 1\frac{3}{4}$$

or

$$1\frac{3}{4} = \frac{7}{4}$$

It is important to know how to change mixed numbers to improper fractions and improper fractions to mixed numbers.

One of the most important reasons for the need to know how to change mixed numbers to improper fractions is that it is often easier to apply arithmetic operations to improper fractions than to their corresponding mixed numbers.[91] When solving problems that involve mixed numbers, it is common to convert all the given mixed numbers to improper fractions in preparation for the application of arithmetic operations.

Mixed numbers do have their advantages. One of the strengths of the mixed number notation is in its ability to communicate the overall size with clarity. As an example, it is easier to understand how large $7\frac{1}{4}$ is compared to $\frac{29}{4}$ even though both have the same size (draw diagrams to see for yourself). As a result, answers to problems which happen to be improper are usually converted to a mixed number before the solution is communicated to others.

We will now present an algorithm for the conversion of mixed numbers to improper fractions and another for the conversion of improper fractions to mixed numbers.

Consider the mixed number $2\frac{3}{5}$. This is shown in Figure 4.3.8.

Figure 4.3.8: $2\frac{3}{5}$ as a mixed number

In Figure 4.3.9 we have broken each of the 2 whole pies into 5 pieces.[92]

---

[90] Note that in math we use the equality symbol to imply that the *overall* sizes are equal.

[91] An exception to this is discussed in the section on ordering, addition and subtraction of fractions.

[92] Note that this is the same as the number of pieces the last pie is cut into.

This gives us a total of 13 pieces: 10 from the two whole pies and the addi-

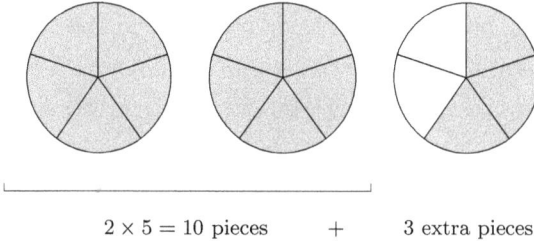

$2 \times 5 = 10$ pieces   +   3 extra pieces

Figure 4.3.9: Conversion of a mixed number to an improper fraction

tional 3 pieces. Therefore, we have 13 pieces each of which is a fifth. This means we have $\frac{13}{5}$. We write

$$2\frac{3}{5} = \frac{13}{5}$$

A mnemonic for quick conversion of a mixed number to an improper fraction is given in Figure 4.3.10. In this mnemonic, Step 1 corresponds to the division of the 2 whole pies above into 5 pieces each to arrive at 10 pieces. Step 2 corresponds to adding the 3 extra pieces to get a total of 13 fifths.

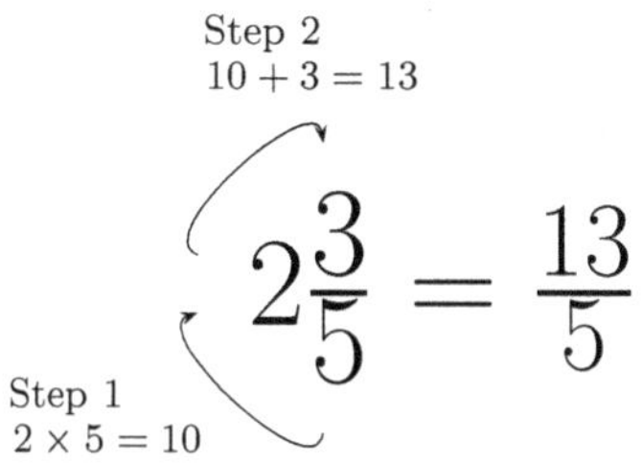

Step 2
$10 + 3 = 13$

$$2\frac{3}{5} = \frac{13}{5}$$

Step 1
$2 \times 5 = 10$

Figure 4.3.10: Mnemonic for quick conversion of a mixed number to an improper fraction

Examples

Convert each of the following to an improper fraction.

1. $6\frac{2}{5}$ 2. $2\frac{1}{2}$ 3. $237\frac{3}{5}$

Answers

a. $\frac{32}{5}$ b. $\frac{5}{2}$ c. $\frac{1188}{5}$

Algorithm for Conversion of Mixed Numbers to Improper Fractions

1. Work out the numerator of the equivalent, improper fraction. To do so, multiply the denominator of the fractional part of the mixed number by its whole part and add its numerator to the product.
2. The denominator of the equivalent, improper fraction remains the same as the denominator of the fractional part of the mixed number.

We will now explain how one can convert an improper fraction to a mixed number.

Consider the improper fraction $\frac{13}{5}$. Here we have 13 pieces each of which is a fifth of a whole. The fact that each piece is a fifth means that with 5 pieces we can make a whole. To convert $\frac{13}{5}$ to a mixed number we must answer the following two questions.

1. How many whole pies can be made from 13 fifths? and
2. How many fifths will be left?

These are the kinds of questions long division can answer. As illustrated in Figure 4.3.11, we divide 13 by 5 to arrive at the mixed formulation of $\frac{13}{5}$.

As you can see from the result of the long division, 2 whole pies (5 pieces each) can be formed from 13 fifths and we will be left with 3 fifths. We write:

$$\frac{13}{5} = 2\frac{3}{5}$$

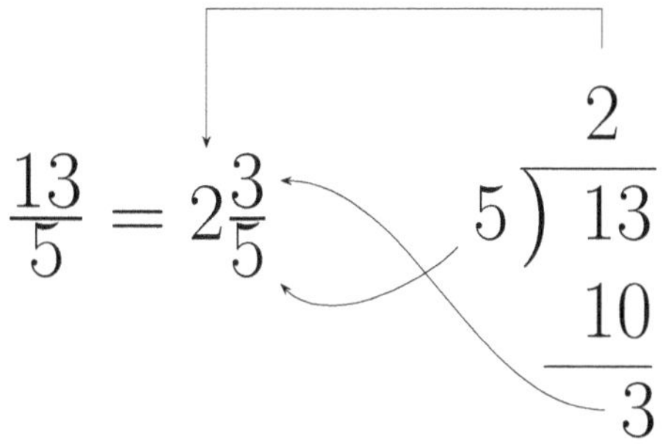

$$\frac{13}{5} = 2\frac{3}{5} \qquad 5\overline{)\,13} \\ \,\underline{10} \\ 3$$

Figure 4.3.11: Mnemonic for quick conversion of an improper fraction to a mixed number

Examples

Convert each of the following to a mixed number.

1. $\frac{7}{5}$

2. $\frac{1205}{77}$

3. $\frac{19}{3}$

Answers

1. $1\frac{2}{5}$

2. $15\frac{50}{77}$

3. $6\frac{1}{3}$

Algorithm for Conversion of Improper Fractions to Mixed Numbers

Divide the numerator of the improper fraction by its denominator and

1. Set the quotient as the whole part of the equivalent mixed number.
2. If the remainder is 0, the result is a whole number and we can stop otherwise set the remainder as the numerator of the fractional part of the equivalent mixed number and set the divisor as the denominator of the fractional part of the equivalent mixed number.

Equivalence

As shown in Figure 4.3.12 below, the fractions $\frac{6}{8}$ and $\frac{3}{4}$ have the same overall size.

(a) Graph of $\frac{6}{8}$: We have 6 shaded pieces   (b) Graph of $\frac{3}{4}$: We have 3 shaded pieces
out of a total of 8 pieces   out of a total of 4 pieces

Figure 4.3.12: Equivalence of the fractions $\frac{6}{8}$ and $\frac{3}{4}$

Fractions whose overall sizes are the same are called **equivalent fractions**. As Figure 4.3.12 shows, the fractions $\frac{6}{8}$ and $\frac{3}{4}$ are equivalent.

Equivalent fractions are set equal to each other. Of the many fractions that are equivalent to each other, the one with the smallest numerator and denominator is preferred as this feature makes it easier to make immediate sense of the fraction's size.

The process of changing a fraction to an equivalent fraction with a smaller numerator and denominator is called **reduction**. Because reduced fractions are easier to make sense of, we always reduce our answers as much as we can. When a fraction has been reduced as much as possible, we say that it has been *reduced to lowest terms*.

To reduce a fraction we divide the numerator and the denominator by the same number. As an example, to reduce $\frac{6}{8}$ we divide both the numerator and the denominator by 2 to get $\frac{3}{4}$. Figure 4.3.13 shows the steps.

The reason we can divide the numerator and the denominator of the fraction $\frac{6}{8}$ by 2 and still end up with the same overall size is that if we have half as many pieces overall (which is what happens when we divide the denominator of $\frac{6}{8}$ by 2), we end up with half as many shaded pieces (which is what happens when we divide the numerator of $\frac{6}{8}$ by 2). This is shown in Figure 4.3.14 below.

While the above discussion illustrates what happens when we divide the numerator and denominator of $\frac{6}{8}$ by 2, the argument holds for any fraction (and not just $\frac{6}{8}$) and any number (and not just 2). As an example, if we divide

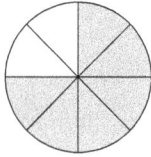

(a) Step 1 in the reduction of $\frac{6}{8}$: Divide the numerator by 2 to get 3

(b) Step 2 in the reduction of $\frac{6}{8}$: Divide the denominator by 2 to get 4

Figure 4.3.13: Steps in the reduction of $\frac{6}{8}$

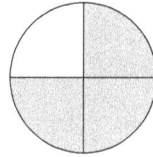

(a) Graph of $\frac{6}{8}$ (before reduction): We have 6 shaded pieces out of a total of 8 pieces

(b) Graph of $\frac{3}{4}$ (after reduction): We have 3 shaded pieces out of a total of 4 pieces

Figure 4.3.14: Reduction of the total number of pieces by one half (i.e., division of the denominator of $\frac{6}{8}$ by 2) results in the reduction of the number of shaded pieces by one half (i.e., division of the numerator of $\frac{6}{8}$ by 2)

the denominator of a fraction by 3, we end up with one third as many pieces in total. This reduces the number of shaded pieces by one third requiring that we divide the numerator by 3 as well. And if we divide the denominator of a fraction by 4, we end up with one quarter as many pieces in total. This reduces the number of shaded pieces by one quarter requiring that we divide the numerator by 4 as well.

Example

Reduce $\frac{6}{21}$ to lowest terms.

Solution

We divide the numerator and the denominator by 3 to get[93]

$$\frac{\overset{2}{\cancel{6}}}{\underset{7}{\cancel{21}}} = \frac{2}{7}$$

We can take multiple steps to reduce a fraction. Here's an example.

Example

Reduce $\frac{28}{42}$ to lowest terms.

Solution

One way to reduce $\frac{28}{42}$ to lowest terms is to divide both the numerator and the denominator by 14 to get $\frac{2}{3}$. We write this as

$$\frac{\overset{2}{\cancel{28}}}{\underset{3}{\cancel{42}}} = \frac{2}{3}$$

Another way to reduce $\frac{28}{42}$ to lowest terms is to divide the numerator and the denominator by 2 first to get $\frac{14}{21}$. Next we divide the numerator and the denominator of this new fraction by 7 to get $\frac{2}{3}$. We write this as

$$\frac{\overset{\overset{2}{\cancel{14}}}{\cancel{28}}}{\underset{\underset{3}{\cancel{21}}}{\cancel{42}}} = \frac{2}{3}$$

Reduction in one step seems neat but it is usually easier, and quite common, to use smaller numbers and reduce fractions in two or more steps. Reduction in one step may also seem somewhat faster as it requires a single division. However, it does take time to find the largest number that divides

---

[93]Note that when we reduce numbers we cross them out. We do this to remind ourselves that the crossed out numbers should no longer be used.

both the numerator and the denominator.[94] In addition, the division itself takes longer as the divisor is larger. In the balance, both approaches seem to be similar in terms of efficiency.

When working with mixed numbers, only the fractional part should be reduced. As an example, the mixed number $2\frac{6}{8}$ reduces to $2\frac{3}{4}$, i.e., the whole part of the mixed number should be left alone. Think in pictures and you should be able to see why.

We have shown that to reduce a fraction we must divide the numerator and the denominator by the same number. Sometimes this number is easy to find. As an example, it is easy to see that we can use 2 to reduce $\frac{6}{8}$. At other times it may be difficult to find this number as is the case with $\frac{57}{95}$. We can use 19 to reduce this fraction *but how do we find 19 in the first place?*

The following algorithm provides an answer and gives a step-by-step reduction guide.

Algorithm for Reducing Fractions to Lowest Terms

1. Cancel out an equal number of 0s in the numerator and denominator. This is the same as repeated division by 10.[95]
2. Use the Divisibility Test for 5 to see if you can divide the numerator and the denominator by 5. If so, reduce the fraction using 5. Repeat this step until you can no longer divide by 5.
3. Use the Divisibility Test for 2 to see if you can divide the numerator and the denominator by 2. If so, reduce the fraction using 2. Repeat this step until you can no longer divide by 2.
4. Use the Divisibility Test for 3 to see if you can divide the numerator and the denominator by 3. If so, reduce the fraction using 3. Repeat this step until you can no longer divide by 3.
5. Prime factorize the numerator or the denominator,[96] whichever seems easier.[97] Every time a new prime number shows up in the prime factorization process, see if that prime number can be used in the reduction of the other number (the numerator or the denominator not being prime factorized). If so, reduce the fraction using that prime number before

---

[94]This number is called the **largest common divisor** of the numerator and the denominator.

[95]If, at any point after Step 1, you happen to think of a number that you can use to reduce the fraction further, reduce the fraction using that number before you continue with the algorithm.

[96]The manner in which this process works will be illustrated through an example shortly. For now the reader should know that the process identifies those primes that the numerator or denominator is divisible by.

[97]If both seem easy to prime factorize or both seem difficult to prime factorize, then you should prime factorize the smaller of the two.

you continue with the algorithm. The process stops when the prime factorization ends, i.e., when the process generates 1.

Example

Reduce $\frac{57}{95}$ to lowest terms.

Solution

We follow the steps listed in the algorithm above:

1. There are no 0s on the right ends of 57 and 95. In addition, it is not easy to see what number works. We move on to the next step.

2. 5 does not work for 57 (57 does not have a 0 or a 5 on its right end). We move on to the next step.

3. 2 does not work for 57 (57 is not even). We move on to the next step.

4. 3 works for 57 (the sum of the digits in 57 is 5 + 7 or 12 which is divisible by 3) but 3 does not work for 95 (the sum of the digits in 95 is 9 + 5 or 14 which is not divisible by 3).

5. Both 57 and 95 seem easy to prime factorize: 57 is divisible by 3 and 95 by 5. We prime factorize 57. The first prime number is 2. 57 is not divisible by 2. The next prime number is 3. 57 is divisible by 3. Divide 57 by 3 to get 19:

$$
\begin{array}{c|c}
3 & 57 \\
  & 19 \\
\end{array}
$$

So 3 works for 57 but it does not work for 95. Continue with the prime factorization of 57. Since 19 is itself a prime number, only 19 goes into 19. We get:

$$
\begin{array}{c|c}
3 & 57 \\
19 & 19 \\
1 & \\
\end{array}
$$

Therefore 19 works for 57. We try dividing 95 by 19 and

we note that we get 5. Therefore, 19 works for 95 as well. Reducing the fraction using 19, we arrive at $\frac{3}{5}$. In addition, since the process of prime factorization has reached 1, we have reached the end of the process and can stop.[98]

Exercise Set 4.3.1

1. a. What does the denominator of a fraction signify?
   b. What does the numerator of a fraction signify?

2. What happens to the size of a piece if the denominator of a fraction increases? What if the denominator decreases?

3. a. Suppose the numerator of a fraction stays the same. What happens to the size of the fraction itself as the denominator of the fraction increases? What if the denominator decreases?
   b. Suppose the denominator of a fraction stays the same. What happens to the size of the fraction itself as the numerator of the fraction increases? What if the numerator decreases?

4. What fraction corresponds to each of the pie graphs below?

a.

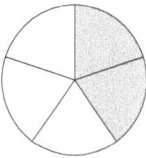

d.

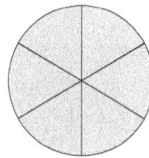

b.

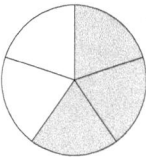

e.

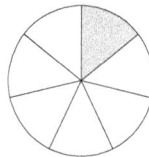

c.

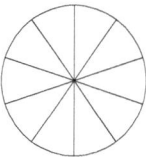

f.

---

[98]Otherwise we would have had to try 19 again and move on to higher primes until the process of prime factorization generates 1.

5. Classify each of the following as either a proper fraction, improper fraction, whole number, a mixed number or none of the above.

a. $\frac{17}{4}$ d. $\frac{14}{1}$ g. $2\frac{3}{20}$ j. $7\frac{21}{8}$ m. $2\frac{7}{7}$

b. $4\frac{3}{10}$ e. 7 h. $5\frac{1}{4}$ k. $\frac{12}{6}$ n. $\frac{18}{27}$

c. $\frac{4}{4}$ f. $\frac{9}{90}$ i. $\frac{1}{6}$ l. $\frac{5}{7}$ o. $\frac{44}{33}$

6. Convert each of the following improper fractions to either a mixed number or a whole number.

a. $\frac{13}{2}$ d. $\frac{20}{3}$ g. $\frac{1117}{18}$ j. $\frac{327}{100}$ m. $\frac{58}{9}$

b. $\frac{18}{9}$ e. $\frac{9}{4}$ h. $\frac{5201}{100}$ k. $\frac{304}{25}$ n. $\frac{16}{16}$

c. $\frac{28}{15}$ f. $\frac{34}{1}$ i. $\frac{34}{17}$ l. $\frac{56}{7}$ o. $\frac{25}{24}$

7. Convert each of the following to an improper fraction.

a. $7\frac{1}{2}$ d. $32\frac{1}{3}$ g. 8 j. $18\frac{4}{5}$ m. $100\frac{29}{120}$

b. $10\frac{2}{15}$ e. $2\frac{1}{4}$ h. $4300\frac{3}{4}$ k. $5\frac{2}{3}$ n. $24\frac{1}{20}$

c. $1\frac{17}{365}$ f. 182 i. 1 l. $20\frac{3}{8}$ o. $7\frac{1}{8}$

8. Use the divisibility test for 10 to decide which of the following can be reduced using 10. Reduce the ones that can be reduced. For some problems you may be able to use 10 more than once.

a. $\frac{40}{50}$ d. $\frac{4}{25}$ g. $\frac{260}{3170}$ j. $\frac{1000}{4900}$

b. $\frac{2}{3}$ e. $\frac{7}{20}$ h. $\frac{170}{15000}$ k. $\frac{2}{5}$

c. $\frac{300}{1100}$ f. $\frac{11000}{130000}$ i. $\frac{8}{9}$ l. $\frac{80000}{86300}$

9. Use the divisibility test for 5 to decide which of the following can be reduced using 5. Reduce the ones that can be reduced. For some problems you may be able to use 5 more than once.

a. $\frac{30}{55}$

b. $\frac{4}{9}$

c. $\frac{10}{35}$

d. $\frac{55}{100}$

e. $\frac{475}{575}$

f. $\frac{75}{205}$

g. $\frac{43}{191}$

h. $\frac{8}{9}$

i. $\frac{35}{20}$

j. $\frac{200}{225}$

k. $\frac{75}{115}$

l. $\frac{250}{375}$

10. Use the divisibility test for 2 to decide which of the following can be reduced using 2. Reduce the ones that can be reduced. For some problems you may be able to use 2 more than once.

a. $\frac{2}{6}$

b. $\frac{2}{5}$

c. $\frac{32}{100}$

d. $\frac{256}{400}$

e. $\frac{4}{5}$

f. $\frac{6}{16}$

g. $\frac{19}{173}$

h. $\frac{1}{23}$

i. $\frac{64}{68}$

j. $\frac{30}{154}$

k. $\frac{26}{31}$

l. $\frac{16}{128}$

11. Use the divisibility test for 3 to decide which of the following can be reduced using 3. Reduce the ones that can be reduced. For some problems you may be able to use 3 more than once.

a. $\frac{15}{21}$

b. $\frac{3}{20}$

c. $\frac{111}{126}$

d. $\frac{45}{504}$

e. $\frac{54}{351}$

f. $\frac{36}{63}$

g. $\frac{14}{29}$

h. $\frac{9}{39}$

i. $\frac{18}{75}$

j. $\frac{18}{35}$

k. $\frac{96}{165}$

l. $\frac{12}{27}$

12. Reduce the following fractions to lowest terms.

a. $\frac{8}{16}$

b. $\frac{10}{16}$

c. $\frac{8}{10}$

d. $\frac{18}{24}$

e. $\frac{56}{72}$

f. $\frac{11}{88}$

g. $\frac{6}{20}$

h. $20\frac{5}{15}$

i. $\frac{6}{42}$

j. $15\frac{9}{12}$

k. $\frac{50}{75}$

l. $\frac{84}{192}$

m. $\frac{22}{40}$

n. $\frac{9}{54}$

o. $\frac{21}{28}$

p. $\frac{10}{25}$

13. Reduce the following fractions to lowest terms.

a. $\frac{24}{32}$

b. $\frac{63}{81}$

c. $\frac{12}{16}$

d. $\frac{6}{15}$

e. $\frac{20}{36}$

f. $\frac{60}{42}$

g. $\frac{18}{45}$

h. $120\frac{21}{84}$

i. $\frac{58}{87}$

j. $\frac{91}{104}$

k. $\frac{150}{600}$

l. $1\frac{25}{100}$

m. $\frac{34}{119}$ n. $\frac{14}{18}$ o. $\frac{33}{48}$ p. $\frac{8}{20}$

14. Reduce the following fractions to lowest terms.

a. $\frac{3}{21}$ f. $\frac{22}{55}$ k. $\frac{11}{24}$ p. $\frac{420}{756}$

b. $\frac{20}{40}$ g. $\frac{35}{75}$ l. $\frac{24}{72}$ q. $\frac{38}{57}$

c. $\frac{15}{45}$ h. $\frac{48}{64}$ m. $\frac{85}{30}$ r. $\frac{25}{1000}$

d. $10\frac{48}{56}$ i. $\frac{80}{200}$ n. $\frac{57}{76}$ s. $\frac{32}{125}$

e. $\frac{35}{49}$ j. $7\frac{21}{56}$ o. $\frac{27}{81}$ t. $\frac{2}{20}$

15. Reduce the following fractions to lowest terms.

a. $\frac{7}{21}$ e. $\frac{126}{175}$ i. $\frac{123}{328}$ m. $\frac{14}{21}$

b. $\frac{2}{5}$ f. $\frac{323}{76}$ j. $\frac{62}{217}$ n. $\frac{7}{35}$

c. $\frac{88}{99}$ g. $\frac{261}{580}$ k. $\frac{44}{11}$ o. $\frac{33}{1639}$

d. $\frac{345}{92}$ h. $\frac{195}{728}$ l. $\frac{39}{52}$ p. $\frac{19}{38}$

4.3.2   Formal Reading

The Size of a Piece

Ordinals can be used to refer to the size of a fractional section of a whole. As an example, if a whole is divided into 5 equal parts, we refer to the size of a piece as a *fifth*. By this we mean that *the size of the piece is one fifth of the size of the whole*.

Although in most cases ordinals are used to refer to the size of a piece in the manner described above, there are some exceptions: If the denominator is 1, we use the word *one* to refer to the size of a piece. If the denominator is 2, we use the word *half* (plural *halves*) to refer to the size of a piece. If the denominator is 4, we use the word *quarter* (in place of the less common *fourth*) to refer to the size of a piece. And finally, If the denominator is 100, 1000, 10 000, etc., we use the words *hundredth, thousandth, ten thousandth*, etc., to refer to the size of a piece. See Figure 4.3.15 for an example.

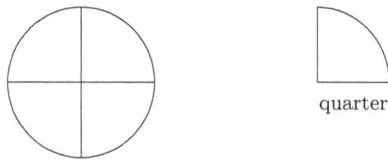

quarter

Figure 4.3.15: An example of referencing the
size of a fractional piece

Writing Fractions

When we write $\frac{2}{4}$ it means we have 2 pieces from a whole that is cut into 4
pieces. Since each piece is a *quarter* and we have two pieces, we can say we
have *two quarters*. This is illustrated in Figure 4.3.16.

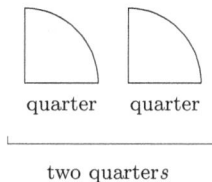

quarter   quarter

two quarter*s*

Figure 4.3.16: Referencing many pieces of equal size

As a second example consider $\frac{3}{4}$. We have three pieces each of which is a
quarter. Therefore we have *three quarters*. Figure 4.3.17 provides an analogy.

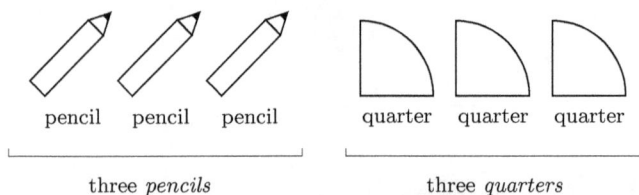

Figure 4.3.17: An analogy for referring to a collection of many pieces of equal size

Note that when we have more than one piece we use the plural 's' as in *three quarters*. If there is only one piece then the plural 's' should not be used. So, we read $\frac{1}{4}$ as *one quarter* and not *one quarters*.

To write a fraction in words

1. Write the numerator as a cardinal.
2. Write the denominator as an ordinal.
3. Attach a plural 's', or else use the plural form in place of the singular, unless you have a single piece.

We read $\frac{2}{10}$ as *two tenths*, $\frac{1}{12}$ as *one twelfth* and $\frac{3}{2}$ as *three halves*. See Figure 4.3.18 for how we use the steps above to read $\frac{3}{5}$.

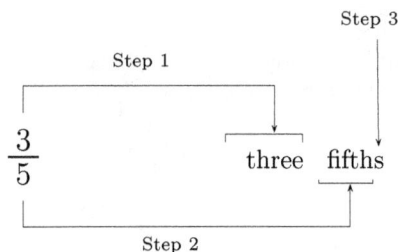

Figure 4.3.18: Steps to write $\frac{3}{5}$ using words

The reading of negative rationals is preceded with the word *negative* and, if the + sign is present, that of positive rationals by the word *positive*. As an example, we read $-\frac{3}{4}$ as *negative three quarters* and we read $+\frac{3}{4}$ as *positive three quarters.* Other than appending the word *negative* or the word *positive*, the reading of the fraction is not affected and follows the rules that we have listed above.

To write a mixed number in words

1. Write the whole part of the mixed number as a whole number as outlined in the section on whole numbers.
2. Write *and* to separate the whole part from the fractional part.
3. Write the fractional part of the mixed number as outlines above.

As an example, we read $1\frac{3}{4}$ as *one and three quarters.*[99]

The reading of negative mixed numbers is preceded with the word *negative* and, if the + sign is present, that of positive mixed numbers is preceded with the word *positive.* We read $-2\frac{3}{4}$ as *negative two and three quarters* and we read $+4\frac{1}{3}$ as *positive four and one third.*

Exercise Set 4.3.2

1. Write each of the following using words.

   a. $\frac{5}{6}$ e. $\frac{0}{2}$ i. $\frac{1}{3}$ m. $\frac{46}{100}$ q. $-\frac{8}{15}$

   b. $\frac{1}{10}$ f. $\frac{18}{18}$ j. $\frac{2}{16}$ n. $\frac{15}{1000}$ r. $-\frac{4}{5}$

   c. $\frac{25}{100}$ g. $\frac{12}{1}$ k. $\frac{9}{7}$ o. $\frac{7}{58}$ s. $-\frac{3}{3}$

   d. $\frac{3}{2}$ h. $\frac{7}{4}$ l. $\frac{12}{12}$ p. $-\frac{5}{3}$ t. $-\frac{1}{24}$

2. Write each of the following as a rational number.

   a. four fifths
   b. three tenths
   c. eighteen thirds
   d. forty-two sixths
   e. zero hundredths
   f. one fifteenth
   g. two halves
   h. seventy-eight hundredths
   i. twenty-five tenths
   j. positive one thousandth
   k. one sixth
   l. nine fifty-fifths

---

[99]Note that the word *and* is used to separate the whole part of a mixed number from its fractional part and, as we will see later, will also be used to separate the whole part of a number written in decimal notation from its fractional part. This explains our insistence earlier in the text that the word *and* should *not* be used in the reading of a whole number or that of an integer.

m. negative eight sevenths

n. six ones

o. twenty-one thirty-ninths

p. positive forty sevenths

q. negative sixteen sixteenths

r. one half

s. sixty-nine quarters

t. negative eighty fourths

3. Write each of the following in words.

a. $4\frac{3}{5}$

b. $\frac{3}{5}$

c. $\frac{14}{3}$

d. $7\frac{2}{3}$

e. $120\frac{1}{10}$

f. $15\frac{1}{2}$

g. $8\frac{7}{100}$

h. $\frac{9}{10\,000}$

4. Write each of the following as a mixed number.

a. nine and three fifths

b. two and a half

c. fifteen and three quarters

d. thirty-two and one eighth

e. negative fourteen and twenty-seven hundredths

f. forty-four and eighteen nineteenths

g. positive five and two thirds

h. five and two thirds

i. negative five and two thirds

j. negative two and one fifth

k. seventeen and three tenths

l. positive four hundred thirty-five and seventy-nine hundredths

m. negative one and five sevenths

4.3.3   Order

Fractions whose denominators are equal are called **like fractions**. As an example, the fractions $\frac{2}{5}$, $\frac{3}{5}$ and $\frac{9}{5}$ are like fractions. Since like fractions have the same denominators, they divide a unit into the same number of pieces. As an example, each fraction listed above divides a unit into 5 equal parts. This means that a piece from any of these fractions has the same size as a piece from any other one of these fractions. For this reason, as we will see below, like fractions are easy to order.[100]

Fractions whose denominators are different are called **unlike fractions**. As an example, the fractions $\frac{2}{3}$, $\frac{1}{4}$ and $\frac{3}{5}$ are unlike fractions. Since unlike fractions have different denominators, they divide a unit into a different number of pieces. This means that a piece from any one of these fractions has a size that is different from the size of a piece from any other one of these

---

[100] As we will see in the next subsection, they are also easy to add and subtract.

fractions. For this reason, as we will see below, unlike fractions are more difficult to order.[101] In fact, to order unlike fractions, we convert them to their equivalent, like fractions and then proceed to order these like fractions.[102] Please read Appendix I on how to convert a given set of unlike fractions into their equivalent, like fractions before reading further.

Ordering Like Fractions

Like fractions cut a unit into an equal number of pieces so that, in a list of like fractions, a piece from any fraction has the same size as a piece from any other fraction. Since the pieces are all of the same size, the fraction with the fewest pieces is the smallest fraction in the set and the fraction with the most pieces is the largest fraction in the set. Since the number of pieces is given by the numerators of the fractions, we can conclude that the fraction with the smallest numerator is the smallest fraction in the set and the fraction with the largest numerator is the largest fraction in the set. We can, therefore, order like fractions according to the size of their numerators.

Example

Arrange

$$\frac{7}{10}, \frac{3}{10}, \frac{9}{10}$$

in increasing order.

Solution

We begin by listing the fractions vertically.

$$\frac{7}{10}$$
$$\frac{3}{10}$$
$$\frac{9}{10}$$

Each of these fractions cuts a unit into 10 pieces so that a piece from one of these fractions has the same size as a piece from any of

---

[101] As we will see in the next subsection, they are also more difficult to add and subtract.
[102] As we will see in the next subsection, the same strategy works when adding and subtracting unlike fractions.

the other fractions. Since the pieces are of equal size, the fraction with the fewest pieces is the smallest fraction. The number of pieces are given by the numerators: The fraction $\frac{7}{10}$ has 7 pieces, the fraction $\frac{3}{10}$ has 3 pieces and the fraction $\frac{9}{10}$ has 9 pieces. Since 3 is less than 7 and 9, the fraction $\frac{3}{10}$ is the smallest fraction in the list. We note this as follows.

$$\frac{7}{10}$$
$$\frac{3}{10} \qquad 1$$
$$\frac{9}{10}$$

The next smallest fraction in the list is $\frac{7}{10}$. We note this as follows.

$$\frac{7}{10} \qquad 2$$
$$\frac{3}{10} \qquad 1$$
$$\frac{9}{10}$$

The fraction $\frac{9}{10}$ is the largest fraction in the list. This leads to

$$\frac{7}{10} \qquad 2$$
$$\frac{3}{10} \qquad 1$$
$$\frac{9}{10} \qquad 3$$

The original fractions can now be listed in increasing order:

$$\frac{3}{10}, \frac{7}{10}, \frac{9}{10}$$

To order negative rationals, associate the negative sign to the numerator and order the rationals according to their numerators using the guidelines given for the ordering of integers.

Example

Arrange

$$-\frac{3}{8}, \frac{-1}{8}, \frac{5}{-8}$$

in increasing order.

Solution

We begin by listing these rational numbers vertically.

$$-\frac{3}{8}$$

$$\frac{-1}{8}$$

$$\frac{5}{-8}$$

Next, we associate the negative signs to the numerators:

$$-\frac{3}{8} = \frac{-3}{8}$$

$$\frac{-1}{8} = \frac{-1}{8}$$

$$\frac{5}{-8} = \frac{-5}{8}$$

We now order these rational numbers by ordering their numerators.

$$-\frac{3}{8} = \frac{-3}{8} \qquad 2$$

$$\frac{-1}{8} = \frac{-1}{8} \qquad 3$$

$$\frac{5}{-8} = \frac{-5}{8} \qquad 1$$

The original fractions can now be listed in increasing order:

$$\frac{5}{-8}, \ -\frac{3}{8}, \ \frac{-1}{8}$$

A mix of positive and negative rationals can be ordered following the same guidelines given above for ordering negative rationals.

Mixed numbers and integers may be converted to improper fractions before they are ordered.

Example

Arrange

$$2\frac{1}{8}, \frac{15}{8}, \frac{19}{8}$$

in increasing order.

Solution

We begin by listing the fractions vertically.

$$2\frac{1}{8}$$
$$\frac{15}{8}$$
$$\frac{19}{8}$$

Next, we convert the mixed numbers and integers in the list to improper fractions. This yields

$$2\frac{1}{8} = \frac{17}{8}$$
$$\frac{15}{8} = \frac{15}{8}$$
$$\frac{19}{8} = \frac{19}{8}$$

We now order these fractions by ordering their numerators.

$$2\frac{1}{8} = \frac{17}{8} \qquad 2$$
$$\frac{15}{8} = \frac{15}{8} \qquad 1$$
$$\frac{19}{8} = \frac{19}{8} \qquad 3$$

The original fractions can now be listed in increasing order:

$$\frac{15}{8}, 2\frac{1}{8}, \frac{19}{8}$$

Ordering Unlike Fractions

To compare unlike fractions convert them to their equivalent, like fractions and *then* compare them.[103]

---

[103] We cannot make a decision on order based on a comparison of the numerators of unlike fractions. As an example, the fraction $\frac{3}{10}$ is smaller than the fraction $\frac{1}{2}$ even though 3 is larger than 1. This is because, although the fraction $\frac{3}{10}$ has 3 pieces and the fraction $\frac{1}{2}$ has only 1 piece, the size of each of the 3 pieces is a tenth and, therefore, they add up to less than the size of the 1 piece which is a half.

Example

Arrange

$$\frac{1}{4}, \frac{3}{5}, \frac{1}{2}$$

in increasing order.

Solution

We begin by listing the fractions vertically.

$$\frac{1}{4}$$
$$\frac{3}{5}$$
$$\frac{1}{2}$$

Next, we convert these fractions to their equivalent, like fractions using 20 as the least common denominator.

$$\frac{1}{4} = \frac{5}{20}$$
$$\frac{3}{5} = \frac{12}{20}$$
$$\frac{1}{2} = \frac{10}{20}$$

We now compare these like fractions by comparing their numerators and order them from smallest to largest as follows.

$$\frac{1}{4} = \frac{5}{20} \qquad 1$$
$$\frac{3}{5} = \frac{12}{20} \qquad 3$$
$$\frac{1}{2} = \frac{10}{20} \qquad 2$$

We now list the original fractions in increasing order.

$$\frac{1}{4}, \frac{1}{2}, \frac{3}{5}$$

Here is an example involving improper fractions, mixed numbers, whole numbers and integers.

Example

Arrange

$$\frac{2}{3}, \frac{5}{2}, 2\frac{1}{4}, 2$$

in increasing order.

Solution

We begin by listing the fractions vertically.

$$\frac{2}{3}$$
$$\frac{5}{2}$$
$$2\frac{1}{4}$$
$$2$$

Next, we convert the mixed numbers, whole numbers and integers in the list to improper fractions. This yields

$$\frac{2}{3} = \frac{2}{3}$$
$$\frac{5}{2} = \frac{5}{2}$$
$$2\frac{1}{4} = \frac{9}{4}$$
$$2 = \frac{2}{1}$$

We now convert the fractions in the list to their equivalent, like fractions using 12 as the least common denominator. This yields

$$\frac{2}{3} = \frac{2}{3} = \frac{8}{12}$$
$$\frac{5}{2} = \frac{5}{2} = \frac{30}{12}$$
$$2\frac{1}{4} = \frac{9}{4} = \frac{27}{12}$$
$$2 = \frac{2}{1} = \frac{24}{12}$$

We now compare these like fractions by comparing their numerators and order them from smallest to largest as follows.

$$\frac{2}{3} = \frac{2}{3} = \frac{8}{12} \qquad 1$$

$$\frac{5}{2} = \frac{5}{2} = \frac{30}{12} \qquad 4$$

$$2\frac{1}{4} = \frac{9}{4} = \frac{27}{12} \qquad 3$$

$$2 = \frac{2}{1} = \frac{24}{12} \qquad 2$$

We now list the original fractions in increasing order.

$$\frac{2}{3}, \, 2, \, 2\frac{1}{4}, \, \frac{5}{2}$$

Our next example involves negative rationals.

Example

Arrange

$$-\frac{2}{3}, \, \frac{1}{-4}, \, \frac{-5}{8}$$

in increasing order.

Solution

We begin by listing the rationals vertically.

$$-\frac{2}{3}$$

$$\frac{1}{-4}$$

$$\frac{-5}{8}$$

We next convert these to their equivalent, like rationals using 24 as the least common denominator. In addition, we associate the negative signs with the numerators. This yields

$$-\frac{2}{3} = \frac{-16}{24}$$

$$\frac{1}{-4} = \frac{-6}{24}$$

$$\frac{-5}{8} = \frac{-15}{24}$$

We now compare these like rationals by comparing their numerators and order them from smallest to largest as follows.

$$-\frac{2}{3} = \frac{-16}{24} \qquad 1$$

$$\frac{1}{-4} = \frac{-6}{24} \qquad 3$$

$$\frac{-5}{8} = \frac{-15}{24} \qquad 2$$

We now list the original rationals in increasing order.

$$-\frac{2}{3}, \frac{-5}{8}, \frac{1}{-4}$$

When a mix of positive and negative rationals are present, we recommend that you order the positive rationals separately from the negative rationals. This does require the use of two least common denominators (one for the positive rationals and one for the negative rationals) as opposed to one but it is usually the case that the two least common denominators are smaller in size and, therefore, easier to work with, than the one that would work for the full set.

Example

Arrange

$$\frac{7}{5}, -\frac{3}{4}, \frac{8}{3}, -1\frac{1}{2}$$

in increasing order.

Solution

We will order the positive rationals separately from the negative rationals.

For the positive rationals in the list, we begin by listing them vertically.

$$\frac{7}{5}$$

$$\frac{8}{3}$$

We now convert these to their equivalent, like fractions using 15 as the least common denominator. This yields

$$\frac{7}{5} = \frac{21}{15}$$
$$\frac{8}{3} = \frac{40}{15}$$

We can now order these equivalent, like fractions by comparing their numerators.

$$\frac{7}{5} = \frac{21}{15} \qquad 1$$
$$\frac{8}{3} = \frac{40}{15} \qquad 2$$

We now move on to order the negative rationals in the given list. A vertical listing of these negative rationals yields

$$-\frac{3}{4}$$
$$-1\frac{1}{2}$$

Next, we convert the mixed numbers, whole numbers and integers in the list to improper fractions, associating the negative signs to the numerators. This yields

$$-\frac{3}{4} = \frac{-3}{4}$$
$$-1\frac{1}{2} = \frac{-3}{2}$$

We now convert these rationals into their equivalent, like rationals using 4 as the least common denominator.

$$-\frac{3}{4} = \frac{-3}{4} = \frac{-3}{4}$$
$$-1\frac{1}{2} = \frac{-3}{2} = \frac{-6}{4}$$

We now order these like rationals by ordering their numerators. This yields

$$-\frac{3}{4} = \frac{-3}{4} = \frac{-3}{4} \qquad 2$$
$$-1\frac{1}{2} = \frac{-3}{2} = \frac{-6}{4} \qquad 1$$

The rational numbers in the original list can now be ordered as

$$-1\frac{1}{2}, -\frac{3}{4}, \frac{7}{5}, \frac{8}{3}$$

Algorithm for Ordering Rational Numbers

To order a set of rational numbers

1. Convert all mixed numbers and integers to improper fractions.[104]
2. If the fractions in the list are unlike fractions, convert them to their equivalent like fractions. Associate any negative signs to the numerators.
3. Order the like fractions by ordering their numerators using the ordering guidelines given in the section on integers.
4. Order the original numbers in the list accordingly.

Exercise Set 4.3.3

1. Insert $<$ or $>$ as appropriate.

a. $\dfrac{3}{4}$ $\dfrac{2}{4}$ e. $\dfrac{-3}{2}$ $\dfrac{5}{-2}$ i. $-\dfrac{8}{5}$ $\dfrac{-7}{5}$

b. $-\dfrac{1}{5}$ $-\dfrac{4}{5}$ f. $\dfrac{32}{2}$ $\dfrac{12}{2}$ j. $\dfrac{20}{100}$ $\dfrac{40}{100}$

c. $\dfrac{0}{3}$ $\dfrac{2}{3}$ g. $\dfrac{25}{50}$ $\dfrac{32}{50}$ k. $\dfrac{4}{4}$ $\dfrac{3}{4}$

d. $\dfrac{520}{121}$ $\dfrac{519}{121}$ h. $\dfrac{6}{1}$ $\dfrac{8}{1}$ l. $\dfrac{50}{3}$ $\dfrac{18}{3}$

2. Insert $<, =$ or $>$ as appropriate.

a. $\dfrac{3}{4}$ $\dfrac{2}{3}$ e. $\dfrac{-2}{2}$ $-\dfrac{3}{6}$ i. $\dfrac{8}{9}$ $\dfrac{3}{5}$

b. $\dfrac{1}{5}$ $\dfrac{3}{4}$ f. $\dfrac{10}{4}$ $\dfrac{25}{7}$ j. $\dfrac{16}{5}$ $\dfrac{15}{2}$

c. $-\dfrac{2}{5}$ $-\dfrac{3}{4}$ g. $\dfrac{-1}{2}$ $\dfrac{2}{-3}$ k. $-\dfrac{1}{4}$ $\dfrac{3}{-8}$

d. $\dfrac{8}{20}$ $\dfrac{4}{10}$ h. $\dfrac{0}{4}$ $\dfrac{2}{5}$ l. $\dfrac{50}{3}$ $\dfrac{230}{12}$

---

[104]For an alternative algorithm where the rational numbers in the list are converted to mixed numbers and integers before they are compared please see Appendix J. In some cases, this alternative algorithm can be very powerful and, ideally, you should know both alternatives and choose one or the other depending on the particular rational numbers that you wish to order. As an example, the algorithm given in the body of the text may work faster if the majority of the rational numbers in the given list are improper fractions while the alternative algorithm given in Appendix J may be faster if the majority of the rational numbers in the given list are mixed numbers or integers.

The main advantage of the approach given in the body of the text is that it can be extended easily later when we take up the study of algebraic expressions and equations.

3. Arrange in order from smallest to largest.

a. $\frac{1}{5}, \frac{3}{4}, \frac{2}{5}$

b. $\frac{2}{3}, \frac{1}{4}, \frac{3}{7}$

c. $\frac{3}{5}, \frac{1}{2}, \frac{1}{3}$

d. $-\frac{11}{20}, \frac{-2}{5}, \frac{7}{-15}, \frac{-8}{15}$

e. $-\frac{1}{5}, -\frac{2}{3}, -\frac{3}{4}, \frac{-5}{6}$

f. $\frac{3}{2}, \frac{5}{11}, \frac{3}{15}$

g. $-\frac{1}{4}, -\frac{4}{9}, -\frac{3}{4}$

h. $\frac{1}{3}, \frac{4}{3}, \frac{4}{5}, \frac{7}{3}$

i. $\frac{-7}{9}, -\frac{2}{7}, \frac{-8}{15}$

j. $\frac{1}{2}, \frac{2}{5}, \frac{3}{8}$

k. $\frac{11}{10}, \frac{13}{15}, \frac{12}{11}, \frac{2}{3}$

l. $\frac{3}{4}, \frac{4}{20}, \frac{2}{5}$

4. Insert $<, =$ or $>$ as appropriate.

a. $3\frac{1}{4}$ $2\frac{3}{5}$

b. $14\frac{1}{5}$ $6\frac{3}{4}$

c. $-2\frac{3}{5}$ $-2\frac{4}{7}$

d. $6\frac{1}{2}$ $6\frac{2}{3}$

e. $10\frac{14}{20}$ $10\frac{7}{10}$

f. $1\frac{3}{15}$ $1\frac{5}{20}$

g. $20\frac{1}{4}$ $12\frac{4}{9}$

h. $-13\frac{2}{3}$ $-40\frac{2}{5}$

i. $5\frac{7}{9}$ $5\frac{2}{7}$

j. $6\frac{1}{20}$ $6\frac{2}{25}$

k. $-2\frac{1}{5}$ $-2\frac{3}{10}$

l. $7\frac{8}{10}$ $7\frac{4}{5}$

5. Arrange in ascending order.

a. $\frac{9}{10}, \frac{3}{10}, \frac{7}{10}$

b. $\frac{1}{8}, \frac{7}{8}, \frac{3}{8}$

c. $2\frac{3}{5}, 2\frac{1}{5}, 2\frac{4}{5}$

d. $-\frac{1}{4}, -\frac{1}{5}, -\frac{3}{10}$

e. $\frac{1}{10}, \frac{2}{25}, \frac{3}{20}$

f. $8\frac{1}{2}, 8\frac{1}{4}, 8\frac{3}{4}$

g. $\frac{-4}{5}, \frac{9}{-10}, \frac{-19}{20}, -\frac{3}{4}$

h. $\frac{2}{25}, \frac{1}{5}, \frac{9}{50}, \frac{3}{10}$

i. $\frac{9}{20}, -\frac{3}{5}, \frac{3}{4}, -\frac{9}{10}$

j. $\frac{1}{5}, \frac{3}{25}, \frac{4}{5}, \frac{2}{5}$

k. $4\frac{2}{3}, 1\frac{1}{2}, 2\frac{1}{5}, 3\frac{1}{4}$

l. $3\frac{2}{5}, -5\frac{1}{3}, -5\frac{2}{3}, 3\frac{2}{3}$

m. $18\frac{1}{20}, 18\frac{3}{20}, 18\frac{3}{100}, 18\frac{1}{4}$

n. $9\frac{3}{4}, 9\frac{1}{2}, 9\frac{4}{5}, 9\frac{1}{5}, 9\frac{2}{3}$

o. $3\frac{1}{3}, 3\frac{1}{5}, 3\frac{3}{4}, 3\frac{2}{3}$

p. $2, 2\frac{2}{3}, 2\frac{3}{4}, 2\frac{1}{2}$

6. Arrange in decreasing order.

a. $\frac{9}{25}, \frac{3}{25}, \frac{14}{25}$

b. $\frac{3}{100}, \frac{9}{100}, \frac{7}{100}$

c. $4\frac{11}{17}, 4\frac{3}{17}, 4\frac{1}{17}, 4\frac{8}{17}$

d. $\frac{1}{2}, \frac{1}{5}, \frac{1}{10}, \frac{1}{3}$

e. $\frac{-2}{3}, \frac{1}{-2}, -\frac{3}{5}$

f. $\frac{4}{5}, \frac{1}{5}, \frac{2}{3}$

g. $-3\frac{4}{5}, -3\frac{1}{3}, -3\frac{1}{6}$

h. $8\frac{1}{2}, 8\frac{3}{4}, 8\frac{2}{5}$

i. $\frac{9}{10}, \frac{1}{2}, \frac{3}{4}, \frac{4}{5}$

j. $91\frac{1}{2}, 92\frac{1}{2}, 92\frac{1}{3}, 92\frac{3}{4}$

k. $-\frac{3}{10}, \frac{4}{5}, -\frac{1}{5}, 3\frac{1}{2}$ m. $2\frac{1}{5}, 3\frac{2}{5}, 2\frac{3}{10}, 3\frac{1}{2}$

l. $40\frac{2}{3}, 40\frac{1}{2}, 40\frac{3}{4}$ n. $18\frac{1}{3}, -18, -18\frac{2}{5}$

4.3.4   Addition, Subtraction, Multiplication and Division

In this subsection we will present algorithms for addition, subtraction, multiplication and division of rational numbers. We will begin with addition and subtraction.

Addition and Subtraction

There are two cases to consider: addition of like fractions and addition of unlike fractions.

Addition of Like Fractions

Consider the problem

$$\frac{2}{9} + \frac{5}{9}$$

As shown in Figure 4.3.19, since all pieces have the same size we can add the 2 pieces and the 5 pieces to get a total of 7 pieces. This means we have $\frac{7}{9}$. We write all of this as follows.

$$\frac{2}{9} + \frac{5}{9} = \frac{2+5}{9}$$
$$= \frac{7}{9}$$

To add like fractions add their numerators to find the total number of pieces. The denominator, which sets the size of a piece, stays the same.

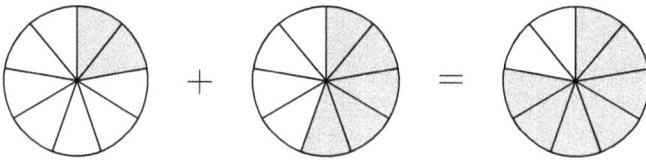

Figure 4.3.19: Addition of like fractions

Addition of Unlike Fractions

When we try to use the logic above to add *unlike* fractions, we run into trouble. Consider, as an example, the sum of $\frac{2}{5}$ and $\frac{3}{8}$. This is written as

$$\frac{2}{5} + \frac{3}{8}$$

The sum is shown in Figure 4.3.20. Now we have a problem: It makes no

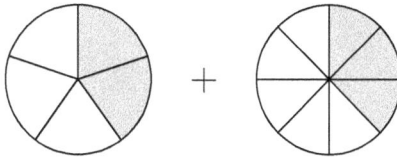

Figure 4.3.20: Addition of unlike fractions

sense to add the 2 pieces on the left and the 3 pieces of the right to say we have 5 pieces. 5 pieces of what size? The pieces are not all of the same size.

To add unlike fractions, convert them to their equivalent, like fractions[105] to make sure all the pieces are of the same size. We can then add the numerators to find the total number of pieces. Here's an example.

$$\begin{aligned}
\frac{2}{5} + \frac{3}{8} &= \frac{16}{40} + \frac{15}{40} \\
&= \frac{16 + 15}{40} \\
&= \frac{31}{40}
\end{aligned}$$

---

[105]See Appendix I on how to convert a set of fractions to their equivalent, like fractions.

The following shortcut is highly recommended.[106]

$$\frac{2}{5} + \frac{3}{8} = \frac{16 + 15}{40}$$
$$= \frac{31}{40}$$

This is the manner in which you should record your solutions. To assist the reader in writing the first line, we will present the thought processes that go into the making of it. We begin by finding the least common denominator. For the problem at hand, the least common denominator is 40. We write

$$\frac{2}{5} + \frac{3}{8} = \frac{}{40}$$

We now follow the mnemonic given in Figure I.2.3 in Appendix I to work out the numerators. We divide 40 by 5 to get 8. Multiply 8 by 2 to get 16. We now write

$$\frac{2}{5} + \frac{3}{8} = \frac{16}{40}$$

Next we write the + sign:

$$\frac{2}{5} + \frac{3}{8} = \frac{16 +}{40}$$

We now divide 40 by 8 to get 5. Multiply 5 by 3 to get 15. We write

$$\frac{2}{5} + \frac{3}{8} = \frac{16 + 15}{40}$$

The rest follows.

As usual improper results should be converted to mixed numbers. Here's an example.

Example

$$\frac{2}{3} + \frac{4}{5} = \frac{10 + 12}{15}$$
$$= \frac{22}{15}$$
$$= 1\frac{7}{15}$$

Reduction may be in order as the following example shows.

---

[106]There is more to this shortcut than efficiency. It sets the style needed for working with algebraic equations at higher levels of math as it dispenses with overuse of brackets and naturally emphasizes the division.

Example

$$\frac{5}{12} + \frac{1}{4} = \frac{5 + 3}{12}$$
$$= \frac{8}{12}$$
$$= \frac{2}{3}$$

And one may have to reduce *and* convert to a mixed number. In such cases it makes no difference whether one reduces the fraction first and *then* converts to a mixed number, or converts to a mixed number and *then* reduces the fractional part. Both approaches are equally efficient and of course both generate the same answer.

When whole numbers and mixed numbers are present we can change them to improper fractions before we proceed. Here's an example.

Example

$$1\frac{3}{4} + \frac{8}{3} = \frac{7}{4} + \frac{8}{3}$$
$$= \frac{21 + 32}{12}$$
$$= \frac{53}{12}$$
$$= 4\frac{5}{12}$$

Subtraction of fractions is similar to addition of fractions. The only difference is that instead of adding pieces we subtract them.

Subtraction of Like Fractions

We illustrate the idea using an example.

Example

$$\frac{9}{25} - \frac{2}{25} = \frac{9 - 2}{25}$$
$$= \frac{7}{25}$$

The pie graph of each of the fractions $\frac{9}{25}$ and $\frac{2}{25}$ is cut into 25 pieces. Therefore, a piece from one pie graph has the same size as a piece from the other pie graph. We have 9 pieces. From this we wish to subtract 2 pieces. We end up with 7 pieces. The denominator, which sets the size of a piece, stays the same.

To subtract like fractions subtract their numerators. The denominator stays the same.

Subtraction of Unlike Fractions

To subtract unlike fractions, convert them to their equivalent, like fractions and *then* subtract the numerators. The denominator stays the same. The following example illustrates the idea.

Example

$$\frac{5}{6} - \frac{3}{4} = \frac{10 - 9}{12}$$
$$= \frac{1}{12}$$

When mixed numbers, whole numbers and integers are present, we can convert them to improper fractions and then proceed to subtract them. Here's an example.

Example

$$6\frac{4}{5} - 2\frac{2}{3} = \frac{34}{5} - \frac{8}{3}$$
$$= \frac{102 - 40}{15}$$
$$= \frac{62}{12}$$
$$= \frac{31}{6}$$
$$= 5\frac{1}{6}$$

Example

$$6 - 2\frac{1}{5} = \frac{6}{1} - \frac{11}{5}$$
$$= \frac{30 - 11}{5}$$
$$= \frac{19}{5}$$
$$= 3\frac{4}{5}$$

Algorithm for Evaluation of Addition and Subtraction of Two Rational Numbers

1. Convert any mixed numbers and whole numbers to improper fractions[107] and associate any leading negative sign to the numerator of the leading rational number.
2. Convert unlike fractions to their equivalent, like fractions.
3. Add and/or subtract the numerators. To do so, follow the *Algorithm for Evaluation of Addition and subtraction of Two Integers*. The denominator, which sets the size of a piece, stays the same.

Multiplication and Division

We will start by presenting the algorithm for working out multiplication and division of rational numbers. Following this, we will provide justification for each of the steps in the algorithm.

---

[107]For an alternative algorithm where the rational numbers involved are converted to mixed numbers and integers before they are added and subtracted please see Appendix J. In some cases, this altrenative algorithm can be very powerful and, ideally, you should know both alternatives and choose one or the other depending on the particular rational numbers that you wish to add or subtract. As an example, the algorithm given in the body of the text may work faster if the majority of the rational numbers involved are improper fractions while the alternative algorithm given in Appendix J may be faster if the majority of the rational numbers involved are mixed numbers or integers.

The main advantage of the approach given in the body of the text is that it can be extended easily later when we take up the study of algebraic expressions and equations.

Algorithm for Evaluation of Multiplication and Division of Two Rational Numbers

1. Work out the sign following the guidelines given for working out the sign of chain multiplication/division of integers. If the answer is negative, write the negative sign down. From this point on, ignore all signs.
2. Convert all mixed numbers and whole numbers to improper fractions.
3. Convert all divisions to multiplication and flip the fraction that follows each division.
4. Work out the magnitude. To do so

   a. reduce any numerator with any denominator and repeat this reduction until it is no longer possible to do so.
   b. multiply the numerators and multiply the denominators.

Step 1 extends the logic behind the multiplication and division of integers.

Step 2 rewrites the numbers in a format that is required in Step 4.

Step 3 reduces the number of different operations involved to one: multiplication. The argument here is that division by a rational number, i.e., $\div \frac{m}{n}$, can be rewritten as multiplication by its **reciprocal**, i.e., $\times \frac{n}{m}$. To explain why, consider the semantics of $\div \frac{m}{n}$. Since division is the inverse of multiplication, we expect $\div \frac{m}{n}$ to undo what $\times \frac{m}{n}$ does. Since $\times \frac{m}{n}$ may be viewed as multiplication by m followed by division by n, its inverse, as shown in Figure 4.3.21, is multiplication by n followed by division by m which may be written as $\times \frac{n}{m}$. This shows that the opposite of $\times \frac{m}{n}$, i.e., $\div \frac{m}{n}$, is the same as $\times \frac{n}{m}$.

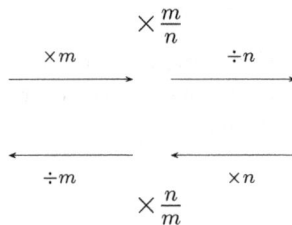

$$\times \frac{m}{n}$$

$$\times m \qquad \div n$$

$$\longrightarrow \qquad \longrightarrow$$

$$\longleftarrow \qquad \longleftarrow$$

$$\div m \qquad \times n$$

$$\times \frac{n}{m}$$

Figure 4.3.21: Equivalence of $\div \frac{m}{n}$ and $\times \frac{n}{m}$

Step 4 relies on a combination of four key types of reasoning. These lines of reasoning will be illustrated using four key examples.

Key Example 1

Evaluate[108] $2 \times \frac{3}{8}$

Solution

Steps 1, 2, and 3 do not apply. We move on to Step 4(a) and reduce 2 and 8 to get 1 and 4. Next we follow Step 4(b), multiply the numerators to get 3 and multiply the denominators to get 4. This leads to the fraction $\frac{3}{4}$ as the result:

$$\overset{1}{\cancel{2}} \times \frac{3}{\underset{4}{\cancel{8}}} = \frac{3}{4}$$

The argument here is that, to double the size of $\frac{3}{8}$, we can double the size of a piece, from an *eighth* to a *quarter*. The crossing out of 8 and its replacement with 4 doubles the size of a piece and the crossing out of 2 indicates that, since the size of a piece has doubled, we only need half as many pieces, i.e., in place of 2 things, each of which has a size of $\frac{3}{8}$, we take 1 thing which has a size of $\frac{3}{4}$.

Key Example 2

Evaluate[109] $2 \times \frac{3}{5}$

Solution

Steps 1, 2, 3, and 4(a) do not apply. We move on to Step 4(b) and multiply the numerators to get 6 and multiply the denominators to get 5. This leads to $\frac{6}{5}$ or $1\frac{1}{5}$:

$$2 \times \frac{3}{5} = \frac{6}{5}$$
$$= 1\frac{1}{5}$$

In this problem we seek to double the size of the fraction $\frac{3}{5}$. Unlike the previous problem where we achieved this effect by doubling the size of a piece

---

[108]Note that this expression may be seen as $\frac{2}{1} \times \frac{3}{8}$.
[109]Note that this expression may be seen as $\frac{2}{1} \times \frac{3}{5}$.

through the reduction of 8 and 2, here we are not able to do so as 5 is not divisible by 2. Therefore, we switch to a different logic and achieve the desired effect by doubling the number of pieces, from 3 to 6, while keeping the size of a piece unchanged. Multiplying the numerators, i.e., 2×3, doubles the number of pieces while multiplying the denominators, i.e., 1×5, keeps the size of a piece unchanged.

Key Example 3

Evaluate $\frac{1}{2} \times \frac{4}{5}$

Solution

Steps 1, 2 and 3 do not apply. We move on to Step 4(a) and reduce 2 and 4 as shown. Next we move on to Step 4(b) and multiply the numerators to get 2 and multiply the denominators to get 5. This leads to the fraction $\frac{2}{5}$ as the result:

$$\frac{1}{\cancel{2}} \times \frac{\cancelto{2}{4}}{5} = \frac{2}{5}$$

To understand why this kind of reduction works, we appeal to the semantics of $\frac{1}{2} \times$ which translates to $\frac{1}{2}$ *of*.[110] Since $\frac{4}{5}$ is made up of 4 pieces each of which is a *fifth*, to find half of this amount we can simply divide the total number of pieces by 2, i.e., take 2 pieces each of which is a *fifth*. The crossing out of 4 in the numerator and the rewriting of it as 2 reflects the fact that the total number of pieces was divided by 2. The crossing out of the 2 in the denominator indicates that, in place of $\frac{1}{2}$ of the thing whose size is $\frac{4}{5}$, we take 1 piece that has double the size, i.e., $\frac{2}{5}$.

---

[110]The equivalence between the use of the word *of* and multiplication is also seen in our dealings with whole numbers. Note that 2 *of something* implies $2\times$ *something*. For more on the relationship between the word *of* and multiplication please see the companion textbook *Semantics and the Syntax of Algebra* by the author. Here we simply note that the association between the word *of* and multiplication in working with rational numbers is an extension of the association between the word *of* and multiplication in working with whole numbers.

Key Example 4

Evaluate $\frac{1}{2} \times \frac{3}{5}$

Solution

Steps 1, 2, 3, and 4(a) do not apply. We move on to Step 4(b) and multiply the numerators to get 3 and multiply the denominators to get 10. This leads to the fraction $\frac{3}{10}$ as the result:

$$\frac{1}{2} \times \frac{3}{5} = \frac{3}{10}$$

Unlike the previous problem where we were able to divide the numerator 4 by 2, here we are not able to do so as 3 is not divisible by 2. To work out the problem we switch to a different logic: $\frac{3}{5}$ indicates that we have 3 pieces each of which is a *fifth*. To find $\frac{1}{2}$ of this, we keep the 3 pieces unchanged which results from the multiplication of the numerators, i.e., 1×3, but we reduce the size of each piece by a half, i.e., from a *fifth* to a *tenth*. The 10, of course, comes from the doubling of 5 in the denominator, i.e., from 2×5.

The four key examples above explain the logic behind Step 4 in the *Algorithm for Evaluation of Multiplication and Division of Two Rational Numbers*. The logic in these four key examples can be applied repeatedly to calculations that involve a mix of the scenarios above. Here are some examples.

Example

Evaluate $\frac{4}{5} \times \frac{1}{6}$

Solution

Steps 1, 2 and 3 do not apply. We move on to Step 4. We reduce 4 and 6 using 2 and then multiply the numerators and multiply the denominators to arrive at $\frac{2}{15}$:

$$\frac{\overset{2}{\cancel{4}}}{5} \times \frac{1}{\underset{3}{\cancel{6}}} = \frac{2}{15}$$

Example

Evaluate $\frac{8}{35} \times \frac{14}{6}$

Solution

Steps 1, 2 and 3 do not apply. We move on to Step 4. Reduce as much as possible and then multiply the numerators and the denominators to arrive at $\frac{8}{15}$:

$$\frac{\overset{4}{\cancel{8}}}{\underset{5}{\cancel{35}}} \times \frac{\overset{2}{\cancel{14}}}{\underset{3}{\cancel{6}}} = \frac{8}{15}$$

Here we reduced 8 and 6 using 2. Next we reduced 35 and 14 using 7. The new values can no longer be reduced. We multiply the numerators to get 8 and we multiply the denominators to get 15.[111]

Example

Evaluate

$$\frac{2}{5} \div \frac{-3}{10}$$

Solution

Step 1 requires that we work out the sign. Since there are an odd number of negative signs, the result will be negative. We write

$$\frac{2}{5} \div \frac{-3}{10} = -\frac{2}{5} \div \frac{3}{10}$$

Step 2 does not apply. We move on to Step 3 and convert $\div \frac{3}{10}$ to $\times \frac{10}{3}$.

---

[111]There are other ways to reduce the fractions in this multiplication. We could have started by reducing 14 and 6 as an example. However, no matter how you reduce the fractions, the final result will be the same. Try it yourself.

$$= -\frac{2}{5} \times \frac{10}{3}$$

Next we move on to Step 4(a) and reduce as much as possible.

$$= -\frac{2}{\cancel{5}} \times \frac{\cancel{10}^{2}}{3}$$
$$\phantom{= -\frac{2}{5}} {}_{1}$$

And finally we move on to Step 4(b), multiply the numerators and multiply the denominators.

$$= -\frac{4}{3}$$
$$= -1\frac{1}{3}$$

In the example above we used the symbol $\div$ to represent division with. As we have noted earlier, the formal notation to represent division with is the horizontal line. The use of the horizontal line to represent division of rational numbers leads to a structural setup called a **complex fraction**. A complex fraction is a fraction whose numerator and/or denominator contains one or more fractions. Such expressions arise from the use of the horizontal line to denote nested division. An example of a complex fraction is the following.

$$\frac{4\frac{1}{5}}{\frac{3}{4}}$$

The numerator of this fraction is $4\frac{1}{5}$ and its denominator is $\frac{3}{4}$. The main fraction is represented by the longest line present.[112] When writing complex fractions, the main fraction line should be drawn first[113], followed by the numerator and the denominator. This adheres to the discipline of good writing style and naturally emphasizes the fraction line as the main operation

---

[112]Confusion can arise if this convention is not followed. The expression

$$\frac{\frac{1}{2}}{3} \qquad \text{(Divide } \frac{1}{2} \text{ into 3 parts)}$$

is equal to $\frac{1}{6}$ whereas the expression

$$\frac{1}{\frac{2}{3}} \qquad \text{(How many } \frac{2}{3} \text{ fit into a 1?)}$$

evaluates to $1\frac{1}{2}$.

[113]Estimate the length.

which strengthens analysis. In addition, style requires that the numerator and denominator should be centred.

The key structure in working with *all* complex fractions is the one that has the form

$$\frac{\frac{a}{b}}{\frac{c}{d}}$$

This complex fraction has a single fraction in its numerator and a single fraction in its denominator. Recall that the fraction line is based on the horizontal line that denotes division (of the numerator by the denominator). Division of the numerator of the complex fraction above by its denominator yields

$$\frac{\frac{a}{b}}{\frac{c}{d}} = \frac{a}{b} \div \frac{c}{d}$$
$$= \frac{a}{b} \times \frac{d}{c}$$
$$= \frac{a \times d}{b \times c}$$
$$= \frac{ad}{bc}$$

Since the steps listed above in the evaluation of the key structure need to be performed frequently, it pays to memorize the following shortcut:

$$\frac{\frac{a}{b}}{\frac{c}{d}} = \frac{ad}{bc}$$

You may wish to keep the mnemonic in Figure 4.3.22 in mind.[114]

---

[114]Shortcuts and mnemonics are useful in that they speed up otherwise time-consuming tasks, especially when such tasks need to be performed repeatedly. However, the user of such shortcuts and mnemonics should always be prepared to defend the logic behind them if called upon to do so.

$$\left(\frac{\dfrac{a}{b}}{\dfrac{c}{d}} = \frac{ad}{bc}\right.$$

Figure 4.3.22: Mnemonic for quick evaluation of $\dfrac{\frac{a}{b}}{\frac{c}{d}}$

Example

Simplify the complex fraction

$$\frac{\dfrac{2}{3}}{\dfrac{5}{7}}$$

Solution

Application of the shortcut introduced above yields the following solution.

$$\frac{\dfrac{2}{3}}{\dfrac{5}{7}} = \frac{2 \times 7}{3 \times 5}$$

$$= \frac{14}{15}$$

Note that we insist on writing the expression

$$\frac{2 \times 7}{3 \times 5}$$

We do not recommend that you make a jump to the last line by multiplying the various numbers to get $\frac{14}{15}$ in one step which would be the case if we wrote

$$\frac{\dfrac{2}{3}}{\dfrac{5}{7}} = \frac{14}{15}$$

The reason is that often reduction may be in order as the following example shows.

Example

Simplify the complex fraction

$$\frac{\frac{4}{15}}{\frac{2}{35}}$$

Solution

Application of the shortcut introduced above with intermediate reduction yields the following solution.

$$\frac{\frac{4}{15}}{\frac{2}{35}} = \frac{\overset{2}{\cancel{4}} \times \overset{7}{\cancel{35}}}{\underset{3}{\cancel{15}} \times \underset{1}{\cancel{2}}}$$

$$= \frac{14}{3}$$

$$= 4\frac{2}{3}$$

As we noted earlier, complex fractions of the form

$$\frac{\frac{a}{b}}{\frac{c}{d}}$$

are key to simplifying *all* complex fractions. In fact, given any complex fraction, we first simplify its numerator and denominator to a single fraction, and then apply the shortcut given above. Application of this algorithm to the case where mixed numbers or integers are present requires that they should be converted to improper fractions before the shortcut is applied. Here is an example:

Example

Simplify the complex fraction

$$\frac{8}{3\frac{1}{2}}$$

Solution

We simplify the numerator and denominator of the complex fraction into a single fraction and then apply the shortcut presented above.

$$\frac{8}{3\frac{1}{2}} = \frac{\frac{8}{1}}{\frac{7}{2}}$$

$$= \frac{8 \times 2}{1 \times 7}$$

$$= \frac{16}{7}$$

$$= 2\frac{2}{7}$$

If negative rationals are present, the sign is worked out following the rules that govern the multiplication and division of integers. Here is an example:

Example

Simplify the complex fraction

$$\frac{\frac{-3}{4}}{\frac{-1}{-5}}$$

Solution

$$\frac{\frac{-3}{4}}{\frac{-1}{-5}} = -\frac{\frac{3}{4}}{\frac{1}{5}}$$

$$= -\frac{3 \times 5}{4 \times 1}$$

$$= -\frac{15}{4}$$

$$= -3\frac{3}{4}$$

Exercise Set 4.3.4

1. Evaluate.

    a. $\frac{2}{5} + \frac{1}{5}$ d. $\frac{2}{3} + \frac{1}{3}$ g. $\frac{1}{2} + \frac{1}{2}$ j. $\frac{4}{10} + \frac{9}{10}$

    b. $\frac{1}{8} + \frac{2}{8}$ e. $\frac{4}{9} + \frac{2}{9}$ h. $\frac{1}{10} + \frac{7}{10}$ k. $\frac{1}{3} + \frac{1}{3}$

    c. $\frac{3}{4} + \frac{2}{4}$ f. $\frac{3}{4} + \frac{3}{4}$ i. $\frac{12}{18} + \frac{15}{18}$ l. $\frac{0}{9} + \frac{3}{9}$

2. Evaluate.

    a. $\frac{1}{3} + \frac{2}{7}$ d. $\frac{4}{10} + \frac{7}{20}$ g. $\frac{32}{40} + \frac{17}{30}$ j. $\frac{11}{34} + \frac{4}{51}$

    b. $\frac{2}{5} + \frac{2}{10}$ e. $\frac{4}{7} + \frac{1}{9}$ h. $\frac{1}{25} + \frac{1}{20}$ k. $\frac{1}{15} + \frac{15}{32}$

    c. $\frac{3}{4} + \frac{1}{8}$ f. $\frac{11}{50} + \frac{6}{25}$ i. $\frac{12}{33} + \frac{19}{22}$ l. $\frac{11}{19} + \frac{2}{17}$

3. Evaluate.

    a. $2\frac{1}{3} + 1\frac{3}{5}$ d. $12 + 1\frac{3}{8}$ g. $182\frac{3}{4} + \frac{425}{3}$ j. $-6 + 3\frac{2}{5}$

    b. $8\frac{1}{4} + \frac{1}{2}$ e. $4 + \frac{2}{3}$ h. $-\frac{3}{8} + \frac{2}{5}$ k. $42 + 57\frac{1}{3}$

    c. $2\frac{3}{5} + 3$ f. $\frac{1}{9} + 2$ i. $-4\frac{1}{5} + 2\frac{1}{2}$ l. $-2\frac{2}{5} + 7$

4. Evaluate.

    a. $\frac{4}{5} - \frac{3}{5}$ d. $\frac{30}{25} - \frac{10}{25}$ g. $\frac{19}{3} - \frac{8}{3}$ j. $\frac{14}{5} - \frac{8}{5}$

    b. $\frac{15}{22} - \frac{7}{22}$ e. $\frac{20}{105} - \frac{14}{105}$ h. $\frac{3}{8} - \frac{1}{8}$ k. $\frac{305}{6} - \frac{292}{6}$

    c. $\frac{3}{4} - \frac{1}{4}$ f. $\frac{32}{5} - \frac{18}{5}$ i. $\frac{9}{8} - \frac{3}{8}$ l. $\frac{46}{9} - \frac{40}{9}$

5. Evaluate.

    a. $\frac{3}{4} - \frac{1}{2}$ d. $\frac{4}{5} - \frac{1}{2}$ g. $\frac{8}{3} - \frac{4}{5}$ j. $\frac{102}{7} - \frac{206}{15}$

    b. $\frac{9}{16} - \frac{2}{12}$ e. $\frac{3}{10} - \frac{1}{8}$ h. $\frac{19}{15} - \frac{11}{20}$ k. $\frac{42}{5} - \frac{18}{25}$

    c. $\frac{3}{10} - \frac{4}{15}$ f. $\frac{9}{10} - \frac{8}{25}$ i. $\frac{3}{10} - \frac{1}{20}$ l. $\frac{19}{3} - \frac{17}{4}$

6. Evaluate.

a. $4\frac{3}{4} - 1\frac{1}{3}$ f. $25 - 18\frac{3}{4}$ k. $3 - 4\frac{1}{9}$ p. $-9\frac{5}{8} - 23\frac{1}{6}$

b. $2\frac{4}{5} - \frac{3}{4}$ g. $12\frac{1}{5} - 10$ l. $5\frac{3}{4} - 8\frac{1}{3}$ q. $-2 - \frac{3}{8}$

c. $16 - 5\frac{4}{5}$ h. $87\frac{5}{12} - 63\frac{7}{8}$ m. $2\frac{9}{10} - \frac{25}{4}$ r. $-5 - 2\frac{1}{2}$

d. $21\frac{1}{5} - 14\frac{2}{3}$ i. $95\frac{1}{3} - 809\frac{2}{5}$ n. $-\frac{43}{10} - \frac{27}{6}$ s. $-4\frac{2}{7} - \frac{6}{1}$

e. $18\frac{1}{3} - 12$ j. $\frac{3}{4} - \frac{5}{6}$ o. $-3\frac{1}{5} - \frac{2}{7}$ t. $5\frac{2}{3} - 8$

7. Evaluate.

a. $\frac{2}{9} \times \frac{3}{4}$ e. $\frac{14}{25} \times \frac{25}{14}$ i. $\frac{1}{2}\left(\frac{2}{1}\right)$ m. $\frac{32}{11}\left(\frac{22}{16}\right)$

b. $\frac{3}{16} \times \frac{4}{3}$ f. $\frac{24}{18} \times \frac{12}{20}$ j. $\frac{14}{3}\left(\frac{2}{7}\right)$ n. $\frac{13}{10}\left(\frac{1}{26}\right)$

c. $\frac{30}{52} \times \frac{4}{3}$ g. $\frac{5}{3} \times \frac{9}{8}$ k. $\frac{5}{18}\left(\frac{3}{20}\right)$ o. $\frac{5}{80}\left(\frac{30}{2}\right)$

d. $\frac{3}{2} \times \frac{2}{3}$ h. $\frac{12}{12} \times \frac{3}{4}$ l. $\frac{3}{4}\left(\frac{16}{20}\right)$ p. $\frac{7}{10}\left(\frac{5}{14}\right)$

8. Evaluate.

a. $3\frac{1}{2} \times \frac{1}{4}$ e. $-3\left(\frac{2}{15}\right)$ i. $\frac{-14}{3}(-2)$ m. $-\frac{11}{2} \times \frac{4}{-33}$

b. $2\frac{1}{3} \times 1\frac{1}{7}$ f. $-1\frac{1}{3}\left(-\frac{1}{4}\right)$ j. $\frac{7}{-12} \times \frac{6}{7}$ n. $\frac{-34}{-15} \times \frac{-20}{14}$

c. $-4\frac{2}{3} \times 2\frac{1}{2}$ g. $2\left(\frac{-5}{4}\right)$ k. $-2\frac{1}{2}(3)$ o. $14 \times \frac{1}{-21}$

d. $-\frac{1}{5} \times \frac{5}{6}$ h. $3\frac{1}{8}(-16)$ l. $\frac{-2}{-15}\left(\frac{-3}{11}\right)$ p. $-5\frac{1}{3} \times \frac{-1}{-4}$

9. Evaluate.

a. $\frac{4}{3} \div \frac{6}{7}$ e. $\frac{12}{15} \div \frac{8}{25}$ i. $\frac{9}{10} \div \frac{27}{100}$ m. $\frac{6}{20} \div \frac{8}{10}$

b. $\frac{3}{10} \div \frac{3}{5}$ f. $\frac{42}{13} \div \frac{21}{26}$ j. $\frac{6}{5} \div \frac{12}{5}$ n. $\frac{14}{15} \div \frac{7}{20}$

c. $\frac{2}{3} \div \frac{8}{6}$ g. $\frac{3}{4} \div \frac{3}{4}$ k. $\frac{3}{25} \div \frac{6}{100}$ o. $\frac{16}{15} \div \frac{20}{105}$

d. $\frac{16}{21} \div \frac{8}{35}$ h. $\frac{8}{3} \div \frac{16}{9}$ l. $\frac{18}{15} \div \frac{6}{10}$ p. $\frac{32}{15} \div \frac{7}{30}$

10. Evaluate.

a. $\frac{-5}{14} \div \frac{1}{-7}$ e. $4\frac{2}{3} \div \left(-3\frac{1}{2}\right)$ i. $\frac{4}{3} \div \left(-1\frac{2}{3}\right)$ m. $-2\frac{1}{3} \div 14$

b. $\frac{3}{-4} \div 1\frac{1}{8}$ f. $-3\frac{1}{2} \div \frac{-1}{-8}$ j. $14 \div \frac{7}{8}$ n. $-\frac{18}{21} \div (-6)$

c. $-1\frac{3}{8} \div \frac{22}{3}$ g. $-7\frac{1}{2} \div \frac{-3}{10}$ k. $-2 \div \frac{4}{5}$ o. $-8\frac{1}{2} \div 4$

d. $-1\frac{1}{2} \div \left(-\frac{5}{6}\right)$ h. $\frac{14}{5} \div \left(-\frac{7}{10}\right)$ l. $16 \div \left(-\frac{1}{2}\right)$ p. $-\frac{18}{7} \div 6$

11. In each case identify the numerator and denominator of the main fraction.

a. $\dfrac{\frac{2}{3}}{\frac{4}{5}}$ b. $\dfrac{\frac{3}{2}}{\frac{2}{5}}$ c. $\dfrac{\frac{4}{5}}{1}$ d. $\dfrac{6\frac{1}{2}}{\frac{1}{2}}$

12. simplify.

a. $\dfrac{\frac{2}{3}}{\frac{4}{5}}$ e. $\dfrac{\frac{4}{15}}{\frac{3}{20}}$ i. $\dfrac{\frac{-2}{3}}{-\frac{5}{-8}}$ m. $\dfrac{\frac{1}{2}}{\frac{2}{3}}$

b. $\dfrac{\frac{9}{5}}{\frac{7}{15}}$ f. $\dfrac{\frac{14}{35}}{\frac{9}{5}}$ j. $\dfrac{\frac{4}{-5}}{\frac{-3}{4}}$ n. $\dfrac{-4}{5\frac{1}{2}}$

c. $\dfrac{\frac{3}{8}}{\frac{3}{4}}$ g. $\dfrac{\frac{-18}{15}}{\frac{27}{25}}$ k. $\dfrac{\frac{1}{23}}{\frac{3}{46}}$ o. $\dfrac{4\frac{2}{3}}{3\frac{1}{4}}$

d. $\dfrac{\frac{1}{2}}{\frac{10}{3}}$ h. $\dfrac{\frac{2}{3}}{\frac{8}{-9}}$ l. $\dfrac{\frac{-3}{2}}{-\frac{2}{5}}$ p. $\dfrac{3\frac{1}{4}}{2}$

4.3.5   Expressions

We will begin with an algorithm for the evaluation of expressions that involve chain addition/subtraction of rational numbers. Next we will present an algorithm for the evaluation of expressions that involve chain multiplication/division of rational numbers. We will end this subsection with the extension of the analysis-synthesis algorithm given in the section on integers for rapid evaluation of expressions that involve rational numbers with a mix of the four basic arithmetic operations.

Once the terms have been evaluated and the signs simplified, expressions involving rational numbers degenerate into chain addition/subtraction of fractions with the possibility of a leading negative number. The algorithm for the evaluation of such chains is similar to the *Algorithm for Evaluation of Addition and Subtraction of Two Rational Numbers* introduced in the previous section. Here is an example.

Example

Evaluate

$$\frac{2}{3} + \frac{4}{5} - \frac{3}{4} + \frac{1}{2}$$

Solution

We will provide detailed explanations for the steps in the evaluation of this expression.

Since there are no mixed numbers or integers present, we will compute the least common denominator for the fractions involved. The least common denominator for the fractions in the expression above is 60. We write

$$\frac{2}{3} + \frac{4}{5} - \frac{3}{4} + \frac{1}{2} = \frac{\rule{3cm}{0.4pt}}{60}$$

We now divide 60 by the denominators of the fractions in the expression in turn and multiply the result by the corresponding numerators beginning with the first fraction in the expression: 60 divided by 3 is 20. 20 times 2 is 40. We write

$$\frac{2}{3} + \frac{4}{5} - \frac{3}{4} + \frac{1}{2} = \frac{40}{60}$$

This is followed by the symbol for addition. We write

$$\frac{2}{3} + \frac{4}{5} - \frac{3}{4} + \frac{1}{2} = \frac{40 +}{60}$$

We move on to the next fraction: 60 divided by 5 is 12. 12 times 4 is 48. We write

$$\frac{2}{3} + \frac{4}{5} - \frac{3}{4} + \frac{1}{2} = \frac{40 + 48}{60}$$

This is followed by the symbol for subtraction. We write

$$\frac{2}{3} + \frac{4}{5} - \frac{3}{4} + \frac{1}{2} = \frac{40 + 48 -}{60}$$

We move on to the next fraction: 60 divided by 4 is 15. 15 times 3 is 45. We write

$$\frac{2}{3} + \frac{4}{5} - \frac{3}{4} + \frac{1}{2} = \frac{40 + 48 - 45}{60}$$

This is followed by the symbol for addition. We write

$$\frac{2}{3} + \frac{4}{5} - \frac{3}{4} + \frac{1}{2} = \frac{40 + 48 - 45 +}{60}$$

We move on to the last fraction in the list: 60 divided by 2 is 30. 30 times 1 is 30. We write

$$\frac{2}{3} + \frac{4}{5} - \frac{3}{4} + \frac{1}{2} = \frac{40 + 48 - 45 + 30}{60}$$

We now work out the numerator to get.

$$= \frac{73}{60}$$

This can be converted to a mixed number.

$$= 1\frac{13}{60}$$

The full solution without the intervening comments and explanations is given below.

$$\frac{2}{3} + \frac{4}{5} - \frac{3}{4} + \frac{1}{2} = \frac{40 + 48 - 45 + 30}{60}$$
$$= \frac{73}{60}$$
$$= 1\frac{13}{60}$$

In case mixed numbers and/or whole numbers are present, we can convert them to improper fractions and continue as above.

Example

$$\frac{1}{6} + 1\frac{3}{4} + 2\frac{2}{3} + 2 = \frac{1}{6} + \frac{7}{4} + \frac{8}{3} + \frac{2}{1}$$
$$= \frac{2 + 21 + 32 + 24}{12}$$
$$= \frac{79}{12}$$
$$= 6\frac{7}{12}$$

Example

$$2\frac{1}{2} + 6 - \frac{1}{4} - 3\frac{2}{3} = \frac{5}{2} + \frac{6}{1} - \frac{1}{4} - \frac{11}{3}$$
$$= \frac{30 + 72 - 3 - 44}{12}$$
$$= \frac{55}{12}$$
$$= 4\frac{7}{12}$$

Algorithm for Evaluation of Expressions Involving Chain Addition/Subtraction of Rational Numbers

If you have access to a calculator, then go from left to right and perform the operations as you see them[115] otherwise

1. Convert any mixed numbers and whole numbers to improper fractions[116] and associate any leading negative sign to the numerator of the leading rational number.
2. Convert unlike fractions to their equivalent, like fractions.
3. Add and/or subtract the numerators. To do so, follow the *Algorithm for Evaluation of Addition and Subtraction of Two Integers*. The denominator, which sets the size of a piece, stays the same.

The algorithm for working out multiplication and division of two rational numbers can also be extended to work out expressions that contain chain multiplication/division of rational numbers.

---

[115]This approach also works well by hand if the numbers involved are small.

[116]For an alternative algorithm where the rational numbers involved are converted to mixed numbers and integers before they are added and subtracted please see Appendix J. The main advantage of the approach given in the body of the text is that it is often more efficient and is generally the preferred approach when working with algebraic expressions and equations.

Algorithm for Evaluation of Expressions Involving Chain Multiplication/Division of Rational Numbers

If you have access to a calculator, then go from left to right and perform the operations as you see them[117] otherwise

1. Work out the sign following the guidelines given for working out the sign of chain multiplication/division of integers. If the answer is negative, write the negative sign down. From this point on, ignore all signs.
2. Convert all mixed numbers and whole numbers to improper fractions.
3. Convert all divisions to multiplication and flip the fraction that follows each division.
4. Work out the magnitude. To do so

   a. reduce any numerator with any denominator and repeat this reduction until it is no longer possible to do so.

   b. multiply the numerators and multiply the denominators.

Example

Evaluate

$$-4\frac{2}{3} \div \frac{7}{5} \times 6 \times \frac{-1}{-18}$$

Solution

We will show in detail how this multiplication is worked out. We begin with Step 1. Since there are three negative signs, the result will be negative. We write

$$-4\frac{2}{3} \div \frac{7}{5} \times 6 \times \frac{-1}{-18} = -$$

We now move on to Step 2 and convert all integers and mixed numbers to improper fractions.

$$-4\frac{2}{3} \div \frac{7}{5} \times 6 \times \frac{-1}{-18} = -\frac{14}{3} \div \frac{7}{5} \times \frac{6}{1} \times \frac{1}{18}$$

We move on to Step 3 and convert $\div\frac{7}{5}$ to $\times\frac{5}{7}$.

$$= -\frac{14}{3} \times \frac{5}{7} \times \frac{6}{1} \times \frac{1}{18}$$

---

[117]This approach also works well by hand if the numbers involved are small.

We now move on to Step 4(a) and reduce as much as we can.

$$= -\frac{\overset{2}{\cancel{14}}}{3} \times \frac{5}{\cancel{7}_{1}} \times \frac{\cancel{6}^{1}}{1} \times \frac{1}{\cancel{18}_{3}}$$

We now apply Step 4(b), multiply the numerators and multiply the denominators.

$$= -\frac{10}{9}$$

Finally, we convert the result to a mixed number.

$$= -1\frac{1}{9}$$

The full solution without the intervening comments and explanations is given below.

$$-4\frac{2}{3} \div \frac{7}{5} \times 6 \times \frac{-1}{-18} = -\frac{14}{3} \div \frac{7}{5} \times \frac{6}{1} \times \frac{1}{18}$$

$$= -\frac{\overset{2}{\cancel{14}}}{3} \times \frac{5}{\cancel{7}_{1}} \times \frac{\cancel{6}^{1}}{1} \times \frac{1}{\cancel{18}_{3}}$$

$$= -\frac{10}{9}$$

$$= -1\frac{1}{9}$$

Example

Evaluate

$$\frac{-2}{-5} \div 1\frac{3}{5} \times (-6) \div (-2)$$

Solution

$$\frac{-2}{-5} \div 1\frac{3}{5} \times (-6) \div (-2) = \frac{2}{5} \div 1\frac{3}{5} \times 6 \div 2$$

$$= \frac{2}{5} \div \frac{8}{5} \times \frac{6}{1} \div \frac{2}{1}$$

$$= \frac{2}{5} \times \frac{5}{8} \times \frac{6}{1} \times \frac{1}{2}$$

$$= \frac{3}{4}$$

We let the reader provide the detail for the reduction at the last step.

The *Analysis-Synthesis Algorithm* can be extended to deal with expressions that involve rational numbers.

The Analysis-Synthesis Algorithm for Evaluation of Expressions Involving Addition, Subtraction, Multiplication and Division of Rational Numbers

1. Analyze:

    a. Analyze the expression into terms using additions and subtractions *outside* brackets.

    b. Analyze each term into factors using multiplications.

    c. Analyze each factor further if needed.

2. Synthesize:

    a. Evaluate the factors.

    b. Multiply the factors to evaluate the terms. To do so

        i. Determine whether the term should be added or subtracted. Use the *Algorithm for Evaluation of Expressions Involving Chain Multiplication/Division of Integers*. From this point on, ignore all signs.

        ii. Determine the *size* of the term. Use the *Algorithm for Evaluation of Expressions Involving Chain Multiplication/Division of Rational Numbers*.

    c. Add and subtract the terms to evaluate the expression. Use the *Algorithm for Evaluation of Expressions Involving Chain Addition/ Subtraction of Rational Numbers*.

We illustrate the use of the above algorithm using examples.[118]

---

[118]In what follows, we will no longer show the details of reduction when we reduce a chain multiplication and leave it to the reader to work out such detail.

Example

Evaluate

$$\frac{2}{3} + \frac{1}{2} \times \frac{3}{4} - \frac{1}{10}$$

Solution

Application of the analysis-synthesis algorithm yields the following solution.

$$\boxed{\frac{2}{3}} + \boxed{\frac{1}{2} \times \frac{3}{4}} - \boxed{\frac{1}{10}} = \boxed{\frac{2}{3}} + \boxed{\frac{3}{8}} - \boxed{\frac{1}{10}}$$

$$= \frac{80 + 45 - 12}{120}$$

$$= \frac{113}{120}$$

Example

Evaluate

$$\frac{3}{4} \div \frac{2}{3} + \frac{1}{4} \times \frac{2}{3} \times \frac{1}{3} + \frac{1}{2}$$

Solution

$$\boxed{\frac{3}{4} \div \frac{2}{3}} + \boxed{\frac{1}{4} \times \frac{2}{3} \times \frac{1}{3}} + \boxed{\frac{1}{2}} = \boxed{\frac{3}{4} \times \frac{3}{2}} + \boxed{\frac{1}{18}} + \boxed{\frac{1}{2}}$$

$$= \boxed{\frac{9}{8}} + \boxed{\frac{1}{18}} + \boxed{\frac{1}{2}}$$

$$= \frac{81 + 4 + 36}{72}$$

$$= \frac{121}{72}$$

$$= 1\frac{49}{72}$$

Example

Evaluate

$$\frac{4}{5} \times 3\frac{2}{5} \div \frac{34}{35} \;-\; \frac{3}{4} \div \frac{3}{20} \div 2 \;-\; \frac{3}{4} \times \frac{1}{3}$$

Solution

$$\boxed{\frac{4}{5} \times 3\frac{2}{5} \div \frac{34}{35}} \;-\; \boxed{\frac{3}{4} \div \frac{3}{20} \div 2} \;-\; \boxed{\frac{3}{4} \times \frac{1}{3}}$$

$$= \boxed{\frac{4}{5} \times \frac{17}{5} \times \frac{35}{34}} \;-\; \boxed{\frac{3}{4} \times \frac{20}{3} \times \frac{1}{2}} \;-\; \boxed{\frac{1}{4}}$$

$$= \boxed{\frac{14}{5}} \;-\; \boxed{\frac{5}{2}} \;-\; \boxed{\frac{1}{4}}$$

$$= \frac{56 \;-\; 50 \;-\; 5}{20}$$

$$= \frac{1}{20}$$

Example

Evaluate

$$\frac{3}{4} \times \frac{2}{3} + \frac{1}{2}\left(\frac{2}{3} - \frac{3}{5} \times \frac{1}{3}\right) - \frac{3}{8}$$

Solution

$$\boxed{\frac{3}{4} \times \frac{2}{3}} + \boxed{\frac{1}{2}\left(\frac{2}{3} - \frac{3}{5} \times \frac{1}{3}\right)} - \boxed{\frac{3}{8}}$$

$$= \boxed{\frac{1}{2}} + \boxed{\frac{1}{2}\left(\frac{2}{3} - \frac{1}{5}\right)} - \boxed{\frac{3}{8}}$$

$$= \boxed{\frac{1}{2}} + \boxed{\frac{1}{2}\left(\frac{10 - 3}{15}\right)} - \boxed{\frac{3}{8}}$$

$$= \boxed{\frac{1}{2}} + \boxed{\frac{1}{2}\left(\frac{7}{15}\right)} - \boxed{\frac{3}{8}}$$

$$= \boxed{\frac{1}{2}} + \boxed{\frac{7}{30}} - \boxed{\frac{3}{8}}$$

$$= \frac{60 + 28 - 45}{120}$$

$$= \frac{43}{120}$$

Example

Evaluate

$$\frac{-2}{-5} \times \frac{5}{-3} - \frac{-1}{4} \times \frac{-4}{-3} \times \frac{1}{2} - \frac{1}{-2}$$

Solution

In the solution given below, we work out the sign of each term before we work out its size.

$$\boxed{\frac{-2}{-5} \times \frac{5}{-3}} - \boxed{\frac{-1}{4} \times \frac{-4}{-3} \times \frac{1}{2}} - \boxed{\frac{1}{-2}} = \boxed{\frac{-2}{3}} + \boxed{\frac{1}{6}} + \boxed{\frac{1}{2}}$$

$$= \frac{-4 + 1 + 3}{6}$$

$$= \frac{0}{6}$$

$$= 0$$

Using the horizontal line to represent division with can lead to the presence of complex fractions.

Example

Evaluate

$$\frac{\frac{2}{3} + \frac{1}{4}}{\frac{1}{2} - \frac{3}{5}}$$

Solution

As usual, the longest line points to the main fraction. The numerator of the main fraction above is

$$\frac{2}{3} + \frac{1}{4}$$

and its denominator is

$$\frac{1}{2} - \frac{3}{5}$$

To simplify such complex fractions, we work out the numerator, work out the denominator, and then take the shortcut[119]

$$\frac{\frac{a}{b}}{\frac{c}{d}} = \frac{ad}{bc}$$

Applying this algorithm to the problem above yields the following solution.

$$\frac{\frac{2}{3} + \frac{1}{4}}{\frac{1}{2} - \frac{3}{5}} = \frac{\frac{8 + 3}{12}}{\frac{5 - 6}{10}}$$

$$= \frac{\frac{11}{12}}{\frac{-1}{10}}$$

$$= -\frac{11 \times 10}{12 \times 1}$$

$$= -\frac{55}{6}$$

$$= -9\frac{1}{6}$$

While we may rightly insist on the use of the horizontal line to represent division with, others might choose to use other division symbols along with the horizontal line and, therefore, we need to know how to work with such expressions. Here is an example:

---

[119]The logic behind this shortcut was discussed in the previous subsection.

Example

$$\frac{4 + 2 \times 15 \div 3}{5 \times 2 - 14 \div 7} = \frac{4 + 10}{10 - 2}$$

$$= \frac{14}{8}$$

$$= \frac{7}{4}$$

$$= 1\frac{3}{4}$$

We present one more example.

Example

$$\frac{\frac{2}{3} + 3 \times 4}{1 - \frac{1}{3} \div \frac{3}{4}} = \frac{\frac{2}{3} + 12}{1 - \frac{1}{3} \times \frac{4}{3}}$$

$$= \frac{\frac{2}{3} + \frac{12}{1}}{1 - \frac{4}{9}}$$

$$= \frac{\frac{2 + 36}{3}}{\frac{1}{1} - \frac{4}{9}}$$

$$= \frac{\frac{38}{3}}{\frac{9 - 4}{9}}$$

$$= \frac{\frac{38}{3}}{\frac{5}{9}}$$

$$= \frac{38 \times 9}{3 \times 5}$$

$$= \frac{114}{5}$$

$$= 22\frac{4}{5}$$

Exercise Set 4.3.5

1. Evaluate

   a. $\frac{2}{5} + \frac{5}{6} + \frac{1}{3}$

   b. $\frac{1}{4} + \frac{2}{5} + \frac{2}{3}$

   c. $\frac{3}{8} + \frac{1}{4} - \frac{1}{2}$

   d. $\frac{5}{12} - \frac{1}{8} - \frac{1}{6}$

   e. $\frac{1}{12} + \frac{2}{3} - \frac{1}{5} + \frac{5}{6}$

   f. $-\frac{1}{15} - \frac{5}{12} + \frac{5}{8}$

   g. $-\frac{1}{17} + \frac{5}{34} - \frac{3}{2}$

   h. $\frac{3}{8} + \frac{1}{2} - \frac{1}{4}$

   i. $-\frac{9}{14} + \frac{2}{21} - \frac{3}{7}$

   j. $\frac{3}{16} - \frac{1}{4} + \frac{2}{3}$

   k. $\frac{4}{25} + \frac{1}{15} + \frac{2}{3}$

   l. $-\frac{7}{10} - \frac{1}{2} - \frac{3}{4} - \frac{1}{8}$

   m. $\frac{1}{3} + \frac{3}{4} - \frac{1}{4}$

   n. $\frac{11}{10} + \frac{1}{2} + \frac{1}{4}$

   o. $-\frac{2}{3} - \frac{2}{5} + \frac{15}{4}$

   p. $\frac{3}{9} + \frac{1}{6} + \frac{4}{3} + \frac{2}{3}$

2. Evaluate

   a. $\frac{4}{5} + 5\frac{1}{2} - 1\frac{3}{4}$

   b. $\frac{9}{5} - 1\frac{3}{4} - 2\frac{1}{5}$

   c. $4\frac{2}{3} + 1\frac{1}{2} - \frac{3}{4}$

   d. $6\frac{2}{3} - 1 + 3\frac{1}{2} - 2\frac{1}{4}$

   e. $\frac{3}{4} - 2\frac{1}{2} + 4$

   f. $14\frac{2}{3} + 1\frac{1}{3} - 3\frac{2}{5}$

   g. $15\frac{2}{3} + 6\frac{1}{2} - 12\frac{3}{4}$

   h. $4\frac{3}{7} - 2\frac{1}{14} + 1 - 3\frac{1}{4}$

   i. $2\frac{1}{5} + 3\frac{3}{4} - 1\frac{1}{2} + 4\frac{2}{3}$

   j. $-3\frac{2}{5} + 1\frac{1}{2} - 2\frac{1}{4}$

   k. $14\frac{3}{4} - 11\frac{1}{2} + 2\frac{3}{5}$

   l. $-6\frac{1}{3} - 2\frac{3}{4} - 1\frac{1}{3}$

   m. $12 - 10\frac{3}{5} + 8\frac{1}{2}$

   n. $-160\frac{2}{3} + 200 + 360\frac{1}{3}$

   o. $-2\frac{1}{3} - 4\frac{2}{5} + 5 + 3\frac{2}{3}$

   p. $8\frac{1}{5} + 2\frac{1}{3} + 3\frac{1}{4} - 5\frac{1}{2}$

3. Evaluate

   a. $\frac{2}{3} \times \frac{1}{2} \times \frac{3}{4}$

   b. $\frac{4}{5} \times 1\frac{1}{2} \times 15$

   c. $-2\frac{1}{3} \times 9 \times 4\frac{1}{2}$

   d. $4 \times 3\frac{1}{4} \times \frac{5}{26}$

   e. $16 \times \frac{-1}{144} \times 12$

   f. $-4\frac{2}{3} \times \frac{1}{7} \times 24 \times 3\frac{1}{6}$

   g. $9 \times \frac{4}{3} \times 2\frac{1}{3} \times \frac{3}{4}$

   h. $3\frac{4}{5} \times \frac{18}{19} \times 2\frac{1}{6}$

i. $1\frac{5}{6} \times 1\frac{1}{2} \times 8 \times \frac{-4}{33}$

j. $-9 \times \frac{-2}{-3} \times 3\frac{1}{4}$

k. $5\frac{1}{2} \times 34 \times \frac{3}{17}$

l. $1\frac{2}{15} \times \frac{2}{7} \times \frac{14}{34}$

m. $-32 \times 4\frac{1}{8} \times 2 \times \frac{-1}{-11}$

n. $\frac{3}{4} \times 14 \times \frac{1}{21} \times 5 \times \frac{1}{15}$

4. Evaluate

a. $\frac{3}{14} \div \frac{5}{7} \times \frac{2}{9}$

b. $\frac{4}{5} \div 1\frac{1}{6} \times 10\frac{1}{2}$

c. $\frac{3}{4} \div \frac{3}{2} \div \frac{5}{4} \times \frac{1}{3}$

d. $4\frac{4}{5} \div \frac{14}{15} \div \frac{6}{21}$

e. $\frac{1}{4} \div \frac{2}{5} \times 5$

f. $2\frac{2}{3} \div 3\frac{1}{2} \times 2\frac{4}{5} \times \frac{3}{4}$

g. $4\frac{2}{3} \div \frac{1}{2} \div \frac{1}{3}$

h. $-16 \times 1\frac{1}{2} \div 3$

i. $\frac{-1}{3} \times \frac{-1}{-4} \div \frac{-2}{5} \div \frac{1}{2}$

j. $\frac{4}{3} \div 4\frac{2}{3} \times 1\frac{1}{5} \div \frac{2}{7}$

k. $-1\frac{1}{15} \div \frac{-7}{21} \times \frac{5}{-13}$

l. $1\frac{2}{3} \times 1 \div \frac{-3}{4} \div \frac{1}{-2}$

m. $-36 \div \frac{-18}{-21} \times \frac{1}{-7} \times 2$

n. $8\frac{1}{2} \div 4 \div 3$

o. $\frac{3}{4} \div 1\frac{1}{2} \times 7\frac{1}{2}$

p. $4\frac{2}{3} \div \frac{3}{2} \times 1\frac{1}{2} \div 2$

5. Evaluate

a. $\frac{1}{2} + \frac{2}{3} \times \frac{3}{8} + \frac{1}{3}$

b. $\frac{2}{3} \times \frac{1}{2} + \frac{3}{5}$

c. $\frac{1}{8} + \frac{2}{3} \times \frac{3}{5} \div \frac{3}{5}$

d. $\frac{2}{15} \times \frac{3}{2} \times \frac{1}{4} + 1\frac{2}{3} \div 3 \times \frac{1}{2}$

e. $3 \div \frac{4}{7} - 2\frac{1}{4} + \frac{2}{5} \times 4\frac{1}{2} \div 8$

f. $\frac{3}{4} \div \frac{2}{3} - \frac{1}{2} \times \frac{3}{5} + \frac{2}{3} \times \frac{1}{5}$

g. $\frac{1}{3} \times \frac{2}{5} + 2 - \frac{3}{4} \div \frac{5}{4}$

h. $\frac{1}{2}\left(2 - \frac{1}{3}\right) + \frac{3}{4}\left(4 + \frac{1}{2}\right)$

i. $4\frac{1}{5} \times \frac{5}{7} \div \frac{3}{4} - \frac{2}{3}\left(\frac{1}{5} + \frac{3}{4} \times \frac{1}{2}\right)$

j. $2\left(\frac{1}{2}\right)\left(\frac{3}{7} \times \frac{1}{2} + \frac{1}{3}\right)$

k. $\frac{3}{5} \times \frac{2}{3} - \frac{1}{10}\left(\frac{2}{5} - \frac{1}{2} \times \frac{1}{3}\right) + \frac{3}{4} \times \frac{2}{5}$

l. $\frac{1}{3} + \frac{2}{3}\left(\frac{3}{4} + 1\right) + 2\left(\frac{3}{5} \times \frac{2}{3} + 3 \div \frac{12}{7}\right)$

m. $\left(\frac{1}{4} \times \frac{6}{7} + \frac{1}{2}\right)\left(\frac{5}{6} - 2 \times \frac{1}{8}\right)$

n. $\frac{2}{-3} - \frac{-4}{5} \times \frac{-5}{-6} + \frac{-1}{3}$

o. $\frac{-6}{7} \times \frac{-1}{2} + \frac{1}{-4}$

p. $\frac{-2}{-5} - \frac{-2}{3} \times \frac{3}{4} \times \frac{-2}{-3}$

q. $\frac{-5}{6} \times \frac{-3}{2} \times \frac{-1}{-5} - \frac{-3}{5} \div (-6) \times \frac{-5}{2}$

r. $-\frac{1}{3} \div \frac{5}{-6} - 3\frac{1}{2} + \frac{-1}{-3} \times \frac{-3}{1} \div \frac{-2}{5}$

s. $\frac{-5}{-7} \div \frac{1}{-7} - \frac{2}{-5} \times \frac{-5}{3} - \frac{2}{7} \times 1\frac{2}{5}$

t. $\frac{1}{3} \times \frac{-6}{5} - \frac{-2}{-7} + \frac{1}{-4} \div \frac{-5}{4}$

6. In each case determine the numerator and denominator of the main fraction.

a. $\dfrac{8 - \frac{2}{3}}{4}$

b. $\dfrac{\frac{9}{2} - 2}{3\frac{1}{4}}$

c. $\dfrac{2}{\frac{3}{4} \times \frac{1}{3} + \frac{1}{2} \times 5}$

d. $\dfrac{5 + 3\frac{1}{2}}{2 - 1\frac{2}{3}}$

7. Evaluate

a. $\dfrac{8 - \frac{2}{3}}{4}$

h. $\dfrac{6\frac{1}{3} - 4}{\frac{2}{5} + \frac{1}{3} \times 6}$

b. $\dfrac{-8}{2 + \frac{3}{4}}$

i. $\dfrac{-4\frac{1}{3} - 4 \times \frac{3}{8}}{-4 - 1\frac{1}{2}}$

c. $\dfrac{5 + 3\frac{1}{2}}{2 - 1\frac{2}{3}}$

j. $\dfrac{8 - 1}{\frac{3}{4} \times 5}$

d. $\dfrac{1\frac{1}{2} + 3\frac{1}{4}}{2\frac{1}{3} - 1\frac{1}{2}}$

k. $\dfrac{4 - 1\frac{1}{3}}{2 \times \frac{1}{2} + 1}$

e. $\dfrac{\frac{9}{2} - 2}{3\frac{1}{4}}$

l. $\dfrac{2}{\frac{3}{4} + \frac{1}{2} \times 5}$

f. $\dfrac{-6\frac{2}{3} + 4 \times \frac{3}{2}}{2}$

m. $\dfrac{\frac{3}{4} + \frac{1}{2} \times \frac{4}{3}}{2 + \frac{1}{3}}$

g. $\dfrac{5 - 2\frac{1}{2}}{\frac{3}{4} + 2 \times \frac{1}{3}}$

n. $\dfrac{\frac{1}{3} \times 4 + 2}{3 - \frac{1}{2} \times \frac{1}{4}}$

p. $\dfrac{2\frac{1}{4} - \frac{1}{2} \times \frac{1}{3}}{\frac{3}{4} + \frac{1}{2}}$

o. $\dfrac{\frac{1}{2} + \frac{2}{3} \times 6}{\frac{3}{4} \div \frac{1}{2} + 1}$

4.3.6 Graphing

In the section on whole numbers we introduced the real line and extended the notion in the section on integers. The real line is shown in Figure 4.3.23.

Figure 4.3.23: The real line

We have already shown how whole numbers and integers are located on the real line. We will now extend this discussion to rational numbers. We begin with proper fractions.

Graphing Proper Fractions

To locate a proper fraction on the real line:

1. Divide the distance from 0 to 1 into as many equally-sized pieces as the denominator says by placing ticks.[120]
2. Counting from the left find the tick identified by the numerator.[121]

This tick represents the proper fraction. Place a heavy dot on the tick. The coordinate of the point may be written under the dot or above the dot.

As an example, to graph $\frac{2}{3}$, we divide the distance from 0 to 1 into 3 equally-sized pieces, and count 2 ticks from the left. The graph is shown in

---

[120]The number of ticks should be 1 less than the number of required divisions. For example, for a denominator of 4, corresponding to a division of 1 into 4 parts, 3 ticks are needed.

[121]The first tick is assigned 1.

Figure 4.3.24.

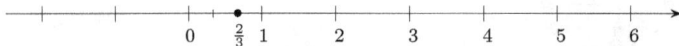

Figure 4.3.24: Graph of $\frac{2}{3}$ on the real line

Graphing Mixed Numbers

We explain how mixed numbers are located on the real line using an example. Consider the mixed number $2\frac{1}{4}$. This is greater than 2 and less than 3. To find $2\frac{1}{4}$ on the real line we divide the space between 2 and 3 into 4 equally-sized pieces and take 1 piece. This is shown in Figure 4.3.25.

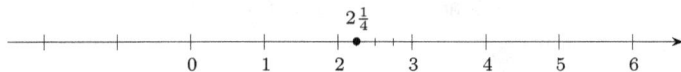

Figure 4.3.25: Graph of $2\frac{1}{4}$ on the real line

Graphing Improper Fractions

To graph an improper fraction convert the fraction to a mixed number and locate the mixed number on the real line as outlined above.[122]

---

[122]While it is possible to graph an improper fraction without conversion to a mixed number, the process can become inefficient. As an example, to graph $\frac{142}{5}$ without conversion to a mixed number, we would have to start at 0 and count 142 fifths. The equivalent mixed formulation of $\frac{142}{5}$ is $28\frac{2}{5}$ which shows that the graph lies at a point which is $\frac{2}{5}$ to the right of 28.

Graphing Negative Rationals

Consider $-\frac{13}{4}$. We convert this to a mixed number to get $-3\frac{1}{4}$. This rational number can be located on the real line by starting at 0 and moving to the left over 3 units and a further $\frac{1}{4}$ of a unit. This is shown in the diagram below.

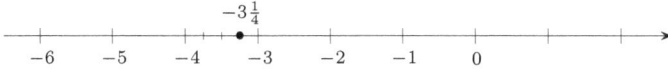

Figure 4.3.26: Graph of $-3\frac{1}{4}$ on the real line

The reader should take note not to read this as $-4\frac{3}{4}$. When reading the coordinate of a point on the line, it may help avoid this error by starting at 0 and then moving toward the point.

Exercise Set 4.3.6

1. Graph each of the following on the real line.

   a. $\frac{3}{6}$

   b. $\frac{0}{3}$

   c. $\frac{5}{8}$

   d. $\frac{3}{4}$

   e. $1\frac{1}{3}$

   f. $\frac{24}{5}$

   g. $\frac{5}{2}$

   h. $3\frac{4}{5}$

   i. $-\frac{1}{3}$

   j. $-\frac{4}{5}$

   k. $-\frac{8}{3}$

   l. $-27\frac{5}{6}$

2. What is the coordinate of each of the points in the graphs below?

(a)

(b)

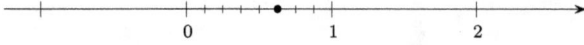

(c)

(d)

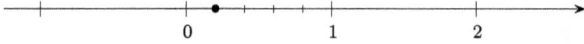

(e)

(f)

(g)

(h)

3. What mixed number should be associated with each graph below?

(a)

(b)

(c)

(d)

(e)

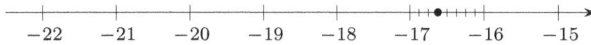

(f)

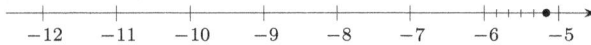

4.4 Real Numbers

The view that we have been promoting in this book attempts to show the manner in which, beginning with the set of natural numbers to which we arrive naturally when we solve problems that require that we count, new numbers are introduced into the system to accommodate our problem-solving need for closure with respect to the many operations, as we move from the simplest of the problems to increasingly more complex problems: We pointed to the lack of closure of the set of natural numbers with respect to subtraction and cited practical applications that require such closure. This led to the expansion of the set of natural numbers to generate the set of whole numbers and the set of integers. To represent these new numbers we used the symbol used to represent subtraction with as it is subtraction with respect to which closure was sought. We pointed to the lack of closure of the set of integers with respect to division and cited practical applications that require such closure. This led us to the expansion of the set of integers to generate the set of rational numbers. To represent these new numbers we used the symbol used to represent division with as it is division with respect to which closure was sought.

The order in which closure is enforced follows the increasing level of the difficulty of the problems that we solve. This starts with direct superposition problems which are problems whose solution relates to the use of addition, inverse superposition problems which are problems whose solution relates to the use of subtraction, direct proportion problems which are problems whose solution relates to the use of multiplication, and inverse proportion problems which are problems whose solution relates to the use of division.

It is this line of thinking that led us to the extension of the set of rational numbers to generate the set of real numbers but, while the previous extensions attempted to impose maximum closure with respect to a single operation, this latest extension aims to maximize closure with respect to all other operations, i.e., those beyond addition, subtraction, multiplication and division, that are used in the solution of problems that are more complex than direct superposition, inverse superposition, direct proportion, and inverse proportion problems.[123]

Such other operations fall into two categories:

The first category relates to those operations that are applied to many numbers, such as roots and logarithms, for which we use the operation symbols, such as $\sqrt{}$ and log, to represent them with.[124] This is in line with our earlier representation of integers that use the subtraction symbol and rationals which use the division symbol for the representation of their elements.

The second category relates to those operations that are unique, involving many stages in their evaluation, such as the number obtained through the division of the circumference of a circle by its diameter, for which we use specific symbols, such as the symbol π.

In this section of the textbook we will represent the most commonly encountered elements of the set of real numbers beyond rationals, i.e., roots and logarithms both of which relate to the lack of closure of the set of rational numbers with respect to the next simplest type of problems that we solve: Exponential problems.

---

[123]We remind the reader that the set of these other operations does not include closure with respect to an even root of a negative number which provides yet another extension from the set of real numbers to the set of complex numbers.

[124]Beyond addition, subtraction, multiplication, division, exponents and roots, we normally use a shortened version of the name of the operation to represent the operation with. Examples of this are the symbol log used to represent logarithms, the symbol sin used to represent sines, cos used to represent cosines, and the like.

4.4.1   Exponents

Notation and Terminology

The notation 2^5 implies that we should multiply 2 by itself 5 times, i.e.,

$$2^5 = 2 \times 2 \times 2 \times 2 \times 2$$
$$= 32$$

We usually skip right to the end result as shown below.[125]

$$2^5 = 32$$

In 2^5, 2 is called the **base** and 5 is called the **exponent**. This is shown in Figure 4.4.1 below.

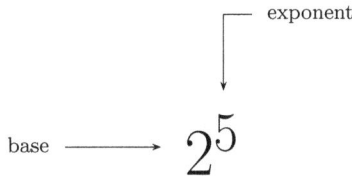

Figure 4.4.1: The base and the exponent

We read 2^5 as 2 *raised to the power of* 5 or 2 *to the power of* 5 for short. When the exponent is 2, we usually use the word *squared* in place of the phrase *to the power of* 2. As an example, we usually read 6^2 as 6 *squared* instead of 6 *to the power of* 2. When the exponent is 3, we usually use the word *cubed* in place of the phrase *to the power of* 3. As an example, we usually read 5^3 as 5 *cubed* instead of 5 *to the power of* 3.[126]

When the exponent is large we leave the expression as it is otherwise we will have to write a very large number. As an example, 2^{104} has some 32 digits so we usually leave 2^{104} as it is although we now see it as the *result* of

---

[125]This is especially desirable when the exponent is large.

[126]Such terminology has its roots in geometric applications of math. In this sense, the reading of 3^2 as 3 *squared* provides a geometric analogy for the abstract concept of 3 *to the power of* 2. The geometric analogy is the area of the square formed by generating a square out of 3, i.e., a square whose side length is 3. A similar analogy exists between the abstract concept 5^3 and the volume of the cube formed by 5, i.e., a cube with a side length of 5.

raising 2 to the power of 104.[127]

Finally, when we speak of *powers of* 2, we mean the list

$$2^1, \ 2^2, \ 2^3, \ 2^4, \ 2^5, \ 2^6, \ \ldots$$

i.e.,

$$2, \ 4, \ 8, \ 16, \ 32, \ 64, \ \ldots$$

Similar meanings are associated with phrases *powers of* 3, *powers of* 4 and the like.

The Base

In this section we will discuss the base by considering cases where the base is a natural number, a whole number, an integer and a rational number. We will limit the discussion to exponents that are whole numbers. Cases where the exponent is an entity other than a whole number will be discussed in the next section.

Natural Numbers as Base

As discussed in the previous section, the notation 2^5 means that 2 must be multiplied by itself 5 times. Although in most cases it is easy to make sense of bases that are natural numbers, there are a few cases that warrant further discussion.

1 As Base

1 raised to any exponent equals 1.

Example

$$1^8 \ = \ 1$$

This is rather obvious once you recall that

$$1^8 \ = \ 1 \times 1 \times 1 \times 1 \times 1 \times 1 \times 1 \times 1$$

which is 1.

---

[127]This last comment is important and parallels the one given earlier on the interpretation of rationals. Recall that the notation $\frac{3}{4}$ may be interpreted as a directive (i.e., divide 3 by 4) or the *result* of that directive (i.e., three quarters). In a similar manner, the notation 2^{104} may be viewed as a directive (i.e., raise 2 to the power of 104) or the result of that directive (2 to the power of 104).

Powers of 10 As Base

The result of raising a power of 10 (i.e., 10, 100, 1000, etc.) to an exponent is a 1 followed by a number of 0s. To find the number of 0s in the result, multiply the number of 0s in the base by the exponent.

Example

$$100^3 = 1\,000\,000$$

There are 2 zeros in 100. The exponent is 3. Multiply 2 by 3 to get 6. The result is a 1 followed by 6 zeros.

Multiples of Powers of 10 As Base

The shortcut above can be extended to calculate the result of raising a multiple of a power of 10 (e.g., 700, 20, 30 000, etc.) to an exponent. We will illustrate the algorithm using an example.

Example

$$4000^2 = 16\,000\,000$$

We raise 4 to the power of 2 to get 16. The number of zeros following 16 can be calculated in exactly the same manner as described above.

Whole Numbers as Base

The only novel case is 0 as base. Other cases map onto natural numbers which we just covered.

0 As Base

0 raised to any exponent equals 0.[128]

---

[128]This result is true except for 0^0 which is indeterminate. See Appendix A for a discussion on indeterminate expressions.

Example

$$0^6 = 0$$

This is rather obvious once you recall that

$$0^6 = 0 \times 0 \times 0 \times 0 \times 0 \times 0$$

which is 0.

Integers as Base

When raising a negative integer to an exponent brackets must be used. Here's an example.

Example

$$(-3)^4 = (-3)(-3)(-3)(-3)$$
$$= 81$$

If we drop the brackets then only 3 will be raised to the power of 4 as the following shows.

Example

$$-3^4 = -3 \times 3 \times 3 \times 3$$
$$= -81$$

What the two examples above show is that, if brackets are used, the number within brackets is multiplied by itself repeatedly. However, if brackets are not used, then, by convention, we take this to mean that the number without the negative sign should be multiplied repeatedly and the result should be made negative.

Both expressions are valid. You use one or the other depending on what it is that you want to communicate.

It is a good idea to keep in mind that the result of raising a negative number to an even exponent is positive. This is because an even exponent implies multiplication of an even number of negatives which results in a positive answer. As an example, $(-3)^4$ equals 81.

In addition, the result of raising a negative number to an odd exponent is negative. This is because an odd exponent implies multiplication of an odd number of negatives which results in a negative answer. As an example, $(-2)^3$ equals -8.

Rational Numbers as Base

When exponents are applied to rational numbers, the rational number should be placed inside brackets.[129] Here are a couple of examples.

Example

$$\left(\frac{2}{3}\right)^4 = \left(\frac{2}{3}\right)\left(\frac{2}{3}\right)\left(\frac{2}{3}\right)\left(\frac{2}{3}\right)$$

$$\text{Shortcut} \quad = \frac{2 \times 2 \times 2 \times 2}{3 \times 3 \times 3 \times 3}$$

$$= \frac{2^4}{3^4}$$

$$= \frac{16}{81}$$

As the shortcut shows, to raise a fraction to an exponent, we can raise the numerator and the denominator to that exponent.

Mixed numbers should be converted to improper fractions before the shortcut is applied.

Any negative signs can be processed using the argument given above for the case when the base is an integer: An even exponent implies a positive result and an odd exponent implies a negative result.

---

[129]This is irrespective of whether the rational number is positive or negative. If brackets are not used, then the numerator should be raised to the exponent but not the denominator or any negative signs. As examples, $\frac{2^4}{3} = \frac{2 \times 2 \times 2 \times 2}{3}$, and $-\frac{2^4}{3} = -\frac{2 \times 2 \times 2 \times 2}{3}$. Even $\frac{-2^4}{3} = \frac{-2 \times 2 \times 2 \times 2}{3}$. The placement of the exponent relative to the rational number makes no difference in its interpretation: $\frac{2}{3}^4$ is still seen as $\frac{2 \times 2 \times 2 \times 2}{3}$. Such strict guidelines help avoid miscommunication.

Example

To evaluate

$$\left(\frac{-2}{3}\right)^3$$

we first argue that the result will be negative as there are an odd number of negatives and the exponent is odd. We write

$$\left(\frac{-2}{3}\right)^3 = -$$

We now ignore any negative signs and raise the numerator and denominator of the rational number to the power of 3:

$$\left(\frac{-2}{3}\right)^3 = -\frac{2^3}{3^3}$$

This yields the following:

$$\left(\frac{-2}{3}\right)^3 = -\frac{2^3}{3^3}$$
$$= -\frac{8}{27}$$

Example

$$\left(\frac{2}{-3}\right)^3 = -\frac{2^3}{3^3}$$
$$= -\frac{8}{27}$$

When the exponent is even, the result will be positive. Unless there is reason to emphasize that the result is positive, the positive sign may be dropped.

Example

$$\left(-\frac{2}{3}\right)^4 = \frac{2^4}{3^4}$$
$$= \frac{16}{81}$$

Example

$$\left(\frac{2}{-3}\right)^4 = \frac{2^4}{3^4}$$
$$= \frac{16}{81}$$

The Exponent

We now shift our focus and discuss the exponent. We will begin with natural numbers as exponent and extend the discussion to cases where the exponent is a whole number, an integer and a rational number.

Natural Numbers as Exponent

This case has been covered already. A natural exponent implies repeated multiplication of the base. In this sense, an exponent of 2 implies that the base must be multiplied by itself twice. Another way to think about this is that the expanded expression will involve two factors. Similarly, an exponent of 3 implies that the base must be multiplied by itself 3 times, i.e., the expanded expression has three factors.

This alternate view of exponents as the number of factors implies that an exponent of 1 will involve a single factor.

1 as Exponent

If the exponent equals 1, the answer equals to the base.

Example

$$26^1 = 26$$

Example

$$(-3)^1 = -3$$

Example

$$\left(\frac{3}{4}\right)^1 = \frac{3}{4}$$

Whole Numbers as Exponent

The only novel case is when the exponent is 0.

0 as Exponent

Any base other than 0 raised to the power of 0 equals 1. This is usually written as

$$x^0 = 1, \qquad x \neq 0$$

In addition, 0^0 is indeterminate. See Appendix K for a discussion on this.

Example

$$18^0 = 1$$

Example

$$\left(\frac{3}{5}\right)^0 = 1$$

Integers as Exponent

A nonzero base raised to a negative exponent can be written as the reciprocal of the base to the power of a positive exponent. This is usually written as

$$x^{-n} = \frac{1}{x^n}, \qquad n \in \mathbb{N}, \ x \neq 0$$

If the base is 0, the expression is undefined. See Appendix L for a discussion on this.

Example

$$5^{-2} = \frac{1}{5^2}$$
$$= \frac{1}{25}$$

Example

$$(-3)^{-5} = \frac{1}{(-3)^5}$$
$$= \frac{1}{-243}$$
$$= -\frac{1}{243}$$

When the base is a fraction and the exponent is negative, we can use the following result to evaluate the expression.[130]

$$\left(\frac{a}{b}\right)^{-n} = \left(\frac{b}{a}\right)^{n}, \qquad n \in \mathbb{N}, \quad a \neq 0, \quad b \neq 0$$

The result follows from the fact that the reciprocal of a nonzero fraction $\frac{a}{b}$ is equal to $\frac{b}{a}$.

Example

$$\left(\frac{2}{5}\right)^{-3} = \left(\frac{5}{2}\right)^{3}$$
$$= \frac{5^3}{2^3}$$
$$= \frac{125}{8}$$
$$= 15\frac{5}{8}$$

---

[130]If a is 0 but not b, then both the expression on the left and the one on the right are undefined. If b is 0 but not a, then the expression on the left is undefined while the one on the right is 0. If both a and b are 0, then both the expression on the left and the one on the right are indeterminate.

Example

$$\left(-\frac{4}{3}\right)^{-2} = \left(-\frac{3}{4}\right)^{2}$$
$$= \frac{3^2}{4^2}$$
$$= \frac{9}{16}$$

Rational Numbers as Exponent

The root notation is commonly used to represent rational exponents. As an example

$$3^{\frac{2}{5}} = \sqrt[5]{3}^2$$

See the section ahead on roots for detail.

Exercise Set 4.4.1

1. Read each of the following in English.

   a. 1^8 b. 0^{12} c. 4^2 d. 18^3 e. 9^1

2. Write each of the following using mathematical notation.

   a. 4 to the power of 7   c. 12 raised to the power of 5
   b. 6 squared   d. 2 cubed

3. In each case identify the base and the exponent.

   a. 7^2 b. 1^{75} c. 15^6

4. List the first four powers of

   a. 2   b. 5

5. Evaluate each of the following.

   a. 2^3 d. 14^2 g. 100^3 j. 2^8

   b. 3^5 e. 24^1 h. 0^3 k. 300^2

   c. 1^6 f. 10^4 i. 15^{28} l. 20^3

6. Evaluate each of the following.

   a. 0^2 d. 100^2 g. 0^{18} j. 1200^2

   b. 10^5 e. 140^3 h. 30^4 k. 1^0

   c. 1^5 f. 1^{125} i. 1000^2 l. 0^0

7. Evaluate each of the following.

   a. $(-3)^4$ d. -7^2 g. -5^3

   b. $(-1)^{17}$ e. -10^3 h. $(-3)^3$

   c. $(-5)^3$ f. -2^4 i. $(-2)^6$

8. Evaluate each of the following.

   a. $\left(\frac{2}{5}\right)^4$ d. $\left(\frac{12}{5}\right)^2$ g. $\left(3\frac{1}{2}\right)^3$ j. $\left(\frac{3}{2}\right)^6$

   b. $\left(\frac{3}{2}\right)^5$ e. $\left(\frac{1}{10}\right)^4$ h. $\left(\frac{3}{10}\right)^4$ k. $\left(2\frac{7}{10}\right)^3$

   c. $\left(5\frac{1}{3}\right)^2$ f. $\left(1\frac{2}{3}\right)^2$ i. $\left(\frac{7}{300}\right)^3$ l. $\left(\frac{5}{14}\right)^2$

9. Evaluate each of the following.

   a. $\frac{2^3}{7}$ c. $\frac{2^5}{3}$ e. $\frac{1}{10^2}$ g. $\frac{1^8}{2}$ i. $\frac{10^3}{3^4}$

   b. $\frac{1}{2^4}$ d. $\frac{4^2}{5^3}$ f. $\frac{3^4}{8}$ h. $\frac{9^2}{10}$

10. Evaluate each of the following.

   a. $\left(\frac{-2}{3}\right)^4$ d. $-\left(\frac{4}{5}\right)^2$ g. $\left(-\frac{7}{30}\right)^3$

   b. $\left(\frac{-11}{20}\right)^1$ e. $\left(\frac{-2}{5}\right)^4$ h. $-\left(\frac{1}{100}\right)^2$

   c. $\left(\frac{3}{5}\right)^3$ f. $\left(\frac{5}{-6}\right)^3$ i. $\left(\frac{-4}{7}\right)^5$

11. Evaluate each of the following.

a. 4^0

b. 12^1

c. $\left(\frac{1}{5}\right)^1$

d. $\left(\frac{2}{3}\right)^0$

e. $\left(\frac{4}{5}\right)^1$

f. $(-3)^0$

g. $(-2)^1$

h. $\left(2\frac{1}{4}\right)^0$

i. 1^0

j. 1^1

k. 0^0

l. 0^1

12. Evaluate each of the following.

a. 4^{-2}

b. 8^{-1}

c. 2^{-5}

d. $(-3)^{-2}$

e. -3^{-2}

f. -2^{-1}

g. $(-2)^{-1}$

h. 4^{-3}

13. Evaluate each of the following.

a. $\left(\frac{2}{3}\right)^{-3}$

b. $\left(\frac{3}{10}\right)^{-2}$

c. $\left(\frac{1}{4}\right)^{-1}$

d. $-\left(\frac{4}{7}\right)^{-4}$

e. $\left(\frac{-1}{5}\right)^{-3}$

f. $\left(\frac{-3}{2}\right)^{-1}$

g. $\left(-\frac{1}{2}\right)^{-2}$

h. $\left(\frac{-2}{7}\right)^{-1}$

4.4.2   Roots

Terminology and Notation

We begin with an example. Suppose we are given $\sqrt[5]{32}$. As a directive, this asks us to find the number that, when raised to the power of 5 (i.e., when multiplied by itself 5 times), generates 32. This number is 2, i.e.,

$$\sqrt[5]{32} = 2$$

Keep the following in mind.

$$\sqrt[5]{32} \quad \text{means} \quad ?^5 = 32$$

As a directive, then, the root notation is used to find a missing base in an exponent problem.

In $\sqrt[5]{32}$, the number 5 is called the **root number** and 32 is referred to as the **radicand**. The symbol $\sqrt{}$ is called the **radical**. The terminology is illustrated in Figure 4.4.2.

root number

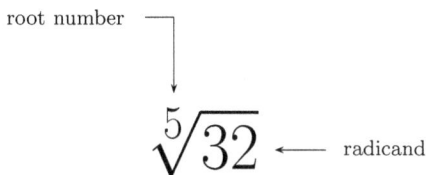

radicand

Figure 4.4.2: The root number and the radicand

We read $\sqrt[5]{32}$ as *the 5th root of* 32. When the root number is 2, we usually use the phrase *square root* in place of *2nd root*. As an example, we usually read $\sqrt[2]{25}$ as *the square root of* 25. And when the root number is 3, we usually use the phrase *cube root* in place of *3rd root*. As an example, we usually read $\sqrt[3]{8}$ as *the cube root of* 8.

When the root number is 2, it is usually not written down. As an example, we usually write $\sqrt{25}$ in place of $\sqrt[2]{25}$.

Depending on the root number and the radicand, it may be possible to simplify certain roots but not others. As an example, $\sqrt{9}$ simplifies to 3, $\sqrt[3]{-64}$ simplifies to -4 and $\sqrt[4]{\frac{1}{16}}$ simplifies to $\frac{1}{2}$. However, as shown in Appendix B, there is no rational number (and therefore no integer, whole number or natural number either) that corresponds to $\sqrt{2}$. In line with our earlier adoption of the operation symbol to represent the integers and the rationals, we will use the root notation to refer to the *results* of roots that do not simplify to a rational number.

When the root number is even and the radicand is negative, the expression is undefined. As an example, $\sqrt[4]{-16}$ is undefined as there are no numbers in $\mathbb{R}$ whose power of 4 is negative.[131]

The root notation may be converted to a fractional exponent:

$$\sqrt[n]{a^{m}} = a^{\frac{m}{n}} \quad \text{with } a \text{ nonnegative if } n \text{ is even and } m \text{ is odd}$$

and

$$\sqrt[n]{a}^{m} = a^{\frac{m}{n}} \quad \text{with } a \text{ nonnegative if } n \text{ is even}$$

The side conditions above ensure that the expressions on the left are defined in $\mathbb{R}$.

---

[131] This is because an even power applied to any real number (whether positive or negative) generates a positive result.

To explain where the first equality comes from, consider the expression

$$5^3 \times 5^3$$

This evaluates to 5^6 as shown below:

$$5^3 \times 5^3 = 5 \times 5 \times 5 \times 5 \times 5 \times 5$$
$$= 5^6$$

This implies that the square root of 5^6 is 5^3. Note that, in going from 5^6 to 5^3 the exponent halves so that we can write

$$\sqrt[2]{5^6} = 5^{\frac{6}{2}}$$
$$= 5^3$$

While in this case the exponent 6 is divisible by 2, we can extend the idea to any exponent. As an example, we can write

$$\sqrt[2]{5^3} = 5^{\frac{3}{2}}$$

and of course

$$\sqrt{5} = \sqrt[2]{5^1}$$
$$= 5^{\frac{1}{2}}$$

To explain where the second equality comes from, consider the expression

$$\sqrt[2]{5}^6$$

The expanded form of this expression is

$$\sqrt[2]{5}^6 = \sqrt[2]{5} \times \sqrt[2]{5} \times \sqrt[2]{5} \times \sqrt[2]{5} \times \sqrt[2]{5} \times \sqrt[2]{5}$$

or

$$\sqrt[2]{5}^6 = \sqrt[2]{5} \times \sqrt[2]{5} \times \sqrt[2]{5} \times \sqrt[2]{5} \times \sqrt[2]{5} \times \sqrt[2]{5}$$
$$= 5^{\frac{1}{2}} \times 5^{\frac{1}{2}} \times 5^{\frac{1}{2}} \times 5^{\frac{1}{2}} \times 5^{\frac{1}{2}} \times 5^{\frac{1}{2}}$$

Since $5^{\frac{1}{2}}$ is the square root of 5, then $\left(5^{\frac{1}{2}}\right)^2$, i.e., $5^{\frac{1}{2}} \times 5^{\frac{1}{2}}$, evaluates to 5 so that the right side of the equality above may be seen as

$$\sqrt[2]{5}^6 = \sqrt[2]{5} \times \sqrt[2]{5} \times \sqrt[2]{5} \times \sqrt[2]{5} \times \sqrt[2]{5} \times \sqrt[2]{5}$$
$$= 5^{\frac{1}{2}} \times 5^{\frac{1}{2}} \times 5^{\frac{1}{2}} \times 5^{\frac{1}{2}} \times 5^{\frac{1}{2}} \times 5^{\frac{1}{2}}$$
$$= 5 \times 5 \times 5$$
$$= 5^3$$

which shows that

$$\sqrt[2]{5}^6 = 5^{\frac{6}{2}}$$
$$= 5^3$$

The Root Number

We will limit the discussion to cases where the root number is a natural number.[132]

Square Roots

It is easy to calculate the square root of a perfect square.[133] As an example, $\sqrt{16}$ is 4.

Perfect square may also appear as the numerator and denominator of fractions. In such cases the result will be equal to the root of the numerator over the root of the denominator. As an example, $\sqrt{\frac{36}{49}}$ is $\frac{6}{7}$.

If the radicand is not a perfect square or a fraction whose numerator and denominator are both perfect squares, then it may still be possible to simplify the root but we will not discuss such scenarios in this book.

Higher Roots

The case for higher roots is similar to the case for square roots. Some radicands may simplify to a natural number or a whole number or an integer or a rational number.

Example

$$\sqrt[3]{-125} = -5$$

Example

$$\sqrt[4]{81} = 3$$

---

[132]It is highly uncommon to encounter root numbers that are not natural. However, in case they appear, one can always rewrite the expression using rational exponents and simplify the root number to a natural number.

[133]Perfect squares are 1, 4, 9, 16, 25, 36, etc. They arise from 1^2, 2^2, 3^2, 4^2, 5^2, 6^2, etc. It is a good idea to memorize the list of perfect squares up to at least 100.

Example

$$\sqrt[5]{\frac{1}{32}} = \frac{1}{2}$$

The Radicand

In this section we will discuss the radicand. We will begin with natural numbers as radicands and then extend the discussion to radicands that are whole numbers, integers and rational numbers.

Natural Numbers as Radicand

This scenario has been discussed in detail already. Here we will add a note on the special case when the radicand is 1.

1 as Radicand

Any root of 1 equals 1.

Example

$$\sqrt[3]{1} = 1$$

Whole Numbers as Radicand

The only novelty here is when the radicand is 0.

0 as Radicand

Any root of 0 equals 0.

Example

$$\sqrt[6]{0} = 0$$

Integers as Radicand

The root of a positive number is positive. The root of a negative number is negative if the root number is odd and it is undefined if the root number is even.[134]

Example

$$\sqrt{9} = 3$$

And not -3 which is represented as $-\sqrt{9}$.

Example

$$\sqrt[3]{8} = 2$$

Example

$$\sqrt{-9} \quad \text{is undefined.}$$

There is no real number whose square is negative.

Rational Numbers as Radicand

To find the root of a fraction, find the roots of the numerator and the denominator.

Example

$$\sqrt[3]{\frac{8}{125}} = \frac{\sqrt[3]{8}}{\sqrt[3]{125}}$$
$$= \frac{2}{5}$$

---

[134]This is true provided we limit ourselves to the real number system which is the case in this book.

Example

$$\sqrt[5]{\frac{-1}{32}} = \frac{\sqrt[5]{-1}}{\sqrt[5]{32}}$$
$$= \frac{-1}{2}$$
$$= -\frac{1}{2}$$

Exercise Set 4.4.2

1. Read each of the following in English.

   a. $\sqrt[3]{64}$ c. $\sqrt{36}$ e. $\sqrt{9}$

   b. $\sqrt[4]{625}$ d. $\sqrt[7]{20}$ f. $\sqrt[6]{1}$

2. Write each of the following using mathematical notation.

   a. the square root of 25   d. the cube root of 8
   b. the 5th root of 1   e. the square root of 4
   c. the fourth root of 81   f. the cube root of 125

3. In each case identify the root number and the radicand.

   a. $\sqrt[3]{8}$ b. $\sqrt{49}$ c. $\sqrt[4]{10\,000}$ d. $\sqrt[3]{1}$

4. Rewrite each of the following using fractional exponents.

   a. $\sqrt[3]{64}$ c. $\sqrt[4]{24}$ e. $\sqrt{9}^{\,7}$ g. $\sqrt{400}$

   b. $\sqrt{6}$ d. $\sqrt[5]{2}^{\,4}$ f. $\sqrt[10]{12}$ h. $\sqrt[4]{5}^{\,3}$

5. Rewrite each of the following using the root notation.

   a. $5^{\frac{1}{2}}$ c. $1^{\frac{3}{5}}$ e. $6^{\frac{1}{4}}$ g. $5^{\frac{4}{3}}$

   b. $12^{\frac{2}{3}}$ d. $0^{\frac{1}{3}}$ f. $100^{\frac{12}{5}}$ h. $7^{\frac{1}{5}}$

6. Evaluate each of the following.

a. $\sqrt[3]{64}$

b. $\sqrt[4]{16}$

c. $\sqrt{169}$

d. $\sqrt[3]{512}$

e. $\sqrt[4]{0}$

f. $\sqrt[3]{8}$

g. $\sqrt[3]{27}$

h. $\sqrt[3]{1000}$

i. $\sqrt[4]{10\,000}$

j. $\sqrt[3]{8}$

k. $\sqrt[3]{-8}$

l. $\sqrt[4]{16}$

m. $\sqrt[4]{-16}$

n. $\sqrt[5]{-32}$

o. $\sqrt[3]{-64}$

p. $\sqrt{100}$

q. $\sqrt{-100}$

r. $\sqrt[3]{1000}$

s. $\sqrt[3]{-1000}$

t. $\sqrt[3]{\dfrac{1}{27}}$

u. $\sqrt[3]{\dfrac{8}{27}}$

v. $\sqrt[4]{\dfrac{16}{81}}$

w. $\sqrt[4]{-\dfrac{1}{81}}$

x. $\sqrt[5]{\dfrac{1}{32}}$

y. $\sqrt[4]{\dfrac{-81}{16}}$

4.4.3 Logarithms

Terminology and Notation

We begin with an example. Suppose we are given $\log_2 32$. As a directive, this asks us to find the number that, when 2 is raised to the power of that number, generates 32. The number that we seek is 5, i.e.,

$$\log_2 32 = 5$$

Keep the following in mind.

$$\log_2 32 \qquad \text{means} \qquad 2^? = 32$$

As a directive, then, the logarithm notation is used to find a missing exponent in an exponent problem.

In $\log_2 32$, the number 2 is called the base in line with our definition of exponents. We refer to 32 as the result. The symbol log is the logarithm notation. The terminology is illustrated in Figure 4.4.3.

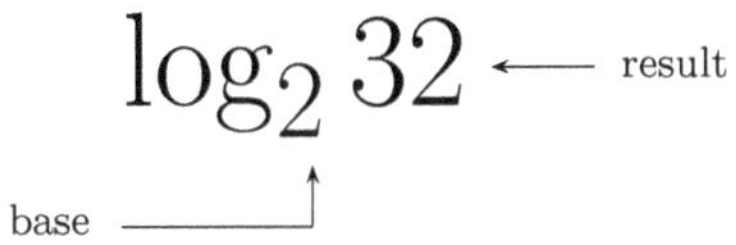

$$\log_2 32 \longleftarrow \text{result}$$

base

Figure 4.4.3: The base and the result

We read $\log_2 32$ as *log to the base 2 of 32.*

When the base is 10, it is usually not written down. As an example, we usually write $\log 100$ in place of $\log_{10} 100$.

Depending on the base and the result, it may be possible to simplify certain logarithms to a natural number, a whole number, an integer or a rational number. As an example, $\log_3 9$ simplifies to 2 and $\log_4 \frac{1}{64}$ simplifies to -3. However, there is no rational number (and therefore no integer, whole number or natural number either) that corresponds to $\log_3 2$ which implies that $\log_3 2$ is an irrational number. In line with our earlier adoption of the operation symbol to represent the integers, the rationals, and irrationals that involve roots, we will use the logarithm notation, i.e., log, to refer to the *results* of logarithms that do not simplify to a rational number.

Exercise Set 4.4.3

1. Read each of the following in English.

   a. $\log_6 36$ c. $\log_2 7$ e. $\log 100$

   b. $\log_3 2$ d. $\log_{10} 150$ f. $\log 24$

2. Write each of the following using mathematical notation.

   a. log to the base 3 of 81   d. log to the base 7 of 7

   b. log to the base 10 of 34   e. log to the base 2 of 1

   c. log to the base 4 of 16   f. log to the base 5 of 125

3. In each case identify the base.

   a. $\log_6 36$ b. $\log_{10} 1000$ c. $\log 52$ d. $\log_3 18$

4. Evaluate each of the following.

   a. $\log_6 36$ d. $\log_4 16$ g. $\log_5 5$

   b. $\log_3 \frac{1}{81}$ e. $\log_2 \frac{1}{8}$ h. $\log_4 \frac{1}{4}$

   c. $\log 100$ f. $\log_{10} 1000$ i. $\log_8 1$

4.4.4   Expressions

When we evaluate an expression the first thing we do is analyze it into its terms and the terms into their factors. The factors may be as simple as a number or they may involve brackets, exponents, roots, or any other mathematical constructs. To synthesize the expression we need to multiply the factors but before we can do so, we need to know what these factors are, i.e., we need to work them out.

Example

To evaluate

$$5 \times 6 + 4 \times 2^3$$

we break the expression up into two terms as shown below.

$$\boxed{5 \times 6} + \boxed{4 \times 2^3}$$

The first box contains two factors: 5 and 6. We will be multiplying these shortly. The second box contains two factors also. The first is 4 and the second is 2^3. Before we can carry out the multiplication we need to work out the exponent. This yields the following.

$$\boxed{5 \times 6} + \boxed{4 \times 2^3} = \boxed{30} + \boxed{4 \times 8}$$

The rest is simple.

$$\boxed{5 \times 6} + \boxed{4 \times 2^3} = \boxed{30} + \boxed{4 \times 8}$$
$$= 30 + 32$$
$$= 64$$

Example

To evaluate

$$5\left(-6\right)^2 - \left(-2\right)^3\left(-4\right)\left(-3\right)$$

we break it up into terms as shown below.

$$\boxed{5\left(-6\right)^2} - \boxed{\left(-2\right)^3\left(-4\right)\left(-3\right)}$$

Next we break the terms into factors. The first term contains two factors: 5 and $\left(-6\right)^2$. The second term contains three factors: $\left(-2\right)^3$, $\left(-4\right)$, and $\left(-3\right)$. We begin by working out the factors.

$$\boxed{5\left(-6\right)^2} - \boxed{\left(-2\right)^3\left(-4\right)\left(-3\right)}$$
$$= \boxed{5\left(36\right)} - \boxed{\left(-8\right)\left(-4\right)\left(-3\right)}$$

Having evaluated the factors, we can now multiply them to evaluate the terms. The first term evaluates to 180. For the second term we combine the three negative signs inside the box with the operation that sits to the left of the box to get a '+' followed by $8 \times 4 \times 3$ which evaluates to 96.

$$\boxed{5\left(-6\right)^2} \; - \; \boxed{\left(-2\right)^3\left(-4\right)\left(-3\right)}$$
$$= \boxed{5\left(36\right)} \; - \; \boxed{\left(-8\right)\left(-4\right)\left(-3\right)}$$
$$= 180 + 96$$

The rest is simple.

$$\boxed{5\left(-6\right)^2} \; - \; \boxed{\left(-2\right)^3\left(-4\right)\left(-3\right)}$$
$$= \boxed{5\left(36\right)} \; - \; \boxed{\left(-8\right)\left(-4\right)\left(-3\right)}$$
$$= 180 + 96$$
$$= 276$$

Here's an example involving fractions.

Example

$$\boxed{\tfrac{-2}{3}\left(\tfrac{3}{5}\right)^2} \; + \; \boxed{\left(-4\right)\left(-\tfrac{1}{2}\right)^2} = \boxed{\tfrac{-2}{3}\left(\tfrac{9}{25}\right)} \; + \; \boxed{\left(-4\right)\left(\tfrac{1}{4}\right)}$$
$$= \frac{-6}{25} - 1$$
$$= \frac{-6}{25} - \frac{1}{1}$$
$$= \frac{-6 - 25}{25}$$
$$= \frac{-31}{25}$$
$$= -1\frac{6}{25}$$

The following example shows what to do when exponents are applied to brackets.

Example

$$\boxed{5\left(2\times4-4\times3\right)^2} + \boxed{\left(-3\right)^2\left(5\right)} = \boxed{5\left(8-12\right)^2} + \boxed{9\left(5\right)}$$

$$= \boxed{5\left(-4\right)^2} + \boxed{45}$$

$$= \boxed{5\left(16\right)} + \boxed{45}$$

$$= 80 + 45$$

$$= 125$$

We now present examples that involve roots.

Example

To evaluate

$$5\sqrt{16} + 2\times6$$

we break the expression up into two terms as shown below.

$$\boxed{5\sqrt{16}} + \boxed{2\times6}$$

The first term contains $5\sqrt{16}$. This is the same as $5\times\sqrt{16}$. There are, therefore, two factors in the first term: 5 and $\sqrt{16}$. Before we can multiply these we have to work out the root. The second box evaluates to 12.

$$\boxed{5\sqrt{16}} + \boxed{2\times6} = \boxed{5\times4} + \boxed{12}$$

The rest is simple.

$$\boxed{5\sqrt{16}} + \boxed{2\times6} = \boxed{5\times4} + \boxed{12}$$

$$= 20 + 12$$

$$= 32$$

Example

Evaluate

$$5\sqrt[4]{-81} - \left(-2\right)^5\left(17\right)$$

Solution

The expression is undefined (it is meaningless) as it contains an even root of a negative radicand, i.e., $\sqrt[4]{-81}$.

Here's an example with fractions.

Example

$$\boxed{\frac{5}{12}\sqrt[3]{\frac{-27}{1000}}} + \boxed{(-4)\left(-\frac{1}{2}\right)} = \boxed{\frac{5}{12} \times \frac{-3}{10}} + \boxed{\frac{2}{1}}$$

$$= \frac{-1}{8} + \frac{2}{1}$$

$$= \frac{-1 + 16}{8}$$

$$= \frac{15}{8}$$

$$= 1\frac{7}{8}$$

And here is an example of a root whose radicand needs some work.

Example

Evaluate

$$3\sqrt{53 - 4 \times 7} + 5\sqrt[3]{8}$$

Solution

$$\boxed{3\sqrt{53 - 4 \times 7}} + \boxed{5\sqrt[3]{8}} = \boxed{3\sqrt{53 - 28}} + \boxed{5 \times 2}$$

$$= \boxed{3\sqrt{25}} + \boxed{10}$$

$$= \boxed{3 \times 5} + \boxed{10}$$

$$= 15 + 10$$

$$= 25$$

Here is an example of an expression that involves logarithms.

Example

Evaluate

$$5 \log_3 9 \; - \; 4 \log_2 32$$

Solution

The expression breaks into two terms: $5 \log_3 9$ and $4 \log_2 32$. The first term contains two factors: 5 and $\log_3 9$. The second term contains two factors as well: 4 and $\log_2 32$. To work out the expression, we start by working out the factors, multiply the factors to evaluate the terms, and add and/or subtract the terms to evaluate the expression. Here is the full solution.

$$\boxed{5 \log_3 9} \; - \; \boxed{4 \log_2 32} \; = \; \boxed{5 \times 2} \; - \; \boxed{4 \times 5}$$
$$= \; 10 \; - \; 20$$
$$= \; -10$$

We end this section with one more example on logarithms.

Example

$$\boxed{-3 \log_3 \tfrac{1}{3}} \; - \; \boxed{2 \log \tfrac{1}{100}} \; = \; \boxed{-3\,(-1)} \; - \; \boxed{2\,(-2)}$$
$$= \; 3 \; + \; 4$$
$$= \; 7$$

Exercise Set 4.4.4

1. Evaluate each of the following.

   a. $3 \times 2 + 4^2 \times 3$

   b. $5 \times 3^2 + 8 \times 2$

   c. $-7 \times 4^2 + 3^3 \times 2$

   d. $2^3 \times 5^2 + 8$

   e. $15 + 2^5 \times 3^3 - 6^2 \div 9$

   f. $-4 + (-3)^2 - (-3)\,(4)$

   g. $2\,(-3)\,(4^2) + 3^2\,(-2)\,(-1)$

   h. $2\,(4^2)\,(-1) - (2^3)\,(-3)$

   i. $3\,(2 + 3^2)\,(1 \times 3 - 2)$

   j. $2\,(4 - 2^3) - 3 \times 4^2$

k. $18 \left(4 \times 2 - 3^2\right)^2$

n. $8 \left(\frac{1}{4}\right)^2 - \frac{3}{5} \times 4 + (-3) \left(\frac{2}{3}\right)^3$

l. $\left(\frac{2}{3}\right)^2 \times 3 - (-5) \left(\frac{1}{3}\right)^3$

o. $4 \times 3^2 + 2 \left(-1 + 2 \times 3\right)$

m. $\frac{2}{3} \left(\frac{3}{4}\right)^2 + \frac{1}{2} \div \frac{4}{5} - \frac{1}{3}$

p. $\frac{1}{3} \times 2 + \left(\frac{3}{4}\right)^2$

q. $-(-3)^2 (5) (2) - 4 \left(-2^3\right) (4) - 4^3 \div 8$

r. $-2 \left(4 + 2 \times 3\right) + 2 \left(3^4 - 2 \times 5\right)$

s. $(2 + 3 \times 4 - 18)^3 + 4 \left(5 - 3 \times 2\right)^2$

2. Evaluate each of the following.

a. $4\sqrt{25} + 3\sqrt{4}$

h. $8\sqrt{1} - 2\sqrt{9} - 5\sqrt{16}$

b. $\sqrt{36} + \sqrt{25}$

i. $5\sqrt{16} + 2\sqrt{0} - 3\sqrt{4} + \sqrt{1}$

c. $3\sqrt{1} + 2\sqrt{16} - 3\sqrt{4}$

j. $2\sqrt{3} - 5\sqrt{3} + \sqrt{3}$

d. $8 + 3\sqrt{4} - 2 \times 5$

k. $4\sqrt{5} - 2\sqrt{5} - 7\sqrt{5}$

e. $4\sqrt{\frac{1}{36}} + \sqrt{\frac{9}{4}} - \frac{3}{5}$

l. $-\sqrt{2} - 4\sqrt{2} + 5\sqrt{2}$

f. $6\sqrt{9} - \sqrt{16} + 5\sqrt{4}$

m. $-3\sqrt[3]{2} + 4\sqrt[3]{2} - \sqrt[3]{2}$

g. $\sqrt{4} \times 5 + 3\sqrt{25} - 3 \times 4$

n. $\sqrt[4]{10} + 2\sqrt[4]{10} - 6\sqrt[4]{10}$

o. $-\sqrt{5} + 2\sqrt{5} - 3\sqrt{5}$

3. Evaluate each of the following.

a. $\log_2 8 + \log_2 16$

f. $\log_3 81 - 2\log_2 2$

b. $\log_3 9 + \log_2 8$

g. $\log_5 \frac{1}{25} - (-3)\log_2 \frac{1}{2}$

c. $\log_5 25 + \log_3 81$

h. $-\log \frac{1}{100} + 2\log_3 3$

d. $-4\log_2 4 + \log_3 9$

i. $-3\log 100 - \log_2 2 - 2\log_3 27$

e. $-\log 100 + \log_6 36$

j. $4\log_3 3 + 5\log_2 32 - \log 10$

4. Evaluate each of the following.

a. $4^2 - 3\log_2 8 + 5\sqrt{16}$

c. $4^2 - 2\log 100 + 3\sqrt{25}$

b. $3\log_4 16 - \sqrt[3]{8}$

d. $-2\log_5 5 + \log_3 9 - 2^3$

e. $-\sqrt[3]{81} - 2^4 + \sqrt{36}$

f. $4^2 + 3\log_2 8 - 3\sqrt[3]{-1}$

g. $\sqrt[3]{-8} - \log_2 \frac{1}{4} + 1$

h. $-4 + 2\log \frac{1}{100} + 3^2$

i. $-2^3 - \log_2 2 + \log_4 1$

j. $2\sqrt{-4} + 2\log_2 8 - 3^3$

k. $\sqrt{2} + 3\sqrt{2}$

l. $-2\sqrt[3]{5} - 4\sqrt[3]{5}$

m. $4\log_2 3 + \log_2 3 - 3\log_2 3$

n. $3\sqrt{2} - 2\log_2 5 + 5\sqrt{2} - 3\log_2 5$

Part III
Applications and Base-Ten Notations

In this part of the book we shift our focus from theory to applications of theory. The treatment begins by providing justification for use of base-ten notations in the expression of measured values; expressions that are often present in discussions that pertain to applications of scientific theories. Following this, we present and discuss structural, fratures of, and algorithmic tools for working with, two such base-ten notations: the decimal notation and scientific notation.

Chapter 5
Communication Needs of Applications

One of the drawbacks of the formal notation is its complexity. While the notation used to represent the natural numbers and whole numbers may seem simple enough,[1] the extension of the symbol set to represent such irrational numbers as the number pi, π, Euler's number, e, the golden ratio, ϕ and the like is not so intuitive. Note that for a typical adult it is not easy to work with, say, π, without resorting to its approximate decimal representation of 3.14. In addition, the formal notation's use of the various operations to represent the integers, which use the notations for addition and subtraction to represent the positives and the negatives, the use of the division symbol, i.e., the fraction line, to represent the rational numbers, and the use of other symbols used to represent the various operations that give rise to the irrationals, e.g., $\sqrt{\ }$, log, etc., is enough to discourage the average person from the study of basic math.[2]

The decimal notation represents an attempt to use a unified notation to represent all the numbers in $\mathbb{R}$, whether natural, whole, integral, rational or irrational. Realization of such unified notation would be a feat as it would free us from having to study separate algorithms to work with (compare, add, subtract, multiply, divide, etc.) the various symbols introduced by the formal notation. In fact, it would render the formal notation obsolete.

The problem, however, is that, so long as one insists on representing the various numbers in $\mathbb{R}$ *exactly*, the attempt fails to deliver on its promise. Let us explain why.

---

[1]This is disputable. The apparent simplicity has its root in familiarity: Keep in mind that most adults are introduced to the notation in early childhood and it is the repetitive use of the notation since our early school years that makes it seem simple.

[2]Note that with the addition of each symbol comes a host of algorithms that teach us how to work with them: There are algorithms that tell us how to work with integers; others that explain how we can work with the rationals; and still others that show us how to work with roots, logarithms and the like.

The decimal notation starts by adopting the symbolism and the place value mechanism to represent the natural numbers and whole numbers. However, as its main feature is that it extends the place value mechanism to represent the *sizes* of the various numbers in $\mathbb{R}$, it is inherently incapable of representing the negative integers (and, indeed, all negatives). The notation, therefore, follows the formal lead and adopts the subtraction symbol for the representation of the negatives as one sees in such expressions as -15.9.

Furthermore, the notation is only capable of expressing the rationals exactly. Even then, in most cases, its representations are awkward. More often than not exact representation of rationals run into infinite sequences of repeating digits which are very unpleasant to work with. An example of this is the decimal representation of $\frac{3}{7}$ which is $0.\overline{428\,571}$ with the line over the digits pointing to a repeating block. It is not immediately clear how one can add, subtract, multiply or divide such representations.

Beyond rationals, the notation becomes useless in practice as its infinite-digit representations do not exhibit repeating behaviour. As an example, the decimal representation of $\sqrt{2}$ is $1.414\,213\,562\ldots$ a sequence of digits that never ends and no part of which ever repeats. How does one work with such representations?

Another weakness of base-ten notations is in their lack of clarity. There is something to be said about the clarity communicated by the notation $\frac{3}{7}$ which immediately brings 3 *out of* 7 *pieces* to mind. The decimal equivalent $0.428\,571\ldots$ is incapable of doing so. As a second example, the use of 2π makes it clear that the number π plays a significant role in the discussion whereas the fact is fully hidden in the decimal representation $6.283\,185\,307\ldots$.

Given the strengths and weaknesses of base-ten notations, it should be obvious that they should only be used under conditions where exactness or descriptiveness are not of concern. The expression of values that are inherently approximate and have no need for clarity, such as measured values, provides one such area of use for base-ten notations. Measurements are rarely exact[3] and one would be hard-pressed trying to convince us that the height of some person has as much significance as the value of π. Given that measured values are rounded and are, therefore, never exact, we will never end up with repeating digits[4] and given that the specific height of an individual is hardly significant, we do not need symbols to represent them with. In such situations the shortcomings of base-ten notations are of no consequence whereas their advantages come shining through.

Limiting the use of the formal notation to the expression of theoretical discussions and that of base-ten notations to the expression of applied dis-

---

[3]The notable exception is a measurement that results from the act of counting.

[4]Other than during intermediate calculations that involve decimals which is something that can be relatively easily dealt with.

cussions has the added advantage that it makes it clear whether a given discussion is theoretical or applied. As stated in Chapter 3, theoretical relationships are general statements that can be used to solve specific problems. The theoretical statement $E = \frac{1}{2}mv^2$, as an example, can be used to solve any specific problem in which the values of kinetic energy, mass and speed are of interest.[5] An application of this theory (an example of a problem that uses the theoretical formula $E = \frac{1}{2}mv^2$ in its solution) is the problem *An object, moving at a speed of* 27.7 m/s *has a kinetic energy of* 1700 J. *Calculate its mass.* The solution to this problem can be communicated as follows:

$$E = \frac{1}{2}mv^2 \qquad \text{Theory}$$

$$m = \frac{2}{v^2}E$$

$$m = \frac{2}{27.7^2} \times 1700 \qquad \text{Application}$$

$$m = \frac{2}{767.29} \times 1700$$

$$m = 4.43 \text{ kg}$$

Note once again the clean separation of the theoretical argument and the application of the theory to solve the specific problem under consideration. So long as formal notation is used, the discussion is theoretical. Appearance of base-ten notations signal the shift from theory to applications of that theory.

Exercise Set 5

1. What notation is suitable for use in the communication of applications of scientific theories? Why?
2. Explain why base-ten notations are *not* used in the communication of the values of constants in science formulas.
3. What does the appearance of a numerical value in base-ten form in a science formula imply?
4. How can you tell whether a discussion is theoretical or applied?

---

[5] Formulas that result from observed relationships between quantities are considered to be correct within the range of experimental verification. Such relationships may or may not hold when applied outside the range within which they have been verified.

Chapter 6
Working with Base-Ten Notations

In this chapter we will discuss two base-ten notations: The decimal notation and scientific notation both of which are used extensively in the expression of the values of quantities in the applications of scientific theories. For each notation, we will discuss the structure of that notation, the formal reading of numbers written in that notation, algorithms for performing arithmetic operations on numbers written in that notation, the evaluation of expressions involving numbers written in that notation, and the graphing of numbers written in that notation. Our focus will be on the development of tools that are meaningful and efficient as it is this combination of features that makes a mathematical tool practical.

6.1 The Decimal Notation

6.1.1 Structure

The decimal notation extends the place value mechanism to represent the sizes of the various numbers in $\mathbb{R}$. The extension is implemented by introducing a point, called the **decimal point**, to the right end of a natural number, a whole number or an integer followed by more places for the placement of more digits with place values that are sub-powers of ten.

Apart from the sign, the left side of the decimal point is referred to as the **whole part** of the decimal number and the right side of the decimal point is referred to as the **fractional part** of the decimal number. The terminology reflects the fact that the digits on the left side of the decimal point represent the number of wholes that are present in the decimal number while the digits on the right side of the decimal point represent bits and pieces whose overall size is always less than 1. As an example, the whole part of 45.3 is 45 indicating that there are 45 wholes[1] in 45.3 while its fractional side

---

[1] Or ones or units, if you like.

corresponds to the fraction $\frac{3}{10}$; and in 62.97 the whole side is 62 indicating that there are 62 wholes in 62.97 while its fractional side corresponds to the fraction $\frac{97}{100}$ and so on.

Note that a single digit on the fractional side represents a value out of 10. The largest digit that one can place in this location is a 9 which results in $\frac{9}{10}$ which is less than 1. If there are two digits on the fractional side, the value is out of 100. The largest digits that can occupy the two places are 9s resulting in $\frac{99}{100}$; still less than 1. This argument can be extended to any number of digits on the fractional side of the decimal point and this shows that the total value represented by any and all digits on the right side of the decimal point is always less than 1 in size.

The digits on the right side of the decimal point are referred to as **decimal digits**. The terminology is used quite frequently in many contexts, as in the rounding of decimals when one is asked to round a decimal number to a specific number of decimal digits, or in the multiplication of decimal numbers where one needs to keep track of the number of decimal digits in the numbers being multiplied. Figure 6.1.1 below illustrates the terminology introduced so far.

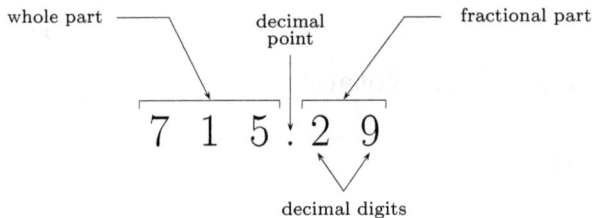

Figure 6.1.1: Structure of a decimal number

Place Values

As we discussed above, the decimal notation extends the place value scheme by adding place values to the right side of the decimal point. These place values are sub-powers of ten and are illustrated in Figure 6.1.2.

Note the presence of *th* in the naming of the place values on the right side of the decimal point. This indicates a sub-power of ten. As an example, the first place value on the right side of the decimal point indicates that we are dealing with a size that is a *tenth* of 1 so that the presence of, say, digit 7

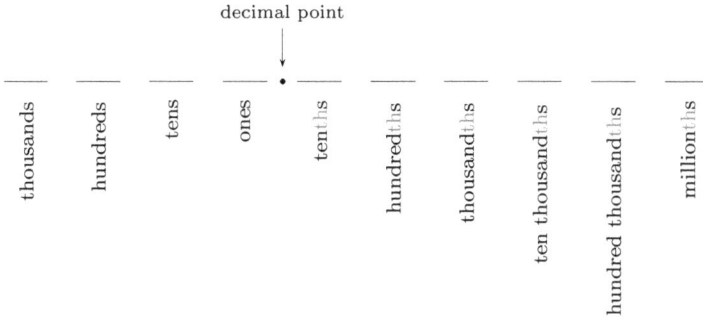

Figure 6.1.2: The more common place values

in this place indicates that we have 7 tenths, or $\frac{7}{10}$. This is an extension of the scheme used to make sense of the place values in natural numbers, whole numbers and integers: The *tens* place value, as an example, refers to *tens of* 1 and the presence of, say, digit 7 in that place indicates that we have 7 tens, or 70.

Place Value Size Relationships

The name of a place value relates the size of that place value to that of 1. As an example, the interpretation of the place value *tenths* speaks of a piece whose size is a tenth *of one*. An example of such a relationship between *tenths* and *ones* using a measure of length is given in Figure 6.1.3 below.

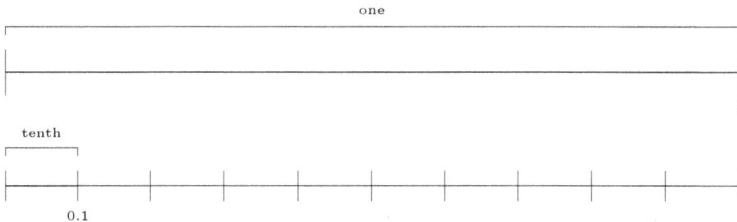

Figure 6.1.3: Illustration of the relationship between the size of a *tenth* and the size of a *one* using a measure of length

Similarly, the name of the place value *hundredths* speaks of a piece whose size is a hundredth *of one*. An example of such a relationship between *hun-*

dredths and *ones* using a measure of length is given in Figure 6.1.4 below.

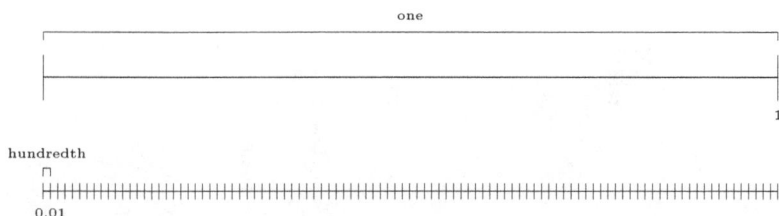

Figure 6.1.4: Illustration of the relationship between the size of a *hundredth* and the size of a *one* using a measure of length

The discussion above shows the manner in which one should view the 1s in the decimal number 1.11, i.e., 1 *one*, 1 *tenth*, and 1 *hundredth*.[2]

In addition to comparing the sizes of the various place values to that of a *one*, an activity that, as shown above, is facilitated by the reading of the name of the place values, it is also possible to relate the sizes of two place values neither of which may be *one*. As an example we can say that there are 10 *hundredths* in a *tenth*. An example of this relationship between *tenths* and *hundredths* using a measure of length is given in Figure 6.1.5 below.

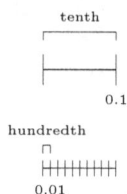

Figure 6.1.5: Illustration of the relationship between the size of a *hundredth* and the size of a *tenth* using a measure of length

The relationships between the sizes of *ones* and *tenths*, as well as those of *ones* and *hundredths*, and *tenths* and *hundredths* can be generalized to relate the sizes of any two place values. At the root of all such relationships is the key structural feature of the decimal notation which states that each place

---

[2]Note how quickly the sizes of these 1s decrease as we go from the 1 in the place value *ones* to the 1 in the place value *tenths* to the 1 in the place value *hundredths*. Following this pattern, a 1 in the next place value. i.e., *thousandths*, would represent a piece whose size is a tenth of a *hundredth*.

value is 10 times smaller than the one on its left and 10 times larger than the one on its right. This implies the following:

Each place value is

- 10 times larger than the one on its right.
- 100 times larger than the second place value on its right.
- 1000 times larger than the third place value on its right.
- etc.

or, equivalently, each place value is

- 10 times smaller than the one on its left.
- 100 times smaller than the second place value on its left.
- 1000 times smaller than the third place value on its left.
- etc.

As an example of such relationships, we can say that a *tenth* is

- 10 times larger than a *hundredth*.
- 100 times larger than a *thousandth*.
- 1000 times larger than a *ten thousandth*.
- etc.

or we can say that a *hundredth* is

- 10 times smaller than a *tenth*.
- 100 times smaller than a *one*.
- 1000 times smaller than a *ten*.
- etc.

Note that the relationship between the sizes of place values in decimal numbers is an extension of the relationship between the sizes of the place values in natural numbers, whole numbers and integers.

To relate the sizes of two place values we recommend that you divide the larger place value by the smaller place value. As an example, to relate the sizes of *thousandths* and *tenths* we can perform the division $0.1 \div 0.001$ which should immediately bring 100 to mind.[3] This is a direct extension of the recommended technique for establishing the relationship between the place values in natural numbers, whole numbers and integers and is the one that, from the point of view of semantics, makes most sense.

---

[3]See the section on division of decimals to see how you can perform such divisions mentally.

A somewhat less efficient, but still acceptable, technique for the determination of the relative sizes of two place values is to place a 1 in the larger place value and then fill up the place values to the right side of this 1 with 0s until we arrive at the smaller place value. The number generated by the 1 and the 0s that follow it tells us how many times the place value on the left is larger than the one on the right, or, equivalently, how many times the place value on the right is smaller than the place value on the left. As an example of this, consider the task of determining how many *tenths* fit in a *hundred*. We identify the place values *hundreds* and *tenths* as shown below:

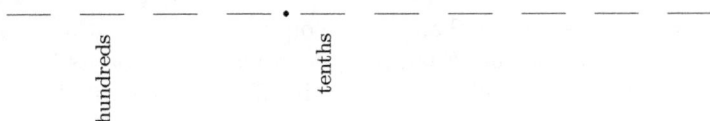

hundreds   tenths

We now place a 1 in the larger place value *hundreds* and fill those on its right with 0s until we arrive at the place value *tenths*. This is illustrated below:

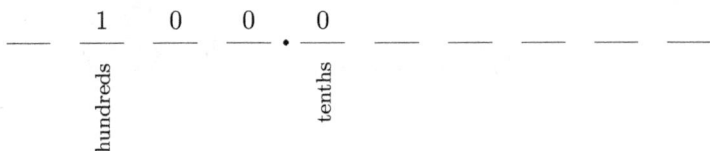

1   0   0 . 0

hundreds   tenths

This tells us that a *hundred* is 1000 times larger than a *tenth*, or, equivalently, a *tenth* is 1000 times smaller than a *hundred*, i.e., 1000 *tenths* make a *hundred*.

The size relationship between the various place values in the decimal notation is exploited in the design of units in the International System of Units, SI, to ease the task of unit conversion. It figures in the design of algorithms that facilitate the task of working with, not just decimals, but also scientific notation. It explains why in basic science courses students are asked to keep a few (typically three to five) significant digits in the results of their calculations. It shows up again when we relate changes in the value of the exponent of 10 to the size of the value of the number itself, a feature seen in such contexts as the definition of the pH scale in determining the acidity of solutions. These are just a few of the applications of the structural feature of decimals that relates the sizes of various place values. The reader will do well to make sure that she or he feels comfortable with this structural feature before continuing further.

Interpretation of Decimal Digits

It is also important that the reader should have two views of the digits that are written on the right side of the decimal point: One takes account of each digit separately from the others and the other deals with all the decimal digits together.

Separately, each digit refers to the number of pieces of a certain size, with the size of a piece set by the place value of that digit. As an example, in 0.627, we have 6 tenths or $\frac{6}{10}$, 2 hundredths or $\frac{2}{100}$ and 7 thousandths or $\frac{7}{1000}$, a view that is often written as

$$\frac{6}{10} + \frac{2}{100} + \frac{7}{1000}$$

This is the **expanded form** of the decimal number 0.627.

In addition to the view that interprets the value of each digit on its own, the reader should also be able to make sense of the decimal digits together, e.g., as 627 thousandths in the example above, a view that is written as

$$\frac{627}{1000}$$

The equivalence between the two views can be established by adding the fractions in the expanded form using the least common denominator as follows:

$$\frac{6}{10} + \frac{2}{100} + \frac{7}{1000} = \frac{600 + 20 + 7}{1000}$$
$$= \frac{627}{1000}$$

The latter view provides an algorithm for writing a decimal number as a fraction or a mixed number. The whole part of the decimal number, if not 0, appears as the whole part of a mixed number followed by the fractional side placed over 10, 100, 1000, etc. as the case may be. Here are a few examples:

Example

Write 42.3 as a fraction.

Solution

$$42.3 = 42\frac{3}{10}$$

Example

Write 0.26 as a fraction.

Solution

$$0.26 = \frac{26}{100}$$
$$= \frac{13}{50}$$

Note that it may be possible to reduce the fractional part as shown in the example above.

Exercise Set 6.1.1

1. In each case determine the whole part and the fractional part of the decimal number.

   a. 926.4   c. 95   e. 0.004
   b. 3.068   d. 0.159   f. 420

2. How does the size of the fractional part of a decimal number compare with the unit, i.e., 1?

3. How many decimal digits are there in each of the following?

   a. 45.6   d. 28.0   g. 325
   b. 0.008   e. 427.05   h. 0.046
   c. 95.40   f. 0.49   i. 9000

4. How many tenths are there in a unit?
5. How many hundredths are there in a unit?
6. How many thousandths are there in a unit?
7. How many hundredths are there in a tenth?
8. How many thousandths are there in a tenth?

9. How many thousandths are there in a hundredth?
10. How many tenths are there in a ten?
11. How many times is a tenth larger than a hundredth?
12. How many times is a tenth larger than a thousandth?
13. How many times is a unit larger than a hundredth?
14. How many times is a ten larger than a thousandth?
15. How many times is a ten larger than a unit?
16. How many times is a hundred larger than a tenth?

17. Convert to a fraction or a mixed number.

   a. 4.18   d. 0.45   g. 0.30

   b. 60.07   e. 75.20   h. 0.300

   c. 0.05   f. 0.3   i. 0.3000

18. Consider the diagram below. In each case determine the value of the
    location pointed to by the arrow.

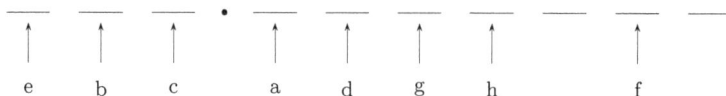

19. What is the place value of 2 in 17.429?
20. What is the place value of the digit 9 in 0.0092?
21. What is the place value of 1 in 0.01?
22. What is the place value of the digit 6 in 94.62?
23. What is the place value of the digit 7 in 0.000072?
24. What is the place value of the digit 0 in 2.405?
25. What is the place value of 8 in 0.169857?

26. Write each of the following decimals in expanded form and interpret the
    result.

   a. 0.457   c. 0.1284   e. 0.32185

   b. 0.07   d. 92.79   f. 128.91

6.1.2   Formal Reading

The manner in which we formally read a number written in decimal format
depends on whether or not the whole part of that number is 0.

To formally read a number written in decimal format when the whole part is *not* 0

1. If a sign is present, read the sign as *positive* or *negative*.
2. Read the whole part following the conventions listed in the formal reading of whole numbers.
3. Use the word *and* to separate the whole part from the fractional part.[4]
4. Read the fractional part as a whole number following the conventions listed in the formal reading of whole numbers.
5. Read the place value of the rightmost digit on the fractional side. Use a plural 's' if the fractional side implies that we have more than one piece.

If the whole part is 0, skip Steps 2 and 3, i.e., do *not* say *zero and*.

Following the guidelines above, we read 4.2 as *four and two tenths*, -34.01 as *negative thirty-four and one hundredth*, and $+0.076$ and *positive seventy-six thousandths*. The steps in the formal reading of a number written using the decimal notation with a nonzero whole part are illustrated in Figure 6.1.6 below in the reading of 12.025.

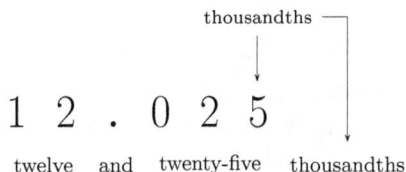

thousandths

$$1 \; 2 \; . \; 0 \; 2 \; 5$$

twelve   and   twenty-five   thousandths

Figure 6.1.6: Formal reading of a number written in decimal format with a nonzero whole part

And the steps in the formal reading of a number written using the decimal notation with zero on the whole part are illustrated in Figure 6.1.7 below in the reading of 0.018.

In going from the wording of the reading of a number written using the

---

[4]This has led some to claim that the word *and* implies the presence of the decimal point. This is not necessarily so. The formal interpretation of the word *and* in the reading of a number separates the whole part from the fractional part of that number regardless of what notation is used to represent the number. As an example, the word *and* is used in the formal reading of mixed numbers such as $3\frac{1}{2}$ where there is no decimal point present. When a number is written using the decimal notation, however, the word *and* maps onto the decimal point.

thousandths

$$0 \,.\, 0 \ 1 \ 8$$

eighteen   thousandths

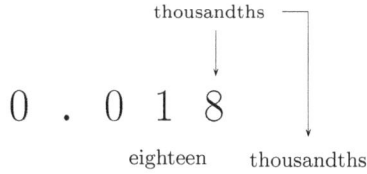

Figure 6.1.7: Formal reading of a number written in decimal
format with a zero whole part

decimal notation to its mathematical representation using digits, one should
first hunt down the special word *and* and the reading of the place value of
the rightmost digit in that number.

Example

Write *fifty-three and four hundredths* using digits.

Solution

We underline the word *and* and the reading of the place value of
the rightmost digit.

fifty-three <u>and</u> four <u>hundredths</u>

We now write 53, followed by the decimal point in place of the
word *and*. On the fractional side we notice that the rightmost
digit must be in the *hundredths* place, meaning that we need two
decimal digits. Since *four* has a single digit, we place a 0 on its
left to arrive at 53.04.

In case the word *and* is missing, it means that the whole part is 0.

Example

Write *thirty-six ten thousandths* using digits.

Solution

The word *and* is not present so we underline the reading of the place value of the rightmost digit:

thirty-six <u>ten thousandths</u>

Since there is no *and*, we write a 0 followed by the decimal point. The place value *ten thousandths* indicates that we need four digits on the fractional part. Since *thirty-six* has two digits, we place two 0s on its left to arrive at 0.0036.

Informal readings of decimals, as in the reading of 4.3 as *four point three* may be acceptable in cases when one is mindlessly processing numbers. The formal reading is, however, much more appealing from the point of view of semantics. As an example, the statement *seven tenths of the people voted* makes immediate sense whereas *zero point seven of the people voted* does not.[5]

The formal reading of a decimal also maps onto the formal reading of a rational number or a mixed number. As an example, both 4.3 and $4\frac{3}{10}$ read formally as *four and three tenths*[6] whereas their informal readings as *four point three* and *four and three over ten* sound quite different. Repeated use of the formal reading of decimals and rationals trains the brain to remember the equivalency between the two different ways of representing the same value.[7]

Exercise Set 6.1.2

1. Write each of the following decimals in words.

   a. 120.27   d. 45.719   g. 0.1   j. −0.18
   b. −1.7   e. 0.034   h. 0.03   k. 72.429
   c. 14.06   f. −0.751   i. −920.005   l. 0.0075

2. Write the following using digits.

   a. negative fifty-six and five tenths
   b. seven and twenty-four hundredths

---

[5] And who talks like this?

[6] And this *should* be the case as the two numbers represent the same value.

[7] It should be noted that, while the reading of a number written in decimal format is unique, it is possible to write two numbers corresponding to the reading of some decimals. As an example, both 0.110 and 0.0100 correspond to the reading *one hundred ten thousandths*. This is unfortunate but is not a problem that is commonly encountered.

    c. nine hundred twenty-five and seventy-six thousandths

    d. negative twenty-five hundredths

    e. eight hundred forty thousandths

    f. nine hundred fifteen and seven hundred fifty-nine thousandths

    g. negative four and one hundredth

    h. nine thousand four ten thousandths

6.1.3   Order

We will focus on size comparison of positive decimals. To order negative decimals, apply the same rules as those that were given for the ordering of integers.

Algorithm for Ordering Numbers written in Decimal Notation

1. Compare the whole parts of the two numbers. If the whole parts are different, the one with the larger whole part is the larger of the two numbers.

2. If the whole parts are the same, then compare the digits on the fractional side place value by place value, starting with the digits in the tenths and moving on to the right. The larger number can be identified as soon as a difference in the digits in corresponding place values is seen: The one with the larger digit is the larger of the two numbers.

The structural feature of the decimal notation that relates the sizes of the many place values can be used to explain why the algorithm above works.

Consider, for the moment, the whole part of a decimal number. On the whole side, we know from past experience that the value represented by a digit in the *hundreds* place is larger than the value represented by a digit in the *tens* place and that this is true irrespective of what the actual digits are. Even if the digit in the *tens* place is larger than the digit in the *hundreds* place, the *value* represented by the digit in the *hundreds* place is larger. As an example, a 1 in the *hundreds* results in a value of 100 and this is larger than a 9 in the *tens* which has a value of 90.

This conclusion is based on the fact that *tens* are smaller than *hundreds* and that we need 10 *tens* to make 1 *hundred*. Given that the *tens* place can, at most, host a single digit, even the largest such single digit, i.e., 9, in the *tens* place will be less than the required 10 *tens* that equal to 1 *hundred*.

The same argument holds on the fractional side of a decimal number. As an example, 0.1 is larger than 0.09 because we need 10 *hundredths* to get 1 *tenth* whereas the 9 in 0.09 represents 9 *hundredths*, 1 fewer *hundredth* than needed to make 1 *tenth*.

This fact is also apparent from the interpretation of the sizes of 0.1 and 0.09 as lengths as shown on the ruler in Figure 6.1.8 below.

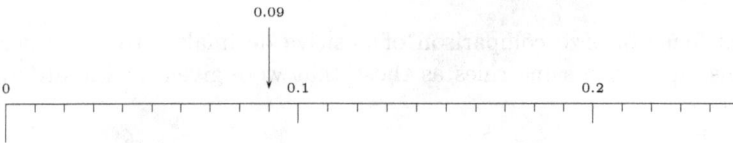

Figure 6.1.8: Comparison of the sizes of the numbers 0.1 and 0.09

As you can see it is the comparison of the digits in the *tenths* place that matters in the ordering of 0.1 and 0.09. The digit in the *hundredths* makes no difference to the order of the sizes of the two numbers.

As a second example, consider the sizes of 0.2 and 0.193. Some might be tempted to think that 0.193 is larger than 0.2 as 193 is larger than 2. However, note that in 0.2, digit 2 has a value of 2 *tenths* whereas 0.193 represents a totality of 1 *tenth*, 9 *hundredths*, and 3 *thousandths*. The total value represented by the digits 1, 9 and 3 in 0.193, then, is smaller than the total value represented by digit 2 in 0.2. Figure 6.1.9 below illustrates the idea.

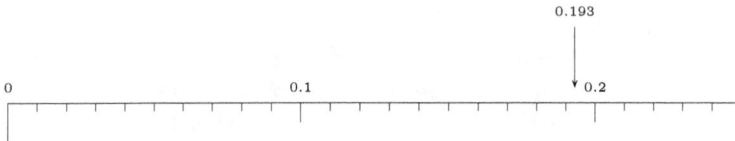

Figure 6.1.9: Comparison of the sizes of the numbers 0.2 and 0.193

So a larger digit in the *tenths* means a larger number overall regardless of what happens in the *hundredths* and the *thousandths*. This observation can be extended to all place values and it is this observation that figures in the formulation of the algorithm for the ordering of decimals given above.

Example

Which is larger: 15.8 or 9.99?

Solution

15.8 is larger as its whole part is larger.

Example

Which is larger: 0.0856 or 4.01?

Solution

4.01 is larger as its whole part is larger.

Example

Which is larger: 0.052 or 0.0378?

Solution

Since the whole sides are equal, we move on to the fractional sides and compare digits place value by place value, starting with the digits in the tenths place. The digits in the tenths place are the same. We move one place value to the right and compare the digits in the hundredths place. Since 0.052 has a 5 in the hundredths place and 0.0378 has a 3 in the hundredths place, 0.052 is larger.

Example

Which is larger: 2.3 or 2.27?

Solution

Since the whole sides are equal, we move on to the fractional sides and compare digits place value by place value. We start with the tenths place. Since 2.3 has a 3 in the tenths place and 2.27 has a 2 in the tenths place, 2.3 is larger.

As stated earlier, if negative decimals are present they can be ordered according to the same rules that were given for the ordering of integers. Here is an example.

Example

Arrange 0.56, −0.245, 0.245, 0, −0.56 in increasing order.

Solution

$-0.56, -0.245, 0, 0.245, 0.56$

Exercise Set 6.1.3

1. Arrange 42.75, 45.994, 40.25 and 41.7 in increasing order.
2. Arrange 0.068, 0.608, 0.68 and 0.0068 in descending order.
3. Arrange 570.245, 570.254, 570.2504, 570.2054 and 570.2405 in decreasing order.
4. Arrange 0.168, 0.1096, 0.1993, 0.21 in ascending order.
5. Arrange −2.4, −3.8, −4.1 and −1.5 in increasing order.
6. Arrange 15.4, 14.2, −13.8 and −10.1 in decreasing order.
7. Arrange −0.045, −0.2, −0.43, 0 and −0.08 in descending order.
8. Arrange 1.25, 1.2, −1.34, −1.3 and −2.6 in ascending order.

6.1.4   Addition, Subtraction, Multiplication and Division

In this section of the text we will present efficient algorithms for the addition, subtraction, multiplication and division of numbers written using the decimal notation.

Addition

When adding numbers written using the decimal notation, we must make sure that corresponding place values line up. The reason for this should be apparent: One must add the hundredths together and the tenths together and the ones together and so on.

We can force the lining of the place values by lining the decimal points up. Here is an example.

Example

Evaluate 14.56 + 2.091

Solution

We line up the decimal points as shown below

```
    1  4  .  5  6
+      2  .  0  9  1
   _____
```

We now add the digits in the columns, carrying whatever may be necessary as we would with whole numbers.[8] When we get to the decimal point we move it straight down. This results in

```
              1
    1  4  .  5  6
+      2  .  0  9  1
   _____
    1  6  .  6  5  1
```

More than two numbers can be handled in a similar manner.

Example

4.9 + 13.2 + 0.41 = 18.51

When one of the numbers involved does not have a decimal point, the representation indicates that there are only whole pieces present. This implies

---

[8]The reason we can carry such digits is due to the relationship between the sizes of adjacent place values: Each place value is 10 times smaller than the one on its left which means every time we end up with a sum of 10 in the adding of the digit in a given column, we can pass on a 1 to the column on its left.

that its digits should be lined up with the digits on the whole side of the other number.

Example

Evaluate 19.3 + 423

Solution

We line up the digits as shown below and add column by column:

```
      1  9 . 3
  +   4  2   3
  ─────────────
      4  4  2 . 3
```

Subtraction

Similar to addition, place values must be lined up when subtracting numbers that are written using the decimal notation. The reasons for this requirement are similar to those given for addition: One must subtract hundredths from hundredths, tenths from tenths, ones from ones and so on. To make sure that place values are lined up properly, we line up the decimal points.

Example

Subtract 2.16 from 4.51

Solution

We line up the decimal points as shown below:

```
      4 . 5  1
  -   2 . 1  6
  ─────────────
```

We Start with the rightmost column. Since we cannot subtract 6 from 1, we need to borrow. Borrowing a tenth and breaking it into 10 hundredths gives us[9]

$$
\begin{array}{r}
{}^{4}\ \ {}^{11} \\
4\ .\ \not{5}\ \ \not{1} \\
-\ 2\ .\ 1\ \ 6 \\
\hline
\end{array}
$$

Subtracting along the hundredths column and then moving to the left one column at a time, we arrive at the following answer:

$$
\begin{array}{r}
{}^{4}\ \ {}^{11} \\
4\ .\ \not{5}\ \ \not{1} \\
-\ 2\ .\ 1\ \ 6 \\
\hline
2\ .\ 3\ \ 5
\end{array}
$$

When the top number has fewer decimal digits than the bottom number, one must add 0s to the right end of the top number and borrow to subtract. The following example illustrates the point.

Example

Evaluate $24.8 - 1.47$

Solution

We line up the decimal points as shown below:

$$
\begin{array}{r}
2\ 4\ .\ 8 \\
-\ \ \ 1\ .\ 4\ \ 7 \\
\hline
\end{array}
$$

---

[9]Note how the relationships between the sizes of adjacent place values explains the borrowing step.

Note that we cannot move the 7 down and write

```
    2  4  .  8
  -    1  .  4  7
  _____
               7  ⟵── WRONG
```

This would have been correct if the 7 was on top in which case one would subtract 0 from it to get 7 back. However, in the present case, 7 should be subtracted from 0 on top as shown below:

```
    2  4  .  8  0
  -    1  .  4  7
  _____
```

This requires that we borrow as shown below:

```
              7  10
    2  4  .  8̸  0̸
  -    1  .  4  7
  _____
```

The rest follows:

```
              7  10
    2  4  .  8̸  0̸
  -    1  .  4  7
  _____
    2  3  .  3  3
```

We end the discussion on subtraction of decimals with an example where one of the numbers involved in the subtraction does not have a decimal point.

Example

Evaluate $96 - 3.8$

Solution

Note that 96 should be lined up with the whole side of 3.8. This should be apparent if we recall that if there is no decimal point, then the digits represent the number of wholes that are present in the number.

$$
\begin{array}{r}
9\ 6 \\
-\ \ 3\ .\ 8 \\
\hline
\end{array}
$$

Placing the decimal point in 96 and adding a 0 in the tenth place, we have:

$$
\begin{array}{r}
9\ 6\ .\ 0 \\
-\ \ 3\ .\ 8 \\
\hline
\end{array}
$$

Borrowing from the ones and breaking it down into 10 tenths results in:

$$
\begin{array}{r}
{}^{5}\ {}^{10} \\
9\ \not6\ .\ \not0 \\
-\ \ 3\ .\ 8 \\
\hline
\end{array}
$$

We can now finish the subtraction:

$$
\begin{array}{r}
{}^{5}\ {}^{10} \\
9\ \not6\ .\ \not0 \\
-\ \ 3\ .\ 8 \\
\hline
9\ 2\ .\ 2
\end{array}
$$

Algorithm for Evaluation of Addition and Subtraction of Two Numbers Written in Decimal Notation

1. Arrange the numbers vertically making sure corresponding place values line up. To do so, line up the decimal points.
2. Ignore the decimal point and add and subtract as you would with whole numbers. Carry the decimal point down before moving to the ones column.

Multiplication

We will begin by presenting an algorithm for the multiplication of two numbers written using the decimal notation and present an example before justifying it.

Algorithm for Evaluation of Multiplication of Two Numbers Written in Decimal Notation

1. Drop the decimal point in both numbers.
2. Multiply the resulting whole numbers.
3. Place the decimal point so that the result has the same number of decimal digits as the total number of decimal digits in the numbers that you multiplied. Add 0s on the left if you need to.

Example

Evaluate 0.23×0.7

Solution

Following step 1 in the algorithm above, we drop the decimal points in 0.23 and 0.7 to get 23 and 7.

Following Step 2, we multiply 23 by 7 to get 161.

Following Step 3, we find the total number of decimal digits in 0.23 and 0.7. 0.23 has two decimal digits and 0.7 has one decimal digit. The total number of decimal digits is three. The result must have three decimal digits. We place the decimal point in 161 so that we end up with three decimal digits and arrive at 0.161.

The key structural feature of the decimal notation that relates the sizes of the many place values may be used to explain how this algorithm works: In the multiplication of 0.23 and 0.7, when we drop the decimal point in the 0.23, we make the number 100 times larger. In addition, when we drop the decimal point in 0.7, we make the number 10 times larger. The product 23×7 that is computed in Step 2 is, therefore, 1000 times larger than it should be. If we make the product of 23 and 7, 1000 times smaller, then we will arrive at an answer for the product of 0.23 and 0.7. This is what Step 3 achieves.

Example

Evaluate 2.6×3.18

Solution

Following step 1 in the algorithm above, we drop the decimal points in 2.6 and 3.18 to get 26 and 318.

Following Step 2, we multiply 26 by 318 to get 8268.

Following Step 3, we find the total number of decimal digits in 2.6 and 3.18. 2.6 has one decimal digit and 3.18 has two decimal digits. The total number of decimal digits is three. The result must have three decimal digits. We place the decimal point in 8268 so that we end up with three decimal digits and arrive at 8.268.

The algorithm given above works even in the case when one of the numbers involved does not have a decimal point. Here is an example:

Example

Evaluate 0.0024×17

Solution

Following step 1 in the algorithm above, we drop the decimal points in 0.0024 and 17 to get 24 and 17.

Following Step 2, we multiply 24 by 17 to get 408.

Following Step 3, we find the total number of decimal digits in 0.0024 and 17. 0.0024 has four decimal digit and 17 has no decimal digits. The total number of decimal digits is four. The result must have four decimal digits. We place the decimal point in 408 so that we end up with four decimal digits. This requires that we add a 0 to the left of 408 arriving at 0.0408.

Multiplication by powers of 10

There is a fast algorithm for multiplying a decimal number by powers of 10 (i.e., 10, 100, 1000, etc.). The algorithm requires that we move the decimal point in the given number to the right over as many digits as the number of 0s in the power of 10.

Example

Evaluate 4.523×100

Solution

Since 100 has two 0s, we move the decimal point in 4.523 over two digits to the right to get 452.3.

Example

Evaluate 3.2×1000

Solution

Since 1000 has three 0s, we move the decimal point in 3.2 over three digits to the right to get 3200. Note that this requires that we add 0s to the right end of 3.2.

The logic behind this algorithm is rooted in the relationship between the sizes of the many place values. Since each place value is 10 times smaller than the one on its left and 10 times larger than the one on its right, movement of the decimal point to the right over three digits reassigns the place values so that the value of each digit becomes 1000 times larger. This, in turn, implies that the value of the number itself becomes 1000 times larger.

Multiplication by Sub-Powers of 10

There is also a fast algorithm for multiplying a decimal number by sub-powers of 10 (i.e., 0.1, 0.01, 0.001, etc.). The algorithm requires that we move the decimal point in the given number to the left over as many digits as there are decimal digits in the sub-power of 10.

Example

Evaluate 0.01×783.1

Solution

Since 0.01 has two decimal digits, we move the decimal point in 783.1 over two digits to the left to get 7.831.

Example

Evaluate 0.1×275

Solution

Since 0.1 has one decimal digit, we move the decimal point in 275 (which is on its right end) over one digit to the left to get 27.5.

Example

Evaluate 0.001×3.15

Solution

Since 0.001 has three decimal digits, we move the decimal point in 3.15 over three digits to the left to get 0.003 15. Note that this requires that we add 0s to the left of 3.15.

The justification for this movement of the decimal point is based on the relationships between the sizes of the many place values and the semantics of multiplication by subpowers of 10. The expression

$$0.001 \times 3.15$$

requires that we calculate *one-thousandth* of 3.15. We would, of course, expect that the answer will be 1000 times smaller than 3.15. Since each place value is 10 times smaller than the one on its left and 10 times larger than the one on its right, the movement of the decimal point to the left over three digits, which reassigns the place values, makes each digit 1000 times smaller. This, in turn, makes the decimal number itself 1000 times smaller than it originally was.

Division[10]

There are two cases to consider when discussing the division of numbers that are written in decimal form. The first is easier and concerns divisions where the fractional part of the divisor[11] is 0. The second case involves divisions where the fractional part of the divisor is not 0. Note that it makes no difference whether the fractional part of the dividend[12] is 0 or not.

In the case where the fractional part of the divisor is 0, we can start the division right away. The steps in the algorithm for such divisions are identical to those given for the division of whole numbers with three important new features: First, begin with the whole side even if this results in a 0 on the whole side. Second, right before you pull down the first decimal digit, place a decimal point on the right side of the quotient. And third, if you run out of digits and the remainder is not 0, add 0s to the right side of the dividend and continue with the division until at some point the remainder becomes 0 or else you note a repeating pattern in the quotient. As with the case of division of whole numbers it is very important that you pull down one digit at a time and following this move answer the question *how many divisors go into the remainder* otherwise you may miss intervening 0s.

We will illustrate these point using examples.

Example

Evaluate $41.4 \div 3$

---

[10] We will begin with a note of caution: It may be tempting to make certain assumptions that may seem obvious such as the assumption that the result of the division of two numbers that have nonzero fractional parts must have a nonzero fractional part, or that the result of the division of two numbers with zero fractional parts should have a zero fractional part, etc. Such assumptions may not necessarily be true: The division $5 \div 2$ equals to 2.5 and the division $4.6 \div 0.2$ equals to 23. Furthermore, the division $4 \div 0.3$ equals to $13.\overline{3}$ whereas the division $4 \div 0.2$ equals to 20. The only case where one can claim that the answer will have a nonzero fractional part is the division of a number that has a nonzero fractional part by a number that has a zero fractional part. In such a case the result can be proven to have a nonzero fractional part.

[11] The **divisor** is the number you divide by. As an example, in $5 \div 2$ the divisor is 2.

[12] The **dividend** is the number being divided. As an example, in $5 \div 2$, the dividend is 5.

Solution

Step 1.
Since the divisor, 3, does
not have a decimal point, we can
start the division right away.

$$3\overline{)\,4\,1\,.\,4}$$

Step 2.
We start with 4 and ask: How
many 3s fit into a 4? The
answer is 1. Write 1 as shown.

$$3\overline{)\,\overset{\textstyle 1}{4\,1\,.\,4}}$$

Step 3.
Multiply 1 by 3 to get 3.
Write 3 as shown.

$$3\overline{)\,\overset{\textstyle 1}{4\,1\,.\,4}}$$
$$\mathbf{3}$$

Step 4.
Subtract 3 from 4 to get 1.
Write 1 as shown.

$$3\overline{)\,\overset{\textstyle 1}{4\,1\,.\,4}}$$
$$-\,3$$
$$\overline{\mathbf{1}}$$

Step 5.
Bring 1 down as shown.

$$3\overline{)\,\overset{\textstyle 1}{4\,1\,.\,4}}$$
$$-\,3$$
$$\overline{1\,\mathbf{1}}$$

Step 6.
Ask: How many 3s go into 11?
Answer: 3. Write 3 as shown

$$3\overline{)\,\overset{\textstyle 1\,\mathbf{3}}{4\,1\,.\,4}}$$
$$-\,3$$
$$\overline{1\,1}$$

Step 7.
Multiply 3 by 3 to get 9.
Write 9 down as shown.

$$3\overline{)\,\overset{\textstyle 1\,3}{4\,1\,.\,4}}$$
$$-\,3$$
$$\overline{1\,1}$$
$$\mathbf{9}$$

Step 8.
Subtract 9 from 11 to.
get 2. Write 2 as shown.

```
        1 3
   3 ) 4 1 . 4
     - 3
     ─────
       1 1
      - 9
      ─────
         2
```

Step 9.
Write the decimal point
in the quotient.

```
        1 3 .
   3 ) 4 1 . 4
     - 3
     ─────
       1 1
      - 9
      ─────
         2
```

Step 10.
Bring 4 down as shown.

```
        1 3 .
   3 ) 4 1 . 4
     - 3
     ─────
       1 1
      - 9
      ─────
         2 4
```

Step 11.
Ask: How many 3s fit
into 24? Answer: 8.
Write 8 as shown.

```
        1 3 . 8
   3 ) 4 1 . 4
     - 3
     ─────
       1 1
      - 9
      ─────
         2 4
```

Step 12.
Multiply 8 by 3 to
get 24. Write 24
as shown.

```
        1 3 . 8
   3 ) 4 1 . 4
     - 3
     ─────
       1 1
      - 9
      ─────
         2 4
         2 4
```

Step 13.
Subtract 24 from 24 to
get 0. Write 0 as shown.

```
        1 3 . 8
   3 ) 4 1 . 4
     - 3
     ─────
       1 1
      - 9
      ─────
         2 4
         2 4
         ─────
           0
```

At this point we can stop as there are no more digits to pull down and the remainder is 0. We can now write

$$41.4 \div 3 \ = \ 13.8$$

In case the remainder is *not* 0, then we should add 0s to the right end of the dividend and pull them down until the remainder becomes 0 or we notice a repeating pattern in the quotient. Here is an example.

Example

Evaluate $0.11 \div 40$

Solution

Step 1.
Since the divisor, 40, does
not have a decimal point, we can
start the division right away.

$$40\overline{)\,0\,.\,1\,1}$$

Step 2.
We start with the whole side,
i.e., the 0, and ask: How
many 40s fit into 0?
The answer is 0. Write
0 down as shown.

$$40\overline{)\,0\,.\,1\,1}^{\,\,0}$$

Step 3.
Multiply 0 by 40 to get 0.
Write 0 down as shown.
Note: This step and the next
are optional and are often
skipped as they return the same
value after subtraction.

$$\begin{array}{r} 0 \\ 40\overline{)\,0\,.\,1\,1} \\ 0 \end{array}$$

Step 4.
Subtract 0 from 0 to get 0.
Write 0 down as shown.

$$\begin{array}{r} 0 \\ 40\overline{)\,0\,.\,1\,1} \\ -\,0 \\ \hline 0 \end{array}$$

Step 5.
Write the decimal point as shown.

$$\begin{array}{r} 0\,. \\ 40\overline{)\,0\,.\,1\,1} \\ -\,0 \\ \hline 0 \end{array}$$

Step 6.
Bring 1 down as shown.

$$\begin{array}{r} 0\,. \\ 40\overline{)\,0\,.\,1\,1} \\ -\,0 \\ \hline 0\ \,1 \end{array}$$

Step 7.
Ask: How many 40s fit
into 1? Answer: 0. Write
0 as shown.

$$
\begin{array}{r}
0\,.\,\mathbf{0} \\
40\,\overline{)\,0\,.\,1\,1} \\
-\,0 \\
\hline
0\ \ 1
\end{array}
$$

Step 8.
Multiply 0 by 40 to get 0.
Write 0 as shown.
Note: This step and the
next are often skipped
as they return the same
value after subtraction.

$$
\begin{array}{r}
0\,.\,0 \\
40\,\overline{)\,0\,.\,1\,1} \\
-\,0 \\
\hline
0\ \ 1 \\
\mathbf{0}
\end{array}
$$

Step 9.
Subtract 0 from 1 to get 1.
Write 1 down as shown.

$$
\begin{array}{r}
0\,.\,0 \\
40\,\overline{)\,0\,.\,1\,1} \\
-\,0 \\
\hline
0\ \ 1 \\
-\,0 \\
\hline
\mathbf{1}
\end{array}
$$

Step 10.
Bring 1 down as shown.

$$
\begin{array}{r}
0\,.\,0 \\
40\,\overline{)\,0\,.\,1\,1} \\
-\,0 \\
\hline
0\ \ 1 \\
-\,0 \\
\hline
1\,\mathbf{1}
\end{array}
$$

Step 11.
Ask: How many 40s fit into 11?
Answer: 0. Write 0 as
shown.

$$
\begin{array}{r}
0\,.\,0\,\mathbf{0} \\
40\,\overline{)\,0\,.\,1\,1} \\
-\,0 \\
\hline
0\ \ 1 \\
-\,0 \\
\hline
1\ 1
\end{array}
$$

Step 12.
Multiply 0 by 40 to get
0. Write 0 as shown.
Note: This step and the next
are often skipped as they
return the same value
after subtraction.

```
              0 . 0 0
        ─────────────
    4 0 ) 0 . 1 1
        − 0
        ─────────
          0   1
          − 0
          ───────
            1 1
             0
```

Step 13.
Subtract 0 from 11 to
get 11. Write 11 as shown.

```
              0 . 0 0
        ─────────────
    4 0 ) 0 . 1 1
        − 0
        ─────────
          0   1
          − 0
          ───────
            1 1
           − 0
           ───────
            1 1
```

Step 14.
Since remainder is not 0,
we cannot stop. Add a 0
as shown.

```
              0 . 0 0
        ───────────────
    4 0 ) 0 . 1 1 0
        − 0
        ─────────
          0   1
          − 0
          ───────
            1 1
           − 0
           ───────
            1 1
```

Step 15.
Bring the 0 down as shown.

```
              0 . 0 0
        ───────────────
    4 0 ) 0 . 1 1 0
        − 0
        ─────────
          0   1
          − 0
          ───────
            1 1
           − 0
           ───────
            1 1 0
```

Step 16.
Ask: How many 40s fit
into 110? Answer: 2. Write
2 as shown.

```
              0 . 0 0 2
        ───────────────
    4 0 ) 0 . 1 1 0
        − 0
        ─────────
          0   1
          − 0
          ───────
            1 1
           − 0
           ───────
            1 1 0
```

Step 17.
Multiply 2 by 40 to get 80.
Write 80 as shown.

```
              0 . 0 0 2
      40 ) 0 . 1 1 0
          − 0
          ─────
            0   1
              − 0
              ─────
                1 1
                − 0
                ─────
                1 1 0
                  8 0
```

Step 18.
Subtract 80 from 110 to get.
30. Write 30 as shown.

```
              0 . 0 0 2
      40 ) 0 . 1 1 0
          − 0
          ─────
            0   1
              − 0
              ─────
                1 1
                − 0
                ─────
                1 1 0
                  8 0
                ─────
                  3 0
```

Step 19.
Since the remainder is not 0,
we cannot stop. Add a 0 as
shown.

```
              0 . 0 0 2
      40 ) 0 . 1 1 0 0
          − 0
          ─────
            0   1
              − 0
              ─────
                1 1
                − 0
                ─────
                1 1 0
                  8 0
                ─────
                  3 0
```

Step 20.
Bring the 0 down as shown.

```
              0 . 0 0 2
      40 ) 0 . 1 1 0 0
          − 0
          ─────
            0   1
              − 0
              ─────
                1 1
                − 0
                ─────
                1 1 0
                  8 0
                ─────
                  3 0 0
```

Step 21.
Ask: How many 40s fit into
300? Answer: 7. Write 7
down as shown.

$$
\begin{array}{r}
0.002\,\mathbf{7} \\
40\,\overline{)\,0.1100} \\
-\,0 \\
\hline
0\ \ 1 \\
-\,0 \\
\hline
1\ 1 \\
-\,0 \\
\hline
1\ 1\ 0 \\
8\ 0 \\
\hline
3\ 0\ 0
\end{array}
$$

Step 22.
Multiply 7 by 40 to get
280. Write 280 as shown.

$$
\begin{array}{r}
0.002\,7 \\
40\,\overline{)\,0.1100} \\
-\,0 \\
\hline
0\ \ 1 \\
-\,0 \\
\hline
1\ 1 \\
-\,0 \\
\hline
1\ 1\ 0 \\
8\ 0 \\
\hline
3\ 0\ 0 \\
\mathbf{2\ 8\ 0}
\end{array}
$$

Step 23.
Subtract 280 from 300 to get
20. Write 20 as shown.

$$
\begin{array}{r}
0.002\,7 \\
40\,\overline{)\,0.1100} \\
-\,0 \\
\hline
0\ \ 1 \\
-\,0 \\
\hline
1\ 1 \\
-\,0 \\
\hline
1\ 1\ 0 \\
8\ 0 \\
\hline
3\ 0\ 0 \\
-\,2\ 8\ 0 \\
\hline
\mathbf{2\ 0}
\end{array}
$$

Step 24.
Since the remainder is not 0
we cannot stop. Add a 0 as
as shown.

```
          0 . 0 0 2 7
4 0 ) 0 . 1 1 0 0 0
    − 0
      0  1
       − 0
         1 1
         − 0
           1 1 0
             8 0
             3 0 0
           − 2 8 0
               2 0
```

Step 25.
Bring the 0 down as shown.

```
          0 . 0 0 2 7
4 0 ) 0 . 1 1 0 0 0
    − 0
      0  1
       − 0
         1 1
         − 0
           1 1 0
             8 0
             3 0 0
           − 2 8 0
             2 0 0
```

Step 26.
Ask: How many 40s fit into
200? Answer: 5. Write 5
down as shown.

```
          0 . 0 0 2 7 5
4 0 ) 0 . 1 1 0 0 0
    − 0
      0  1
       − 0
         1 1
         − 0
           1 1 0
             8 0
             3 0 0
           − 2 8 0
             2 0 0
```

Step 27.
Multiply 5 by 40 to get
200. Write 200 as shown.

$$
\begin{array}{r}
0.00275 \\
40\,\overline{)\,0.11000} \\
-0 \\[-2pt]
\hline
0\;1 \\
-0 \\[-2pt]
\hline
1\;1 \\
-0 \\[-2pt]
\hline
1\;1\;0 \\
8\;0 \\[-2pt]
\hline
3\;0\;0 \\
-2\;8\;0 \\[-2pt]
\hline
2\;0\;0 \\
\mathbf{2\;0\;0}
\end{array}
$$

Step 28.
Subtract 200 from 200 to get 0.
Write 0 as shown.

$$
\begin{array}{r}
0.00275 \\
40\,\overline{)\,0.11000} \\
-0 \\[-2pt]
\hline
0\;1 \\
-0 \\[-2pt]
\hline
1\;1 \\
-0 \\[-2pt]
\hline
1\;1\;0 \\
8\;0 \\[-2pt]
\hline
3\;0\;0 \\
-2\;8\;0 \\[-2pt]
\hline
2\;0\;0 \\
-2\;0\;0 \\[-2pt]
\hline
\mathbf{0}
\end{array}
$$

The remainder is 0 and there are no more digits to pull down. We can now stop and write

$$0.11 \div 40 = 0.00275$$

It is frequently the case that the division never ends as the remainder continues to be nonzero. In such cases, at some point we find that the digits in the quotient repeat. An example of this is the division $21 \div 11$. This division will generate the quotient $0.636363\ldots$[13] As the reader notes, the digits 6 and 3 repeat (try the division to see for yourself). A shorthand for writing

---

[13]The repeating pattern does not have to start in the tenths. Nor does the whole side have to be 0: In $42.1565656\ldots$ the repeating pattern starts in the hundredths and the whole side is 42. It is also possible for the repeating pattern to start on the whole side. An example of this is $1424.24242\ldots$ where the repeating pattern starts in the hundreds.

the results of such divisions is to place a line over the digits that repeat, e.g., to write $0.\overline{63}$.[14,15]

The question naturally arises as to the manner in which one works with such representations. If exact calculations involving decimals with repeating digits are sought, then, unless the case is somewhat clear as in $0.\overline{1} + 0.\overline{2}$ which equals to $0.\overline{3}$, then one would have to convert the decimal representation to formal notation, i.e., a rational number, and then process the values using formal notation.

However, recall that in the sciences the decimal notation is used in cases where we cannot be exact as in the case when we work on applications that involve measured values. Measured values always involve a level of uncertainty and are, for this reason, rounded. As such, they do not involve repeating digits. In addition, the results of calculations performed on measured values are subject to rounding rules. For us, then, the repeating pattern in the representation of some decimals will not be of major concern.[16]

We will present one more example.

Example

Evaluate $15 \div 2$

---

[14]Another convention is to place a dot over each digit that repeats. In the example above, this alternative notation yields $0.\dot{6}\dot{3}$.

[15]Similar notations are used when the repeating pattern starts elsewhere, e.g., $42.1\overline{56}$. In the case of repeating patterns that start on the whole side, we place the line over the first repeating block on the fractional side, e.g., $1424.\overline{24}$.

[16]See Chapter 9 on working with measured values for more on this.

Solution

Step 1.
Since the divisor, 2, does
not have a decimal point, we can
start the division right away.

$$2 \overline{)\ 1\ 5}$$

Step 2.
We start with 1 and ask: How
many 2s fit into a 1? The
answer is 0. Write 0 as shown.
Note: This step and the next two
are often skipped as the 0 in
the quotient is a leading 0
and the subtraction that
follows returns the same value.

$$\overset{\textbf{0}}{2 \overline{)\ 1\ 5}}$$

Step 3.
Multiply 0 by 2 to get 0.
Write 0 down as shown.

$$\begin{array}{r} 0 \\ 2 \overline{)\ 1\ 5} \\ \textbf{0} \end{array}$$

Step 4.
Subtract 0 from 1 to get 1.
Write 1 as shown.

$$\begin{array}{r} 0 \\ 2 \overline{)\ 1\ 5} \\ -\ 0 \\ \hline 1 \end{array}$$

Step 5.
Bring 5 down as shown.

$$\begin{array}{r} 0 \\ 2 \overline{)\ 1\ 5} \\ -\ 0 \\ \hline 1\ \textbf{5} \end{array}$$

Step 6.
Ask: How many 2s go into 15?
Answer: 7. Write 7 as shown

$$\begin{array}{r} 0\ \textbf{7} \\ 2 \overline{)\ 1\ 5} \\ -\ 0 \\ \hline 1\ 5 \end{array}$$

Step 7.
Multiply 7 by 2 to get 14.
Write 14 down as shown.

$$\begin{array}{r} 0\ 7 \\ 2 \overline{)\ 1\ 5} \\ -\ 0 \\ \hline 1\ 5 \\ \textbf{1 4} \end{array}$$

Step 8.
Subtract 14 from 15 to get
1. Write 1 down as shown.

$$
\begin{array}{r}
0\ 7\ \ \ \\
2\ )\overline{1\ 5\ \ \ } \\
-\ 0\ \ \ \\
\hline
1\ 5\ \ \ \\
-\ 1\ 4\ \ \ \\
\hline
1\ \ \ \\
\end{array}
$$

Step 9.
Since the remainder is not
0, we cannot stop. Place
the decimal point in 15
as shown.

$$
\begin{array}{r}
0\ 7\ \ \ \\
2\ )\overline{1\ 5\ .} \\
-\ 0\ \ \ \\
\hline
1\ 5\ \ \ \\
-\ 1\ 4\ \ \ \\
\hline
1\ \ \ \\
\end{array}
$$

Step 10.
Place the decimal point in
the quotient as shown.

$$
\begin{array}{r}
0\ 7\ .\\
2\ )\overline{1\ 5\ .} \\
-\ 0\ \ \ \\
\hline
1\ 5\ \ \ \\
-\ 1\ 4\ \ \ \\
\hline
1\ \ \ \\
\end{array}
$$

Step 11.
Add a 0 as shown.

$$
\begin{array}{r}
0\ 7\ .\ \ \\
2\ )\overline{1\ 5\ .\ \mathbf{0}} \\
-\ 0\ \ \ \ \\
\hline
1\ 5\ \ \ \ \\
-\ 1\ 4\ \ \ \ \\
\hline
1\ \ \ \ \\
\end{array}
$$

Step 12.
Move the 0 down as shown.

$$
\begin{array}{r}
0\ 7\ .\ \ \\
2\ )\overline{1\ 5\ .\ 0} \\
-\ 0\ \ \ \ \\
\hline
1\ 5\ \ \ \ \\
-\ 1\ 4\ \ \ \ \\
\hline
1\ \ \mathbf{0}\\
\end{array}
$$

Step 13.
Ask: How many 2s fit into 10?
Answer: 5. Write 5 down as
shown.

$$
\begin{array}{r}
0\ 7\ .\ \mathbf{5}\\
2\ )\overline{1\ 5\ .\ 0} \\
-\ 0\ \ \ \ \\
\hline
1\ 5\ \ \ \ \\
-\ 1\ 4\ \ \ \ \\
\hline
1\ \ 0\\
\end{array}
$$

Step 14.
Multiply 5 by 2 to get 10.
Write 10 down as shown.

$$
\begin{array}{r}
0\,7\,.\,5 \\
2\,\overline{)\,1\,5\,.\,0} \\
-\,0 \\
\hline
1\,5 \\
-\,1\,4 \\
\hline
1\quad 0 \\
\mathbf{1\quad 0}
\end{array}
$$

Step 15.
Subtract 10 from 10 to get 0.
Write 0 down as shown.

$$
\begin{array}{r}
0\,7\,.\,5 \\
2\,\overline{)\,1\,5\,.\,0} \\
-\,0 \\
\hline
1\,5 \\
-\,1\,4 \\
\hline
1\quad 0 \\
-\,1\quad 0 \\
\hline
\mathbf{0}
\end{array}
$$

Since the remainder is 0 and there are no more digits to pull down, we can stop and write

$$15 \div 2 \;=\; 7.5$$

In case the divisor has a nonzero fractional side, we use the following logic to turn the problem into an equivalent one, i.e., one with the same quotient, where the divisor has a zero fractional side. The logic that we have in mind states that in a division such as $20 \div 5$, one can multiply both the dividend and the divisor by the same number, e.g., turn $20 \div 5$ into $200 \div 50$ by multiplying both 20 and 5 by 10, and still get the same quotient. In addition, as noted earlier, multiplication of a decimal number by 10 moves the decimal point to the right over one digit. Therefore, to generate an equivalent division in which the divisor has a zero fractional side we can move the decimal point in the divisor to the right end of the divisor[17] and count the number of digits the decimal point moves over[18] and move the decimal point in the dividend to the right over as many digits[19] adding 0s if necessary. We can now apply the steps given earlier for the case where the divisor does not have a decimal point to this new division. The answer to this division is the same as the answer to the original division.

---

[17]This corresponds to the multiplication of the divisor by 10 or 100 or 1000, etc., depending on how many digits the decimal point needs to move over to get to the right end of the divisor.

[18]This reminds us whether we multiplied the divisor by 10 or 100 or 1000, etc.

[19]This corresponds to the multiplication of the dividend by the same value, i.e., 10, 100, 1000, etc., that the divisor was multiplied by.

We will present a few examples to illustrate the steps in the algorithm above.

Example

Evaluate $14.2 \div 7.1$

Solution

The divisor, 7.1, has a decimal point. We move the decimal point over one digit to the right to get 71. This corresponds to the multiplication of the divisor by 10. Since the decimal point moves over one digit in 7.1, we move the decimal point in 14.2 over one digit to the right as well to get 142. This corresponds to the multiplication of the dividend by 10. Our division has now turned to

$$142 \div 71$$

which evaluates to 2 (try the division yourself). The answer to the original division, $14.2 \div 7.1$, is also 2.

Example

Evaluate $1.008 \div 0.02$

Solution

The divisor, 0.02, has a decimal point. We move the decimal point over two digits to the right to get 2. This corresponds to the multiplication of the divisor by 100. Since the decimal point moves over two digits in 0.02, we move the decimal point in 1.008 over two digits to the right as well to get 100.8. This corresponds to the multiplication of the dividend by 100. Our division has now turned to

$$100.8 \div 2$$

which evaluates to 50.4 (try the division yourself). The answer to the original division, $1.008 \div 0.02$, is also 50.4.

Example

Evaluate $7.5 \div 0.015$

Solution

The divisor, 0.015, has a decimal point. We move the decimal point over three digits to the right to get 15. This corresponds to the multiplication of the divisor by 1000. Since the decimal point moves over three digits in 0.015, we move the decimal point in 7.5 over three digit to the right as well to get 7500.[20] This corresponds to the multiplication of the dividend by 1000. Our division has now turned to

$$7500 \div 15$$

which evaluates to 500 (try the division yourself). The answer to the original division, $7.5 \div 0.015$, is also 500.

Algorithm for Evaluation of Division of Two Numbers Written in Decimal Notation

1. If the divisor has no decimal digits, skip this step otherwise

   a. Move the decimal point in the divisor to its right end[21] and count the number of digits the decimal point moves over.[22]

   b. Move the decimal point in the dividend to the right over as many digits.[23]

2. Set up the division. Start on the whole side of the dividend and work with one digit at a time as in the case of the division of whole numbers. Before you pull the digit in the tenths down, place the decimal point on the right end of the quotient. Add 0s to the right end of the dividend as needed until the reminder is 0 and there are no more digits to pull down.

---

[20] Note that this requires that we add two 0s to the right end of 7.5.

[21] This corresponds to multiplication of the divisor by 10, 100, 1000, etc.

[22] This reminds us whether we multiplied the divisor by 10 or 100 or 1000, etc.

[23] This corresponds to the multiplication of the dividend by the same value, i.e., 10, 100, 1000, etc., that the divisor was multiplied by.

Division by Powers of 10

There is a fast algorithm for dividing a decimal number by powers of 10 (i.e., 10, 100, 1000, etc.). The algorithm requires that we move the decimal point to the left over as many digits as there are 0s in the power of 10.

Example

Evaluate $45.6 \div 10$

Solution

Since 10 has one 0, we move the decimal point in 45.6 to the left over one digit to get 4.56.

To explain why this works, we appeal to the key structural feature of decimals that states that each place value is 10 times smaller than the one on its left and 10 times larger than the one on its right. This means that the movement of the decimal point reassigns the place values and since it moves to the left over one digit, the place value of each digit becomes 10 times smaller. This implies that the overall size of the number becomes 10 times smaller which is what we seek from the division of a number by 10.

Example

Evaluate $0.089 \div 100$

Solution

Since 100 has two 0s, we move the decimal point in 0.089 to the left over two digits to get 0.000 89. Note that this move requires that we add 0s to the left of 0.089.

The movement of the decimal point to the left over two digits makes the value of each digit 100 times smaller. This means that the overall size of the number itself becomes 100 times smaller which is what we seek from the division of a number by 100.

Example

Evaluate $920 \div 1000$

Solution

Since 1000 has three 0s, we move the decimal point in 920 (which is on its right end) to the left over three digits to get 0.92. Note that we have dropped the right-end 0 in 920 *after* the move.[24]

Division by Sub-Powers of 10

There is also a fast algorithm for dividing a decimal number by sub-powers of 10 (i.e., 0.1, 0.01, 0.001, etc.). The algorithm requires that we move the decimal point in the given number to the right over as many digits as there are decimal digits in the sub-power of 10.

Example

Evaluate $43.89 \div 0.1$

Solution

Since 0.1 has one decimal digit, we move the decimal point in 43.89 to the right over one digit to get 438.9.

Note that a division by 0.1 implies that we need to compute the number of *tenths* that fit in the dividend. A bit of reflection shows that the answer to such a question should be 10 times larger than the dividend as 10 *tenths* fit in a *one*.[25] Since the movement of the decimal point in the dividend increases its size by a factor of 10, we can get the answer by making such a move.

Example

Evaluate $719 \div 0.01$

---

[24]Not always is 0.92 equal to 0.920. See Chapter 9 for cases where the presence of such right-end 0s is semantically significant.

[25]You should think this through and convince yourself that this is the case before you continue further.

Solution

Since 0.01 has two decimal digits, we move the decimal point in 719 to the right over two digits to get 71 900. Note that this requires that we add 0s to the right-end of 719.

Example

Evaluate $0.3 \div 0.001$

Solution

Since 0.001 has three decimal digits, we move the decimal point in 0.3 to the right over three digits to get 300. Note that this requires that we add 0s to the right end of 0.3.

Exercise Set 6.1.4

1. Evaluate the following expressions.

   a. $23.5 + 16.3$
   b. $2.1 + 4.6$
   c. $9.25 + 0.14$
   d. $120.08 + 376.21$
   e. $0.26 + 0.43$
   f. $0.072 + 0.516$
   g. $0.25 + 0.31$
   h. $0.002 + 0.006$
   i. $42.6 + 18.4$
   j. $3.05 + 2.95$
   k. $345.2 + 809.6$
   l. $45.32 + 72.99$
   m. $6.8 + 3.4 + 1.2$

2. Evaluate the following expressions.

   a. $0.02 + 0.451$
   b. $0.96 + 0.004$
   c. $0.026 + 0.57$
   d. $0.589 + 0.2$
   e. $32.4 + 5.37$
   f. $9.25 + 10.024$
   g. $180.6 + 2.14$
   h. $75.1 + 21.58$
   i. $0.82 + 0.106$
   j. $0.057 + 0.23$
   k. $0.004 + 0.5823$
   l. $0.2 + 0.328$
   m. $38.2 + 4.85$
   n. $9.7 + 0.49$
   o. $1.008 + 15.3$
   p. $582.04 + 36.5$
   q. $94.6 + 0.05$

3. Evaluate the following expressions.

   a. $0.9 - 0.2$
   b. $0.07 - 0.01$
   c. $0.047 - 0.006$
   d. $0.5 - 0.4$
   e. $12.73 - 10.28$
   f. $150.2 - 104.7$
   g. $900.3 - 264.9$
   h. $32.47 - 2.93$
   i. $18.93 - 12.483$
   j. $0.462 - 0.29$
   k. $0.0056 - 0.004$
   l. $410.6 - 205$

m. $12.51 - 3$ o. $27 - 4.2$ q. $0.0158 - 0.007$

n. $100 - 29.46$ p. $3241.42 - 423.541$ r. $25.08 - 10.4$

4. Evaluate the following expressions.

a. $8.49 + 0.9 - 0.32$ f. $47 + 14.8 - 12.99 + 4.169 + 200$

b. $0.07 + 3.27 - 0.01$ g. $2.9 + 31 - 1 + 5.8$

c. $0.917 - 0.72 + 0.34$ h. $170.27 - 271.9 + 343.8$

d. $0.261 + 4.281 - 3.25 - 0.2$ i. $31.8 - 42.8 + 31.92$

e. $15.42 - 2.98 + 4.9 - 7.179$

5. Evaluate the following expressions.

a. 2.4×3.2 h. 4000×0.001 o. 329×0.4

b. 5.1×0.8 i. 4000×0.0001 p. 251×0.06

c. $12.9\,(3.7)$ j. $382\,(0.1)$ q. 32×0.6

d. $6.93\,(0.09)$ k. $382\,(0.01)$ r. 9.8×0.5

e. 0.5×72.41 l. $382\,(0.001)$ s. 3.71×15

f. 4000×0.1 m. $382\,(0.0001)$ t. 0.27×528

g. 4000×0.01 n. $382\,(0.000\,01)$

6. Evaluate the following expressions.

a. $358.8 \div 23$ k. $\frac{2}{5}$ s. $\frac{2}{15}$

b. $99.84 \div 39$ l. $\frac{225}{15}$ t. $\frac{15.6}{0.12}$

c. $5601.9 \div 71$ m. $\frac{2}{3}$ u. $\frac{0.3816}{0.72}$

d. $62 \div 0.5$

e. $792 \div 0.06$ n. $\frac{2}{9}$ v. $\frac{42.8}{0.18}$

f. $48 \div 0.012$ o. $\frac{3}{11}$ w. $\frac{320}{0.02}$

g. $2.43 \div 0.15$ p. $\frac{23}{999}$ x. $\frac{2.4}{0.12}$

h. $15.08 \div 0.025$ q. $\frac{1}{7}$ y. $\frac{17.82}{0.3}$

i. $4.92 \div 0.5$

j. $\frac{1}{4}$ r. $\frac{3}{8}$ z. $\frac{0.52}{2.4}$

7. Evaluate the following expressions.

a. $14.2 \times 6 \div 0.5 \times 8$ d. $54.28 \div 0.4 \div 4.2$

b. $19.2 \div 2.5 \times 7.9$ e. $6.38 \times 9 \div 0.03$

c. $6.02 \div 4.5 \times 2.6$ f. $\frac{3.2 \times 4}{2 \times 2.5}$

g. $\dfrac{6.41}{2.4 \times 5}$

h. $\dfrac{15.1 \times 3.4 \times 7.2}{5 \times 0.2}$

i. $\dfrac{9.2 \times 6}{0.004}$

6.1.5   Expressions

We will begin with an algorithm for the evaluation of expressions that involve chain addition/subtraction of decimals. Next we will present an algorithm for evaluating expressions that involve chain multiplication/division of decimals. We will end the section with an adaptation of the analysis-synthesis algorithm given in the section on integers, for rapid evaluation of expressions that involve a mix of the four basic arithmetic operations.

Chain Addition/Subtraction

As in the case for the chain addition/subtraction of integers, two approaches are possible here.

The first approach starts at the left and works its way to the right adding and subtracting the numbers as instructed. This approach follows the manner in which the expression is written and adds and subtracts our gains and losses as sequenced. This approach is the preferred approach if one has access to a calculator otherwise repeated application of the *Algorithm for Evaluation of Additions and Subtractions of Two Numbers Written in Decimal Notation* will slow the pace to the point where it becomes inefficient.

The second approach is based on the algorithm introduced in the section on whole numbers and proceeds by adding the terms that represent gains, adding the terms that represent losses, and then subtracting the latter from the former. This approach is the preferred approach if one evaluates an expression by hand as it minimizes the number of subtractions to one. In addition, the logic is in line with the way we normally reason as in when we add our sources of income, add our expenses, and then subtract the latter from the former.

The observations above are summarized below.

Algorithm for Evaluation of Chain Addition/Subtraction of Decimals

To evaluate an expression involving chain addition/subtraction, go from left to right and perform the operations as you see them if you have access to a calculator otherwise

1. Add the terms that represent gains (those with a + on their left side).
2. Add the terms that represent losses (those with a − on their left side).
3. Subtract the sum of the losses from the sum of the gains.

Example

Evaluate

$$-2.5 + 1.8 - 7.2 - 1.2 + 1.9 - 3.1$$

Solution

Following the first approach we write:

$$-2.5 + 1.8 - 7.2 - 1.2 + 1.9 - 3.1 = -10.3$$

Following the second approach we write:

$$
\begin{aligned}
-2.5 + 1.8 &- 7.2 - 1.2 + 1.9 - 3.1 \\
&= 1.8 + 1.9 - (2.5 + 7.2 + 1.2 + 3.1) \\
&= 3.7 - 14 \\
&= -10.3
\end{aligned}
$$

Chain Multiplication/Division

The algorithm developed in the section on integers on evaluating the products and quotients of nonzero integers can be adapted for the evaluation of expressions that contain chain multiplication and/or division of decimals.

Algorithm for Evaluation of Chain Multiplication/Division of Decimals

1. Determine the *sign* of the result: If there are an even number of negative signs in the chain, the result is positive otherwise the result is negative.
2. Determine the *size* of the result: Ignore the signs and multiply and/or divide the numbers involved.

Here is an example:

Example

Evaluate

$$4.6 \, (-2.3) \, (7.1)$$

Solution

$$4.6 \, (-2.3) \, (7.1) \; = \; -75.118$$

Example

Evaluate

$$\frac{2.46 \, (-1.8)}{-0.36 \, (-1.23)}$$

Solution

$$\frac{2.46 \, (-1.8)}{-0.36 \, (-1.23)} \; = \; -10$$

In working out the expression above, we first work out the sign and then multiply and divide the numbers as indicated to arrive at the answer.

Addition, Subtraction, Multiplication, and Division

The *Analysis-Synthesis Technique* can be extended to deal with expressions that involve decimals.

The Analysis-Synthesis Algorithm for Evaluation of Expressions Involving Addition, Subtraction, Multiplication and Division of Decimals

To evaluate an expression involving the addition, subtraction, multiplication and division of decimals

1. Analyze:
   a. Analyze the expression into terms using additions and subtractions *outside* brackets.
   b. Analyze each term into factors using multiplications.
   c. Analyze each factor further if needed.
2. Synthesize:
   a. Work out the factors.
   b. Multiply the factors to evaluate the terms. To do so
      i. Work out the sign: Count the number of negative signs within the term. If the terms is being subtracted (i.e., the operation on the left of the term is subtraction), then count this subtraction as a negative sign in addition to the ones within the term itself. An even number of negative signs implies that the term will be added (write +). An odd number of negative signs implies that the term will be subtracted (write −).
      ii. Work out the magnitude: Ignore all signs and apply the *Algorithm for Evaluation of Chain Multiplication/Division of Decimals* given earlier to the expression within the term.
   c. Evaluate the expression. To do so, use the *Algorithm for Evaluation of Chain Addition/Subtraction of Decimals* given earlier.

We illustrate the use of the above algorithm using examples.

Example

Evaluate
$$4.2 \times 3.1 \div 0.6 + 5.2 \div 4 \times 2$$

Solution

$$4.2 \times 3.1 \div 0.6 + 5.2 \div 4 \times 2 = \boxed{4.2 \times 3.1 \div 0.6} + \boxed{5.2 \div 4 \times 2}$$
$$= \boxed{21.7} + \boxed{2.6}$$
$$= 24.3$$

Note that in going from line 1 to line 2 we perform the operations in order from left to right.

Example

Evaluate

$$32.8\,(-3.1)\,(-1.7) \;-\; (-7.5)\,(-1.8)\,(-5)$$

Solution

$$32.8\,(-3.1)\,(-1.7) \;-\; (-7.5)\,(-1.8)\,(-5)$$

$$= \boxed{32.8\,(-3.1)\,(-1.7)} \;-\; \boxed{(-7.5)\,(-1.8)\,(-5)}$$

$$= \boxed{172.856} \;+\; \boxed{67.5}$$

$$= 240.356$$

Example

Evaluate

$$8.3\,(1.2 \,-\, 9.5) \;+\; (-3.9)\,(5.3 \,+\, 2.4)$$

Solution

$$8.3\,(1.2 \,-\, 9.5) \;+\; (-3.9)\,(5.3 \,+\, 2.4)$$

$$= \boxed{8.3\,(1.2 \,-\, 9.5)} \;+\; \boxed{(-3.9)\,(5.3 \,+\, 2.4)}$$

$$= \boxed{8.3\,(-8.3)} \;+\; \boxed{(-3.9)\,(7.7)}$$

$$= \boxed{-68.89} \;-\; \boxed{30.03}$$

$$= -98.92$$

Example

Evaluate

$$\frac{3.4\,(-1.9)}{-1.7} \;-\; 4.7 \times \frac{3.5}{-5}$$

Solution

$$\frac{3.4\,(-1.9)}{-1.7} - 4.7 \times \frac{3.5}{-5} = \boxed{\frac{3.4\,(-1.9)}{-1.7}} - \boxed{4.7 \times \frac{3.5}{-5}}$$
$$= \boxed{3.8} + \boxed{3.29}$$
$$= 7.09$$

We will now consider cases where higher order operations such as exponents, roots and logarithms are present in the expression.

Example

Evaluate

$$1.4^2\,(-3.8) - 8.1\sqrt{5.3}$$

Solution

The expression

$$1.4^2\,(-3.8) - 8.1\sqrt{5.3}$$

analyzes into two terms: $1.4^2\,(-3.8)$ and $8.1\sqrt{5.3}$.

The first term breaks into two factors: 1.4^2 and -3.8. We work out the first factor by multiplying the base 1.4 by itself to get 1.96. The second factor is already simplified.

The second term breaks into two factors as well: 8.1 and $\sqrt{5.3}$. The first is already simplified. The second evaluates to $2.302\,172\ldots$.[26] The full solution follows:

$$1.4^2\,(-3.8) - 8.1\sqrt{5.3} = 1.96\,(-3.8) - 8.1\,(2.302\,172\ldots)$$
$$= -7.448 - 18.647\,600\ldots$$
$$= -26.095\,600\ldots$$

---

[26] A few issues need to be clarified. First, when the root notation is used with decimals, the notation should be taken as a directive. In addition, since decimals are involved, we can apply algorithms for evaluating the results of such directives as a decimal. However, we find that beyond addition, subtraction, multiplication, division and exponents, the algorithms for the evaluation of other directives are complex and time-consuming. As a result, we normally use a calculator to perform such calculations.

Example

Evaluate

$$2\sqrt{4.8} + (-3)\log_{2.3} 17.5$$

Solution

$$
\begin{aligned}
2\sqrt{4.8} + (-3)\log_{2.3} 17.5 &= 2\,(2.190\,890\ldots) + (-3)\,(3.436\,390\ldots) \\
&= 4.381\,780\ldots - 10.309\,171\ldots \\
&= -5.927\,390\ldots
\end{aligned}
$$

Exercise Set 6.1.5

Evaluate each expression.

1. a. $9.4 \times 3.2 + 2.8 \times 4$

   b. $3.6 + 2.7 \times 9.1$

   c. $42.8 - 3.5 \times 2 + 6 \div 1.5$

   d. $-18 - 2.46 \times 42 \div 1.5 + (-2.4) \div 0.3$

   e. $42.5 \div 0.5 \times 4 + 6.2 \times 4.5 \div 2.5$

   f. $\dfrac{18.6 \times 1.4}{2.1} + 8.2 \times 5.6$

   g. $\dfrac{-19.4\,(4.6)}{-2.3\,(-0.5)} + \dfrac{-2.3\,(8)}{4\,(0.2)}$

   h. $-4.5\,(-3.6 + 2.8 \times 1.4)$

   i. $6.2\,(5.6 \times 2 + 1.5) - 2.1\,(0.6 - 5 \times 0.1)$

   j. $\dfrac{18.6 \times 2.5}{5} + \dfrac{3.2}{4}\left(9.5 + \dfrac{3}{0.5}\right)$

   k. $\dfrac{16.2}{-2.5} - 3\left(4 \times \dfrac{2}{0.5} - 3 \times 2.6\right)$

   l. $3.4\sqrt{1.8} - 2.1\sqrt{5.7}$

   m. $-1.7\sqrt{9.93} + 1.17\sqrt{1.1}$

   n. $-2.5\sqrt{53.2} - (-4.6)\sqrt{3.1}$

   o. $4.4 \times 1.7^2 + 2.7 \times 3.8^2$

   p. $-7.4 \times 4.6^3 + 3.2^2 \times 1.6^2$

   q. $1.2 \times 3.4^2 \times \sqrt{3.2}$

r. $\frac{2.5}{3.1} \times \sqrt{4.6} + \frac{1.5}{3.2^2} \times 1.8^3$

s. $\frac{1.4^3 + 3.2^2}{3.8 \times 4.1^2}$

t. $3\log_{2.3} 8.4 - 2\log_{4.5} 17.8$

u. $-4\log_{6.3} 12.7 + (-3)\log_{1.4} 0.8$

v. $-\log 132.7 + 3\sqrt{5.9}$

6.1.6 Rounding

When we ask you to round a decimal number to a certain place value, it means find the integral multiple of that place value that is closest to the given number.

To explain what this means, let us start with a discussion on the manner in which positive decimals are rounded to a certain place value and, in doing so, we will take on the task of rounding a positive decimal to a place value on the whole side before we tackle the task of rounding a decimal number to a place value on the fractional side. Following this we will discuss the manner in which one rounds negative decimals to a given place value.

Rounding to the Nearest Ten

When we ask you to round a decimal number to the nearest ten, it means find the whole multiple of 10 that is closest to the given number. The whole multiples of 10 are

0, 10, 20, 30, 40, 50, 60, 70, 80, 90, 100, 110, 120, ...

As an example, suppose we ask you to round 342 to the nearest ten. To do so, we must find the whole multiple of 10 that is closest to 342. As the diagram on the left side in Figure 6.1.10 shows, the closest whole multiple of 10 to 342 is 340.

This visual tool is quite descriptive but a faster technique for use in practice is preferred. To explain how this faster technique works, see the diagram on the right side in Figure 6.1.10. Here we have magnified the interval from 340 to 350 with 345 as the halfway point. Since the 2 in 342 is less than the 5 in 345, 342 is below the halfway point and is, therefore, closer to 340. This leads to the technique in Figure 6.1.11 below.

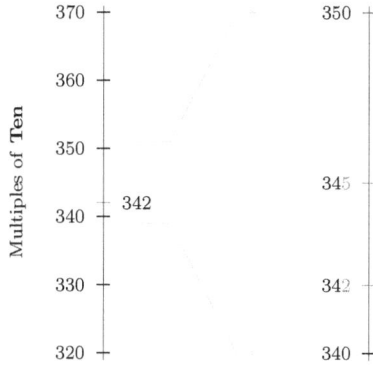

Figure 6.1.10: Rounding 342 to the nearest **ten**:
Which multiple of 10 is 342 closest to?

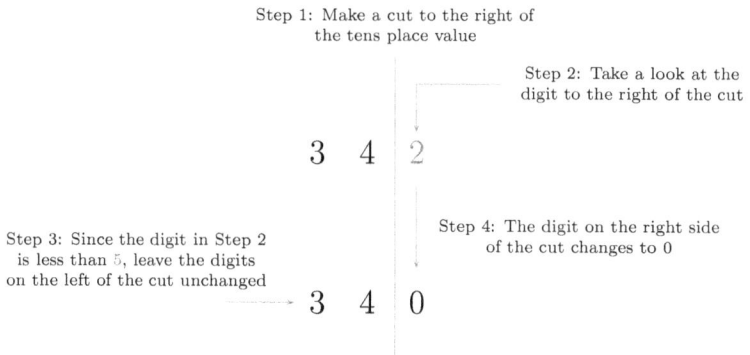

Figure 6.1.11: An efficient technique for
rounding 342 to the nearest ten

Suppose now that we are asked to round 348 to the nearest ten. To do so, we must find the whole multiple of 10 that is closest to 348. As the diagram on the left in Figure 6.1.12 shows, the closest whole multiple of 10 to 348 is 350.

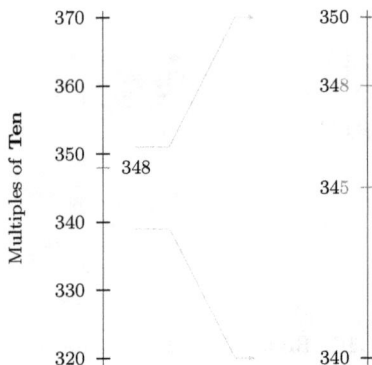

Figure 6.1.12: Rounding 348 to the nearest **ten:**
Which multiple of 10 is 348 closest to?

The diagram on the right side in Figure 6.1.12 provides a more efficient way of rounding 348 to the nearest ten. Note once again that 345 points to the point halfway between 340 and 350. Since the 8 in 348 is greater than the 5 in 345, 348 is above the halfway point and is, therefore, closer to 350. This leads to the technique in Figure 6.1.13 below.

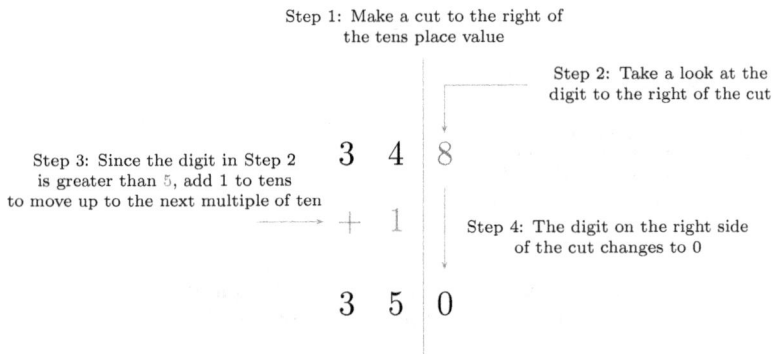

Step 1: Make a cut to the right of the tens place value

Step 2: Take a look at the digit to the right of the cut

Step 3: Since the digit in Step 2 is greater than 5, add 1 to tens to move up to the next multiple of ten

3 4 8

+ 1

Step 4: The digit on the right side of the cut changes to 0

3 5 0

Figure 6.1.13: An efficient technique for rounding 348 to the nearest ten

The question naturally arises as to what one should do if one is asked to round 345 to the nearest ten. The problem of course is that both 340 and 350 are equally close to 345. This is shown in Figure 6.1.14 below.

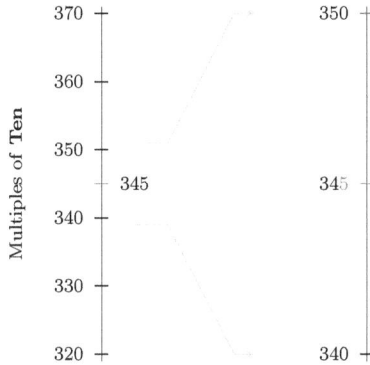

Figure 6.1.14: Rounding 345 to the nearest **ten**:
Which multiple of 10 is 345 closest to?

Unless otherwise stated, we round all halfway values up to the next whole multiple.[27] The answer to the problem at hand, therefore, is 350; see below.

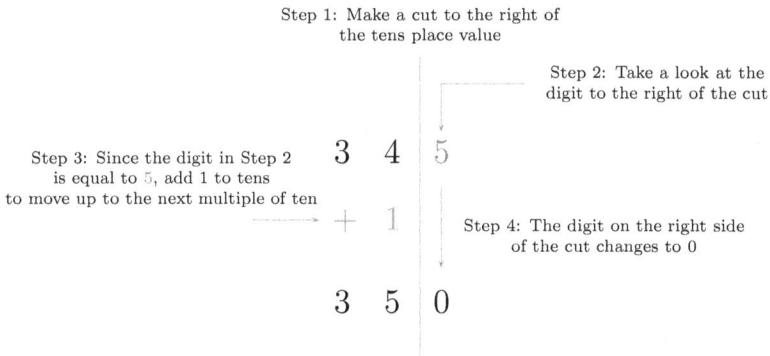

Figure 6.1.15: An efficient technique for rounding 345 to the nearest ten

---

[27]In most contexts such an approach is acceptable – hence the reason why it is the de facto approach to rouding halfway points. In situations when this appraoch is not suitable, other approaches are used. Such special situations are not discussed in this textbook.

We now turn our attention to three special cases. To illustrate the first, suppose we are asked to round 3 to the nearest ten. To do so, we must find the whole multiple of 10 that is closest to 3. The following diagram illustrates the idea.

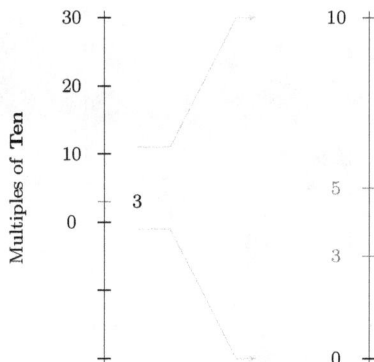

Figure 6.1.16: Rounding 3 to the nearest **ten**:
Which multiple of 10 is 3 closest to?

As the diagram on the left in Figure 6.1.16 clearly shows, the whole multiple of ten closest to 3 is 0. The answer, therefore, is 0. As Figure 6.1.17 below shows, the faster technique works in this case as well.

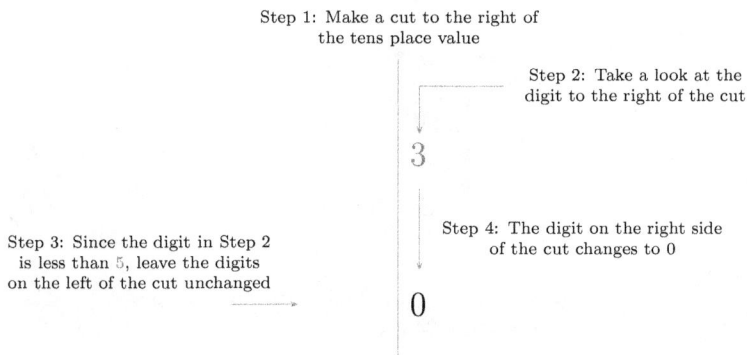

Step 1: Make a cut to the right of the tens place value

Step 2: Take a look at the digit to the right of the cut

3

Step 3: Since the digit in Step 2 is less than 5, leave the digits on the left of the cut unchanged

Step 4: The digit on the right side of the cut changes to 0

0

Figure 6.1.17: An efficient technique for rounding 3 to the nearest ten

To illustrate the second special case, suppose we are asked to round 3496 to the nearest ten. As Figure 6.1.18 shows, the answer is 3500.

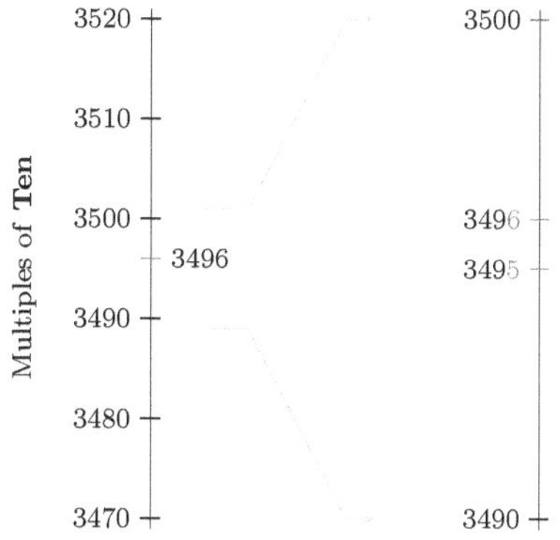
3520
3510
3500
3496
3490
3480
3470
Multiples of Ten
3500
3496
3495
3490

Figure 6.1.18: Rounding 3496 to the nearest **ten**:
Which multiple of 10 is 3496 closest to?

The following diagram shows that the faster technique introduced earlier also works in this situation. Note that we need to use addition with carrying in such a case.

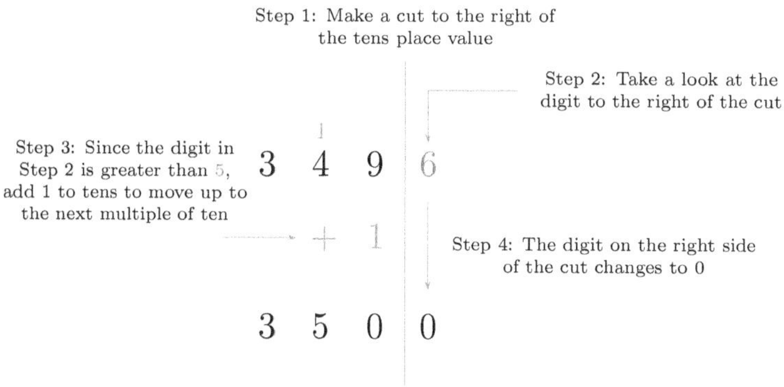
Step 1: Make a cut to the right of the tens place value
Step 2: Take a look at the digit to the right of the cut
Step 3: Since the digit in Step 2 is greater than 5, add 1 to tens to move up to the next multiple of ten
1
3 4 9 6
+ 1
Step 4: The digit on the right side of the cut changes to 0
3 5 0 0

Figure 6.1.19: An efficient technique for rounding 3496 to the nearest ten

To illustrate the third special case, suppose we are asked to round 470 to the nearest ten. As Figure 6.1.20 shows, the answer is 470.

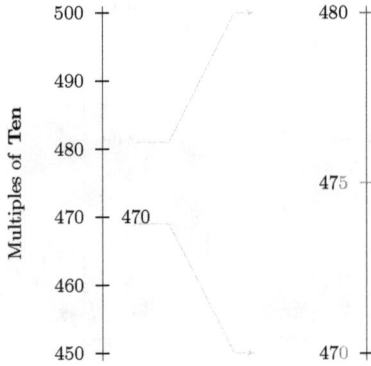

Figure 6.1.20: Rounding 470 to the nearest **ten**: Which multiple of 10 is 470 closest to?

The following diagram shows that the faster technique introduced earlier also works in this situation.

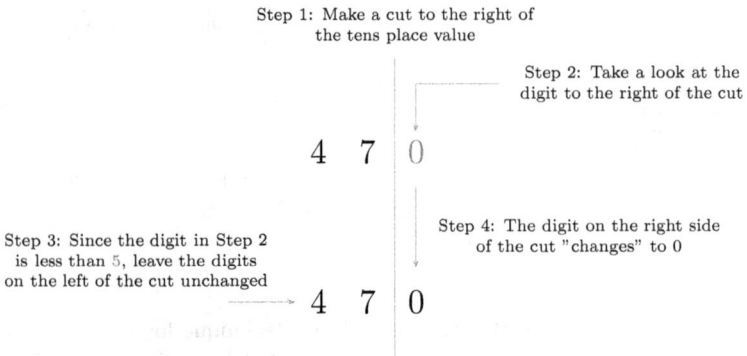

Figure 6.1.21: An efficient technique for rounding 470 to the nearest ten

Examples

Round each of the following to the nearest ten.

1. 84
2. 65

3. 95
4. 424

5. 399
6. 920

Answers

1. 80
2. 70

3. 100
4. 420

5. 400
6. 920

Rounding to the Nearest Hundred

When we ask you to round a decimal number to the nearest hundred, it means find the whole multiple of 100 that is closest to the given number. The whole multiples of 100 are

0, 100, 200, 300, 400, 500, 600, 700, 800, 900, 1000, 1100, 1200, ...

As an example, suppose we ask you to round 627 to the nearest hundred. To do so, we must find the whole multiple of 100 that is closest to 627. As the diagram on the left in Figure 6.1.22 shows, the closest whole multiple of 100 to 627 is 600.

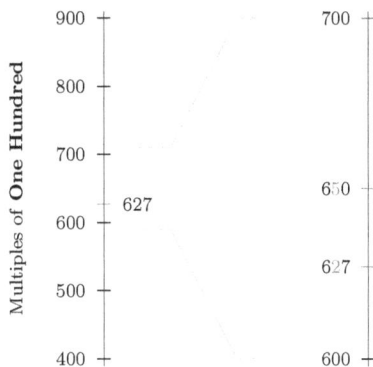

Figure 6.1.22: Rounding 627 to the nearest **hundred**: Which multiple of 100 is 627 closest to?

The faster technique for rounding 627 to the nearest hundred is shown in 6.1.23. Note that in this case the midpoint between 600 and 700 is 650.

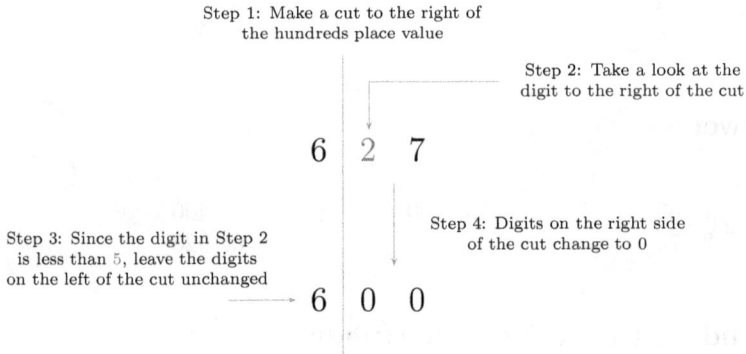

Step 1: Make a cut to the right of
the hundreds place value

Step 2: Take a look at the
digit to the right of the cut

6 | 2 | 7

Step 3: Since the digit in Step 2
is less than 5, leave the digits
on the left of the cut unchanged

Step 4: Digits on the right side
of the cut change to 0

6 | 0 | 0

Figure 6.1.23: An efficient technique for
rounding 627 to the nearest hundred

Note that what matters is digit 2. Digit 7 in the ones will make no difference as the reader can readily tell that 620, 621, 622, 623, ..., 629 are all less than 650 (the halfway point).

As a second example, suppose we ask you to round 1793 to the nearest hundred. To do so, we must find the whole multiple of 100 that is closest to 1793. As the diagram on the left in Figure 6.1.24 shows, the closest whole multiple of 100 to 1793 is 1800.

The faster technique for rounding 1793 to the nearest hundred is shown in 6.1.25. Note that in this case the midpoint between 1700 and 1800 is 1750.

Consider now the process of rounding 347 to the nearest hundred. According to the technique that we have presented, the answer is 300. A question arises as to whether one should instead first round 347 to the nearest ten to arrive at 350, and then round 350 to the nearest hundred to arrive at 400. To answer this question, we draw the reader's attention to Figure 6.1.26 and the semantics of rounding to a given place value. Based on the semantics of rounding to the nearest hundred, we understand that, to round 347 to the nearest hundred, we must find the whole multiple of 100 that is closest to 347. As one can see in the diagram on the left side of Figure 6.1.26, the closest whole multiple of 100 to 347 is, indeed, 300. The diagram on the right side of

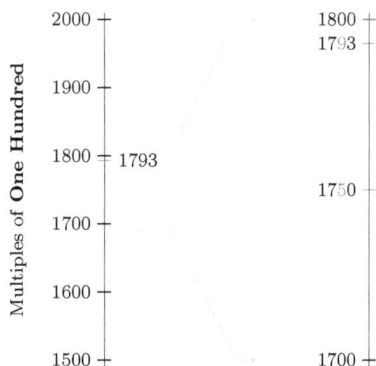

Figure 6.1.24: Rounding 1793 to the nearest **hundred**: Which multiple of 100 is 1793 closest to?

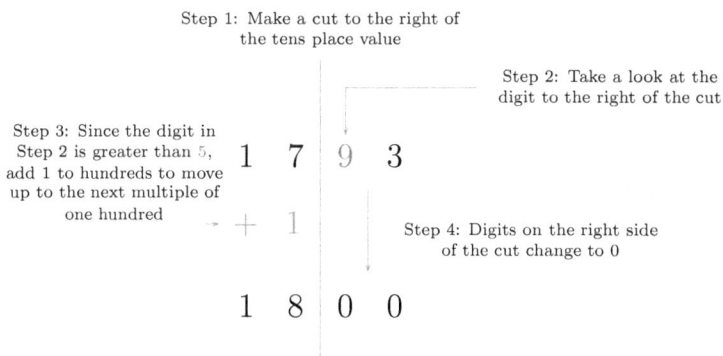

Figure 6.1.25: An efficient technique for rounding 1793 to the nearest hundred

Figure 6.1.26 shows that, when we round in stages (in rounding first to the tens and then to hundreds), in effect we push 347 up to the halfway point (i.e., 350), and then push this up again to 400. Clearly this is not what we intend to do when we round 347 to the nearest hundred.

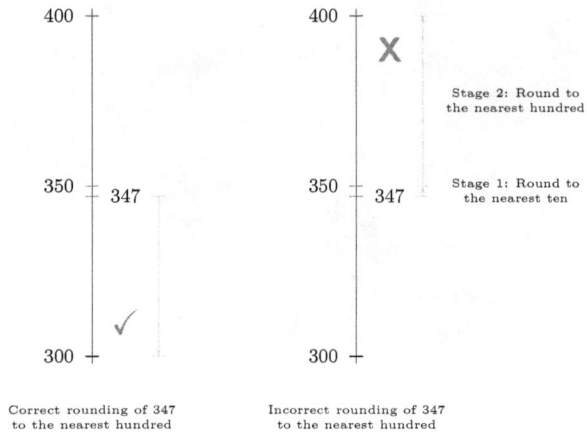

Correct rounding of 347
to the nearest hundred

Incorrect rounding of 347
to the nearest hundred

Figure 6.1.26: Correct and incorrect roundings
of 347 to the nearest hundred

Examples

Round each of the following to the nearest hundred.

1. 781   3. 2419   5. 9962
2. 618   4. 42   6. 500

Answers

1. 800   3. 2400   5. 10 000
2. 600   4. 0   6. 500

Rounding to Higher Place Values

When we ask you to round a decimal number to the nearest thousand, it means find the whole multiple of 1000 that is closest to the given number. The whole multiples of 1000 are

0, 1000, 2000, 3000, 4000, 5000, 6000, 7000, 8000, 9000, 10 000, ...

And when we ask you to round a decimal number to the nearest ten thousand, it means find the whole multiple of 10 000 that is closest to the given number. The whole multiples of 10 000 are

0, 10 000, 20 000, 30 000, 40 000, 50 000, 60 000, ...

Rounding to the nearest hundred thousand, million, ten million, etc. have similar meanings.

The usefulness of the fast technique that we presented earlier in our discussion is not just in the fact that it can be used under all circumstances and special cases, but that it can be extended for use in rounding to any place value. The manner in which we round a decimal number to higher place values is similar to the manner in which we round a decimal number to the nearest ten or hundred. It is only the location of the cut in Step 1 that is different. As an example, suppose we ask you to round 4268 to the nearest thousand. To do so, we must find the whole multiple of 1000 that is closest to 4268. As the diagram on the left in Figure 6.1.27 shows, the closest whole multiple of one thousand to 4268 is 4000.

The faster technique for rounding 4268 to the nearest thousand is shown in 6.1.28.

We have seen that the fast algorithm that we have presented for rounding a given decimal number to a given place value works under all circumstances so far.

Examples

Round to the specified place value.

1. 4529 (thousands)
2. 710 (thousands)
3. 19 920 (thousands)
4. 54 610 (ten thousands)
5. 875 271 (ten thousands)
6. 9 480 192 (millions)

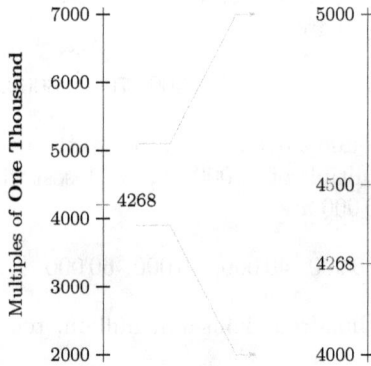

Figure 6.1.27: Rounding 4268 to the nearest **thousand**: Which multiple of 1000 is 4268 closest to?

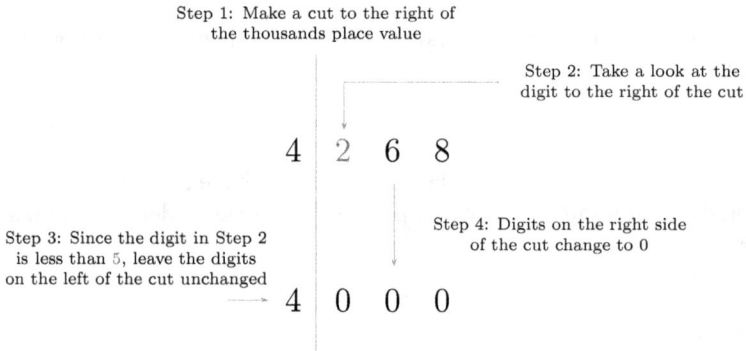

Step 1: Make a cut to the right of the thousands place value

Step 2: Take a look at the digit to the right of the cut

4 | 2 6 8

Step 3: Since the digit in Step 2 is less than 5, leave the digits on the left of the cut unchanged

Step 4: Digits on the right side of the cut change to 0

4 | 0 0 0

Figure 6.1.28: An efficient technique for rounding 4268 to the nearest thousand

Answers

| 1. 5000 | 3. 20 000 | 5. 880 000 |
| 2. 1000 | 4. 50 000 | 6. 9 000 000 |

The semantics of rounding negative decimals to a given place value is similar to the one that we presented for rounding positive decimals. As an example, to round -480 to the nearest hundred, we must find the nearest *integral* multiple of 100 to -480. The integral multiples of 100 are

$$\ldots, -400, -300, -200, -100, 0, 100, 200, 300, 400, \ldots$$

To round a negative decimal to a given place value, round the absolute value of that decimal to the given place value following the rules for rounding positive decimal numbers, preceding the result with a $-$. As an example, to round -480 to the nearest hundred, we must find the integral multiple of 100 that is closest to -480. As shown in Figure 6.1.29 below, the answer is -500.

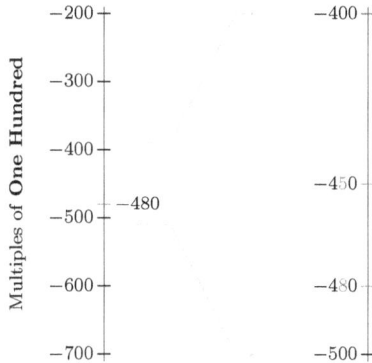

Figure 6.1.29: Rounding -480 to the nearest **hundred**: Which multiple of 100 is -480 closest to?

An efficient technique for rounding -480 to the nearest hundred is to round its absolute value (i.e., 480) to the nearest hundred as shown in Figure 6.1.30 below, and attach a negative sign to the result.

Step 1: Make a cut to the right of
the hundreds place value

Step 2: Take a look at the
digit to the right of the cut

Step 3: Since the digit in Step 2
is greater than 5, add 1 to
hundreds to move up to the next
multiple of one hundred

4 | 8   0

$+$ 1

Step 4: Digits on the right side
of the cut change to 0

5 | 0   0

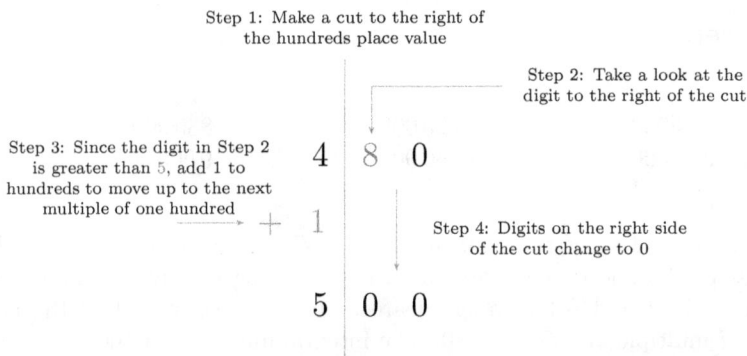

Figure 6.1.30: An efficient technique for rounding
-480 to the nearest hundred: Round the absolute
value of -480 (i.e., 480) to the nearest hundred as
shown, and attach a negative sign to the result.

It is instructive to see how the rounding of -420 to the nearest hundred
compares to the rounding of 420 to the nearest hundred. The comparison is
shown in Figure 6.1.31.

Examples

Round to the indicated place value.

1. -731 (tens)
2. -850 (hundreds)
3. -4145 (hundreds)

4. $+318$ (tens)
5. -1 (hundreds)
6. -1192 (thousands)

Answers

1. -730
2. -900

3. -4100
4. 320

5. 0
6. -1000

Rounding to the Nearest Tenth

When we ask you to round a decimal number to the nearest tenth, it means
find the whole multiple of 0.1 that is closest to the given number. The whole
multiples of 0.1 are

0, 0.1, 0.2, 0.3, 0.4, 0.5, 0.6, 0.7, 0.8, 0.9, 1.0, 1.1, 1.2, ...

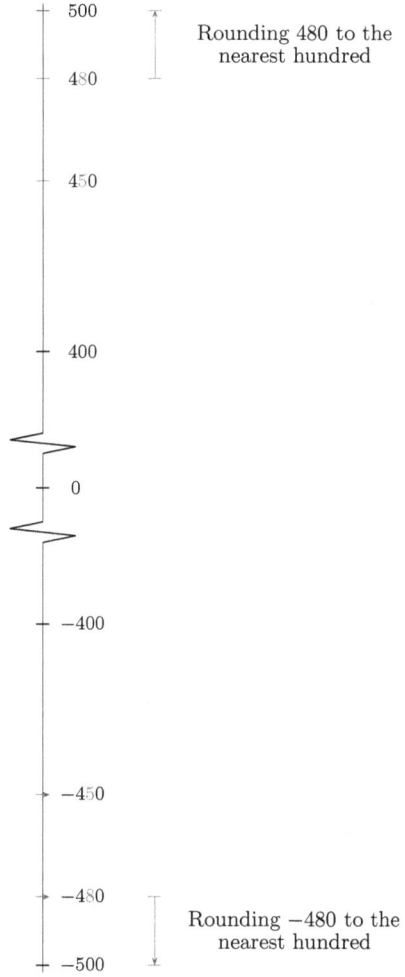

Figure 6.1.31: Comparison of rounding 480 and
−480 to the nearest hundred

As an example, suppose we ask you to round 3.42 to the nearest tenth. To do so, we must find the whole multiple of tenth that is closest to 3.42. As the diagram on the left in Figure 6.1.32 shows, the closest whole multiple of tenth to 3.42 is 3.4.

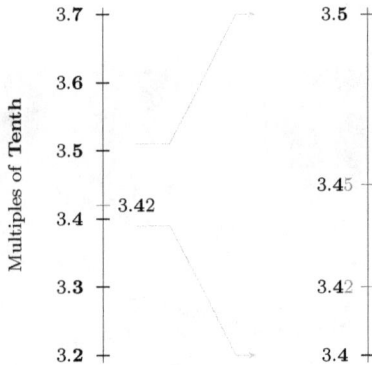

Figure 6.1.32: Rounding 3.42 to the nearest **tenth**: Which multiple of 0.1 is 3.42 closest to?

This visual tool is quite descriptive but a faster technique for use in practice is preferred. To explain how this faster technique works, see the diagram on the right side in Figure 6.1.32. Here we have magnified the interval from 3.4 to 3.5 with 3.45 as the halfway point. Since the 2 in 3.42 is less than the 5 in 3.45, 3.42 is below the halfway point and is, therefore, closer to 3.4. This leads to the technique in Figure 6.1.33 below.

Suppose now that we are asked to round 3.48 to the nearest tenth. To do so, we must find the whole multiple of tenth that is closest to 3.48. As the diagram on the left in Figure 6.1.34 shows, the closest whole multiple of tenth to 3.48 is 3.5.

The diagram on the right side in Figure 6.1.34 provides a more efficient way of rounding 3.48 to the nearest tenth. Note once again that 3.45 points to the point halfway between 3.4 and 3.5. Since the 8 in 3.48 is greater than the 5 in 3.45, 3.48 is above the halfway point and is, therefore, closer to 3.5. This leads to the technique in Figure 6.1.35 below.

Step 1: Make a cut to the right of
the tenths place value

Step 2: Take a look at the
digit to the right of the cut

$$3 \ . \ 4 \ | \ 2$$

Step 4: Drop the digits on the
right side of the cut

Step 3: Since the digit in Step 2
is less than 5, leave the digits
on the left of the cut unchanged $3 \ . \ 4$

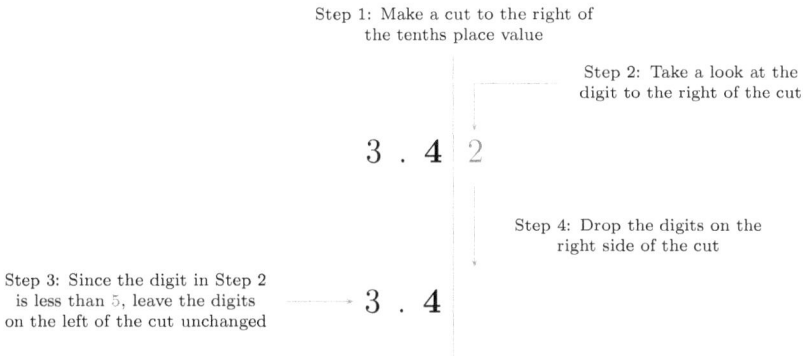

Figure 6.1.33: An efficient technique for
rounding 3.42 to the nearest tenth

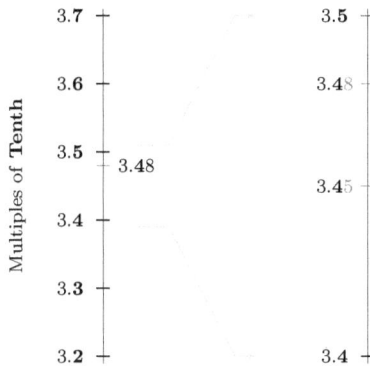

Multiples of Tenth

3.7 —
3.6 —
3.5 — 3.48
3.4 —
3.3 —
3.2 —

3.5 —
3.48 —
3.45 —
3.4 —

Figure 6.1.34: Rounding 3.48 to the nearest
tenth: Which multiple of 0.1 is 3.48 closest to?

Step 1: Make a cut to the right of
the tenths place value

Step 2: Take a look at the
digit to the right of the cut

Step 3: Since the digit in Step 2 is
greater than 5, add 1 to tenths to
move up to the next multiple of tenth

$$3 \; . \; 4 \; | \; 8$$

$$+ \; 1$$

Step 4: Drop the digits on the
right side of the cut

$$3 \; . \; 5$$

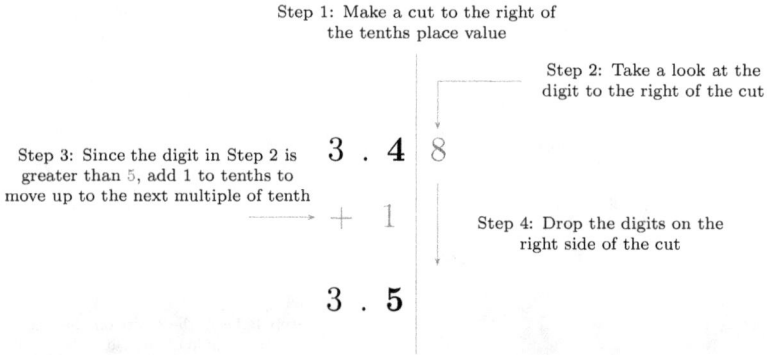

Figure 6.1.35: An efficient technique for
rounding 3.48 to the nearest tenth

The question naturally arises as to what one should do if one is asked to
round 3.45 to the nearest tenth. The problem of course is that both 3.4 and
3.5 are equally close to 3.45. This is shown in Figure 6.1.36 below.

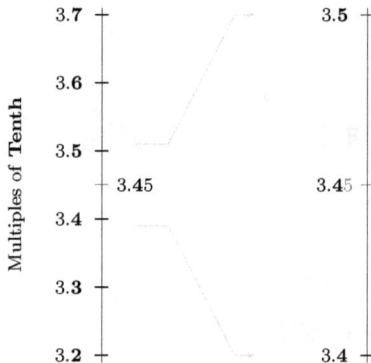

Figure 6.1.36: Rounding 3.45 to the nearest
tenth: Which multiple of 0.1 is 3.45 closest to?

Unless otherwise stated, we round all halfway values up to the next whole
multiple.[28] The answer to the problem at hand, therefore, is 3.5; see below.

---

[28] In most context such an approach is acceptable – hence the reason why it is the de
facto approach to rounding halfway points. In situations when this approach is not suitable,
other approaches are used. Such special situations are not discussed in this textbook.

Step 1: Make a cut to the right of
the tenths place value

Step 2: Take a look at the
digit to the right of the cut

Step 3: Since the digit in Step 2
is equal to 5, add 1 to tenths
to move up to the next multiple
of tenth

$$3 \ . \ 4 \ \mid 5$$

$$+ \ 1$$

Step 4: Drop the digits on the
right side of the cut

$$3 \ . \ 5$$

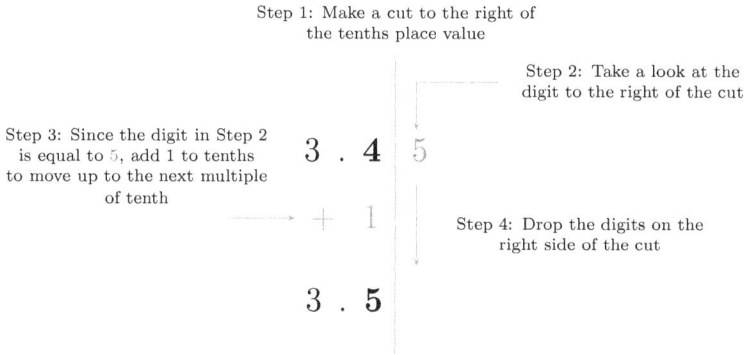

Figure 6.1.37: An efficient technique for
rounding 3.45 to the nearest tenth

We now turn our attention to three special cases. To illustrate the first,
suppose we are asked to round 0.03 to the nearest tenth. To do so, we must
find the whole multiple of 0.1 that is closest to 0.03. The following diagram
illustrates the idea.

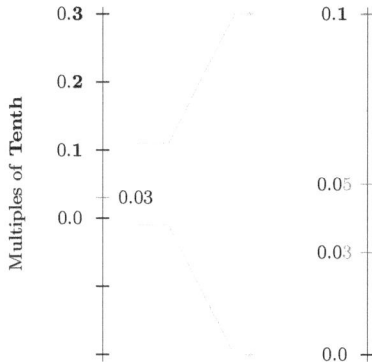

Figure 6.1.38: Rounding 0.03 to the nearest
tenth: Which multiple of 0.1 is 0.03 closest to?

As the diagram on the left in Figure 6.1.38 clearly shows, the whole multiple of tenth closest to 0.03 is 0.0. The answer, therefore, is 0.0.[29] As Figure 6.1.39 below shows, the faster technique works in this case as well.

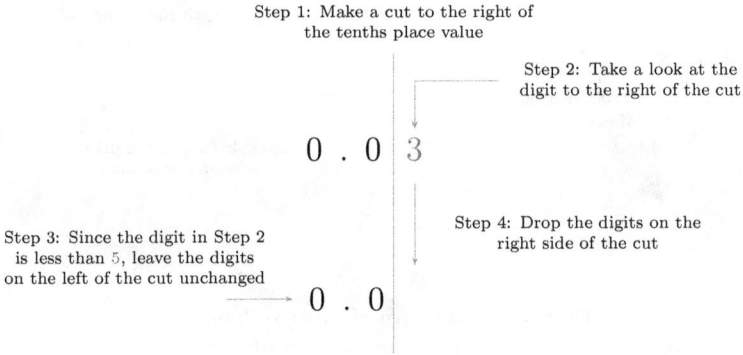

Step 1: Make a cut to the right of
the tenths place value

Step 2: Take a look at the
digit to the right of the cut

$$0 \,.\, 0 \,\big|\, 3$$

Step 3: Since the digit in Step 2
is less than 5, leave the digits
on the left of the cut unchanged

Step 4: Drop the digits on the
right side of the cut

$$0 \,.\, 0$$

Figure 6.1.39: An efficient technique for rounding 0.03 to the nearest tenth

To illustrate the second special case, suppose we are asked to round 34.96 to the nearest tenth. As Figure 6.1.40 shows, the answer is 35.0.

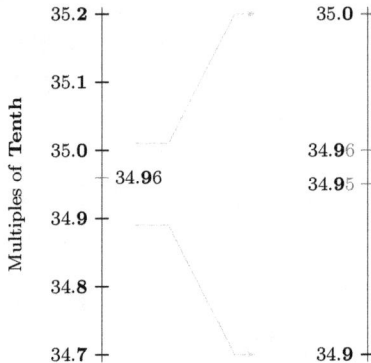

Figure 6.1.40: Rounding 34.96 to the nearest **tenth**: Which multiple of 0.1 is 34.96 closest to?

---

[29]The 0 in the tenths must be shown to make it clear that rounding was done to the nearest tenth, not the nearest unit.

The following diagram shows that the faster technique introduced earlier also works in this situation. Note the use of addition with carrying.

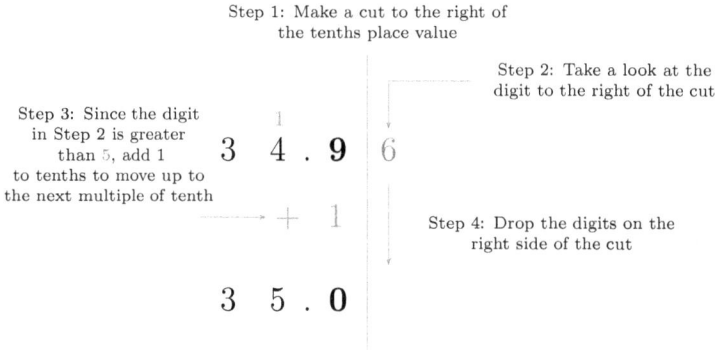

Step 1: Make a cut to the right of
the tenths place value

Step 2: Take a look at the
digit to the right of the cut

Step 3: Since the digit
in Step 2 is greater
than 5, add 1
to tenths to move up to
the next multiple of tenth

1

3   4 . 9   6

+   1

Step 4: Drop the digits on the
right side of the cut

3   5 . 0

Figure 6.1.41: An efficient technique for
rounding 34.96 to the nearest tenth

To illustrate the third special case, suppose we are asked to round 4.7 to the nearest tenth. As Figure 6.1.42 shows, the answer is 4.7.

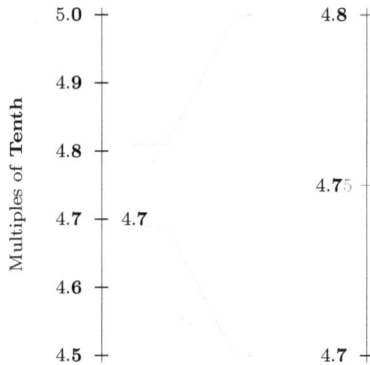

5.0 ┤   4.8 ┤

4.9 ┤

Multiples of Tenth

4.8 ┤

                4.75 ┤

4.7 ┤ 4.7

4.6 ┤

4.5 ┤   4.7 ┤

Figure 6.1.42: Rounding 4.7 to the nearest
tenth: Which multiple of 0.1 is 4.7 closest to?

The following diagram shows that the faster technique introduced earlier also works in this situation.

Step 1: Make a cut to the right of
the tenths place value

Step 2: Take a look at the
digit to the right of the cut
(which is 0)

4 . 7

Step 3: Since the digit in Step 2
is less than 5, leave the digits
on the left of the cut unchanged

Step 4: "Drop" the digits on the
right side of the cut

4 . 7

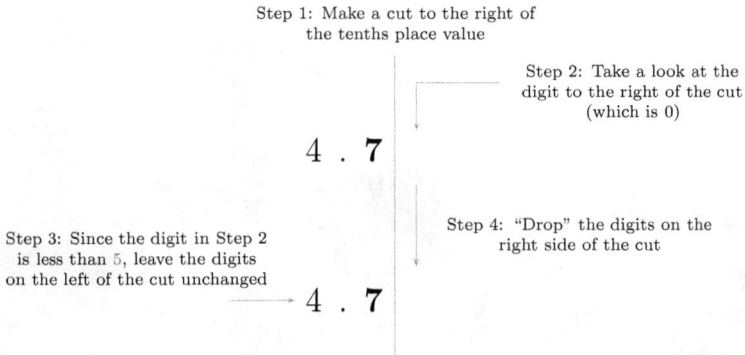

Figure 6.1.43: An efficient technique for
rounding 4.7 to the nearest tenth

Examples

Round each of the following to the nearest tenth.

1. 7.18   3. 4.251   5. 19
2. 22.419   4. 19.0   6. 1.8

Answers

1. 7.2   3. 4.3   5. 19.0
2. 22.4   4. 19.0   6. 1.8

Rounding to the Nearest Hundredth

When we ask you to round a decimal number to the nearest hundredth, it means find the whole multiple of 0.01 that is closest to the given number. The whole multiples of 0.01 are

0, 0.01, 0.02, 0.03, 0.04, 0.05, 0.06, 0.07, ...

As an example, suppose we ask you to round 0.0627 to the nearest hundredth. To do so, we must find the whole multiple of one hundredth that is closest to

0.0627. As the diagram on the left in Figure 6.1.44 shows, the closest whole multiple of one hundredth to 0.0627 is 0.06.

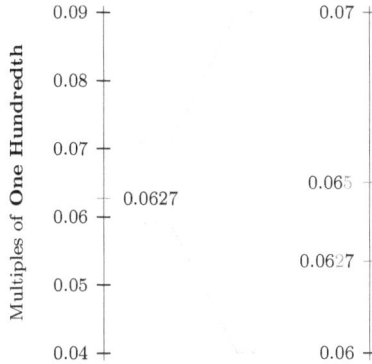

Figure 6.1.44: Rounding 0.0627 to the nearest **hundredth**: Which multiple of 0.01 is 0.0627 closest to?

The faster technique for rounding 0.0627 to the nearest hundredth is shown in 6.1.44. Note that in this case the midpoint between 0.06 and 0.07 is 0.065.

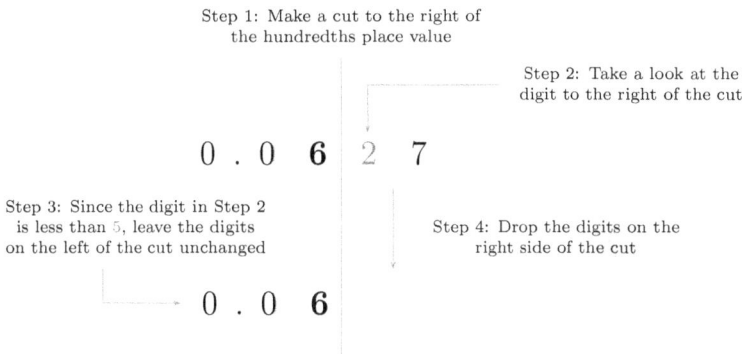

Figure 6.1.45: An efficient technique for rounding 0.0627 to the nearest hundredth

Note that what matters is digit 2. Digit 7 in the thousandths will make no difference as the reader can readily tell that 0.062, 0.0621, 0.0622, 0.0623, ..., 0.0629 are all less than 0.065 (the halfway point).

As a second example, suppose we ask you to round 4.1793 to the nearest hundredth. To do so, we must find the whole multiple of one hundredth that is closest to 4.1793. As the diagram on the left in Figure 6.1.46 shows, the closest whole multiple of one hundredth to 4.1793 is 4.18.

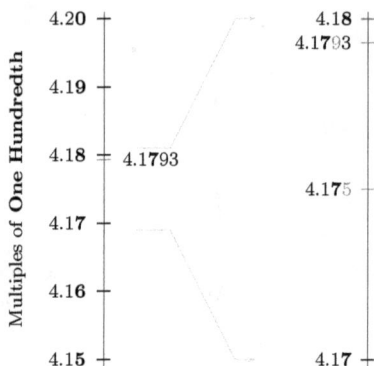

Figure 6.1.46: Rounding 4.1793 to the nearest **hundredth:** Which multiple of 0.01 is 4.1793 closest to?

The faster technique for rounding 4.1793 to the nearest hundredth is shown in 6.1.47. Note that in this case the midpoint between 4.17 and 4.18 is 4.175.

Consider now the process of rounding 5.0347 to the nearest hundredth. According to the technique that we have presented, the answer is 5.03.

A question arises as to whether one should instead first round 5.0347 to the nearest thousandth to arrive at 5.035, and then round 5.035 to the nearest hundredth to arrive at 5.04.

To answer this question, we draw the reader's attention to Figure 6.1.48 and the semantics of rounding to the a given place value. Based on the semantics of rounding to the nearest hundredth, we understand that, to round 5.0347 to the nearest hundredth, we must find the whole multiple of 0.01 that is closest to 5.0347. As one can see in the diagram on the left side of Figure 6.1.48, the closest whole multiple of 0.01 to 5.0347 is, indeed, 5.03.

Step 1: Make a cut to the right of
the hundredths place value

Step 2: Take a look at the
digit to the right of the cut

Step 3: Since the digit
in Step 2 is greater than
5, add 1 to hundredths
to move up to the next
multiple of one hundredth

4 . 1 7 9 3

Step 4: Drop the digits on the
right side of the cut

+ 1

4 . 1 8

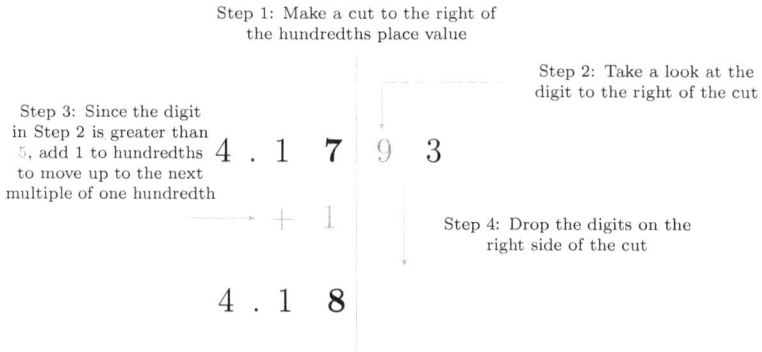

Figure 6.1.47: An efficient technique for
rounding 4.1793 to the nearest hundredth

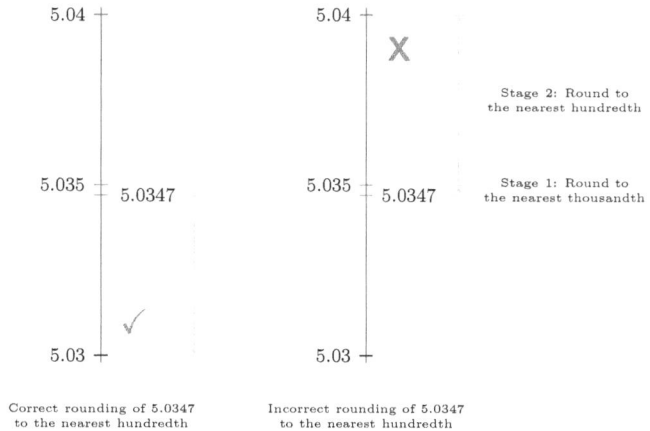

5.04 ┼

5.04 ┼   ✗

Stage 2: Round to
the nearest hundredth

5.035 ┼ 5.0347

5.035 ┼ 5.0347

Stage 1: Round to
the nearest thousandth

✓

5.03 ┼

5.03 ┼

Correct rounding of 5.0347
to the nearest hundredth

Incorrect rounding of 5.0347
to the nearest hundredth

Figure 6.1.48: Correct and incorrect roundings
of 5.0347 to the nearest hundredth

The diagram on the right side of Figure 6.1.48 shows that, when we round in stages (in rounding first to the thousandths and then to hundredths), in effect we push 5.0347 up to the halfway point (i.e., 5.035), and then push this up again to 5.04. Clearly this is not what we intend to do when we round 5.0347 to the nearest hundred.

Examples

Round each of the following to the nearest hundredth.

1. 0.0781	3. 2.500	5. 9.998
2. 42.1749	4. 35	6. 0.004

Answers

1. 0.08	3. 2.50	5. 10.00
2. 42.17	4. 35.00	6. 0.00

Rounding to Lower Place Values

When we ask you to round a decimal number to the nearest thousandth, it means find the whole multiple of 0.001 that is closest to the given number. The whole multiples of 0.001 are

0, 0.001, 0.002, 0.003, 0.004, 0.005, 0.006, 0.007, ...

And when we ask you to round a decimal number to the nearest ten thousandth, it means find the whole multiple of 0.0001 that is closest to the given number. The whole multiples of 0.0001 are

0, 0.0001, 0.0002, 0.0003, 0.0004, 0.0005, 0.0006, ...

Rounding to the nearest hundred thousandth, millionth, ten millionth, etc. have similar meanings.

The usefulness of the fast technique that we presented earlier in our discussion is not just in the fact that it can be used under all circumstances and special cases, but that it can be extended for use in rounding to any place value. The manner in which we round a decimal to lower place values is similar to the manner in which we round a decimal to the nearest tenth or hundredth. It is only the location of the cut in Step 1 that is different. We will present a few examples, leaving it to the reader to draw the relevant diagrams.

Examples

Round to the indicated place value.

1. 0.0247 (thousandths)
2. 0.009 51 (thousandths)
3. 0.001 52 (ten thousandths)
4. 0.000 19 (thousandths)
5. 42.199 32 (thousandths)
6. 0.003 819 (hundred thousandths)

Answers

1. 0.025	3. 0.0015	5. 42.199
2. 0.010	4. 0.000	6. 0.003 82

Rounding to a Place Value on the Whole Side

To round a number written using the decimal notation to a place value on the whole side we can follow the same steps given above. However, the digits in the rounded value will have to be written up to and including the digits in the ones place.

As an example, suppose we ask you to round 142.68 to the nearest ten. To do so, we must find the whole multiple of ten that is closest to 142.68. As the diagram on the left in Figure 6.1.49 shows, the closest whole multiple of ten to 142.68 is 140.

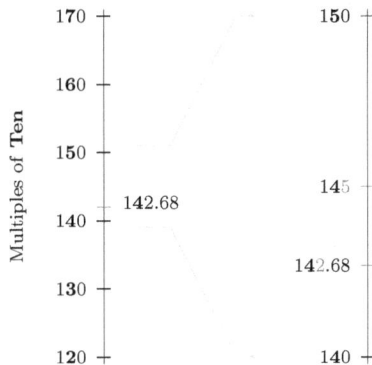

Figure 6.1.49: Rounding 142.68 to the nearest **ten**: Which multiple of 10 is 142.68 closest to?

The faster technique for rounding 142.68 to the nearest ten is shown in Figure 6.1.50.

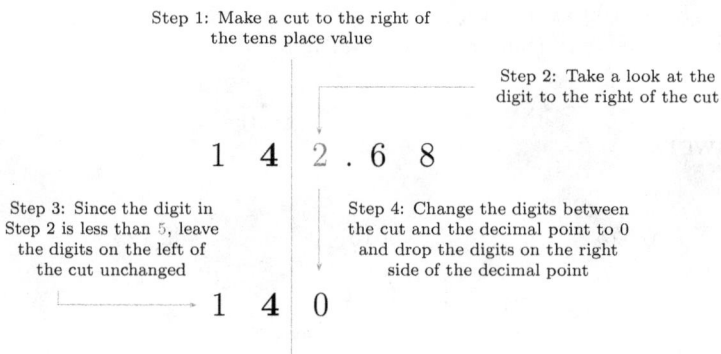

Step 1: Make a cut to the right of
the tens place value

Step 2: Take a look at the
digit to the right of the cut

$$1 \quad 4 \mid 2 \; . \; 6 \quad 8$$

Step 3: Since the digit in
Step 2 is less than 5, leave
the digits on the left of
the cut unchanged

Step 4: Change the digits between
the cut and the decimal point to 0
and drop the digits on the right
side of the decimal point

$$1 \quad 4 \mid 0$$

Figure 6.1.50: An efficient technique for rounding 142.68 to the nearest ten

Examples

Round to the specified place value.

1. 543.2 (tens)
2. 143.9254 (tens)
3. 14 239.8 (thousands)
4. 78 199.9 (tens)
5. 1 951 (hundreds)
6. 112 000 (ten thousands)

Answers

1. 540
2. 140
3. 14 000
4. 78 200
5. 2 000
6. 110 000

The algorithm for rounding a decimal number to a given place value is given below.

Algorithm for Rounding Numbers Written in Decimal Notation

To round a number written in decimal notation to a given place value

1. Make a cut on the right side of the place value that you want to round to.[30]
2. Look at the digit on the right side of the cut. If this digit is greater than or equal to 5, add 1 to the place value to which you want to round to.
3. a. If you are rounding to a place value on the whole side, change all the digits on the right side of the cut to the decimal point to 0, drop the decimal point and drop any digits on the right side of the decimal point.
   b. If you are rounding to a place value on the fractional side, write the decimal digits up to the cut[31] and drop all others.

Exercise Set 6.1.6

1. Round each of the following to the nearest ten.

   a. 46   d. 99   g. 315   j. 3296
   b. 32   e. 320   h. 811   k. 8
   c. 87   f. 428   i. 2492   l. 2

2. Round each of the following to the nearest hundred.

   a. 142   d. 2999   g. 4519   j. 320
   b. 425   e. 4100   h. 7370   k. 8
   c. 3591   f. 3251   i. 12 128   l. 51

3. Round each of the following to the nearest thousand.

   a. 3140   d. 9502   g. 20   j. 4201
   b. 8719   e. 341   h. 3000   k. 471 103
   c. 9199   f. 43 981 i. 14 718   l. 99 999

---

[30] In some cases, as in *Round* 2.4 *to the nearest thousandth*, it may be the case that no digit in the place value that you are asked to round to is given. In such cases you should place 0s in all place values up to and including the one you are asked to round to as well as one on its right and then continue with the algorithm.

[31] These are the digits up to and including the one in the place value that you are rounding to.

4. Round each of the following to the nearest ten thousand.

   a. 30 142   e. 324 192   i. 42 129
   b. 916 921   f. 735 182   j. 10 999
   c. 808 210   g. 9 310   k. 82 130 200
   d. 3 197 010   h. 129   l. 20 000

5. Round to the indicated place value.

   a. 319 (tens)   g. 71 819 (hundreds)
   b. 42 610 (thousands)   h. 319 128 (hundred thousands)
   c. 1920 (tens)   i. 987 291 (hundred thousands)
   d. 1920 (hundreds)   j. 4999 (thousands)
   e. 1920 (thousands)   k. 132 891 989 (ten millions)
   f. 1920 (ten thousands)   l. 99 200 (millions)

6. Round each of the following to the nearest ten.

   a. −46   d. −99   g. −315   j. −3296
   b. 32   e. +320   h. 811   k. 8
   c. −87   f. +428   i. 2492   l. +2

7. Round each of the following to the nearest hundred.

   a. 142   d. −2999   g. −4519   j. +320
   b. −425   e. −4100   h. 7370   k. +8
   c. 3591   f. 3251   i. −12 128   l. −51

8. Round each of the following to the nearest thousand.

   a. −3140   d. 9502   g. −20   j. −4201
   b. −8719   e. −341   h. 3000   k. −471 103
   c. −9199   f. +43 981   i. −14 718   l. −99 999

9. Round each of the following to the nearest ten thousand.

   a. −30 142   e. −324 192   i. 42 129
   b. −916 921   f. −735 182   j. −10 999
   c. 808 210   g. 9 310   k. −82 130 200
   d. −3 197 010   h. −129   l. −20 000

10. Round to the indicated place value.

a. -319 (tens)
b. $+42\,610$ (thousands)
c. -1920 (tens)
d. 1920 (tens)
e. -1920 (hundreds)
f. 1920 (hundreds)
g. -1920 (thousands)
h. 1920 (thousands)

i. -1920 (ten thousands)
j. 1920 (ten thousands)
k. $71\,819$ (hundreds)
l. $-319\,128$ (hundred thousands)
m. $-987\,291$ (hundred thousands)
n. -4999 (thousands)
o. $-132\,891\,989$ (ten millions)
p. $-99\,200$ (millions)

11. Round each of the following to the nearest unit.

a. 35.8
b. 35.1
c. 35.5

d. 35.0
e. 472.9
f. 1.5

g. 0.049
h. 9.816
i. 2.49

j. 7.1
k. 2.51
l. 199.51

12. Round each of the following to the nearest tenth.

a. 0.37
b. 0.35
c. 0.34

d. 0.30
e. 0.78
f. 0.95

g. 2.86
h. 47.561
i. 32.445

j. 1.532
k. 9.99
l. 62

13. Round each of the following to the nearest hundredth.

a. 3.1415
b. 270.8924
c. 0.08

d. 0.052
e. 0.1
f. 20

g. 0.037
h. 0.195
i. 0.03481

j. 0.3245
k. 18.2
l. 0.0006

14. Round each of the following to the nearest thousandth.

a. 0.2451
b. 0.2459
c. 0.2450

d. 0.245
e. 0.24549
f. 0.0003

g. 0.32
h. 42.8199
i. 5.462

j. 270
k. 1.8264
l. 2.435 09

15. Round each of the following to the nearest ten thousandth.

a. 0.004 35
b. 0.004 37
c. 0.004 33
d. 0.004 30

e. 0.0043
f. 0.329 952
g. 3.141 592 65
h. 0.04

i. 0.008
j. 0.000 001
k. 0.000 049
l. 0.000 051

16. Round each of the following to the nearest ten.

   a. 25.6   c. 142.8   e. 99.941   g. 49.99
   b. 32.81   d. 0.099   f. 100.00   h. 52.95

17. Round each of the following to the nearest hundred.

   a. 314.2   b. 1297.42   c. 7.82   d. 1290.371

18. Round each of the following to the nearest thousand.

   a. 5743.7   b. 3099.99   c. 55 091.7   d. 2529.69

19. Round to the indicated place value.

   a. 45.68 (tenths)   h. 1920 (thousandths)
   b. 32.4 (hundreds)   i. 1920 (ten thousandths)
   c. 42 610 (thousandths)   j. 126.737 (hundreds)
   d. 1920 (tenths)   k. 0.641 7 (hundredths)
   e. 241.9 (units)   l. 72.8199 (thousandths)
   f. 1920 (hundredths)   m. 0.006 375 (ten thousandths)
   g. 24.3 (tens)   n. 0.028 26 (thousandths)

6.1.7   Graphing

The number 10 features prominently in the decimal notation. This is not
unexpected: The decimal notation, as its name implies, was designed with
the number 10 in mind. Note the addition properties of decimals: Each time
we reach 10 in a column, we add a 1 over to the place value on the left or the
subtraction properties of decimals: When we borrow we pass on a 10 to the
place value on the right. Note how easy it is to multiply or divide a number
written using the decimal notation by 10 by simply moving the decimal point
back and forth. Or how each place value is 10 times larger than the one on
its right and 10 times smaller than the one on its left.

These advantages of the decimal notation[32] are so attractive that the International System of Units, SI, chose to relate its units of measure after 10,
its powers and its sub-powers. As an example of this consider the measuring
stick shown in Figure 6.1.51.

---

[32] As well as others: The special status of 10 as base for humans; the ease with which
they can be rounded, etc.

Figure 6.1.51: Tick labels correspond to the
digit in the ones place

We associate the tick labels to the digit in the ones place, e.g., if we choose to write either 1 or 2 down, then the 1 or the 2 should be written in the ones place.

We now divide the distance between the two ticks into 10 parts. Each tick corresponding to the smaller subdivisions can be associated with a digit in the tenths place. Figure 6.1.52 illustrates the idea.

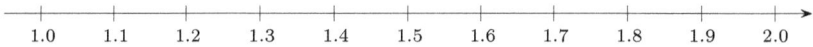

Figure 6.1.52: Tick labels correspond to the
digit in the tenths place

We now take each of the smaller subdivisions and divide those into 10 parts each. These smallest ticks are associated with the digit in the hundredths place as shown in Fig 6.1.53 below.

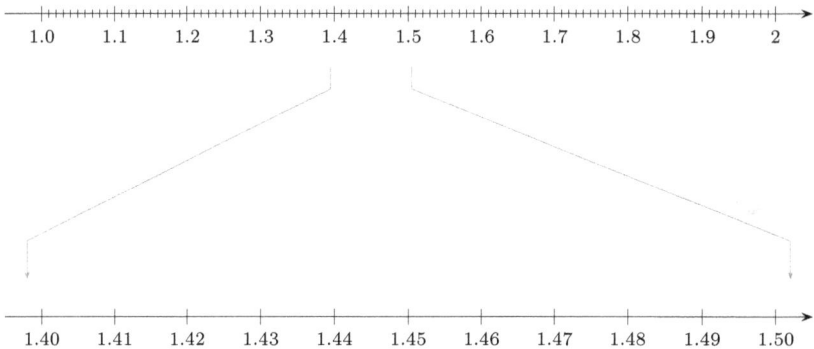

Figure 6.1.53: Tick labels correspond to the
digit in the hundredths place

Such subdivision of an interval into 10 parts continues indefinitely with each new subdivision pushing into the next place value. The repeated subdivision of an interval also helps us estimate the coordinate of a given point. As an example, given the graph below

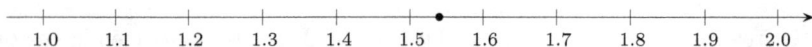

Figure 6.1.54: Estimating the coordinate of
the point as 1.54

we can associate the decimal number 1.54 to the dot. This is done by reading the digits up to and including those on the smallest subdivision and then estimating one more digit by dividing the interval from 1.5 to 1.6 into 10 parts in your mind's eye and estimating the tick that would fall on or near the given point. As we see, then, in reporting a measured value using this type of scale, we can always provide one digit more than those given for the smallest subdivision on the measuring instrument.

Equally, it is important to understand that if a measured value, such as 1.54 above, is given, that the rightmost digit (here, 4 which is in the hundredths) is estimated and that the ruler's smallest divisions increase in steps given by the place value on the left of this (here, the place value to the left of the hundredths which is tenths). A measured value, then, provides us with two pieces of information: One is the smallest subdivision on the ruler used and the estimated digit that further refines the position of the dot.

We will return to these topics in the section on measurement.

Exercise Set 6.1.7

1. In each case estimate the coordinate of the point on the real line as a decimal.

   a.

   b.

c.

d.

e.

f.

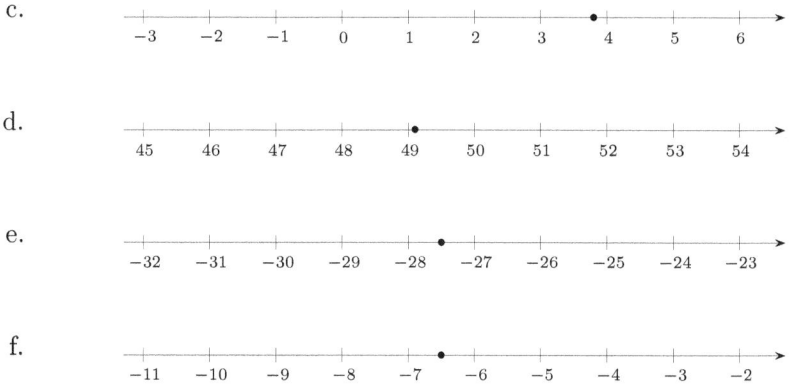

2. In each case estimate the coordinate of the point on the real line as a decimal.

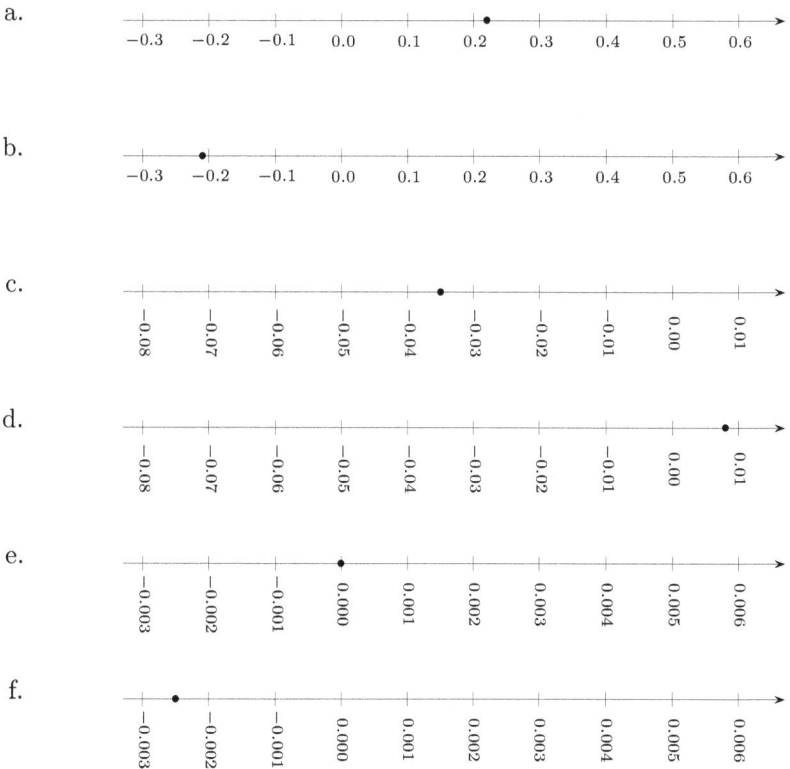

a.

b.

c.

d.

e.

f.

g.

h.

i.

3. Choose an appropriate step size and graph each of the following on the real line.

a. 4.2	d. −34.5	g. −0.42
b. 3.16	e. 3.24	h. 5.6
c. 0.45	f. 0.015	i. −14.0

4. How would you graph 4.263174 on a real line with a step size of 1?
5. How would you graph 173.28 on a real line with a step size of 10?
6. How would you graph 7.298 on a real line with a step size of 0.1?
7. How would you graph 0.0097 on a real line with a step size of 0.01?

6.2   Scientific Notation

6.2.1   Structure

In the sciences we often find that we need to work with certain numbers whose features make the decimal notation a cumbersome notation to use. Examples of such numbers are the particulate mass of electron which is approximately equal to

0.000 000 000 000 000 000 000 000 910 9 g/1

This is an extremely small number with a zero whole side and a long string of fractional-side 0s on the left side of nonzero digits. Another example is Avogadro's number, the number of entities in 1 mole of a substance, which is approximately equal to

602 200 000 000 000 000 000 000 1/mol

This is an extremely large number with a zero fractional side and a long string of whole-side 0s on the right side of nonzero digits. We can calculate the molar mass of electron by multiplying the two numbers above but, in practice, especially if one needs to work with such numbers on a frequent basis, writing all these 0s, counting them and accounting for each can try the patience of the practitioner. Moreover, it is not immediately apparent just how small or how large these numbers are and how such numbers compare in terms of size as they have many 0s. What we need is a notation that makes it easier to work with such numbers and provides an almost immediate feel for their sizes, either in isolation or in comparison.

Scientific notation was invented to make it easier to work with very small numbers with a zero whole side and a string of fractional side 0s on the left side of nonzero digits, as well as very large numbers with a zero fractional side and a string of whole-side 0s on the right side of nonzero digits. We would like to stress that scientific notation was designed specifically to deal with such long strings of 0s. As we will see later, many, though not all, of scientific notation's advantages disappear if such long strings of 0s are not present as in

0.310 241 408 000 210 389 45

or

427 109 290 210 425 670 109

However, since the vast majority of the very small or very large numbers that we may need to work with in the sciences are rounded to a few significant digits,[33] the long strings of 0s *are* present in these very small or very large numbers and, therefore, scientific notation can handle them well.[34]

In addition to the very small and the very large numbers, we do need to work with numbers that are not so large or so small in size such as 42.5, −13.8 and the like. While scientific notation was not designed to work with such numbers, its use to represent them does not pose any issues.[35]

With the introduction above on the practical needs that led to the invention of scientific notation in mind, let us now discuss the notation itself.

A number is said to be written in scientific notation if its representation has the form of a decimal number whose size is between 1 and 10 (inclusive on

---

[33]Typically three to five.

[34]For reasons why very small or very large numbers are rounded to a few significant digits please see Chapter 9 where we discuss the manner in which one works with measured values.

[35]So long as very small numbers or very large numbers are also present otherwise it is easier to work with these not so very small or large numbers using the decimal notation.

1 but not 10) multiplied by a power or subpower of 10. Using mathematical notation, we write this form as

$$s \times 10^e, \quad s \in \mathbb{R} \ \wedge \ 1 \le |s| < 10 \ \wedge \ e \in \mathbb{I}$$

with s written as a decimal.[36]

In the statement above, s is called **significand**[37] and e is the exponent of 10. The side conditions

$$s \in \mathbb{R} \ \wedge \ 1 \le |s| < 10$$

state that s is a real number whose size is between 1 and 10 (inclusive on 1 but not 10). As an example, the numbers 1.62, 9.99, 3, −2.4 and −9 are all acceptable as significands (they are written using decimal notation and their sizes are between 1 and 10, inclusive on 1 but not 10) whereas the numbers −10.5, 10, 193, 0.6, −0.002 and −42 are not acceptable as significands (their sizes are either less than 1 as in the numbers 0.6 and −0.002 in the list above, or greater than or equal to 10 as in the numbers −10.5, 10, 193 and −42 in the list above).[38]

The statement

$$e \in \mathbb{I}$$

states that the exponent must be an integer, i.e., it must belong to the set $\{\ldots, -3, -2, -1, 0, 1, 2, 3, \ldots\}$. As we will soon see this requirement is based on the need to keep track of the location of the decimal point which can be moved back and forth over an integral number of digits and, therefore, the set of integers is the set the exponent naturally belongs to.

Examples

The following numbers are written in scientific notation:

$$4.2 \times 10^3$$

$$1 \times 10^{-2}$$

$$-2.8 \times 10^0$$

---

[36]This is a convention not a requirement but without it many of the advantages of scientific notation will be lost. In practice deviations from this convention are rare.

[37]Also called **mantissa**.

[38]As we will soon see, the requirement that the size of the significand must be between 1 and 10 (inclusive on 1 but not 10) enforces uniformity and makes it easier to compare the sizes of such numbers.

but the following are not:

0.25×10^2 (significand is less than 1)

14.9×10^{-6} (significand is greater than 10)

$\dfrac{2}{3} \times 10^{-4}$ (significand not in decimal format)

2.5×3^2 (base must be 10)

$3.29 \times 10^{1.2}$ (exponent not an integer)

Examples

Which of the following numbers are written in scientific notation?

1. 4.53×10^5
2. 32×10^{-4}
3. 1×10^{-2}
4. $9.82 \times 10^{-2.5}$
5. 0.23×10^4
6. 5×10^0

Answers

1. The number is written in scientific notation.
2. The number is not written in scientific notation. Significand is greater than 10.
3. The number is written in scientific notation.
4. The number is not written in scientific notation. Exponent is not an integer.
5. The number is not written in scientific notation. Significand is less than 1.
6. The number is written in scientific notation.

The best way to understand scientific notation is to look at a few examples in detail. Consider

7.268×10^2

This number is written in scientific notation. We can rewrite this number in decimal notation if we carry out the operations that are listed in the notation 7.268×10^2. Since 10^2 is equal to 100, we can write

7.268×100

We can now carry out the multiplication by moving the decimal point in 7.268 over two digits to the right to get

726.8

So the decimal representation of 7.268×10^2 is 726.8. An exponent of 2 results in multiplication by 100 which has two 0s and this in turn implies that the decimal point should move to the right over two digits. This pattern is true for all positive exponents of 10. We can, therefore, conclude that *a positive exponent of* 10 *implies that the decimal point should move to the right over as many digits as the size of the exponent.*[39]

Note that the exponent is allowed to be negative or 0. We refer the reader to Appendix L on how to make sense of negative exponents. In this appendix we arrive at the understanding that a negative exponent of 10 implies repeated division by 10. So, while $\times 10^3$ implies $\times 10 \times 10 \times 10$ which is repeated multiplication by 10, $\times 10^{-3}$ means $\div 10 \div 10 \div 10$ which is repeated division by 10. This means that the sign of the exponent tells us whether we are multiplying by 10 or dividing by 10 and the size of the exponent tells us the number of times we need to perform the multiplications or the divisions. A number such as

$$5.21 \times 10^{-2}$$

then, implies that we should divide 5.21 by 10 once and then one more time, a chain of divisions that can be executed in one step if we divide 5.21 by 100. This moves the decimal point to the left over two digits arriving at 0.0521 as the decimal representation of the number. This means that *a negative exponent of* 10 *implies that the decimal point should move to the left over as many digits as the size of the exponent.*[40]

We refer the reader to Appendix K on how to make sense of a 0 exponent. In this appendix we show that 10^0 equals to 1 and, therefore, $\times 10^0$ implies $\times 1$ which does not move the decimal point. As an example, 3.99×10^0 is equal to 3.99. We see that *a* 0 *exponent leaves the location of the decimal point unchanged.*

The observations above shed light on the relationship between the movement of the decimal point and changes to the exponent of 10. To explain this relationship consider the number

$$42.947 \times 10^8$$

Note that, in the decimal representation of this number, the location of the decimal point is over 8 digits to the right, i.e., the number, written in decimal

---

[39] One might argue that it would be easier to just stick to 726.8, however, keep in mind that the full power of scientific notation comes to light when we work with very small or very large numbers. While 726.8 may be more conventient to work with than 7.268×10^2, it is easier to work with 6.022×10^{23} than 602 200 000 000 000 000 000 000.

[40] The reader should note once again that, while it may be easier to work with 0.0521 than 5.21×10^{-2}, the full force of scientific notation becomes apparent when we work with very small or very large numbers. While 0.0521 may be more convenient to work with than 5.21×10^{-2}, it is easier to work with 9.109×10^{-28} than 0.000 000 000 000 000 000 000 000 000 910 9.

notation, reads as

    4 294 700 000

Suppose now that, for whatever reason, we wish to move the decimal point to the right over two digits to get 4294.7. Now the decimal point needs to move over only 6 more digits to the right to get to its location in the decimal representation of the number. This means that the exponent of 10 should be changed to 6, i.e., we have

    4294.7×10^6

Indeed, we can say that every time the decimal point in the significand moves over one digit to the right, the exponent should decrease by 1. This makes sense if we note that each time the decimal point in the significand moves over one digit to the right, the size of the significand gets 10 times larger. A decrease in the exponent by 1, which implies one fewer multiplication by 10 or one more division by 10, compensates for this increase and keeps the overall size of the number the same.

Similarly, every time the decimal point in the significand moves over one digit to the left, the exponent should increase by 1. This makes sense if we note that each time the decimal point in the significand moves over one digit to the left, the size of the significand gets 10 times smaller. An increase in the exponent by 1, which implies one more multiplication by 10 or one fewer division by 10, compensates for this decrease and keeps the overall size of the number the same.

Examples

Find the missing significand or the missing exponent.

1. $5.8194 \times 10^6 = 5819.4 \times 10^?$
2. $192.425 \times 10^{-4} = 19.2425 \times 10^?$
3. $-4287.65 \times 10^{143} = -4.287\,65 \times 10^?$
4. $77.1526 \times 10^{-5} = ? \times 10^{-3}$
5. $-15.2 \times 10^{349} = ? \times 10^{346}$
6. $-2 \times 10^0 = ? \times 10^3$

Answers

1. $5.8194 \times 10^6 = 5819.4 \times 10^3$
2. $192.425 \times 10^{-4} = 19.2425 \times 10^{-3}$
3. $-4287.65 \times 10^{143} = -4.287\,65 \times 10^{146}$

4. $77.1526 \times 10^{-5} = 0.771\,526 \times 10^{-3}$
5. $-15.2 \times 10^{349} = -15\,200 \times 10^{346}$
6. $-2 \times 10^0 = -0.002 \times 10^3$

Exercise Set 6.2.1

1. Which of the following numbers are written in scientific notation?

   a. 3.89×10^6 c. -1.0×10^{-2} e. 0.017×10^0

   b. -132×10^{-2} d. $3.42 \times 10^{3.5}$ f. -8.9×10^0

2. a. What does a positive exponent of 10 in scientific notation imply?
   b. What does a 0 exponent of 10 in scientific notation imply?
   c. What does a negative exponent of 10 in scientific notation imply?

3. Find the missing exponent.

   a. $94.53 \times 10^5 = 945.3 \times 10^?$ f. $4.31 \times 10^{-4} = 431 \times 10^?$

   b. $4.923 \times 10^{12} = 492.3 \times 10^?$

   c. $356.2 \times 10^4 = 35620 \times 10^?$ g. $28.712 \times 10^{-7} = 0.028\,712 \times 10^?$

   d. $38.92 \times 10^7 = 3.892 \times 10^?$ h. $0.00236 \times 10^{-4} = 23.6 \times 10^?$

   e. $2.571 \times 10^2 = 0.002571 \times 10^?$ i. $278.912 \times 10^0 = 2.789\,12 \times 10^?$

4. Find the missing significand.

   a. $395.7 \times 10^8 = ? \times 10^5$ f. $987 \times 10^{-2} = ? \times 10^0$

   b. $62.196 \times 10^{19} = ? \times 10^{15}$ g. $190 \times 10^{-4} = ? \times 10^3$

   c. $0.052 \times 10^3 = ? \times 10^{-2}$ h. $0.09 \times 10^{-1} = ? \times 10^{-7}$

   d. $0.001 \times 10^{25} = ? \times 10^{26}$ i. $1 \times 10^0 = ? \times 10^{-1}$

   e. $31.98 \times 10^{-4} = ? \times 10^{-1}$

6.2.2   Formal Reading

To formally read a number written in scientific notation

1. If there is a sign present, use the word *positive* or *negative* as the case may be.
2. Read the single-digit whole part of the significand formally as a whole numebr.
3. If there is a decimal point, append the word *point* for the reading of the decimal point.

4. If there are fractional digits in the significand, read the fractional digits one at a time formally as a whole number.
5. Append the phrase *times ten to the power of.*
6. Read the exponent formally as an integer.

Following the guidelines above, we read 3.84×10^2 as *three point eight four times ten to the power of two,* -8×10^0 as *negative eight times ten to the power of zero,* and -1.0×10^{-2} as *negative one point zero times ten to the power of negative two.*

Exercise Set 6.2.2

1. Write each of the following formally.

   a. 4×10^3 d. 2.4×10^{62} g. -3×10^{21} j. -9.37×10^8
   b. 3×10^0 e. 3.81×10^0 h. -9×10^0 k. -7.0×10^0
   c. 5×10^{-2} f. 2.00×10^{-1} i. -5×10^{-11} l. -1.11×10^{-2}

2. Write the following in scientific notation.

   a. five times ten to the power of eight.
   b. three times ten to the power of zero.
   c. two times ten to the power of four.
   d. two point eight times ten to the power of twelve.
   e. four point two one times ten to the power of zero.
   f. three point zero times ten to the power of negative forty-five.
   g. negative three times ten to the power of fifteen.
   h. negative four times ten to the power of zero.
   i. negative five times ten to the power of negative twenty-seven.
   j. negative eight point zero zero times ten to the power of fifty.
   k. negative seven point one six times ten to the power of zero.
   l. negative six point nine two one times ten to the power of negative seven.

6.2.3   Order

We will begin with an algorithm on how to order numbers that are written in scientific notation. Following this, we will present a number of examples to illustrate the steps in the algorithm.

Algorithm for Ordering Numbers Written in Scientific Notation

To order numbers that are written in scientific notation

1. Compare the signs of the significands. Those with a negative significand are less than those with a positive significand.[41]

2. If the significands have the same sign, then compare the exponents of 10.

   a. If the significands are both positive, then the one with the larger exponent of 10 is the larger number.[42]
   b. If the significands are both negative, then the one with the larger exponent of 10 is the smaller number.[43]

3. If the exponents of 10 are the same, then the one with the larger significand is the larger number.

Example

Which is the larger number: -2.3×10^3 or 1.1×10^1?

Solution

Negative numbers are always less than positive numbers. Therefore, 1.1×10^1 is larger than -2.3×10^3.

---

[41] If the significand is positive, the number as a whole is positive. As an example, the numbers 4.2×10^2, 4.2×10^0, and 4.2×10^{-3} are all positive: They correspond to the numbers 420, 4.2, and 0.0042. If, on the other hand, the significand is negative, the number as a whole is negative. As an example, the numbers -4.2×10^2, -4.2×10^0, and -4.2×10^{-3} are all negative: They correspond to the numbers -420, -4.2, and -0.0042. Since negative numbers are always less than positive numbers, numbers with a negative significand are less than numbers with a positive significand.

[42] As an example, the number 1.2×10^3 is larger than the number 7.3×10^2: The former corresponds to 1200 while the latter corresponds to 730. As another example, the number 3.4×10^{-3} is smaller than the number 2.5×10^{-2}: The former corresponds to the number 0.0034 while the latter corresponds to the number 0.025

[43] As an example, the number -1.4×10^2 is larger than the number -7.3×10^3: The former corresponds to -140 while the latter corresponds to -7300. As another example, the number -6.2×10^{-1} is smaller than the number -2.8×10^{-2}: The former corresponds to the number -0.62 while the latter corresponds to the number -0.028.

Example

Which is the larger number: 4.2×10^3 or 2.7×10^4?

Solution

Both numbers are positive. Therefore, since the exponent of 10 in 2.7×10^4 is larger than the exponent of 10 in 4.2×10^3, the number 2.7×10^4 is larger than the number 4.2×10^3.

Example

Which is the larger number: 3.8×10^3 or 7.5×10^3?

Solution

Both numbers are positive and the exponents of 10 are the same. Therefore, since 7.5 is larger than 3.8, the number 7.5×10^3 is larger than the number 3.8×10^3.

Example

Which is the larger number: -4.2×10^3 or -2.7×10^4?

Solution

Both numbers are negative. Therefore, since the exponent of 10 in -4.2×10^3 is smaller than the exponent of 10 in -2.7×10^4, the number -4.2×10^3 is larger than the number -2.7×10^4.

Example

Which is the larger number: -3.8×10^3 or -7.5×10^3?

Solution

Both numbers are negative and the exponents of 10 are the same. Therefore, since -3.8 is larger than -7.5, the number -3.8×10^3

is larger than the number -7.5×10^3.

Example

Arrange 4.5×10^3, 2.4×10^{-1} and 5.7×10^2 in ascending order.

Solution

All numbers are positive. Therefore, we compare the exponents of 10. The smallest exponent in the set is -1, followed by 2 and then 3. This means that the number 2.4×10^{-1} is the smallest number in the set, followed by 5.7×10^2 and 4.5×10^3. This leads to the following order:

$$2.4 \times 10^{-1}, \ 5.7 \times 10^2, \ 4.5 \times 10^3$$

Example

Arrange -2.7×10^3, -1.6×10^{-2}, 1.7×10^2 and 4.8×10^{-1} in decreasing order.

Solution

Positive numbers are larger than negative numbers. Therefore, the numbers 1.7×10^2 and 4.8×10^{-1} are larger than the numbers -2.7×10^3 and -1.6×10^{-2}.

On the positive side, the number 1.7×10^2 is larger than the number 4.8×10^{-1} as the exponent of 10 in 1.7×10^2 is larger than the exponent of 10 in 4.8×10^{-1}.

On the negative side, the number -1.6×10^{-2} is larger than the number -2.7×10^3 as the exponent of 10 in -1.6×10^{-2} is smaller than the exponent of 10 in -2.7×10^3.

This leads to the following order:

$$1.7 \times 10^2, \ 4.8 \times 10^{-1}, \ -1.6 \times 10^{-2}, \ -2.7 \times 10^3$$

Example

Arrange 2.7×10^2, -5.8×10^2 and 7.7×10^2 in increasing order.

Solution

-5.8×10^2, 2.7×10^2, 7.7×10^2

Order of Magnitude

It is often the case that one wants to know the ratio of two numbers, i.e., how many times one of the numbers is larger or smaller than the other number. Scientific notation makes it easy to get an estimate for the answer to this question. The process requires that we subtract the smaller exponent of 10 from the larger exponent of 10; a difference that tell us whether one number is comparable in size to another, or whether it is 10 times larger than the other, or whether it is 100 times larger than the other, or whether it is 1000 times larger than the other and so on. In this context, each factor of 10 is referred to as an **order of magnitude**.

Given two numbers written in scientific notation, one is n orders of magnitude larger than another if the exponent of 10 in its representation is n units larger than the exponent of 10 in the representation of the other, i.e., if the difference between the larger exponent and the smaller exponent is n.

As an example, the number 1.5×10^5 is 2 orders of magnitude larger than the number 8.8×10^3 as the difference between their exponents, i.e., $5 - 3$, is 2. As another example, the number 7.1×10^{-21} is 3 orders of magnitude larger than the number 2.5×10^{-24} as the difference between their exponents, i.e., $-21 - (-24)$, is 3.[44]

---

[44]We emphasize that orders of magnitude provide a *rough* estimate for the ratio of two numbers. The exact ratio depends on the relative sizes of the significands as well. As an example,1.001×10^4 and 9.99×10^3 are almost equal: The first is 1001 while the second is 999 whereas subtraction of exponents would leave the impression that 1.001×10^4 is 10 times larger than 9.99×10^3. One can say that, if two numbers are n orders of magnitude apart, then their ratio will fall within the range 10^{n-1} and 10^{n+1}. As an exmple, the number 7.8×10^8 is three orders of magnitude larger than the number 2.4×10^5. This implies that their ratio falls between 100 and 10 000, i.e., one is anywhere between 100 to 10 000 times larger than the other. Only when the significands are equal does the difference in orders of magnitude on its own provide an exact value for the ratio of the two numbers involved.

Exercise Set 6.2.3

1. Which is the larger number in the given pair?

   a. 4.5×10^3 and -8.1×10^4 i. -1.7×10^4 and -2.2×10^3

   b. -3.8×10^{-2} and 4.1×10^{-8} j. -6.51×10^0 and -1.1×10^{-2}

   c. -1.7×10^0 and -1.1×10^1 k. -9.0×10^{-3} and -4.2×10^{-7}

   d. 3.8×10^5 and 9.1×10^3 l. -8.2×10^6 and -2.2×10^6

   e. 2.19×10^{-8} and 4.21×10^{-3} m. -1.4×10^0 and -3.8×10^0

   f. 4.99×10^0 and 3.21×10^1 n. -2.450×10^{-22} and -3.21×10^{-22}

   g. 7.2×10^7 and 2.9×10^7 o. -4.7×10^{42} and -1.2×10^{42}

   h. 6.6×10^{-8} and 9.7×10^{-8}

2. In each case list the given numbers in descending order.

   a. 3.7×10^2, 5.7×10^4, 1.8×10^3

   b. 2.4×10^8, 5.8×10^5, 3.3×10^0

   c. 1.8×10^0, 7.2×10^{-1}, 3.4×10^{-2}, 4.7×10^1

   d. 4×10^{-120}, 2×10^{-126}, 8×10^{-119}

   e. 4.2×10^8, 3×10^8, 1.99×10^8, 4.813×10^8

   f. -3.5×10^3, -2.8×10^4, -1.1×10^2

   g. -8×10^{-2}, -5.2×10^{-5}, -8×10^{-4}

3. In each case list the given numbers in ascending order.

   a. -3.81×10^{-8}, -4.2×10^{-8}, -1.9×10^{-8}

   b. -1×10^{14}, -4×10^{14}, -3×10^{14}

   c. -4.8×10^0, -2.71×10^0, -8.600×10^0

   d. -2.44×10^{-3}, -1.8×10^{-3}, -4.5×10^{-3}, -1×10^{-3}

   e. 4.7×10^{-1}, -2.7×10^2, 1.9×10^1, -3.3×10^3

   f. -5×10^3, -2.8×10^4, 3×10^5, -1.4×10^4

   g. 1.99×10^4, -3.8×10^2, 5.9×10^4, 1.9×10^4

4. Calculate the difference in the orders of magnitudes of the given numbers and interpret the result.

a. 3.4×10^9 and 2.4×10^4 e. 3.7×10^{-5} and 7.7×10^0

b. 8.1×10^3 and 4×10^5 f. 8×10^0 and 3.7×10^{-3}

c. 9×10^0 and 3.7×10^1 g. 4.44×10^{-4} and 8.0×10^4

d. 4.88×10^{-3} and 5.7×10^{-8} h. 9×10^{95} and 4.2×10^{-82}

6.2.4   Addition, Subtraction, Multiplication, and Division

We will begin this section with a discussion on the addition and subtraction of numbers written in scientific notation. This is followed by a discussion on the multiplication and division of numbers written in scientific notation.

Addition and Subtraction

To understand how we add and subtract numbers that are written using scientific notation, think back for a moment on how we add and subtract numbers that are written using the decimal notation. When we add and subtract numbers that are written using the decimal notation, we write the numbers one below the other, making sure that the decimal points in the two numbers line up. The reason we line up the decimal points is to make sure that corresponding place values in the two numbers line up which in turn ensures that we will be adding the hundredths to the hundredths, the tenths to the tenths, the ones to the ones and so on.

When we come to consider a number that is written in scientific notation, we notice that the place values in its significand are set by the exponent of 10 in its representation. As an example, in 4.25×10^2, the place value of 4 is not really the apparent *ones*, but *hundreds* as would be apparent if we switched to the decimal notation and wrote the number as 425. Note that the exponent of 10, being 2, pushes the place values to the right over two places, a move that amounts to a reassignment of place values relative to the decimal point. If the exponent was -2 instead, as in 4.25×10^{-2}, it would push the place values to the left over two places so that the 4 would be in the hundredths place as you can see from the decimal representation of the number, i.e., 0.0425.

Now, given two numbers written in scientific notation, so long as the exponents of 10 in their representations are the same, the place values in their significands will be shifted by equal amounts and in the same direction.

This means that, when adding and subtracting such numbers, we can still line up the decimal points in their significands, being certain that the place values line up properly. The exponent of 10 in the result of such calculations remains the same as it simply sets the place values relative to the decimal point.

Example

$$5.31 \times 10^2 + 3.2 \times 10^2 = 8.51 \times 10^2$$

Since the exponents of 10 are the same, place values in the significands are shifted by equal amounts. As an example, the 5 in 5.31 and the 3 in 3.2 are both in the hundreds. This means that if we line up the decimal points in the significands, the place values will be properly lined up. We therefore do so and add the significands to get 8.51. The exponent of 10 remains 2.

Example

$$-9.21 \times 10^{-3} + 5.2 \times 10^{-3} - 8.5 \times 10^{-3} = -12.51 \times 10^{-3}$$
$$= -1.251 \times 10^{-2}$$

The last step rewrites the result in scientific notation by bringing the significand within range and adjusting the exponent of 10.

When the exponents of 10 in the numbers being added or subtracted are different, the shifts in the place values in their significands are different. As an example, in

$$5.48 \times 10^3 + 2.1 \times 10^2$$

the 5 in 5.48 is in the thousands whereas the 2 in 2.1 is in the hundreds. Lining up the decimal points in the significands would end up lining up the thousands with the hundreds.

One way to deal with this problem is to rewrite the various numbers in the expression using a common exponent of 10. This can be done by rewriting the exponent of 10 in each of the numbers involved and then moving the decimal point in their significands accordingly. Once the exponents of 10 are made the same, we can line up the decimal points in the significands of these numbers and carry out the additions and subtractions. As for the choice of the common exponent, while any exponent (including one that is not present in any of the given numbers) can be used, we recommend using the largest exponent of 10 that is present in the given numbers. Often enough this choice results in less work at the end but, more importantly, it provides us with a fast technique for moving the decimal points in the significands in the right direction: Unless the exponent of 10 in an operand is already the same as the largest exponent present, the decimal point in its significand will have to move to the left to compensate for the increase in the value of the exponent of 10 in its representation.

Example

$$5.48 \times 10^3 + 2.1 \times 10^2 = 5.48 \times 10^3 + 0.21 \times 10^3$$
$$= 5.69 \times 10^3$$

In the example above, the largest exponent present is 3. We choose 3 as the common exponent. The first operand, 5.48×10^3, already has this common exponent in its power of 10. It will, therefore, remain unchanged. The second operand, 2.1×10^2, however, has an exponent of 2 on 10. We change this to 3 and since this raises the size of the number ten-fold, we move the decimal point back over 1 digits to compensate for the increase. This gives us 0.21×10^3. We can now line up the decimal points in the significands of these numbers and add them to get 5.69. The exponent of 10 remains 3.

Example

$$9.14 \times 10^{-2} + 3.8 \times 10^1 = 0.009\,14 \times 10^1 + 3.8 \times 10^1$$
$$= 3.809\,14 \times 10^1$$

Note in the example above that the largest exponent of 10 is 1, not -2.

The steps for the chain addition/subtraction of numbers written in scientific notation are given below.

Algorithm for Evaluation of Addition and Subtraction of Numbers Written In Scientific Notation

1. If the exponents of 10 in the given numbers are different, choose the largest exponent present as the common exponent and change all exponents in the given numbers to the common exponent. Move the decimal points in the significands to the left accordingly to compensate for changes in size caused by changes in the exponents.
2. Line up the decimal points in the significands and add and subtract as instructed. Leave the exponent of 10 in the final answer unchanged.

3. If the answer is not in scientific notation, rewrite the answer in scientific notation by moving the decimal point to bring the significand within range and adjusting the exponent accordingly.

This algorithm works well and is quite efficient unless the numbers involved differ greatly in size. As an example of the complexities that arise when we add such disparate numbers, consider the following sum:

$$5.97 \times 10^{24} + 9.109 \times 10^{-31}$$

Following the algorithm above we would choose 24 as the common exponent. This would leave the significand 5.97 the same but it would require that we move the decimal point in 9.109 to the left over fifty-five places which would result in a decimal point followed by fifty-four 0s and then the digits 9, 1, 0, and 9.

One can, of course, show patience and move the decimal point over as many digits as needed. However, practical situations in which we need to add or subtract numbers that differ greatly in size are rare and, if we must add such numbers, the sum will effectively be equal to the larger number, i.e., 5.97×10^{24} in the example above. To put some context to this sum, the number 5.97×10^{24} could be the mass of the Earth while the number 9.109×10^{-31} may be thought of as the mass of an electron, with both values quoted in kilograms. Do we really need this sum? If so, would it not be reasonable to say that the sum of the two masses is effectively equal to the mass of the Earth, i.e., 5.97×10^{24} kg?[45]

In practical situations where the sum or difference of numbers written in scientific notation are sought, the exponents of 10 are typically within one to five units of each other and in these scenarios the number of places over which the decimal points in the significands need to move over will be small.

Example

$$6.21 \times 10^2 + 1.5 \times 10^4 - 2.2 \times 10^3$$
$$= 0.0621 \times 10^4 + 1.5 \times 10^4 - 0.22 \times 10^4$$
$$= 1.3421 \times 10^4$$

---

[45] As we will see in the chapter on working with measured values, application of the rounding rules that will be introduced in that chapter to the sum above will end with 5.97×10^{24} as the result.

Example

$$4.2 \times 10^3 + 4.5 \times 10^3 = 8.7 \times 10^3$$

Example

$$8.2 \times 10^{-4} + 5.6 \times 10^{-4} - 1.1 \times 10^{-4}$$
$$= 12.7 \times 10^{-4}$$
$$= 1.27 \times 10^{-3}$$

Example

$$3.45 \times 10^4 + 2.1 \times 10^2 = 3.45 \times 10^4 + 0.021 \times 10^4$$
$$= 3.471 \times 10^4$$

Example

$$9.22 \times 10^3 + 3.8 \times 10^2 + 6.78 \times 10^{-1}$$
$$= 9.22 \times 10^3 + 0.38 \times 10^3 + 0.000\,678 \times 10^3$$
$$= 9.600\,678 \times 10^3$$

Multiplication and Division

Consider the multiplication

$$2.1 \times 10^3 \times 3.4 \times 10^2$$

Multiplication can be performed in any order so we can choose to rewrite the above expression by listing the significands first, followed by the powers of 10. This is shown below.

$$2.1 \times 3.4 \times 10^3 \times 10^2$$

We now multiply the significands to get 7.14 and we simplify the subexpression involving the powers of 10 by adding their exponents.[46] This yields

---

[46]The reader who is not familiar with exponent rules should read Appendix M before continuing further. In what follows, we will assume this background on the part of the reader.

7.14×10^5 as the answer. The line that shows the rearranged expression is often skipped[47] so that the full solution may be communicated as follows.

$$2.1 \times 10^3 \times 3.4 \times 10^2 = 7.14 \times 10^{3+2}$$
$$= 7.14 \times 10^5$$

Multiple operands may be dealt with at the same time and if negative exponents are present, they are subtracted from the subexpression in the exponent of 10 that precedes them.

Example

$$3.8 \times 10^{21} \times 5.1 \times 10^{40} \times 1.3 \times 10^{34}$$
$$= 25.194 \times 10^{21+40+34}$$
$$= 25.194 \times 10^{95}$$
$$= 2.5194 \times 10^{96}$$

Example

$$4.3 \times 10^5 \times 1.9 \times 10^{-3} = 8.17 \times 10^{5-3}$$
$$= 8.17 \times 10^2$$

Example

$$5.6 \times 10^{-7} \times 2.8 \times 10^{-2} = 15.68 \times 10^{-7-2}$$
$$= 15.68 \times 10^{-9}$$
$$= 1.568 \times 10^{-8}$$

Example

$$-3.9 \times 10^{23} \left(1.7 \times 10^{-18}\right) \left(-2.1 \times 10^{16}\right) \left(-7.2 \times 10^{-10}\right)$$
$$= -100.2456 \times 10^{23-18+16-10}$$
$$= -100.2456 \times 10^{11}$$
$$= -1.002\,456 \times 10^{13}$$

---

[47] And for good reason as otherwise the process would become cumbersome if it involved multiple operands.

As with multiplication, we can break an expression involving the division of two numbers written in scientific notation into a product of two factors. The first factor divides the significands while the second factor divides the expressions that involve powers of 10. As an example, we can rewrite the expression

$$\frac{3.2 \times 10^6}{2.0 \times 10^1}$$

as

$$\frac{3.2}{2.0} \times \frac{10^6}{10^1}$$

The first factor yields 1.6 and the second factor simplifies to 10^{6-1}.[48] This gives us

$$1.6 \times 10^5$$

as the answer.

Example

$$\frac{4.8 \times 10^5}{3.2 \times 10^3} = 1.5 \times 10^{5-3}$$
$$= 1.5 \times 10^2$$

Example

$$\frac{56 \times 10^3}{8 \times 10^{-2}} = 7 \times 10^{3+2}$$
$$= 7 \times 10^5$$

Example

$$\frac{4.5 \times 10^8 \times 4.8 \times 10^{-3}}{-1.5 \times 10^{-4} \times 1.2 \times 10^5} = -12 \times 10^{8-3+4-5}$$
$$= -12 \times 10^4$$
$$= -1.2 \times 10^5$$

The steps for the chain multiplication/division of numbers written in scientific notation are given below.

---

[48]See Appendix M for justification.

Algorithm for Evaluation of Multiplication and Division of Numbers Written In Scientific Notation

1. Multiply and divide the significands as indicated.
2. Add and subtract the exponents of 10. Positive exponents of 10 in the numerator should be added while negative exponents of 10 in the numerator should be subtracted. Positive exponents of 10 in the denominator should be subtracted while negative exponents of 10 in the denominator should be added.
3. If needed, rewrite the result in scientific notation.

Example

$$3.8 \times 10^4 \times 2.1 \times 10^7 = 7.98 \times 10^{4+7}$$
$$= 7.98 \times 10^{11}$$

Example

$$4.5 \times 10^3 \times 5.8 \times 10^{-1} \times 4.7 \times 10^2 = 122.67 \times 10^{3-1+2}$$
$$= 122.67 \times 10^4$$
$$= 1.2267 \times 10^6$$

Example

$$3.4 \times 10^{-3} \times 1.1 \times 10^{-5} = 3.74 \times 10^{-3-5}$$
$$= 3.74 \times 10^{-8}$$

Example

$$\frac{7.9 \times 10^1 \left(2.8 \times 10^{-1}\right)}{4.5 \times 10^3 \left(9.9 \times 10^{-7}\right)} = 0.496\,520\ldots \times 10^{1-1-3+7}$$
$$= 0.496\,520\ldots \times 10^4$$
$$= 4.965\,207\ldots \times 10^3$$

Example

$$\frac{6.2 \times 10^0}{1 \times 10^{12}} = 6.2 \times 10^{0-12}$$

$$= 6.2 \times 10^{-12}$$

Example

$$\frac{9.99 \times 10^{-24}}{8.57 \times 10^{-10}} = 1.165\,694\ldots \times 10^{-24+10}$$

$$= 1.165\,694\ldots \times 10^{-14}$$

Exercise Set 6.2.4

1. Perform the chain addition/subtraction. Rewrite the result in scientific notation.

   a. $4.5 \times 10^3 + 2.1 \times 10^3$

   b. $6.2 \times 10^5 - 7.5 \times 10^5$

   c. $-1.27 \times 10^{23} - 4.7 \times 10^{23}$

   d. $-1 \times 10^{-4} + 2.7 \times 10^{-4}$

   e. $8.235 \times 10^{-4} + 1.8 \times 10^{-2} - 3.9 \times 10^{-3}$

   f. $-2.54 \times 10^3 + 1.6 \times 10^0 - 4.4 \times 10^2$

   g. $-5 \times 10^{35} - 2.1 \times 10^{32} - 6 \times 10^{33}$

   h. $-2.8 \times 10^1 - 3.5 \times 10^2 + 4 \times 10^3$

   i. $4.2 \times 10^{-7} - 3 \times 10^{-5} - 9.92 \times 10^{-5}$

   j. $-9.5 \times 10^0 + 2.4 \times 10^1 - 1.1 \times 10^{-1}$

   k. $3.21 \times 10^4 - 1.9 \times 10^2 + 3 \times 10^2$

   l. $-1 \times 10^1 + 1 \times 10^0 - 1 \times 10^{-1}$

   m. $3.7 \times 10^{410} + 1.5 \times 10^{408} - 1.2 \times 10^{407}$

   n. $-6.3 \times 10^{520} - 3.1 \times 10^{519}$

   o. $2.7 \times 10^{-208} + 3.5 \times 10^{-206} + 2.2 \times 10^{-207}$

2. Perform the chain multiplication/division. Rewrite the result in scientific notation.

   a. $\left(7.9 \times 10^4\right)\left(2.1 \times 10^3\right)$

   b. $\left(1.4 \times 10^{-3}\right)\left(4.9 \times 10^2\right)$

c. $\left(-7.3 \times 10^{-9}\right)\left(4 \times 10^{-2}\right)$

d. $\left(1.5 \times 10^{-6}\right)\left(9 \times 10^{-5}\right)$

e. $\left(-4.4 \times 10^{0}\right)\left(-7.11 \times 10^{4}\right)$

f. $\left(8.4 \times 10^{4}\right)\left(6.2 \times 10^{7}\right)$

g. $\left(-3.66 \times 10^{-2}\right)\left(-5.4 \times 10^{-8}\right)$

h. $\dfrac{3.72 \times 10^{7}}{2.7 \times 10^{2}}$

i. $\dfrac{1.8 \times 10^{3}}{9.4 \times 10^{2}}$

j. $\dfrac{9.21 \times 10^{-1}}{3.7 \times 10^{4}}$

k. $\dfrac{3.214 \times 10^{-5}}{2.7 \times 10^{-2}}$

l. $\dfrac{1.8 \times 10^{7}}{2.59 \times 10^{-5}}$

m. $\left(-9.12 \times 10^{4}\right)\left(-3.87 \times 10^{5}\right)\left(1.8 \times 10^{-1}\right)$

n. $\left(-1.2 \times 10^{-5}\right)\left(3.15 \times 10^{-2}\right)\left(1.02 \times 10^{-6}\right)$

o. $\left(5 \times 10^{-3}\right)\left(1.8 \times 10^{2}\right)\left(4.1 \times 10^{-2}\right)$

p. $\dfrac{\left(4.1 \times 10^{-3}\right)\left(5.7 \times 10^{0}\right)}{\left(7.19 \times 10^{5}\right)\left(2.16 \times 10^{-2}\right)\left(1.1 \times 10^{-1}\right)}$

q. $\dfrac{\left(-9.87 \times 10^{6}\right)\left(-3.8 \times 10^{-6}\right)\left(-4.15 \times 10^{-3}\right)}{\left(1.9 \times 10^{-5}\right)\left(-4.61 \times 10^{-9}\right)\left(-5.6 \times 10^{7}\right)}$

r. $\dfrac{\left(4.1 \times 10^{-2}\right)\left(3.2 \times 10^{-6}\right)\left(4.3 \times 10^{-1}\right)}{9.9 \times 10^{-2}}$

6.2.5   Expressions

The *Analysis-Synthesis Technique* can be extended to deal with expressions
that involve numbers that are written in scientific notation. Here is the
associated algorithm.

The Analysis-Synthesis Algorithm for Evaluation of Expressions Involving Addition, Subtraction, Multiplication and Division of Numbers Written in Scientific Notation

To evaluate an expression involving the addition, subtraction, multiplication and division of numbers written in scientific notation

1. Analyze:

   a. Analyze the expression into terms using additions and subtractions *outside* brackets.
   b. Analyze each term into factors using multiplications.
   c. Analyze each factor further if needed.

2. Synthesize:

   a. Work out the factors.
   b. Multiply the factors to evaluate the terms. To do so, use the *Algorithm for the Evaluation of Chain Multiplication/Division of Numbers Written in Scientific Notation* given earlier.
   c. Add and subtract the terms to evaluate the expression. To do so, use the *Algorithm for Evaluation of Chain Addition/Subtraction of Numbers Written in Scientific Notation* given earlier.

We illustrate the use of the above algorithm using examples.

Example

$$4.5 \times 10^2 \times 3.25 \times 10^0 + 9.8 \times 10^5 \times 2.4 \times 10^{-1}$$

$$= \boxed{4.5 \times 10^2 \times 3.25 \times 10^0} + \boxed{9.8 \times 10^5 \times 2.4 \times 10^{-1}}$$

$$= \boxed{14.625 \times 10^{2+0}} + \boxed{23.52 \times 10^{5-1}}$$

$$= \boxed{14.625 \times 10^2} + \boxed{23.52 \times 10^4}$$

$$= \boxed{1.4625 \times 10^3} + \boxed{2.352 \times 10^5}$$

$$= \boxed{0.014\,625 \times 10^5} + \boxed{2.352 \times 10^5}$$

$$= 2.366\,625 \times 10^5$$

Example

$$2.4 \times 10^1 + \frac{3.5 \times 10^{-4} \times 2.8 \times 10^{-2}}{7 \times 10^{-5}} - \frac{5.6 \times 10^4}{8 \times 10^4}$$

$$= \boxed{2.4 \times 10^1} + \boxed{\frac{3.5 \times 10^{-4} \times 2.8 \times 10^{-2}}{7 \times 10^{-5}}} - \boxed{\frac{5.6 \times 10^4}{8 \times 10^4}}$$

$$= \boxed{2.4 \times 10^1} + \boxed{1.4 \times 10^{-4-2+5}} - \boxed{0.7 \times 10^{4-4}}$$

$$= \boxed{2.4 \times 10^1} + \boxed{1.4 \times 10^{-1}} - \boxed{0.7 \times 10^0}$$

$$= \boxed{2.4 \times 10^1} + \boxed{1.4 \times 10^{-1}} - \boxed{7 \times 10^{-1}}$$

$$= \boxed{2.4 \times 10^1} + \boxed{0.014 \times 10^1} - \boxed{0.07 \times 10^1}$$

$$= 2.344 \times 10^1$$

Example

$$4.2 \times 10^3 \left(1.7 \times 10^2 + 3.9 \times 10^4\right) + 6.6 \times 10^7$$

$$= \boxed{4.2 \times 10^3 \left(1.7 \times 10^2 + 3.9 \times 10^4\right)} + \boxed{6.6 \times 10^7}$$

$$= \boxed{4.2 \times 10^3 \left(0.017 \times 10^4 + 3.9 \times 10^4\right)} + \boxed{6.6 \times 10^7}$$

$$= \boxed{4.2 \times 10^3 \times 3.917 \times 10^4} + \boxed{6.6 \times 10^7}$$

$$= 16.4514 \times 10^{3+4} + 6.6 \times 10^7$$

$$= 16.4514 \times 10^7 + 6.6 \times 10^7$$

$$= 1.64514 \times 10^8 + 6.6 \times 10^7$$

$$= 1.64514 \times 10^8 + 0.66 \times 10^8$$

$$= 2.305\,14 \times 10^8$$

Exercise Set 6.2.5

Evaluate each of the following expressions.

1. $3.6 \times 10^3 \times 1.8 \times 10^{-4} + 1.9 \times 10^1$

2. $-7.4 \times 10^{62} \times 5.9 \times 10^{-33} - 2.4 \times 10^{28}$

3. $6.4 \times 10^{-42} \left(-8.7 \times 10^{-22}\right) + 3.2 \times 10^{-62}$

4. $2.9 \times 10^{-2} - \left(-1.7 \times 10^{-1}\right)\left(3.3 \times 10^{-2}\right)$

5. $7.1 \times 10^{17} - 8.81 \times 10^{42} \left(-6.2 \times 10^{-24}\right)$

6. $-4.925 \times 10^{-8} \left(-3.7 \times 10^{0}\right) - 2.73 \times 10^{-17} \left(-4.5 \times 10^{10}\right)$

7. $3.8 \times 10^{-72} \left(-1.5 \times 10^{29}\right) - 1.6 \times 10^{-80} \left(-1.1 \times 10^{37}\right)$

8. $9.9 \times 10^{0} \left(2.1 \times 10^{-3}\right) - 5.7 \times 10^{-1} \left(8.9 \times 10^{-3}\right)$

9. $3 \times 10^{55} \times 4.1 \times 10^{-23} - 4.22 \times 10^{24} \times 2.6 \times 10^{8}$

10. $4.57 \times 10^{0} \left(-1.9 \times 10^{0}\right) - 3.4 \times 10^{2} \left(1.9 \times 10^{-2}\right)$

11. $3.4 \times 10^{5} \left(-2.1 \times 10^{3}\right) \left(9.2 \times 10^{-2}\right) - 1.8 \times 10^{7}$

12. $9.3 \times 10^{-6} - 2.9 \times 10^{-2} \left(-3.2 \times 10^{-7}\right) \left(5.3 \times 10^{3}\right)$

13. $3.1 \times 10^{-2} \left(1.7 \times 10^{4} - 2.9 \times 10^{3}\right)$

14. $-1.8 \times 10^{3} \left(-5.2 \times 10^{-1} + 3.8 \times 10^{-2}\right)$

15. $4.2 \times 10^{-7} \left(2.1 \times 10^{-3} - 3.8 \times 10^{-3}\right)$

16. $2.2 \times 10^{0} \left(-6.4 \times 10^{-8} - 1.2 \times 10^{-7}\right)$

17. $-1.8 \times 10^{3} \left(2.9 \times 10^{6} - 5.7 \times 10^{7}\right) + 2.4 \times 10^{4} \left(3.2 \times 10^{5} - 2.8 \times 10^{4}\right)$

18. $7.3 \times 10^{-8} \left(1.5 \times 10^{0} + 2.2 \times 10^{-1}\right) - 1.8 \times 10^{-4} \left(9.2 \times 10^{-4} + 3.7 \times 10^{-3}\right)$

19. $1.9 \times 10^{3} + \dfrac{3.7 \times 10^{5}}{5.7 \times 10^{2}}$

20. $-6.3 \times 10^{5} - \dfrac{1.8 \times 10^{2}}{-1.7 \times 10^{-2}}$

21. $\dfrac{1.2 \times 10^{3}}{2.5 \times 10^{-1}} - 2.7 \times 10^{1}$

22. $\dfrac{-3.7 \times 10^{-3}}{-4.7 \times 10^{2}} - \left(-7.5 \times 10^{-4}\right)$

23. $\dfrac{5.5 \times 10^{0}}{-2.1 \times 10^{3}} + 4.3 \times 10^{-2}$

6.2.6   Rounding

To round a number written in scientific notation to a certain place value, we need the actual place values of the digits in the significand which, as discussed earlier, are set by the exponent of 10. As an example, in 4.53×10^{2}, the actual place value of digit 4 is *hundreds* as the equivalent decimal representation of the number, i.e., 453, shows. However, in 4.53×10^{-2}, the actual place value of digit 4 is *hundredths* as the equivalent decimal representation of the number, i.e., 0.0453, shows.

In the scientific notation representation of a number, the exponent of 10 shifts the place values relative to the decimal point. Positive exponents of 10 shift the place values to the left and negative exponents of 10 shift the place values to the right.

Example

Round 7.258×10^3 to the nearest *hundred*.

Solution

The exponent of 10 shifts the place values to the left over three places. Therefore, the digit in the *hundreds* place is digit 2. This results in 7.3×10^3 as the rounded value.

Example

Round 7.258×10^3 to the nearest *ten*.

Solution

The exponent of 10 shifts the place values to the left over three places. Therefore, the digit in the *tens* place is digit 5. This results in 7.26×10^3 as the rounded value.

Example

Round 1.84×10^{-2} to the nearest *thousandth*.

Solution

The exponent of 10 shifts the place values to the right over two places. Therefore, the digit in the *thousandths* place is digit 8. This results in 1.8×10^{-2} as the rounded value.

Example

Round -1.84×10^{-2} to the nearest *thousandth*.

Solution

The exponent of 10 shifts the place values to the right over two places. Therefore, the digit in the *thousandths* place is digit 8. This results in -1.8×10^{-2} as the rounded value.

Exercise Set 6.2.6

Round to the indicated place value.

1. 4.52×10^2 (hundreds)
2. 4.52×10^2 (tens)
3. 4.52×10^2 (ones)
4. 4.52×10^2 (tenths)
5. 3.891×10^0 (ones)
6. 3.891×10^0 (tenths)
7. 3.891×10^0 (hundredths)
8. 6.35×10^{-2} (hundredths)
9. -1.72×10^1 (tens)
10. -1.72×10^1 (ones)
11. -1.72×10^1 (tenths)
12. -1.72×10^1 (hundredths)
13. -7.72×10^{-1} (ones)
14. -7.72×10^{-1} (tenths)
15. -7.72×10^{-1} (hundredths)
16. 3.815×10^3 (thousands)
17. 3.815×10^3 (hundreds)
18. 3.815×10^3 (tens)
19. 1.892×10^2 (hundreds)
20. 1.892×10^2 (tens)
21. 1.892×10^2 (ones)
22. 3.4×10^1 (hundreds)
23. 3.4×10^1 (tens)
24. 9.97×10^{-1} (tens)
25. 9.97×10^{-1} (ones)
26. 9.97×10^{-1} (tenths)
27. 9.97×10^{-1} (hundredths)
28. -8.241×10^3 (thousands)
29. -8.241×10^3 (hundreds)
30. -8.241×10^3 (tens)
31. -1.99×10^1 (hundreds)
32. -1.99×10^1 (tens)
33. -1.99×10^1 (ones)
34. 6.27×10^0 (tens)
35. 6.27×10^0 (ones)
36. 6.27×10^0 (tenths)
37. 3.781×10^3 (thousands)
38. 3.781×10^3 (hundreds)
39. 3.781×10^3 (tens)
40. 9.971×10^{-1} (ones)
41. 9.971×10^{-1} (tenths)
42. 9.971×10^{-1} (hundredths)
43. 1.89×10^0 (tenths)
44. -8.42×10^{-2} (thousandths)
45. -5.731×10^{-1} (ones)
46. -4.86×10^{-2} (hundredths)
47. 4.37×10^3 (hundreds)
48. -8.241×10^{-3} (thousandths)
49. -1.996×10^1 (ones)
50. 6.27×10^0 (tenths)
51. 1.892×10^2 (hundreds)
52. 3.4×10^1 (tens)
53. 4.27×10^2 (tens)
54. 3.182×10^4 (thousands)
55. 4.15×10^0 (hundreds)
56. 9.2715×10^{-1} (hundredths)

6.2.7   Graphing

To graph a number written in scientific notation, graph the significand as a decimal and label the axis as $\times 10^n$ where n is the exponent of 10 in the representation of that number.

As an example, to graph the number 1.47×10^2, we graph 1.47 and label the axis as $\times 10^2$. This is shown in Figure 6.2.1 below.

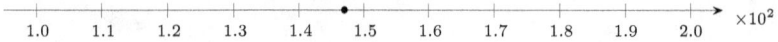

Figure 6.2.1: Graph of 1.47×10^2

To estimate the coordinate of a point in scientific notation, estimate the coordinate of the point as a decimal and append the axis label. As an example, given the graph below

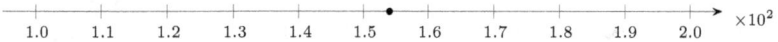

Figure 6.2.2: Estimating the coordinate of the point as 1.54×10^2

we might read the coordinate of the point as 1.54 with digit 4 being estimated. We now append the axis label to this reading and write 1.54×10^2.

As with decimals, it is important to understand that if a measured value, such as 1.54×10^2 above, is given, that the rightmost digit (here, digit 4) is estimated. This means that the ruler's smallest divisions increase in steps given by the place value on the left of the estimated digit. A measured value, then, provides us with two pieces of information: One is the smallest subdivision on the ruler used and the estimated digit that further refines the position of the dot.

Exercise Set 6.2.7

1. In each case estimate the coordinate of the point on the real line in scientific notation.

a.

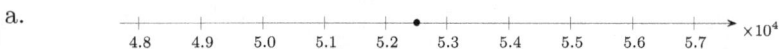

b.

c.

d.

e.

f.

g.

h.

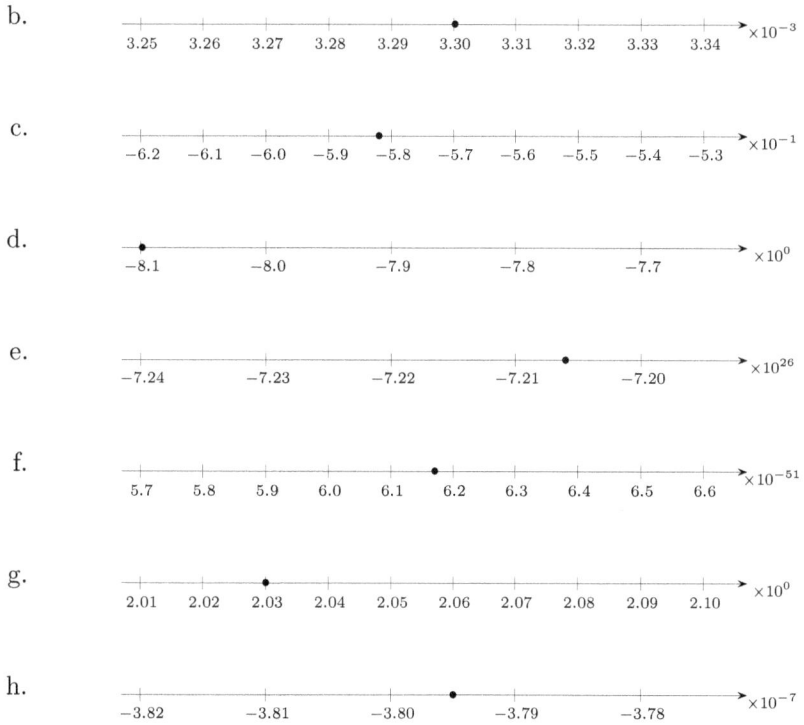

2. Choose an appropriate step size and graph each of the following on the real line.

a. 4.21×10^3

b. 3.168×10^0

c. 4.57×10^{-2}

d. -3.249×10^{32}

e. 1.5×10^{-41}

f. -5.6×10^0

g. -4.400×10^{-2}

h. -3.8×10^0

i. -9.950×10^{-18}

Part IV
Measurement Systems

In this part of the book we introduce the scheme used to formally represent and work with measured values in the sciences. The scheme involves the use of a measurement system, called the *International System of Units*, *SI*, and rules that govern the process of rounding the results of the execution of mathematical operations on measured values.

Chapter 7
Quantities, Units and Values

7.1 Quantities

A **quantity** is a property of an entity that can be measured objectively. Examples of quantities are length, area, volume, mass, time, temperature, speed, force, energy and the like, but not beauty, kindness or serenity, among others. A listing of some of the more common quantities encountered in everyday life activities and basic sciences is given in Table 7.1.1.

Entities

A quantity belongs to an entity. At times the entity to which a given quantity belongs is mentioned explicitly as in

>   The speed of light in vacuum is approximately 300 000 km/s.

Here the quantity is *speed* and the entity the quantity belongs to is *light*. At other times the entity that the given quantity belongs to is implied by the context. As an example, if the context of discussion is the weather, then the sentence

>   The temperature is 20 °C.

associates the quantity *temperature* to the entity *air* at the location where the temperature was measured.

Quantity Name (Symbol)	Unit Name (Symbol)
cost (c)	dollar ($\$$), pound (£), euro (€), etc.
price (p)	dollar ($\$$), pound (£), euro (€), etc.
length (l)	metre (m), kilometre (km), inch (in), etc.
distance (d)	metre (m), kilometre (km), inch (in), etc.
area (A)	square metre (m^2), square inch (in^2), etc.
volume (V)	cubic metre (m^3), litre (l, ℓ, L), etc.
time (t)	second (s), minute (min), hour (h), etc.
speed (s, v)	metres per second (m/s), etc.
acceleration (a)	metres per square second (m/s^2), etc.
mass (m)	gram (g), atomic mass unit (amu), etc.
weight (W)	newton (N), dyne (dyn), etc.
force (F)	newton (N), dyne (dyn), etc.
pressure (p)	pascal (Pa), atmosphere (atm), etc.
temperature (T, t)	kelvin (K), degree Celsius (°C), etc.
energy (E)	joule (J), calorie (cal), Calorie (Cal), etc.
amount of substance (n)	one (1), mole (mol), etc.
electric current (I, i)	ampere (A), etc.

Table 7.1.1: A selection of common quantities and their commonly associated units. The quantity symbol T is used to refer to the thermodynamic temperature, measured using the kelvin as the unit. The quantity symbol t is used to refer to temperature measured using other units, e.g., degree Celsius. The optional quantity symbol for speed, v, relates to the magnitude of the quantity *velocity*. See the discussion ahead on the deviation from SI in classifying the unit *one* as a unit of amount.

Quantity Names and Quantity Symbols

To each quantity we associate a name and a symbol.

Quantity names are words that are used to refer to quantities in English sentences and phrases. Examples of quantity names are length, mass, time, and the like. Quantity names are linguistic entities and, as such, are subject to the rules of English grammar. As an example of use of quantity names, we write

To measure the temperature, press and hold the orange button for approximately 3 seconds.

Quantity symbols are symbols that are used to represent the *values* of quantities in mathematical expressions and equations. Examples of quantity symbols are the symbol m, which is used to represent the value of the quantity *mass*; p, which is used to represent the value of the quantity *pressure*, and the like.

Since quantity symbols represent the *values* of quantities, they are seen as mathematical entities and, as such, are subject to the rules of algebraic grammar. As an example, using t as the quantity symbol to represent the value of the quantity *temperature* in degrees Celsius, we write

$$\frac{9}{5}t + 32$$

to indicate the need to add 32 to $\frac{9}{5}$ of the value of the temperature in degrees Celsius, an activity that converts the value of the temperature in degrees Celsius to the value of the temperature in degrees Fahrenheit.

The understanding that quantity symbols represent the *values* of quantities also implies that they are *not* abbreviations of the names of the quantities whose values they represent. Note that we could have chosen any symbol to represent the value of, say, mass with, e.g., we could have agreed that the symbol ∇ represents the value of mass[1] in which case it would make no sense to refer to the symbol ∇ as an abbreviation of the word *mass*. The reason we prefer to use the symbol m (and not ∇ or some other symbol) to represent the value of mass with is that the *symbol* m reminds us of the *letter* m which may then be seen as an acronym for the word *mass*. Since quantity symbols are not linguistic entities, they are *not* subject to the rules of English grammar. As an example, the period should not be used after quantity symbols.

Exercise Set 7.1

1. What is a quantity? Give an example of a property that is a quantity. Give an example of a property that is not a quantity.
2. What is a quantity name? Give and example of a quantity name.
3. What is a quantity symbol? Give an example of a quantity symbol and name the quantity whose value is represented by the symbol.
4. Explain why quantity symbols may appear in mathematical expressions and equations and why it is not proper to use quantity names in such expressions and equations.

---

[1] Had we chosen to do so, we would write $\nabla = 42.5$ kg in place of $m = 42.5$ kg.

7.2   Units

Units are used to measure the values of quantities. Examples of units are metre, gram, degree Celsius, mole, and the like.

Many different units can be used to measure the value of a given quantity. As an example, we can use the units metre, kilometre, inch, foot and others to measure the value of the quantity *length* with. As another example, we can use the units degree Celsius, degree Fahrenheit, kelvin and others to measure the value of the quantity *temperature* with.

Unit Names and Unit Symbols

To each unit we associate a name and a symbol.

Unit names are words that are used to refer to units in English sentences and phrases. Examples of unit names are metre, gram, second, mole, and the like. Unit names are linguistic entities and, as such, are subject to the rules of English grammar. As an example of use of unit names, we write

> To measure the temperature, press and hold the orange button
> for approximately 3 seconds.

Unit symbols are symbols that are used to represent the *sizes* of units in mathematical expressions and equations. Examples of unit symbols are the symbol m, which is used to represent the size of the unit *metre*; kg, which is used to represent the size of the unit *kilogram*, and the like.

Since unit symbols represent the *sizes* of units, they are seen as mathematical entities and, as such, are subject to the rules of algebraic grammar. As an example, using the symbol m as the unit symbol to represent the size of the unit *metre*, we write

$$d \;=\; 32.6 \; \mathrm{m}$$

The right side of this equation is seen as $32.6 \times \mathrm{m}$. This implies that the equation above may be rearranged and written as

$$\frac{d}{\mathrm{m}} \;=\; 32.6$$

The rules of algebra, then, apply to unit symbols just as they apply to numerical values and quantity symbols.

The understanding that unit symbols represent the *sizes* of units also implies that they are *not* abbreviations of the names of the units whose sizes they represent. Note that we could have chosen any symbol to represent the size of, say, metre with, e.g., we could have agreed that the symbol φ

represents the size of metre[2] in which case it would make no sense to refer to the symbol φ as an abbreviation of the word *metre*. The reason we prefer to use the symbol m (and not φ or some other symbol) to represent the size of metre with is that the *symbol* m reminds us of the *letter* m which may then be seen as an acronym for the word *metre*. Since unit symbols are not linguistic entities, they are *not* subject to the rules of English grammar. As an example, the period should not be used after unit symbols. As another example, the plural s should not be used along with unit symbols, e.g., we write 2.5 m to represent 2.5 metres and not 2.5 ms which is interpreted as 2.5 milliseconds. As a third example, the hyphen should not be used even if the unit symbol acts as an adjective, e.g., we write 'a 2.5 kg object' and not 'a 2.5-kg object'.[3]

Exercise Set 7.2

1. What is a unit? Give an example.
2. What is a unit name? Give an example of a unit name.
3. What is a unit symbol? Give an example of a unit symbol and name the unit whose size is represented by the symbol.
4. Explain why unit symbols may appear in mathematical expressions and equations and why it is not proper to use unit names in such expressions and equations.

7.3   Values

In the sentence

> The height of the patient is 183 cm.

we refer to 183 cm as the value of the quantity *height*. The **value of a quantity** is made up of three parts: The 183 by itself which is called the **numerical value of the quantity**, the empty space between 183 and cm which implies multiplication, and the unit symbol cm which sets the size of the unit used to measure the value of the quantity. The terminology introduced above is illustrated in Figure 7.3.1 below.

Note that the value of a quantity can take on different forms depending on the unit used. As an example, the height of the patient in Figure 7.3.1, i.e., 183 cm, may be expressed as 1.83 m or, approximately, 6 ft.

---

[2]Had we chosen to do so, we would write $d = 32.6\ \varphi$ in place of $d = 32.6$ m.
[3]But we do write *a 2.5-kilogram object*.

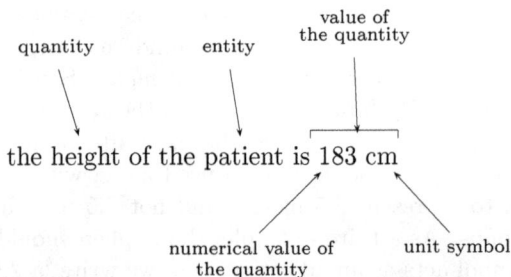

Figure 7.3.1: Illustration of notation and terminology related to measurement

Examples

In each case identify the quantity being measured (list both the quantity name and quantity symbol), the entity to which the quantity belongs, the value of the quantity, the numerical value of the quantity, and the unit used (list both the unit name and unit symbol).

1. The car has a mass of 1243 kg.
2. The reaction generated 5.2 mol of oxygen.
3. There are 760 frogs in the sample.
4. The fridge uses 970 W of power.

Answers

1. Quantity name: mass
   Quantity symbol: m
   Entity: the car
   Value of the quantity: 1243 kg
   Numerical value of the quantity: 1243
   Unit name: kilogram
   Unit symbol: kg

2. Quantity name: amount
   Quantity symbol: n
   Entity: oxygen
   Value of the quantity: 5.2 mol
   Numerical value of the quantity: 5.2

   Unit name: mole
   Unit symbol: mol

3. Quantity name: amount
   Quantity symbol: n
   Entity: the frogs
   Value of the quantity:[4] 760 frogs
   Numerical value of the quantity: 760
   Unit name: one
   Unit symbol: 1

4. Quantity name: power
   Quantity symbol: P
   Entity: the fridge
   Value of the quantity: 970 W
   Numerical value of the quantity: 970
   Unit name: watt
   Unit symbol: W

Exercise Set 7.3

In each case identify the quantity being measured (list both the quantity name and quantity symbol), the entity to which the quantity belongs, the value of the quantity, the numerical value of the quantity, and the unit used (list both the unit name and unit symbol).

1. The test will take 45 minutes.
2. I need a beam that is 5 ft long.

---

[4]The formal notation for the value of this quantity is

   760 1

which is interpreted as seven hundred sixty *ones*. This is similar to 760 g or 760 km, etc., with 1 acting as the unit symbol for *one*, a unit of counting. Further information on the nature of the 1 can be appended to the unit symbol as a subscript, e.g., 760 1_{frog}. This parallels the use of subscripts on other unit symbols as in g_{CO_2} which is interpreted as grams of CO_2.

The use of 1 without a subscript as unit symbol in mathematical expressions and equations is uncommon and can lead to interpretation errors. As an example, it is easy to misinterpret 760 1 (760 ones) as 7601 (seven thousand six hundred one). In practice, the unit symbol 1 is often replaced by a word that describes the nature of the 1, e.g., it is common to write 760 frogs in place of 760 1 or 760 1_{frog} even though this practice is at odds with our earlier insistence that only the unit symbol should be used in mathematical expressions and equations. Note that this exception applies only to measured values that arise as a result of counting with 1 as the unit. In all other cases the use of the unit symbol is required.

In this textbook we follow the common approach and use words that describe the nature of the 1 but, in case we are called upon to name the quanity and its associated unit, we invariably use the quantity name *one* and the corresponding unit symbol, 1.

3. The temperature of the solution is 285 K.
4. These gloves cost $25.
5. Prepare 10 mg of the medication.
6. My son is 4 years old.
7. The operation will take 4.5 hours.
8. I need 1.2 L of paint.
9. The car was travelling at 120 km/h.
10. The sandwich contains 240 Calories.
11. The distance from Toronto to Montreal is 539 km.
12. The reaction generated 45.5 g of CO_2.
13. The book costs $9.20.
14. Rate of tax is 13%.
15. The train was travelling at a speed of 200 km/h.
16. The lot has an area of 4000 ft^2.
17. The temperature of the surface of the sun is 6000 °C.
18. The bullet has a kinetic energy of 2025 J.
19. This BMW can accelerate at 12.66 m/s^2.
20. Charlie[5] has a mass of 32 kg.
21. Charlie weighs 314 N.

---

[5] A dog in my neighborhood :)

Chapter 8
The International System of Units, SI

We begin this chapter with an introduction to features of measurement systems that are of importance to us. Following this, we will focus on the features of the International System of Units, SI; the default measurement system used in the sciences.

8.1 Structure

A **measurement system** consists of a list of quantities and their associated units. Most measurement systems also provide a list of conventions with the aim of establishing a uniform communication protocol.

There are many measurement systems in common use. As examples we have the Imperial System of Units, the US Customary System of Units, the Metric System and the International System of Units, SI.[1]

---

[1]The acronym SI is based on the French version of the title, i.e., *Le Système International d'Unités*. SI is the default measurement system used in the sciences and represents the most recent version of the Metric system. The latter has gone through stages of development, under various names and, since its inception, has been steadily gaining graound in practice. There are notable exceptions in certain branches of pure and applied sciences and industries and among the general population of some English-speaking countries which, for one reason or another, have fallen behind in the drive to convert to the Metric system. The global drive to convert to SI is due to the fact that SI has major advantages over other measurement systems; advantages that we will discuss in the pages ahead as the opportunity to do so presents itself.

Quantity and Unit Types

The quantities on the lists provided by measurement systems are usually classified as either base quantities or derived quantities.

Base quantities are quantities that are given their own units. Units that are associated with base quantities are, in turn, called **base units**. As an example, SI classifies the quantities *length* and *time*, among others, as base quantities with corresponding base units metre and second for measuring their values.

Derived quantities are quantities whose units are given in terms of the units of base quantities. Units that are associated with derived quantities are, in turn, called **derived units**. As an example, SI classifies the quantity *speed* as a derived quantity. Its derived unit, i.e., metres per second, is given in terms of those of base quantities *length*, i.e., metre, and *time*, i.e., second.

Sometimes a special name and an associated special symbol is given to a derived unit. As an example, in SI, the derived unit of derived quantity *energy* in terms of base units is $kg \cdot m^2/s^2$. This combination of base units is given the name *joule* with the corresponding unit symbol J. As a second example, in SI, the derived unit of the derived quantity *pressure* in terms of base units is $kg/(m \cdot s^2)$, a combination of base units that is called the *pascal*, with unit symbol Pa.

One advantage of SI over other systems of units is that in SI the number of base quantities is minimized. This simplifies the expression of science formulas which relate the quantities that are used in the sciences. As an example, in SI, the unit of the derived quantity, speed, i.e., metres per second, naturally relates to the units of base quantities length and time, i.e., the metre and the second. If we were to call speed a base quantity as well, a choice that would force the selection of its own unit of measure, it would complicate science formulas by requiring the formulas to incorporate conversion factors. As an example, using SI, we can write the formula for the speed of an object in terms of distance covered and travel time as

$$v = \frac{d}{t}$$

whereas, if we were to set speed as a base quantity as well and set its unit to, say, knots, we would have to write

$$v = k\frac{d}{t}$$

with the conversion factor, k, roughly equal to 1.944 knot·s/m. In the formula above, the conversion factor k forces the conversion of m/s which arises from the division of distance in metres by time in seconds, to knots.

Exercise Set 8.1

1. What is a measurement system? Give an example of a measurement system.
2. Which measurement system is the default measurement system used in the sciences?
3. Why do most measurement systems provide conventions?
4. What is a base quantity? Give an example of a base quantity in SI.
5. What is a base unit? Give an example of a base unit in SI and the corresponding SI base quantity whose value is measured using the base unit.
6. What is a derived quantity? Give an example of a derived quantity in SI.
7. What is a derived unit? Give an example of a derived unit in SI and the corresponding SI derived quantity whose value is measured using the derived unit.
8. Give an example of a special name that is given to a combination of base units in SI and provide the symbol used to represent its size. Specify the combination of base units in SI that this symbol represents.
9. Why is it advantageous to have the minimum number of base quantities in a measurement system?

8.2   SI Quantity and Unit Types

In this section we will present a list of SI base quantities and their associated base units followed by a selection of commonly used SI derived quantities and their associated derived units.[2]

SI Base Quantities and Units

Table 8.2.1 below lists the base quantities in SI and their associated units.

---

[2]Our main concern in this textbook is to present technical issues that relate to the use of quantities and units. For a detailed discussion of the semantics of quantities and the size of their associated units please see the companion textbook *Semantics and the Syntax of Algebra* by the author.

Quantity	Unit Name	Unit Symbol
length	metre	m
mass	kilogram	kg
time	second	s
electric current	ampere	A
thermodynamic temperature	kelvin	K
amount of substance	mole	mol
luminous intensity	candela	cd

Table 8.2.1: SI base quantities and their associated unit names and unit symbols

The base quantity **mass** with quantity symbol m measures an object's resistance to change in its state of motion, i.e., its velocity.[3] The unit of mass was initially set to be the gram. However, since gram is a very small unit of mass, its use tends to require the use of large numerical values in the representation of the values of masses that we work with on a daily basis. Rather than redefining the gram, the unit *kilogram* was set to replace the gram as the official unit of mass in SI.

The unit of the SI derived quantity **electric charge** with quantity symbol Q is the coulomb with unit symbol C. The base quantity **electric current** with quantity symbol I is a measure of the flow of electric charge and measures the electric charge that passes through a reference point in a unit of time, i.e.,

$$I = \frac{Q}{t}$$

Its unit, coulombs per second, is given the name *ampere* with corresponding unit symbol A.

The unit of **thermodynamic temperature** is the kelvin with unit symbol K. Unlike other units of temperature whose definitions are based on the behaviour of a particular substance,[4] the definition of the kelvin is based on the kinetic energies of the molecules within a substance and is, therefore, more relevant to the physics of heat flow.[5] The definition of kelvin as the unit of temperature sets the 0 at the point where translational movement of molecules within the substance stops, i.e., when the translational speed

---

[3] This is different from an object's **weight** which is a derived quantity in SI with quantity symbol W. An object's weight is a measure of the gravitational force that another object, such as the Earth, exerts on it.

[4] Such as the choice of water to set the 0 and 100 on the Celsius scale.

[5] The SI derived quantity *heat* measures the kinetic energies of the molecules in a substance and is, therefore, related to the random movement of the molecules within that substance.

of molecules within the substance is 0. A temperature of 0 K, then, corresponds to a molecular speed of 0 and, as molecular speed increases, so does the value of temperature. This connection between the kelvin as the unit of temperature and physics of heat flow also implies that equations of physics that involve such quantities will be simpler to write and comprehend.

SI Derived Quantities and Units

We now turn our attention to a discussion of the more commonly used SI derived quantities and their associated derived units.[6]

Area

The SI derived quantity **area** refers to the size of a section of a surface[7] that is contained within a closed curve. The capital letter A is used as the quantity symbol for the size of the area of a plane. The capital letter S is used as the quantity symbol for the size of the area of a curved surface.

The SI unit of area is represented by the area that is enclosed by a **unit square** defined as a square whose side-length measures 1 m. The value of this area is written as 1 m^2 (read *one square metre*); See Figure 8.2.1 for an illustration.[8]

When we say *the area of this garden is* 12 m^2, it means that the garden, *regardless of its actual shape*, covers an area that is equivalent to the area covered by 12 unit squares.

Note that the derived unit of area, i.e., m^2, which is short for m·m, is defined in terms of the base unit of length, i.e., m.

Volume

The SI derived quantity **volume** refers to the size of the space contained within a closed surface. The capital letter V is used as the quantity symbol

---

[6]A proper definition of the derived quantities that are covered in this section of the textbook requires the use of concepts from calculus. Here we adopt the approach that is taken in basic science textbooks and present formulas that relate to average values. As an example, formally, power is defined as the rate of use of energy with respect to time, i.e., $P = \frac{dE}{dt}$. If energy is used at a constant rate over a period of time or else one is interested in the average rate of use of energy with respect to time, the equation above can be simplified to $P = \frac{E}{t}$. It is this latter view of power that is covered in basic science courses and it is the one that we will cover in this textbook.

[7]The surface may be straight or curved. A straight surface is referred to as a **plane**. The area of a curved surface is often referred to as **surface area**.

[8]Note that the drawing is not to scale.

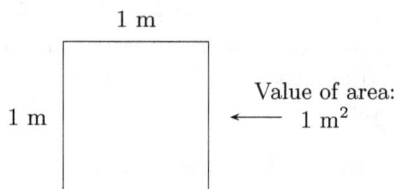

Figure 8.2.1: The SI unit of area is the square metre, with unit symbol m^2, defined as the area enclosed by a unit square

for the size of volume.

The SI unit of volume is represented by the volume that is enclosed by a **unit cube** defined as a cube whose side-length measures 1 m. The value of this volume is written as $1\ m^3$ (read *one cubic metre*); See Figure 8.2.2 for an illustration.[9]

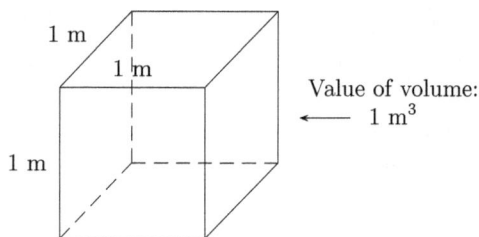

Figure 8.2.2: The SI unit of volume is the cubic metre, with unit symbol m^3, defined as the volume enclosed by a unit cube

When we say *the volume of this container is* $8\ m^3$, it means that the container, *regardless of its actual shape*, covers a volume that is equivalent to the volume enclosed by 8 unit cubes.

Note that the derived unit of volume, i.e., m^3, which is short for m·m·m, is defined in terms of the base unit of length, i.e., m.

---

[9]Note that the drawing is not to scale.

Speed

The SI derived quantity **speed** is defined as distance covered per unit time.[10] This involves the division of distance covered by travel time, i.e.,

$$v = \frac{d}{t}$$

leading to the SI derived unit of *metres per second* as the unit of speed.

Note that the derived unit of speed, i.e., m/s, is defined in terms of the base unit of length, i.e., m, and the base unit of time, i.e., s.

Acceleration

The SI derived quantity **acceleration** is defined as the change in speed in unit time. This involves the division of speed by time, i.e.,

$$a = \frac{v}{t}$$

Its derived unit is $(m/s)/s$, an expression that simplifies to m/s^2.[11]

Note that the derived unit of acceleration, i.e., m/s^2 which is short for $(m/s)/s$, is defined in terms of the base unit of length, i.e., m, and the base unit of time, i.e., s.

---

[10]The SI derived quantity **velocity** is *not* a synonym for speed. Velocity is speed along with direction. As an example, one may say that *the car is travelling at a speed of* 110 km/h but that *the car has a velocity of* 110 km/h *travelling east*. The quantity symbol for velocity is $\vec{v}$ with the arrow over v indicating the inclusion of direction in its value. In this context, the symbol v without the arrow is used to refer to the numerical value of velocity, i.e., speed, without a note on direction of motion. This explains the use of the symbol v to refer to speed in many physics textbooks.

[11]Following the guidelines given earlier in the text on the simplification of complex fractions, we can write

$$\begin{aligned}
(m/s)/s &= \frac{\frac{m}{s}}{s} \\
&= \frac{\frac{m}{s}}{\frac{s}{1}} \\
&= \frac{m \times 1}{s \times s} \\
&= \frac{m}{s^2} \\
&= m/s^2
\end{aligned}$$

This simplified expression is more convenient to work with than the original but not quite as descriptive. When working with equations and formulas the latter form is preferred but if meaning is sought, we recommend the use of the former.

Density

The SI derived quantity **density** is defined as the mass of a substance in a unit volume of that substance. The Greek letter ρ (read *rho*) is used as the quantity symbol for density. By definition

$$\rho = \frac{m}{V}$$

where m represents the mass of the substance and V represents its volume. The SI unit of density is, therefore, kg/m^3.

Note that the derived unit of density, i.e., kg/m^3, is defined in terms of the base unit of mass, i.e., kg, and the base unit of length, i.e., m.

Frequency

The SI derived quantity **frequency** is a measure of the rate at which a cyclic process recurs. As such, it is defined as the number of occurrences of a cycle per unit time. The Greek letter η (read *eta*) is used as the quantity symbol for frequency. By definition

$$\eta = \frac{n}{t}$$

where n is the number of occurrences of the cycle and t is the time during which the cycles occur. The SI unit of frequency is, therefore, $1/s$, a unit that is often referred to as **hertz** with unit symbol Hz.

Note that the derived unit of frequency, i.e., Hz which is short for $1/s$, is defined in terms of the unit of amount,[12] i.e., 1, and the base unit of time, i.e., s.

Force

The SI derived quantity **force** may be viewed as a pull or a push. Newton's formula states that the force exerted on an object causes the object to accelerate in the direction in which force is being applied. For motion on a line, Newton's formula states that

$$F = ma$$

---

[12] Here is an example why the base unit of amount should have been one rather than the mole. One may, of course, rewrite the equation for frequency using the mole as the unit for amount but such an equation would prove awkward for use in practice.

i.e., the relationship between the force exerted on a body and the body's acceleration are related through direct proportion with m acting as the constant of proportionality.[13]

The unit of force, then, is equal to the unit of mass, i.e., kg, times the unit of acceleration, i.e., m/s^2. The unit of force can therefore be written as $kg \cdot m/s^2$, an expression that is often referred to as **newton** with unit symbol N.

Note that the derived unit of force, i.e., N which is short for $kg \cdot m/s^2$, is defined in terms of the base unit of mass, i.e., kg, the base unit of length, i.e., m, and the base unit of time, i.e., s.

Pressure

The SI derived quantity **pressure** with quantity symbol p is defined as force per unit area, i.e.,

$$p = \frac{F}{A}$$

The derived unit of pressure, then, is the unit of force, i.e., N, divided by the unit of area, i.e., m^2. This means that the unit of pressure is N/m^2, an expression that is referred to as **pascal** with unit symbol Pa.

Note that the derived unit of pressure, i.e., Pa which is short for N/m^2 or $kg/(m \cdot s^2)$, is defined in terms of the base unit of mass, i.e., kg, the base unit of length, i.e., m, and the base unit of time, i.e., s.

Work

The SI derived quantity **work** is defined as the product of the value of the component of a force along the line of motion of the object on which the force acts, and the distance the object covers while the force acts on it. For an object moving along a straight line and a force acting along the line of motion of the object, the work done by the force is given by the equation

$$W = Fd$$

The derived unit of work, then, is the product of the unit of force, i.e., N, and the unit of distance, i.e., m. The unit of work can therefore be written as $N \cdot m$, an expression that is referred to as **joule** with corresponding unit symbol J.

---

[13]For more on problems of type *direct proportion* please see the companion textbook *Semantics and the Syntax of Algebra* by the author.

Note that the derived unit of work, i.e., J which is short for N·m or kg·m^2/s^2, is defined in terms of the base unit of mass, i.e., kg, the base unit of length, i.e., m, and the base unit of time, i.e., s.

Energy

The SI derived quantity **energy** can assume many forms. Examples of different forms of energy are **kinetic energy** which refers to the energy carried by an object due to its state of motion, **potential energy** which is energy that is stored in an object due to its interaction with a field such as gravitational or electrostatic fields, **sound energy** which is the energy carried by sound waves, **heat** which is the sum of the energies of the molecules of a substance, and others.

What all the different forms of energy have in common is that they can be converted to work. This equivalence implies that the unit of energy is the same as the unit of work, i.e., joules.

Power

The SI derived quantity **power** is defined as energy used in a unit of time. This relationship can be expressed as

$$P = \frac{E}{t}$$

The unit of power, then, is the same as the unit of energy, i.e., J, divided by the unit of time, i.e., s. This means that the unit of power may be expressed as J/s, an expression that is referred to as **watt** with corresponding unit symbol W.

Note that the derived unit of power, i.e., W which is short for J/s or kg·m^2/s^3, is defined in terms of the base unit of mass, i.e., kg, the base unit of length, i.e., m, and the base unit of time, i.e., s.

Units Outside of SI

In this subsection we will discuss a few of the more commonly used units that are not formally part of SI.

Non-SI Units of Time

The SI unit of time is the second with unit symbol s. One may use metric prefixes with the unit *second* to generate smaller or larger units of time. As an example, the millisecond is an SI unit of time whose size is one thousandth of a second and the kilosecond is a unit of time whose size is one thousand seconds. While some of these units, such as the millisecond, are used in some applications, others are seldom, if ever, used. In place of such units of time, it is more common to use units such as the minute, hour, day, week, month, year, decade, century, millennium and so on. The relationships between such units of time are listed below:

1 min = 60 s
1 h = 60 min
1 day = 24 h
1 week = 7 days
1 year = 52 weeks
1 year = 12 months
1 decade = 10 years
1 century = 100 years
1 century = 10 decades
1 millennium[14] = 1000 years
1 millennium = 100 decades
1 millennium = 10 centuries

Non-SI Units of Mass

While the gram and its SI derivatives such as the kilogram and milligram are commonly used to measure the values of masses of objects that we encounter on a daily basis, they are not suitable for the measurement of masses of such entities as molecules, atoms, and subatomic particles as the masses of such entities are extremely small and, therefore, one would have to use an extremely small numerical value in the expression of the values of their masses. The **atomic mass unit** with unit symbol amu was invented to make it easier to work with the masses of such entities. Using amu as the unit of mass, we can say that the mass of a proton is 1 amu (1.66×10^{-24} g) and that the mass of a C-14 isotope of carbon is 14 amu (2.32×10^{-23} g). The relationship between atomic mass unit and gram is given below

$$1 \text{ g} = 6.023 \times 10^{23} \text{ amu}$$

---

[14]Note that the plural form of *millennium* is *millennia*, not *millenniums*.

Non-SI Units of Volume

In SI, the official unit of volume is the cubic metre with unit symbol m^3 and its derivatives such as the cubic kilometre, km^3, the cubic decimetre, dm^3, and the cubic centimetre, cm^3.

The non-SI unit of volume, the litre, and its derivatives with metric prefixes such as the millilitre are quite common. The **litre**, with unit symbol L,[15] is defined as a unit of volume equal in size to the cubic decimetre, dm^3, i.e.,

$$1\,L = 1\,dm^3$$

Note that this implies that the millilitre, mL, is equal in size to a cubic centimetre, cm^3, i.e.,

$$1\,mL = 1\,cm^3$$

Exercise Set 8.2

1. a. How many base quantities are there in SI?
   b. Name the base quantities in SI and for each state its associated unit name and unit symbol.

2. a. What is the difference between the quantity *mass* and the quantity *weight* in the sciences?
   b. Why was the kilogram, rather than the gram, chosen as the base unit of mass in SI?

3. Why is the SI unit of temperature, i.e., the kelvin, the preferred unit of temperature for use in the sciences?

4. a. What is the SI unit of area?
   b. What does it mean when we say that the area of a shape such as a circle is 2.4 m^2?
   c. Why should one read m^2 as *square metre* but not *metre squared*?

5. a. What is the SI unit of volume?
   b. What does it mean when we say that the volume of an object such as a box is 1.3 m^3?
   c. Why should one read m^3 as *cubic metre* but not *metre cubed*?

---

[15]The official symbol for the litre is the lowercase version of the letter L, i.e., l. However, the symbol l is often confused with the number 1 and, therefore, the uppercase version of the letter, i.e., L, has become the preferred symbol for the representation of the size of the litre. The fancy version of the letter l, i.e., ℓ, is also used as the symbol for the unit *litre* but the symbol may not be readily available on all systems.

6. a. Define the SI derived quantity *speed* and state its SI unit.
   b. Explain how the SI unit of speed relates to the base units in SI.
   c. How is the quantity *speed* different from the quantity *velocity*?

7. a. Define the SI derived quantity *acceleration* and state its SI unit.
   b. Explain how the unit of acceleration in SI can be derived from the definition of acceleration.
   c. Explain how the SI unit of acceleration relates to the base units in SI.

8. a. Define the SI derived quantity *density* and state its SI unit.
   b. Explain how the SI unit of density relates to the base units in SI.

9. a. Define the SI derived quantity *frequency* and state its SI unit.
   b. Explain how the SI unit of frequency relates to the base units in SI.

10. a. Define the SI derived quantity *force* and state its SI unit.
    b. Explain how the SI unit of force relates to the base units in SI.

11. a. Define the SI derived quantity *pressure* and state its SI unit.
    b. Explain how the SI unit of pressure relates to the base units in SI.

12. a. Define the SI derived quantity *work* and state its SI unit.
    b. Explain how the SI unit of work relates to the base units in SI.

13. a. Define the SI derived quantity *energy* and state its SI unit.
    b. Give three examples of different forms of energy.
    c. Explain how the SI unit of energy relates to the base units in SI.

14. a. Define the SI derived quantity *power* and state its SI unit.
    b. Explain how the SI unit of power relates to the base units in SI.

15. What are the commonly used non-SI units of time?

16. a. Why was the unit *atomic mass unit* introduced in the sciences?
    b. What is the relationship between gram and atomic mass unit?

17. a. How is the litre related to the cubic decimetre?
    b. How is the millilitre related to the cubic centimetre?

8.3   SI Prefixes

SI prefixes can be used along with SI units to generate larger or smaller units. Table 8.3.2 provides a list of the more commonly used SI prefixes along with their associated symbols and meanings.

As an example of how we use SI prefixes consider the SI unit of length, the metre. By attaching the SI prefix *kilo* to metre, we generate a new unit

Prefix	Symbol	Meaning
kilo	k	1000
hecto	h	100
deca	da	10
deci	d	0.1
centi	c	0.01
milli	m	0.001

Table 8.3.2: Common SI prefixes, their associated symbols and meanings

of length, the *kilometre*. According to Table 8.3.2, kilo means *one thousand* so that the kilometre is a unit of length whose size is 1000 times larger than that of the metre. We write

$$1 \text{ km} = 1000 \text{ m}$$

As a second example, consider the use of the SI prefix *centi* along with the SI unit of length, the metre. This generates a new unit of length, the *centimetre*. According to Table 8.3.2, centi means *one hundredth* so that the centimetre is a unit of length whose size is one hundredth of the size of the metre. We write

$$1 \text{ cm} = 0.01 \text{ m}$$

Note that the names of the units that arise from the use of SI prefixes consist of one word: We write *kilometre* and not *kilo metre*.

Note also that the use of multiple prefixes in the same unit (e.g., centimillimetre) is strictly forbidden.

Note further that the casing of these prefixes matters. As an example, the symbol mg is used to represent the size of the unit milligram which is equal to one thousandth of a gram, whereas the symbol Mg is used to represent the size of the unit megagram which is equal to one million grams.

The Metric Chart

The metric chart provides a handy tool for rapid conversion of units within SI.[16] The metric chart is shown in Figure 8.3.3 below.

---

[16]We recommend the use of the metric chart when the prefixes used are uncommon. For commonly used prefixes, we recommend that you memorize the relationships and use these relationships to change units. In particular, we recommend that you memorize the

kilo (1000)	hecto (100)	deca (10)	no prefix (1)	deci ($\frac{1}{10}$)	centi ($\frac{1}{100}$)	milli ($\frac{1}{1000}$)
k	h	da		d	c	m

Figure 8.3.3: The metric chart

Figure 8.3.4 shows how SI prefixes can be used along with the metre to generate larger and smaller units of length compared to the metre.

1 km = 1000 m	1 hm = 100 m	1 dam = 10 m	1 m = 1 m	1 dm = 0.1 m	1 cm = 0.01 m	1 mm = 0.001 m
km	hm	dam	m	dm	cm	mm

Figure 8.3.4: SI prefixes with the unit of length, the metre

SI prefixes may be used along with any unit of measure of course, not just the metre. As an example, just as 1 km is a unit of measure for length whose size is equal to 1000 m, 1 kg is a unit of measure for mass whose size is equal to 1000 g and 1 kmol is a unit of measure for amount whose size is equal to 1000 mol. Figure 8.3.5 shows how SI prefixes can be used along with the gram to generate larger and smaller units of mass compared to the gram.

---

following relationships: The relationship between kilo and the unit, e.g., 1 km = 1000 m; the relationship between the unit and centi, e.g., 1 m = 100 cm; the relationship between the unit and milli, e.g., 1 m = 1000 mm; and the relationship between centi and milli, e.g., 1 cm = 10 mm.

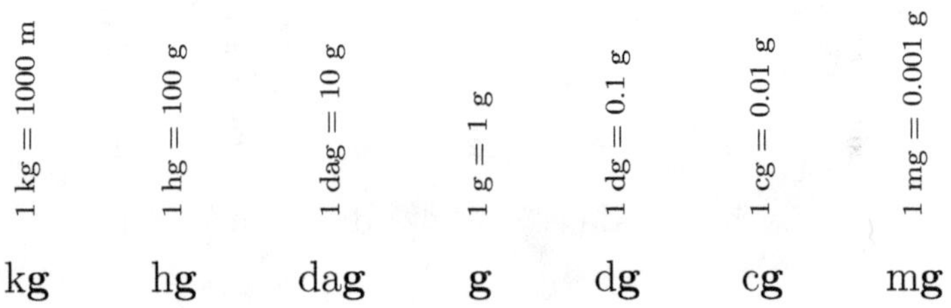

kg	hg	dag	g	dg	cg	mg
1 kg = 1000 m	1 hg = 100 g	1 dag = 10 g	1 g = 1 g	1 dg = 0.1 g	1 cg = 0.01 g	1 mg = 0.001 g

Figure 8.3.5: SI prefixes with the unit of mass, the gram

An important property of the metric chart is that each prefix is 10 times smaller than the one on its left and 10 times larger than the one on its right.[17] As an example, we can say that 1 km is equal to 10 hm or that 1 dm is equal to 10 cm. The relationships between adjacent units on the metric chart using the metre as unit is given below.

1 km = 10 hm
1 hm = 10 dam
1 dam = 10 m
1 m = 10 dm
1 dm = 10 cm
1 cm = 10 mm

Note that the relationships listed above imply that each unit is 100 times larger than the unit that is two steps on its right, 1000 times larger than the unit that is three steps on its right, and so on. Relating each unit to those on its left, we can say that each unit is 100 times smaller than the unit that is two steps on its left, 1000 times smaller than the unit that is three steps on its left, and so on.

Note also that the smaller the size of the unit, the larger the numerical

---

[17]The relationship between the sizes of SI prefixes as described above is intentional and is set to be so to map onto the relationship between place values in the decimal notation. As we will soon see, the correspondence between the relative sizes of SI prefixes and place values in the decimal notation makes it possible to change units within SI by simply moving the decimal point.

Note that the choice of 10 in relating the sizes of SI prefixes is not necessarily *natural*. In the Imperial System, as an example, 1 ft is equal to 12 in.

value that accompanies it. This makes sense: If we make the size of the unit of, say, length, smaller, more of them fit into a given length. As an example, a decimetre is 10 times smaller in size than a metre. Therefore, 10 times as many decimetres fit into a given length compared to the number of metres that fit into that length.[18] This implies that in going from the metre to the decimetre on the metric chart we need to multiply the numerical value by 10 and in going from the decimetre to the metre on the metric chart we need to divide the numerical value by 10. This relationship holds for all adjacent prefixes: In moving from one prefix to the one on its right, we need to multiply the numerical value by 10 and in moving from one prefix to the one on its left, we need to divide the numerical value by 10. Figure 8.3.6 below illustrates the idea.

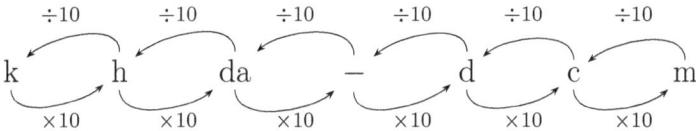

Figure 8.3.6: Change in numerical value corresponding to change in size of unit

The dash at the centre of the chart relates to the case when no prefix is used.

A shortcut for multiplication of a value by 10 is to move the decimal point in that value to the right over one digit and a shortcut for division of a value by 10 is to move the decimal point in that value to the left over one digit. This implies that, in moving from one SI prefix to the one on its right, we need to move the decimal point in the numerical value over one digit to the right and, in moving from one SI prefix to the one on its left, we need to move the decimal point in the numerical value over one digit to the left. These observations form a base for quick conversion of units within SI.

Example

Convert a length of 23.489 dm to cm.

---

[18]The relationship between the numerical value and the size of the unit that accompanies that numerical value is one of inverse proportion. See the companion textbook *Semantics and the Syntax of Algebra* by the author for more on inverse proportion problems.

Solution

We can use the relationship 1 dm = 10 cm to convert the unit in 23.489 dm to cm. Since 1 dm is equal to 10 cm, the conversion of 23.489 dm to cm requires that we multiply the numerical value by 10. A shortcut for multiplication by 10 is to move the decimal point over one digit to the right. This yields

23.489 dm = 234.89 cm

We can also use the metric chart to convert the unit in this example. This is shown in Figure 8.3.7 below.

Figure 8.3.7: Conversion of unit from decimetre to centimetre using the metric chart

As Figure 8.3.7 shows, we begin by associating the current location of the decimal point to the unit dm. The decimal point moves to the right over one digit as we move from dm to cm resulting in a value of 234.89 cm.

Example

Convert a length of 23.489 dm to mm.

Solution

We can use the relationship 1 dm = 100 mm to convert the unit in 23.489 dm to mm. Since 1 dm is equal to 100 mm, the conversion of 23.489 dm to mm requires that we multiply the numerical value by 100. A shortcut for multiplication by 100 is to move the decimal point over two digits to the right. This yields

23.489 dm = 2348.9 mm

We can also use the metric chart to convert the unit in this example. This is shown in Figure 8.3.8 below.

| 2 | 3 . | 4 | 8 | 9 |

dm   cm   mm

Figure 8.3.8: Conversion of unit from decimetre to millimetre using the metric chart

As Figure 8.3.8 shows, we begin by associating the current location of the decimal point to the unit dm. The decimal point moves to the right over one digit as we move from dm to cm and over another digit as we move from cm to mm resulting in a value of 2348.9 mm.

Example

Convert a length of 234.89 cm to dm.

Solution

We can use the relationship 1 dm = 10 cm to convert the unit in 234.89 cm to dm. Since 1 dm is equal to 10 cm, the conversion of 234.89 cm to dm requires that we divide the numerical value by 10. A shortcut for division by 10 is to move the decimal point over one digit to the left. This yields

$$234.89 \, \text{cm} \ = \ 23.489 \, \text{dm}$$

We can also use the metric chart to convert the unit in this example. This is shown in Figure 8.3.9 below.

As Figure 8.3.9 shows, we begin by associating the current location of the decimal point to the unit cm. The decimal point moves to the left over one digit as we move from cm to dm resulting in a value of 23.489 dm.

2   3   4   .   8   9

dm   cm

Figure 8.3.9: Conversion of unit from centimetre to decimetre using the metric chart

Example

Convert a length of 2348.9 mm to dm.

Solution

We can use the relationship 1 dm = 100 mm to convert the unit in 2348.9 mm to dm. Since 1 dm is equal to 100 mm, the conversion of 2348.9 mm to dm requires that we divide the numerical value by 100. A shortcut for division by 100 is to move the decimal point over two digits to the left. This yields

$$2348.9 \text{ mm} = 23.489 \text{ dm}$$

We can also use the metric chart to convert the unit in this example. This is shown in Figure 8.3.10 below.

2   3   4   8   .   9

dm   cm   mm

Figure 8.3.10: Conversion of unit from millimetre to decimetre using the metric chart

As Figure 8.3.10 shows, we begin by associating the current location of the decimal point to the unit mm. The decimal point moves to the left over one digit as we move from mm to cm and over another digit as we move from cm to dm resulting in a value of 23.489 dm.

What the examples above show is that movement on the metric chart is matched by movement of the decimal point in the numerical value: A one-step move to the right on the metric chart implies the movement of the decimal point in the numerical value over one digit to the right, and a two-step move to the right on the metric chart implies the movement of the decimal point in the numerical value over two digits to the right. Furthermore, a one-step move to the left on the metric chart implies the movement of the decimal point in the numerical value over one digit to the left, and a two-step move to the left on the metric chart implies the movement of the decimal point in the numerical value over two digits to the left.

These observations can be generalized: An n-step move to the right on the metric chart implies movement of the decimal point in the numerical value over n digits to the right and an n-step move to the left on the metric chart implies movement of the decimal point in the numerical value over n digits to the left.

Example

Use the metric chart to convert a distance of 0.3457 dam to cm.

Solution

We can use the relationship 1 dam = 1000 cm to convert the unit in 0.3457 dam to cm. Since 1 dam is equal to 1000 cm, the conversion of 0.3457 dam to cm requires that we multiply the numerical value by 1000. A shortcut for multiplication by 1000 is to move the decimal point over three digits to the right. This yields

$$0.3457 \, \text{dam} \; = \; 345.7 \, \text{cm}$$

We can also use the metric chart to convert the unit in this example. This is shown in Figure 8.3.11 below.

As Figure 8.3.11 shows, we begin by associating the current location of the decimal point to the unit dam. The decimal point moves to the right over one digit as we move from dam to m, over another digit as we move from m to dm and over one more digit as we move from dm to cm resulting in a value of 345.7 cm.

If you run out of digits as you move the decimal point to the right or to the left, add as many 0s as needed. Here are two examples.

$$0 \quad . \quad 3 \quad 4 \quad 5 \quad 7$$

dam   m   dm   cm

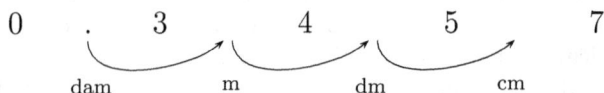

Figure 8.3.11: Conversion of unit from decametre to centimetre using the metric chart

Example

Convert a distance of 3.4 hm to dm.

Solution

We can use the relationship 1 hm = 1000 dm to convert the unit in 3.4 hm to dm. Since 1 hm is equal to 1000 dm, the conversion of 3.4 hm to dm requires that we multiply the numerical value by 1000. A shortcut for multiplication by 1000 is to move the decimal point over three digits to the right. This yields

$$3.4 \text{ hm} \ = \ 3400 \text{ dm}$$

We can also use the metric chart to convert the unit in this example. This is shown in Figure 8.3.12 below.

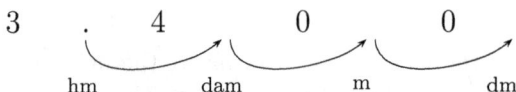

$$3 \quad . \quad 4 \quad 0 \quad 0$$

hm   dam   m   dm

Figure 8.3.12: Conversion of unit from hectometre to decimetre using the metric chart

As Figure 8.3.12 shows, we begin by associating the current location of the decimal point to the unit hm. The decimal point moves to the right over one digit as we move from hm to dam, over another digit as we move from dam to m and over one more digit as we move from m to dm resulting in a value of 3400 dm.

Example

Convert a length of 5.3 dm to dam.

Solution

We can use the relationship 1 dam = 100 dm to convert the unit in 5.3 dm to dam. Since 1 dam is equal to 100 dm, the conversion of 5.3 dm to dam requires that we divide the numerical value by 100. A shortcut for division by 100 is to move the decimal point over two digits to the left. This yields

$$5.3 \text{ dm} = 0.053 \text{ dam}$$

We can also use the metric chart to convert the unit in this example. This is shown in Figure 8.3.13 below.

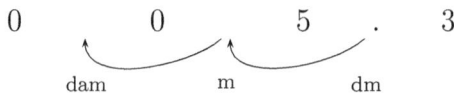

Figure 8.3.13: Conversion of unit from decimetre to decametre using the metric chart

As Figure 8.3.13 shows, we begin by associating the current location of the decimal point to the unit dm. The decimal point moves to the left over one digit as we move from dm to m and over another digit as we move from m to dam resulting in a value of 0.053 dam.

As we noted earlier, metric prefixes can be used along with any unit, not just the metre. Here is an example of conversion of units using the gram as the unit.

Example

Convert a mass of 81 cg to dag.

Solution

We can use the relationship 1 dag = 1000 cg to convert the unit in 81 cg to dag. Since 1 dag is equal to 1000 cg, the conversion of 81 cg to dag requires that we divide the numerical value by 1000. A shortcut for division by 1000 is to move the decimal point over three digits to the left. This yields

$$81 \text{ cg} = 0.081 \text{ dag}$$

We can also use the metric chart to convert the unit in this example. This is shown in Figure 8.3.14 below.

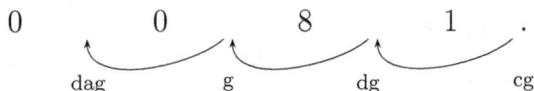

Figure 8.3.14: Conversion of unit from centigram to decagram using the metric chart

As Figure 8.3.14 shows, we begin by associating the current location of the decimal point to the unit cg. The decimal point moves to the left over one digit as we move from cg to dg, over another digit as we move from dg to g and over one more digit as we move from g to dag resulting in a value of 0.081 dag.

Conversion of Units Involving Exponents

Certain quantities involve units that are raised to an exponent. An example of such a unit is the unit of area, the square metre, with unit symbol m^2. Another example of such a unit is the unit of volume, the cubic metre, with unit symbol m^3. Exponents may appear on units other than the unit of length as well as in the unit of acceleration, i.e., m/s^2.[19]

We will explain the manner in which SI units involving exponents can be converted to other SI units using area and volume as examples.

---

[19]Such expressions are shortened versions of longer ones that carry semantics with them and, if meaning is sought, one will have to return to an examination of these longer expressions. As an example, to make sense of an acceleration of 2.5 m/s^2, we return to the origin of the unit symbol as $\frac{m/s}{s}$. The expression 2.5 $\frac{m/s}{s}$ tells us that the speed is increasing by 2.5 m/s every second.

Conversion of Units of Area

The SI unit of area is the square metre which refers to the size of the area covered by a square with a side length of 1 m. Such a square is shown in Figure 8.3.15 below.[20] The square metre is a derived unit and its unit symbol is m^2.

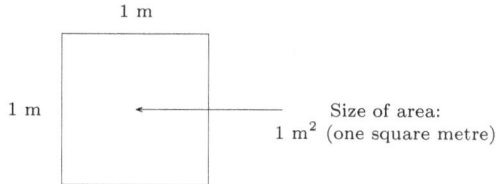

1 m

1 m

Size of area:
$1 \ m^2$ (one square metre)

Figure 8.3.15: The SI unit of area is the square metre, m^2, defined as the area enclosed by a square with a side length of 1 m

When we say *the area of this garden is* 12.5 m^2, it means that, *regardless of its actual shape*, the garden covers an area that is equal in size to the area covered by 12.5 squares of side length 1 m.

In addition to the square metre, SI allows the use of larger and smaller units of area. Examples of these are the square kilometre with unit symbol km^2, the square centimetre with unit symbol cm^2, and the like. The first covers an area equivalent to the area covered by a square of side length 1 km. The second covers an area covered by a square of side length 1 cm. The metric chart for units of area is given in Figure 8.3.16 below.

$$km^2 \qquad hm^2 \qquad dam^2 \qquad m^2 \qquad dm^2 \qquad cm^2 \qquad mm^2$$

Figure 8.3.16: SI units of area

Consider, once again, the square of Figure 8.3.15. The side length of this square is 1 m so that its area is 1 m^2. We now divide each side of the unit square into 10 equally-sized pieces. Since 1 m is equal to 10 dm, each piece will have a length of 1 dm. This is shown below.

---

[20]The drawing is *not* to scale.

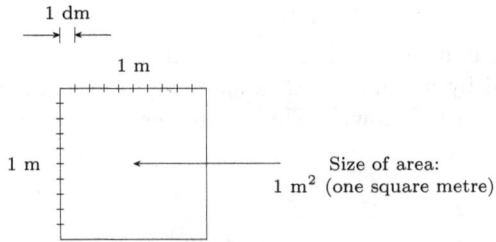

Figure 8.3.17: Subdividing the sides of a 1 m^2
square into 10 parts

We now subdivide the large square into 100 smaller squares as shown below.

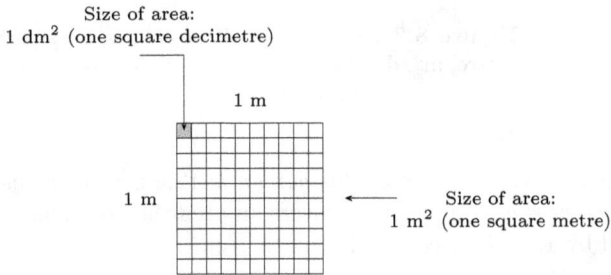

Figure 8.3.18: Subdivision of the unit square into
100 smaller squares

Since the length of the side of each small square is 1 dm, the area of each small square is 1 dm^2. There are 100 small squares in the large square.[21] Therefore

$$1 \text{ m}^2 = 100 \text{ dm}^2$$

---

[21]The correct interpretation of the unit of area for dm^2 is the one that we have given here. It is *incorrect* to view dm^2 as "d" (i.e., deci) followed by m^2. An interpretation that would result in dm^2 being one-tenth of a square metre. When we write dm^2, it is understood to mean dm×dm and not d m×m. Similar comments apply to km^2, hm^2, etc.

Similar examples yield the following relationships between adjacent units of area on the metric chart.

$$1 \text{ km}^2 = 100 \text{ hm}^2$$
$$1 \text{ hm}^2 = 100 \text{ dam}^2$$
$$1 \text{ dam}^2 = 100 \text{ m}^2$$
$$1 \text{ m}^2 = 100 \text{ dm}^2$$
$$1 \text{ dm}^2 = 100 \text{ cm}^2$$
$$1 \text{ cm}^2 = 100 \text{ mm}^2$$

This shows that in order to convert between adjacent units of area on the metric chart, we either multiply by 100 (if we move to the right) or divide by 100 (if we move to the left); See Figure 8.3.19 below.

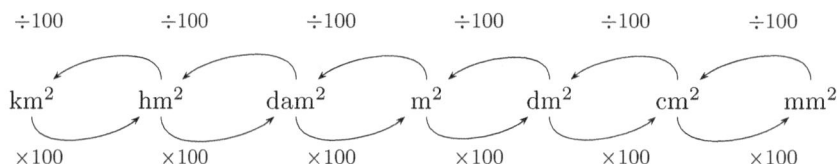

Figure 8.3.19: Change in numerical value corresponding to change in size of unit of area

The relationships listed above also imply that each unit is 10 000 times larger than the unit that is two steps on its right, 1 000 000 times larger than the unit that is three steps on its right, and so on. Relating each unit to those on its left, we can say that each unit is 10 000 times smaller than the unit that is two steps on its left, 1 000 000 times smaller than the unit that is three steps on its left, and so on.

As in the case of length, the decimal point moves in the same direction as the direction of change in the unit on the metric chart. The difference is that now, for each move to the right or left on the metric chart, the decimal point in the numerical value moves in that direction over *two* digits.

Example

Convert 0.475 m^2 to cm^2.

Solution

We can use the relationship $1 \text{ m}^2 = 10\,000 \text{ cm}^2$ to convert the unit in 0.475 m^2 to cm^2. Since 1 m^2 is equal to $10\,000 \text{ cm}^2$, the conversion of 0.475 m^2 to cm^2 requires that we multiply the numerical value by $10\,000$. A shortcut for multiplication by $10\,000$ is to move the decimal point over four digits to the right. This yields

$$0.475 \text{ m}^2 = 4750 \text{ cm}^2$$

We can also use the metric chart to convert the unit in this example. This is shown in Figure 8.3.20 below.

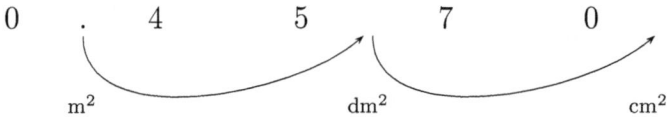

Figure 8.3.20: Conversion of unit of area from square metre to square centimetre using the metric chart

As Figure 8.3.20 shows, we begin by associating the current location of the decimal point to the unit m^2. The decimal point moves to the right over two digits as we move from m^2 to dm^2 and over another two digits as we move from dm^2 to cm^2 resulting in a value of 4570 cm^2.

Example

Convert 34.7 dam^2 to hm^2.

Solution

We can use the relationship $1 \text{ hm}^2 = 100 \text{ dam}^2$ to convert the unit in 34.7 dam^2 to hm^2. Since 1 hm^2 is equal to 100 dam^2, the conversion of 34.7 dam^2 to hm^2 requires that we divide the numerical value by 100. A shortcut for division by 100 is to move the decimal point over two digits to the left. This yields

$$34.7 \text{ dam}^2 = 0.347 \text{ hm}^2$$

We can also use the metric chart to convert the unit in this example. This is shown in Figure 8.3.21 below.

Figure 8.3.21: Conversion of unit of area from square decametre to square hectometre using the metric chart

As Figure 8.3.21 shows, we begin by associating the current location of the decimal point to the unit dam^2. The decimal point moves to the left over two digits as we move from dam^2 to hm^2 resulting in a value of 0.347 hm^2.

Conversion of Units of Volume

The SI unit of volume is the cubic metre which is equivalent to the volume enclosed by a cube with a side length of 1 m. Such a cube is shown in Figure 8.3.22 below.[22] The cubic metre is a derived unit and its unit symbol is m^3.

When we say *the volume of this tank is* 12.5 m³, it means that, *regardless of its actual shape*, the tank has a volume that is equivalent to the volume enclosed by 12.5 cubes of side length 1 m.

In addition to the cubic metre, SI allows the use of larger and smaller units of volume. Examples of these are the cubic kilometre with unit symbol km^3, the cubic centimetre with unit symbol cm^3, and the like. The first encloses a volume equivalent to the volume enclosed by a cube of side length 1 km. The second encloses a volume equivalent to the volume enclosed by a cube of side length 1 cm. The metric chart for units of volume is given in Figure 8.3.23 below.

---

[22]The drawing is *not* to scale.

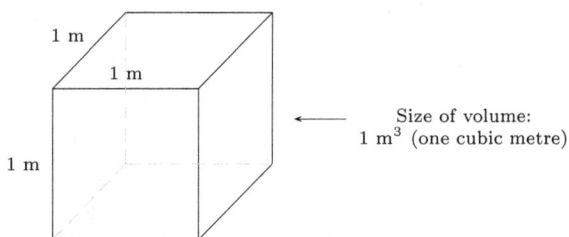

Figure 8.3.22: The SI unit of volume is the cubic metre, m^3, defined as the volume enclosed by a cube with a side length of 1 m

$$km^3 \qquad hm^3 \qquad dam^3 \qquad m^3 \qquad dm^3 \qquad cm^3 \qquad mm^3$$

Figure 8.3.23: SI units of volume

Each unit of volume on the metric chart is 1000 times larger than the unit on its right.[23] This leads to the following relationships between adjacent units of volume on the metric chart.

$$1 \text{ km}^3 = 1000 \text{ hm}^3$$
$$1 \text{ hm}^3 = 1000 \text{ dam}^3$$
$$1 \text{ dam}^3 = 1000 \text{ m}^3$$
$$1 \text{ m}^3 = 1000 \text{ dm}^3$$
$$1 \text{ dm}^3 = 1000 \text{ cm}^3$$
$$1 \text{ cm}^3 = 1000 \text{ mm}^3$$

The practical consequence of this result is that each move on the metric chart will now correspond to the movement of the decimal point over *three* digits.

The relationships listed above also imply that each unit is 1 000 000 times larger than the unit that is two steps on its right, 1 000 000 000 times larger than the unit that is three steps on its right, and so on. Relating each unit to those on its left, we can say that each unit is 1 000 000 times smaller than the unit that is two steps on its left, 1 000 000 000 times smaller than the unit that is three steps on its left, and so on.

---

[23]This can be shown to be true through the subdivision of the unit cube into smaller cubes in a manner similar to the case given for area. We leave the details to you to work out.

Example

Convert $42\,000$ mm^3 to cm^3.

Solution

We can use the relationship 1 cm^3 = 1000 mm^3 to convert the unit in $42\,000$ mm^3 to cm^3. Since 1 cm^3 is equal to 1000 mm^3, the conversion of $42\,000$ mm^3 to cm^3 requires that we divide the numerical value by 1000. A shortcut for division by 1000 is to move the decimal point over three digits to the left. This yields

$$42\,000 \text{ mm}^3 = 42 \text{ cm}^3$$

We can also use the metric chart to convert the unit in this example. This is shown in Figure 8.3.24 below.

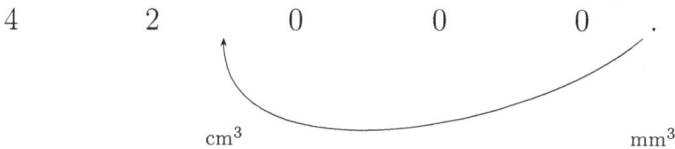

Figure 8.3.24: Conversion of unit of volume from cubic millimetre to cubic centimetre using the metric chart

As Figure 8.3.24 shows, we begin by associating the current location of the decimal point to the unit mm^3. The decimal point moves to the left over three digits as we move from mm^3 to cm^3 resulting in a value of 42 cm^3.

The setting of the relative sizes of adjacent units on the metric chart to match the relative sizes of place values in base-ten notations is one of the greatest achievements of SI. Note the ease with which units can be converted within SI: One simply moves the decimal point back and forth, an activity that does not require the use of pen and paper or even a calculator. Compare this to conversion of feet to inches. Since 1 ft equals 12 in, to convert feet to inches we need to multiply the numerical value of length by 12. This requires the use of pen and paper or a calculator. Conversion of inches to feet is even worse as it requires that we divide the numerical value of length by 12. Conversion of such units gets even more complex as we move from length to area and volume: Since 1 ft^2 equals 144 in^2 (from 12 in × 12 in), to convert square feet to square inches one would have to multiply the numerical value of area by 144. Going from square inches to square feet one would have to divide the numerical value of area by 144. For volume, 1 ft^3 equals 1728 in^3 (from 12 in × 12 in × 12 in), to convert cubic feet to cubic inches one would have to multiply the numerical value of volume by 1728. Going from cubic inches to cubic feet one would have to divide the numerical value of volume by 1728.

Conversions Involving a Mix of Units

Certain units involve a mix of other units. Examples of such units are the unit of speed, i.e., m/s, the unit of force in terms of base units, i.e., kg·m/s^2, and the like. For a detailed discussion of conversions of such units we refer the reader to the companion textbook *Semantics and the Syntax of Algebra* by the author.

Exercise Set 8.3

1. List the SI prefixes by name and state their symbols and sizes relative to the unit.

2. In each case state the relationship between the given pair of units. Choose 1 for the first unit in the given pair.

   a. km and m   e. km and hm   i. cm and hm
   b. m and cm   f. hm and dm   j. dam and km
   c. m and mm   g. hm and cm   k. m and dam
   d. cm and mm   h. mm and dm   l. dm and m

3. In each case state the relationship between the given pair of units. Choose 1 for the first unit in the given pair.

   a. kg and dag   d. mN and cN   g. cW and daW
   b. mol and mmol   e. dacd and cd   h. K and cK
   c. A and cA   f. hJ and dJ   i. cL and mL

4. In each case state the relationship between the given pair of units. Choose 1 for the first unit in the given pair.

a. m^2 and dm^2
b. m^2 and cm^2
c. m^2 and mm^2

d. cm^2 and m^2
e. hm^2 and km^2
f. hm^3 and m^3

g. dam^3 and dm^3
h. dm^3 and dam^3
i. cm^3 and mm^3

5. In each case state the relationship between the given pair of units. Choose 1 for the first unit in the given pair.

a. s^2 and ms^2
b. K^2 and daK^2
c. mol^2 and $dmol^2$

d. cL^2 and dL^2
e. hg^2 and kg^2
f. hs^3 and das^3

g. $mmol^3$ and $cmol^3$
h. mA^3 and dA^3
i. daN^3 and N^3

6. In each case convert the unit as requested.

a. 42.5 m to cm
b. 3.61 m to mm
c. 810 km to dam

d. 0.035 cm to mm
e. 0.241 dm to m
f. 425 dam to km

g. 3100 dm to hm
h. 0.81 m to dam
i. 14.1 dm to km

7. In each case convert the unit as requested.

a. 3.8 dag to cg
b. 24.6 L to mL
c. 35.4 daN to hN

d. 4100 daJ to kJ
e. 1.31 cPa to hPa
f. 0.00236 kW to W

g. 192 mK to dK
h. 0.88 A to hA
i. 13.5 dmol to damol

8. In each case convert the unit as requested.

a. $13.8\ m^2$ to dm^2
b. $4.25\ m^2$ to mm^2
c. $1400\ km^2$ to dam^2

d. $24.9\ cm^2$ to mm^2
e. $0.0712\ dam^2$ to hm^2
f. $1.52\ dam^2$ to dm^2

g. $44.8\ dm^2$ to m^2
h. $47.3\ m^2$ to hm^2
i. $1324\ dm^2$ to hm^2

9. In each case convert the unit as requested.

a. $37.8\ m^3$ to cm^3
b. $0.612\ m^3$ to mm^3

c. $4100\ hm^3$ to dam^3
d. $0.0052\ dm^3$ to cm^3

e. $0.147\ dm^3$ to m^3
f. $637\ dam^3$ to km^3

10. In each case convert the unit as requested.

a. $42.5\ s^2$ to ms^2
b. $0.98\ mol^2$ to $dmol^2$

c. $425\ dam^2$ to km^2
d. $247\ K^3$ to daK^3

e. $1.26\ mA^3$ to cA^3
f. $0.67\ J^3$ to cJ^3

8.4   SI Conventions

SI, like most other measurement systems, provides a list of conventions in addition to the list of quantities and their associated units. For the most part these conventions aim to provide a uniform communication protocol. In this section we will list some of the more important conventions in SI.

In SI, quantity symbols are typed using the italic font style regardless of the font of the surrounding text. This helps differentiate between quantity symbols and unit symbols which are typed using the upright font style. As an example, the symbol m is used to represent the value of the quantity *mass* while the symbol m is used to represent the size of the unit *metre*.

Case matters: The uppercase letter, V, as an example, is used as the quantity symbol for the representation of the value of the quantity *volume* whereas the lowercase version of the letter, v, is used as the quantity symbol for the representation of the value of the quantity *speed*. As a second example, the uppercase letter, P, is used as the quantity symbol for the representation of the value of the quantity *power* while the lowercase version of the letter, p, is used to denote the value of the quantity *pressure*.

Since quantity symbols represent the *values* of quantities, they should not be viewed as abbreviations for the names of the quantities whose values they represent. They are, therefore, not subject to the rules of English grammar. As an example, the period should not be used after a quantity symbol. As another example, the plural s should not be used along with quantity symbols: The representation ms is not interpreted as the plural form of mass but the quantity symbol m multiplied by another quantity symbol s.

If necessary, further specification about a quantity is written as a subscript to the quantity symbol. As an example, in a problem where both the value of the mass of an electron and the value of the mass of a proton are being discussed, we can represent the former as m_e and the latter as m_p. As another example, if values of energies of kinetic and potential varieties are under investigation, we may choose to represent the former as E_k and the latter as E_p. Note that *in all cases the main letter is the quantity symbol and the subscript acts as a qualifier.* Therefore, we write m_e (and not e_m) to refer to the value of the mass of an electron and we write E_k (and not K_e) to refer to the value of kinetic energy.[24] Such subscripts may consist of entire words. As an example, some choose to represent the value of kinetic energy as $E_{kinetic}$ as opposed to E_k.

---

[24]This convention is followed by some but not all practitioners. Physicists tend to follow this convention closely. Others, notably those active in branches of chemistry and the biological sciences, on the other hand, frequently break this convention. As an example, it is common to see K_e, or even worse, KE, or some such similar representations of the value of kinetic energy in chemistry or biology textbooks.

Unit symbols are typed using the upright font style regardless of the font of the surrounding text. This helps differentiate between unit symbols and quantity symbols which, as we saw earlier, are typed using the italic font style. As an example, we write m to refer to the size of the unit *metre*, not *m*.

Case matters: The unit symbol mm, as an example, is used to represent the size of the unit *millimetre* which is a unit of length whose size is one-thousandth of the size of the metre whereas the unit symbol Mm is used to represent the size of the unit *megametre* which is a unit of length whose size is equal to one million metres.

Since unit symbols represent the *sizes* of units, they should not be viewed as abbreviations for the names of units whose sizes they represent. They are, therefore, not subject to the rules of English grammar. As an example, the period should not be used after a unit symbol. As another example, the plural *s* should not be used along with unit symbols: The representation ms is not interpreted as the plural form of metre but as the unit millisecond.

In SI, unit names begin with lowercase letters unless they appear at the beginning of a sentence or in titles where the first letters in the words in the title are capitalized.[25] This helps differentiate between the unit and the individual when the unit name is fashioned after the name of an individual. As an example, the word *kelvin* refers to the unit of temperature whereas *Kelvin* refers to the individual whose name was used in the naming of this particular unit of temperature.

An empty space should separate the numerical value of a quantity and the unit symbol as shown in Figure 8.4.25 below.[26] The empty space is interpreted as multiplication.

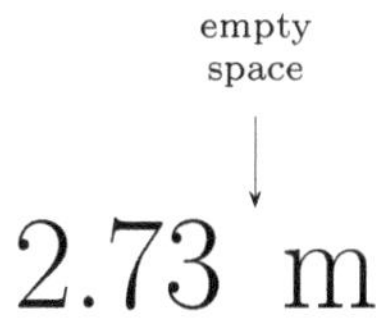

empty
space

$$2.73 \; \text{m}$$

Figure 8.4.25: An empty space should separate the numerical value of a quantity and the unit symbol

---

[25]We emphasize that the conventions listed in this section are those of SI. Other measurement systems may or may not follow these conventions.

[26]Note that 4.5 m is interpreted as 4.5 metres whereas $4.5m$ is interpreted as 4.5 times the value of the mass.

SI encourages the use of base-ten notations and in particular the decimal notation in the expression of the numerical values of quantities. Not only is the decimal notation easier to use (compare, round, add, subtract, multiply, divide, etc.) compared to other notations such as the formal notation, its use also facilitates unit conversion within SI[27] and naturally yields itself to the application of the rules that govern the rounding of the results of the application of arithmetic operations to measured values; a topic that we will cover in the next chapter.

In case the numerical value of a quantity has 0 as its whole part, the 0 must be written down. As an example, we write 0.25 m as opposed to .25 m. This convention forces the reader's attention to the decimal point which might otherwise be easily missed.[28]

When units are multiplied they should be written with an empty space between them otherwise the interpretation changes. Another possibility is to use the multiplication dot between the units being multiplied. As an example, the product of the metre and the second should be written as either

m s

or

m · s

but not ms which represents the size of the unit millisecond.

For division of units we can use either the horizontal line or the forward slash.[29] Of the two, the use of the forward slash is more common as the forward slash is read as *per* as in m/s which is read as metres per second. The use of the forward slash requires the use of brackets to group products in the denominator. As an example, the use of the forward slash in place of the horizontal line in

$$\frac{\text{Pa} \cdot \text{m}^3}{\text{mol} \cdot \text{K}}$$

requires that we write

$$\text{Pa} \cdot \text{m}^3 / (\text{mol} \cdot \text{K})$$

---

[27] One of the main strengths of SI is in its use of the metric chart which allows the movement of the decimal point to convert units within SI. This advantage is lost if the decimal notation is not used.

[28] Missing the decimal point makes the value of a quantity at least 10 times larger. As an example, 0.25 m is equal to a quarter of a metre. Removing the decimal point changes the value of the quantity to 25 m which is 100 times larger than 0.25 m.

[29] Negative exponents are also allowed and present a more systematic presentation. As an example, the unit symbol m/s^2 may be written as m·s^{-2}.

Removal of brackets in the above, i.e.,

$$\mathrm{Pa} \cdot \mathrm{m}^3 / \mathrm{mol} \cdot \mathrm{K}$$

would change the interpretation to

$$\frac{\mathrm{Pa} \cdot \mathrm{m}^3}{\mathrm{mol}} \cdot \mathrm{K}$$

The conventions covered in this section of the textbook represent the most important conventions of SI. There are others however and we invite the interested reader to read the document *SI Brochure* for a study of other conventions of SI.

Exercise Set 8.4

1. What is the reason for having conventions in SI?

2. What font style should be used in the display of quantity symbols in SI? Give an example.

3. Why is the casing of letters used as quantity symbols important in SI? Provide an example to illustrate the point.

4. In each case use subscripts on quantity symbols to denote the value of each of the following quantities.

   a. The mass of a neutron
   b. kinetic energy
   c. mass of a molecule of CO_2
   d. the volume of a cube
   e. the temperature of the sun
   f. water pressure
   g. the current in the wire
   h. energy content of carbohydrates
   i. the width of the lot
   j. the speed of the car

5. What font style should be used in the display of unit symbols in SI? Give an example.

6. Why is the casing of letters used as unit symbols important in SI? Provide an example to illustrate the point.

7. Why do unit names in SI begin with lowercase letters? Give an example to illustrate your answer.

8. What does the empty space between the numerical value of a quantity and the unit symbol that follows it represent in SI? Give an example to illustrate your answer.

9. Why are base-ten notations the preferred notation for the expression of numerical values of quantities in SI?

10. List the different notations used to represent multiplication of units in SI. Give an example.

11. List the different notations used to represent division of units in SI. Give an example.

Chapter 9
Working with Measured Values

In this chapter we will discuss the manner in which one works with measured values. We begin with the classification of measured values as approximate or exact. Following this we will present three different schemes that are commonly used to process mathematical expressions that involve measured values.

9.1 Measured Values

In this section we will discuss the classification of measured values as either approximate or exact. Along the way we will introduce the all-important notions of significant digits and the estimated digit.

9.1.1 Approximate Measured Values

Measured values other than those that result from direct counting bear a measure of uncertainty. To explain why, consider the task of measuring the length of a rod using a measuring tape. To do so, we line up the 0 on the measuring tape with one end of the rod and then read the numerical value that lines up with the other end of the rod. Let us suppose that the measuring tape used is marked in steps of 10 cm and that the end of the rod falls somewhere between the tick that is labelled as 30 and the tick that is labelled as 40. This is shown in Figure 9.1.1 below.

We can immediately tell that the numerical value of the length of the rod is greater than 30 and less than 40. Now, in your mind's eye break the interval from 30 to 40 into 10 parts[1] and guess at the number of ones that fit

---

[1]The choice of 10 and its multiples as the number of subdivisions of a unit is based on the practice promoted by SI. Other measurement systems may or may not follow this convention. For the sake of uniformity we will only use 10 as the number of subdivisions

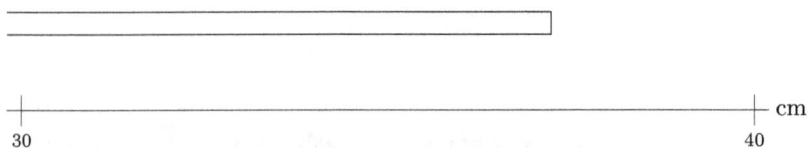

Figure 9.1.1: Measuring the length of a rod using a measuring
tape with tens as its smallest division

into the length extending beyond the 30 cm mark. A reasonable value would
be 7 so that our reading now is 37 cm. Based on the process described above,
digit 3 is certain to be correct while digit 7 is estimated.[2]

We could, of course, get a more accurate reading for the numerical value
of the length of the rod but, to do so, we would need a measuring instrument
with smaller divisions. As an example, we can come up with a more accurate
numerical value for the length of the rod in the example above if we use a
measuring tape that has ones as its smallest division. This is shown in Figure
9.1.2 below.

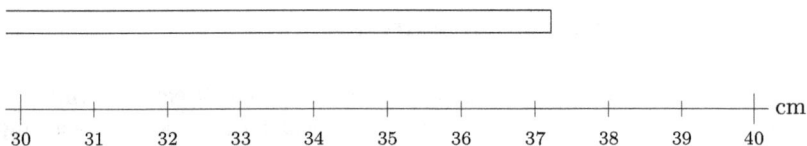

Figure 9.1.2: Measuring the length of a rod using a measuring
tape with ones as its smallest division

---

of a unit in this textbook. Other subdivisions, including those that are based on multiples
of 10, may be treated in a similar manner.

[2]Note that, in the example above, it would be unreasonable to write a numerical value
such as 37.2173 as the numerical value for the length of the rod. The measuring instrument's
smallest divisions set a limit on what digits can be reasonably written down. In the case of
the problem above it would be reasonable to guess at the number of ones that fit into the
numerical value of the length of the rod beyond the 30 cm mark but, with few exceptions,
it would be unreasonable to expect one to guess at the number of tenths that fit into the
numerical value of the length of the rod beyond 37. And it would certainly be questionable
whether there is any validity to digits that pertain to thousandths, ten thousandths, etc.,
of the smallest division on the measuring device.

We can now tell that the numerical value of the length of the rod falls between 37 and 38. Now, in your mind's eye break the interval from 37 to 38 into 10 parts and guess at the number of tenths that fit into the length extending beyond the 37 cm mark. A reasonable value would be 2 so that our reading now is 37.2 cm. Based on the process described above, digits 3 and 7 are certain to be correct while digit 2 is estimated.

By now the reader must have noticed that we can never come up with a value that is an exact measure of the length of the rod. We can use ever more accurate measuring tapes with finer and finer divisions to come up with more and more accurate values for the length of the rod but we will never be able to come up with an exact value for the length of the rod as, at some point, we will have to estimate the rightmost digit.

One might wonder whether it is not possible for the end of the rod to fall right on a tick but note that even if this happens to be so, we would never be certain that it is not off by a tiny amount. As an example, we set the length of the rod in Figure 9.1.3 to be 37.001 cm while the figure seems to imply that the value of the length of the rod is exactly 37 cm.[3]

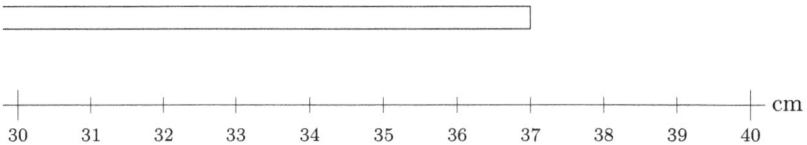

Figure 9.1.3: Does the end of the rod fall exactly on the value 37 cm?

The understanding that at some point we need to estimate a digit as described above allows us to infer the smallest division on the measuring tape used from the numerical value of the length of the rod. As an example, a reading of 37 cm implies that digit 7 is estimated. Since digit 7 is in the ones place, the smallest division on the measuring tape must have been tens. Similarly, the reading 37.2 cm implies that digit 2 is estimated. Since digit 2 is in the tenths place, the smallest division on the measuring tape must have

---

[3]This is not due to the discrete nature of the display in digital devices. The same argument would hold if you were looking at an actual measuring tape and an actual rod.

Note also that physical objects are not infinitely smooth. As we try to refine our reading more and more by using an ever more accurate measuring tape, we will inevitably have to deal with the lack of smoothness of the end of the rod which makes it difficult to decide where that end is.

been ones. The smallest division on the measuring tape, then, is represented by the place value of the digit on the left side of the estimated digit.

Example

The length of a beam is measured as 2.31 m. What is the smallest division on the measuring tape used?

Solution

Since digit 1 is estimated, the smallest division on the measuring tape is represented by the place value of digit 3, i.e., tenths.

The understanding that the place value of the digit on the left side of the estimated digit points to the smallest division on the measuring tape used implies that there is a difference between measured values such as 3.7 and 3.70. The second measured value is more accurate than the first as the measuring tape used to measure the value must have had tenths as its smallest division while the first value must have been measured using a measuring tape that had ones as its smallest division. Note that the reading of 3.7 could have come from any value between 2.65 and 3.74 (inclusive on both ends). The reading of 3.70, on the other hand, could come from any value between 3.695 and 3.704 (inclusive on both ends), a much narrower range of values.

While in performing calculations the rightmost 0 in 3.70 may not matter, from the point of view of the accuracy of the measurement, the 0 plays an important role and, if it happens to be the estimated digit, it should be written down.

Significant Digits

Significant digits are digits in a numerical value that relate to the degree of accuracy in measured values that were used in its calculation. Such digits are present in the measured values themselves but they also appear in the results of performing mathematical operations on measured values.

In this section we will discuss the conventions and associated notations for the identification of significant digits in a numerical value. Following this, we will discuss the reasons behind these conventions.

Conventions on Identification of Significant Digits

1. In a measured value written using the decimal notation

    a. If there is a decimal point in the representation of the numerical value, then all digits other than 0s on the left side of all nonzero digits are significant.

    b. If there is no decimal point in the representation of the numerical value, then all digits other than 0s on the left side or right side of all nonzero digits are significant.

2. In a measure value written using scientific notation all digits in the representation of the significand are significant.[4]

Notwithstanding the above, if there is dot present under a digit, then all digits up to and including the digit with the dot under other than 0s on the left of all nonzero digits are significant.

Convention 1a implies that, in the absence of the dot notation, all nonzero digits are significant. As an example, the numerical value 32.47 has four significant digits. This convention meets the expectation that nonzero digits in a measured value all come from making an act of measurement and, therefore, figure in the accuracy of the measurement. The convention may be overridden through the use of the dot notation. As an example, the numerical value 32.4̣7 has three significant digits (digits 3, 2 and 4). This convention allows us to keep track of digits that reflect the accuracy of the original measurement in numerical values that arise as a result of performing mathematical operations on measured values.[5]

The convention on the 0s on the left side of all nonzero digits effectively excludes them from being classified as being significant. This is reasonable as digits that we care about in a measured value start with the first nonzero digit on the left. As an example, in a measured value such as 4.72 m we hardly care to report that *there were no tens of metres present in the measured value*, an assertion that could be made explicit by writing 4.72 as 04.72, or that *there were no hundreds of metres present in the measured value*, an assertion that could be made explicit by writing 4.72 as 004.72. Such 0s, therefore, are of no interest to us and for this reason we do not consider them as being significant.[6] However, note that the convention also implies that the 0s in a measured value such as 0.0025 m are also not classified as being significant.

---

[4]This explains why the significand is labelled as such. Note also that the structure of the significand does not allow for 0s on the left of all nonzero digits.

[5]As we will see in the next section, certain rounding schemes require that we retain as many digits as possible in numerical values that arise as a result of performing mathematical operations on measured values while at the same time keeping track of those digits in the calculated values that reflect the accuracy of the measured values.

[6]It is for this reason that such left-side 0s are normally not written down.

This is once again due to the fact that the digits that are of importance to us begin with one that is not 0. We do not consider the 0s in the numerical value 0.0025 m as significant as we do not care to report that *there were no tenths of a metre present in the measured value.* However, unlike the previous example, such 0s may not be dropped as this would shift the place values which in turn affects the magnitude of the numerical value.

The convention on the 0s on the right side of the decimal point allows us to exclude such 0s from being significant if they arise as a result of conversion of units in SI. To illustrate, a measured length of 57 km has two significant digits (digits 5 and 7 which come directly from the act of making a measurement). If one were to change units and write the value of this length as 57 000 m, we would still want to count the digits 5 and 7 as significant since these were the original digits that arose as a result of the act of making a measurement but not the 0s that come about as a result of converting the unit from kilometres to metres.[7] In case the 0s in 57 000 m do come from an act of making a measurement as opposed to conversion of units, we can use the dot notation to communicate the fact by writing the measured value as 57 00$\dot{0}$.

Examples

Identify the significant digits in each of the following measured values.

1. 35.29	7. −178.310	13. 5249	19. 8.71×10^{-1}
2. 7.50	8. 3.418 369	14. −2716	20. -6.8×10^{3}
3. −123.08	9. 000 457	15. −124	21. 4.10×10^{-4}
4. 0.0367	10. 0021	16. 1400	22. 3.96×10^{0}
5. 5.20	11. −00573	17. −810	23. -2.95×10^{42}
6. 9.$\dot{2}$0	12. 7150	18. 9260	24. 7.620×10^{51}

Answers

1. 3, 5, 2 and 9	6. 9 and 2
2. 7, 5 and 0	7. 1 and 7
3. 1, 2, 3, 0 and 8	8. 3, 4, 1, 8, 3 and 6
4. 3, 6 and 7	9. 4, 5 and 7
5. 5, 2 and 0	10. 2 and 1

---

[7]In 57 km the measured value is estimated to be between the values 56.5 km and 57.4 km, i.e., it could be in error by 0.5 km. Converting the unit to metres does not make the measurement more accurate. This means that the converted value 57 000 m may be in error by the same amount, i.e., 0.5 km or 500 m.

11. 5, 7 and 3
12. 7, 1 and 5
13. 5, 2, 4 and 9
14. 2, 7, 1 and 6
15. 1, 2 and 4
16. 1 and 4
17. 8 and 1

18. 9, 2 and 6
19. 8, 7 and 1
20. 6 and 8
21. 4, 1 and 0
22. 3 and 9
23. 2, 9 and 5
24. 7, 6, 2 and 0

The Estimated Digit

The estimated digit is the rightmost significant digit.[8]

Examples

In each case identify the estimated digit.

1. 35.29
2. 7.50
3. −123.08
4. 0.0367
5. 5.20
6. 9.2̇0

7. −17̇8.310
8. 3.418 36̇9
9. 000 457
10. 0021
11. −00573
12. 7150

13. 5249
14. −2716
15. −124
16. 1400
17. −810
18. 9260

19. 8.71×10^{-1}
20. -6.8×10^{3}
21. 4.10×10^{-4}
22. $3.9̇6 \times 10^{0}$
23. -2.95×10^{42}
24. 7.620×10^{51}

Answers

1. 9
2. 0
3. 8
4. 7

5. 0
6. 2
7. 7
8. 6

9. 7
10. 1
11. 3
12. 5

13. 9
14. 6
15. 4
16. 4

17. 1
18. 6
19. 1
20. 8

21. 0
22. 9
23. 5
24. 0

9.1.2   Exact Measured Values

As we mentioned at the beginning of this chapter, measured values other than those that result from a direct act of counting have uncertainties associated with them. Measured values that arise from a direct act of counting, on the other hand, have no uncertainties associated with them and are, for this reason, referred to as exact measured values. As an example, when we say

---

[8]Note that this is not the same as *the rightmost digit*. In 4.2̇579 the estimated digit is digit 5 and in 29 000 the estimated digit is digit 9. Neither of these is the rightmost digit but the rightmost *significant* digit.

that there are 20 marbles in the bag, the implication is that there are exactly 20 of them, with no uncertainties. Note that for this claim to exactness to hold, one must make a *direct* count of the number of marbles not calculations that use approximate measured values such as division of the mass of all the marbles by the mass of one marble. Such indirect calculations are subject to uncertainties in the measured values that figure in their calculation.

Some exact values arise from definitions. As an example of such exact values consider the formula for the conversion of the unit of mass from kilograms to grams, i.e.,

$$m_{\mathrm{g}} = 1000 m_{\mathrm{kg}}$$

where m_{g} represents the value of mass in grams and m_{kg} represents the value of mass in kilograms. In this formula the numerical value 1000 is exact. This value comes from the defined relationship between kilo and the unit: By definition there are *exactly* 1000 grams in 1 kilogram. This relationship is not measured but defined and the values involved in its definition are, therefore, exact.

Exact values can also arise from theoretical derivations. As an example, the formula for the perimeter of a square is

$$P = 4s$$

where P represents the value of the perimeter of the square and s represents the value of the length of its side. In this formula the numerical value, 4, is exact:[9] A square has *exactly* 4 sides, not a bit more or less, i.e., it cannot possibly have 3.9 sides or 4.1 sides. In this formula, there are no uncertainties associated with the numerical value 4. By convention, numerical values may only be used in equations if they are exact and arise from theoretical considerations.

Note the difference between an exact value and a constant. Exact values may or may not be constant: The number of toys generated by a toy factory changes as the day progresses but at any point in time we know exactly how many toys are made on that day. Furthermore, constant values, if measured, are not exact: The specific heat capacity of steel, as an example, has a constant value but any measurement of this value such as 490 J/(kg·K) is inherently approximate. In the formula for the force of attraction between two objects of mass m_1 and m_2 at a distance of r from each other, i.e.,

$$F = \frac{G m_1 m_2}{r^2}$$

---

[9]Note that the numerical value arises from counting the number of sides in a square. It may also be seen here as a definition: A square is defined as having 4 sides of equal size with interior angles that are 90°.

G is known to be a constant whose value is measured to be approximately $6.674\,08 \times 10^{-11}$ m^3/(kg·s^2). However, no physics book worthy of its name would write the formula as

$$F = \frac{6.674\,08 \times 10^{-11} m_1 m_2}{r^2}$$

as this leaves the impression that the value is exact.[10] The value $6.674\,08 \times 10^{-11}$ m^3/(kg·s^2) is measured and is, therefore, approximate. The symbol G in the theoretical formula represents the exact value of this constant and it may be replaced by its approximate measured value of $6.674\,08 \times 10^{-11}$ m^3/(kg·s^2) in applications of this formula.

Example

In each case identify the exact values in the given equation or expression.

1. 4.52 kg
2. 12 943 people (measured directly)
3. $d = d_0 + v_0 t + \frac{1}{2} a t^2$
4. 3526 atoms (measured indirectly)
5. $F = \frac{k q_1 q_2}{r^2}$ (k is constant)

Answers

1. The value 4.52 kg is either measured or is based on calculations that use measured values and is, therefore, approximate.
2. The value 12 943 people is exact as it relates to a direct act of counting.
3. The values 1 2 and 2 appear as numerical values in the formula which implies that their derivation is based on theoretical considerations and are, therefore, exact.
4. The value 3526 atoms is based on calculations that use measured values and is, therefore, approximate.
5. 2 appears as a numerical value in the formula which implies that it's derivation is based on theoretical considerations and is, therefore, exact. In addition, k is a constant and is, therefore, exact.

---

[10]This is also at odds with our earlier convention that only the use of formal notation is allowed in the expression of theoretical formulas.

One of the characteristics of an exact value is that every digit in its representation is known. A mass of 4.2 kg may be seen as 4.2??????? kg as the digits in the hundredths, thousandths and so on are unknown. A direct measurement of 120 particles, on the other hand, may be seen as 120.000 000... as we are certain that digits on the right side of the decimal point are all 0s. These rightmost digits in exact values are not normally written down but they are known to be 0. In this sense they are all significant digits and, therefore, an exact value can be said to have an infinite number of significant digits. The estimated digit does not exist as all digits are known to be exact.

Exercise Set 9.1

1. What is the smallest division on the measuring device used to measure each of the following values?

a. 25.8
b. 324
c. 1900
d. −24.18
e. 15.0
f. −8.23
g. 0.00173
h. −0.0413

2. For each measured value below identify the significant digits and the estimated digit.

a. 247
b. 3580
c. 3104
d. −0056
e. 00480
f. −3.24
g. −0.184
h. 0.0029
i. 0.702
j. 41.20
k. 0003.218
l. 2.00
m. 24 300
n. 7200
o. 9420
p. 3000
q. 328
r. 3.2×10^3
s. 1.47×10^{-4}
t. 2.00×10^{-1}

3. What quantity, when measured directly, generates an exact value?

4. Give an example of a defined value that is exact.

5. In each case identify the exact values.

a. $A = \frac{1}{2}bh$
b. $P = 2(l + w)$
c. $E = \frac{1}{2}mv^2$
d. $m_{CO_2} = m_C + 2m_O$

9.2   Rounding Schemes

We now turn our attention to the manner in which measured values are processed. We will begin with an introduction to the scheme from Interval Math, followed by the schemes from the Standard Rule and the Limited-Significant-Digit Rule. These schemes provide guidelines on when and how an act of rounding should be executed as we process expressions that involve approximate and exact measured values.

9.2.1   The Rounding Scheme from Interval Math

The rounding scheme from Interval Math provides a rigorous scheme with rules that are firmly rooted in logic and is capable of providing answers that can be relied on. However, the rules proposed by Interval Math are complex and in most practical situations the level of rigour that they provide seems excessive.

In interval math values of quantities are represented as intervals. As an example, suppose the value of the mass of an object is measured to be 42.7 kg, a value that is ordinarily reported as

$$m = 42.7 \text{ kg}$$

Assuming a metric measuring instrument with the unit as its smallest division has been used to measure the value of this mass, one may argue that the reported value of the mass may be off by 0.05 kg.[11]   This is because, as shown in Figure 9.2.4, any value from $42.7 - 0.05$ to $42.7 + 0.05$, inclusive on $42.7 - 0.05$ but exclusive on $42.7 + 0.05$,[12] would have been estimated as 42.7. The range of values from $42.7 - 0.05$ to $42.7 + 0.05$ is illustrated in Figure 9.2.4 below.

Figure 9.2.4: Range of values corresponding to a measured value of 42.7 kg

In Figure 9.2.4 the range of values that would map onto 42.7 is shown as the heavy line from $42.7 - 0.05$ to $42.7 + 0.05$ with the filled dot on the left indicating inclusion of the value $42.7 - 0.05$ and the open dot on the right indicating exclusion of the value $42.7 + 0.05$.[13]

---

[11] This is equal to half the size of the place value of the estimated digit.

[12] By agreement a value such as $42.7 - 0.05$, i.e., 42.65, would round to 42.7 while a value such as $42.7 + 0.05$, i.e., 42.75, would round to 42.8.

[13] The reader should note the faded appearance of the values 42.6, 42.7, and 42.8 as well as their corresponding ticks in Figure 9.2.4. This is because the subdivisions 42.6, 42.7 and 42.8 appear in the reader's imagination: The measuring instrument has the unit as its smallest division and would only show values such as 42, 43, and the like.

This view requires that the measured value of mass should be reported as a range of values as opposed to a single value. We can report the value of the mass as a range of values using inequalities as shown below.

$$42.7 - 0.05 \text{ kg} \leq m < 42.7 + 0.05 \text{ kg}$$

Simplifying the statement above yields

$$42.65 \text{ kg} \leq m < 42.75 \text{ kg}$$

A more compact notation uses the interval notation[14]

$$m \in [42.65, 42.75) \text{ kg}$$

This statement implies that the measured value of m can range from 42.65 to 42.75 with the square bracket on the left implying inclusion of 42.65 and the parenthesis on the right implying exclusion of 42.75.

Having discussed the manner in which one represents a measured value using the interval notation, we now turn our attention to the manner in which arithmetic operations are performed on such values. We illustrate matters using an example.

Problem

You covered a distance of 18.6 km this morning and a distance of 28 km this afternoon. What total distance did you cover today?

Solution

$$d = [18.55, 18.65) + [27.5, 28.5)$$
$$d = [18.55 + 27.5, 18.65 + 28.5)$$
$$d = [46.05, 47.15) \text{ km}$$

To add the values of two quantities, we add the minimum values and the maximum values. This should make sense as adding the minimum values results in the least possible value for the sum and adding the maximum values results in the greatest possible value for the sum.

As we mentioned earlier, the rounding scheme from Interval Math is quite complex, even for the simple task of adding the measured values of two quantities. Such level of complexity might be needed for more advanced work in the sciences but, at the introductory level, the level of accuracy that they provide is seldom needed and, for this reason, it is often the case that the simpler

---

[14]This is where *Interval Math* gets its name from.

schemes provided by the Standard Rule or the Limited-Significant-Digit Rule are used. However, the reader should note that these simpler schemes are approximate.

9.2.2   The Standard Rule

The scheme from Standard Rule is an approximate alternative to the one from Interval Math. To illustrate how the scheme from the Standard Rule works, we will begin with the case where the expression consists of a chain addition/subtraction of measured values followed by the case where the expression consists of a chain multiplication/division of measured values. We will then combine these rules to present the manner in which expressions that consist of a mix of the four basic arithmetic operations are dealt with. We end the section with a note on the manner in which the scheme from Standard Rule deals with expressions that involve more complex operations on measured values.

Chain Addition/Subtraction

The algorithm for the evaluation of expressions that involve chain addition/subtraction of measured values is given below.

Algorithm for Evaluation of Expressions Involving Chain Addition/Subtraction of Measured Values

1. Finish the chain addition/subtraction exactly with no intermediate roundings and write the result down.
2. Determine the place values of the estimated digits in each measured value in the original expression.[15]
3. Round the result in Step 1 to the largest place value found in Step 2.

Example

Evaluate and round the result according to the scheme from Standard Rule.

1. $45.2 + 3.18$
2. $1500 - 2110 + 1821$

---

[15]Exact values may be ignored at this stage as their representations are infinitely accurate and, therefore, have no estimated digits.

3. $1642 + 3280 - 9819$
4. $2.100 - 1.51 + 120$ 120 is exact
5. $4.3 \times 10^4 + 1.8 \times 10^3$

Solution

1.

$$45.2 + 3.18 = 48.38 \quad \text{Exact}$$
$$= 48.4 \quad \text{Rounded}$$

In going from 48.38 to 48.4 we reasoned as follows: The estimated digit in 45.2 is 2 which is in the tenths place. The estimated digit in 3.18 is 8 which is in the hundredths place. Since a tenth is larger than a hundredth, we round the result to the nearest tenth to get 48.4.

2.

$$1500 - 2110 + 1821 = 1211 \quad \text{Exact}$$
$$= 1200 \quad \text{Rounded}$$

In going from 1211 to 1200 we reasoned as follows: The estimated digit in 1500 is 5 which is in the hundreds place. The estimated digit in 2110 is the rightmost 1 which is in the tens place. The estimated digit in 1821 is the rightmost 1 which is in the ones place. Since a hundred is larger than a ten and a one, we round the result to the nearest hundred to get 1200.

3.

$$1642 + 3280 - 9819 = -4897 \quad \text{Exact}$$
$$= -4900 \quad \text{Rounded}$$

In going from -4897 to -4900 we reasoned as follows: The estimated digit in 1642 is 4 which is in the tens place. The estimated digit in 3280 is 2 which is in the hundreds place. The estimated digit in 9819 is the rightmost 9 which is in the ones place. Since a hundred is larger than a ten as well as a one, we round the result to the nearest hundred to get -4900.

4.

$$2.100 - 1.51 + 120 = 120.59 \quad \text{Exact}$$
$$= 120.59 \quad \text{Rounded}$$

The estimated digit in 2.100 is the rightmost 0 which is in the thousandths place. The estimated digit in 1.51 is the rightmost 1 which is in the hundredths place. We ignore 120 which has been declared to be exact. Since a hundredth is larger than a thousandth, we round the result to the nearest hundredth to get 120.59.

5.

$$4.3 \times 10^4 + 1.8 \times 10^3 = 4.3 \times 10^4 + 0.18 \times 10^4$$
$$= 4.48 \times 10^4 \quad \text{Exact}$$
$$= 4.5 \times 10^4 \quad \text{Rounded}$$

Before we work out the place values of the estimated digits in the measured values being added and subtracted, we must make sure that the exponents of 10 are the same otherwise place values will not line up correctly. Once this is done, we can find the estimated digits in 4.3×10^4 and 0.18×10^4, not the original numbers. With these modified numbers, the estimated digit in 4.3×10^4 is 3 with an apparent place value of tenths. The estimated digit in 0.18×10^4 is 8 with an apparent place value of hundredths. Since a tenth is larger than a hundredth, we round to the nearest tenth to get 4.5×10^4.[16]

The argument provided by the Standard Rule in justifying the scheme for rounding the results of additions and subtractions of measured values relies on the size of the largest estimation in the numerical values involved which masks the effects of other estimations. As an example of this, consider the sum

4400 + 0.0032

According to the scheme given by the Standard Rule, the exact sum, i.e., 4400.0032 should be rounded to 4400. While this may sound strange at first, a bit of reflection will convince the reader that the rounded result is reasonable: Suppose, as an example, that 4400 represents a rough estimate for the distance between your house in Toronto and your friend's house in

---

[16]It is, of course, possible to work with actual place values. However, since what is important is the relative sizes of the place values of the estimated digits in the measured values being added and subtracted, we choose to use apparent place values which are easier to work with.

We emphasize that the exponents of 10 in the numbers being added and subtracted must be the same before we can compare the sizes of the place values of the estimated digits. The need to make the exponents of 10 the same does not add extra work to the work that already needs to be done as the move will have to be made when adding and subtracting numbers that are written in scientific notation anyway.

Vancouver. This rough estimate could be off by 50 km either way. If there is a tree some 3.2 m on the west side of your friend's house, it would be reasonable to say that the distance between your house in Toronto and the tree next to your friend's house is still 4400 km. It would, in fact, be odd to report a distance of 4400.0032 km as the distance between your house in Toronto and the tree next to your friend's house in Vancouver. The additional distance of 3.2 m would get lost in the rough estimate given by 4400 km.

Chain Multiplication/Division

The algorithm for the evaluation of expressions that involve chain multiplication/division of measured values is given below.

Algorithm for Evaluation of Expressions Involving Chain Multiplication/Division of Measured Values

1. Finish the chain multiplication/division exactly with no intermediate roundings and write the result down.
2. Determine the number of significant digits in each measured value in the original expression.[17]
3. Round the result in Step 1 to the minimum number of significant digits found in Step 2.

Example

Evaluate and round according to the scheme from the Standard Rule.

1. 2.4×8.74

2. $\dfrac{3.45}{2.104}$

3. $\dfrac{1400 \times 3210}{1829}$

4. $\dfrac{3 \times 4.1}{5.92 \times 0.0346}$ 3 is exact

5. $\dfrac{5.18 \times 10^3 \times 2.900 \times 10^2}{-9.05 \times 10^6 \times 3.50 \times 10^{-2}}$

---

[17]Exact values may be ignored at this stage as their representations have an infinite number of significant digits. As such, they do not provide the minimum number of significant digits sought in Step 3.

Solution

1.

$$2.4 \times 8.74 = 20.976 \quad \text{Exact}$$
$$= 21 \quad \text{Rounded}$$

In going from 20.976 to 21 we reasoned as follows: The numerical value 2.4 has two significant digits. The numerical value 8.74 has three significant digits. The minimum number of significant digits is two. We round the result to two significant digits to get 21.

2.

$$\frac{3.45}{2.104} = 1.639\,733\ldots \quad \text{Exact}$$
$$= 1.64 \quad \text{Rounded}$$

In going from $1.639\,733\ldots$ to 1.64 we reasoned as follows: The numerical value 3.45 has three significant digits. The numerical value 2.104 has four significant digits. The minimum number of significant digits is three. We round the result to three significant digits to get 1.64.

3.

$$\frac{1400 \times 3210}{1829} = 2457.080\,371\ldots \quad \text{Exact}$$
$$= 2500 \quad \text{Rounded}$$

In going from $2457.080\,371\ldots$ to 2500 we reasoned as follows: The numerical value 1400 has two significant digits. The numerical value 3210 has three significant digits. The numerical value 1829 has four significant digits. The minimum number of significant digits is two. We round the result to two significant digits to get 2500.

4.

$$\frac{3 \times 4.1}{5.92 \times 0.0346} = 60.049\,211\ldots \quad \text{Exact}$$
$$= 60 \quad \text{Rounded}$$

In going from $60.049\,211\ldots$ to 60 we reasoned as follows: We ignore the numerical value 3 as it is exact.[18] The numerical

---

[18] As we discussed earlier, an exact value may be thought of as having an infinite number of significant digits. As such, an exact value cannot be the numerical value with the minimum number of significant digits and it is for this reason that they can be ignored in Step 2 of the algorithm.

value 4.1 has two significant digits. The numerical value 5.92 has three significant digits. The numerical value 0.0346 has three significant digits. The minimum number of significant digits is two. We round the result to two significant digits to get 6Q̣. Note the use of the dot under the 0. This ensures that we have two significant digits.

5.

$$\frac{5.18 \times 10^3 \times 2.900 \times 10^2}{-9.05 \times 10^6 \times 3.50 \times 10^{-2}} = -0.474\,254\ldots \times 10^{3+2-6+2}$$

$$= -0.474\,254\ldots \times 10^1$$

$$= -4.741\,254\ldots \times 10^0 \quad \text{Exact}$$

$$= -4.74 \times 10^0 \quad \text{Rounded}$$

In going from the exact value to the rounded value we reasoned as follows: The numerical value 5.18 has three significant digits. The numerical value 2.900 has four significant digits. The numerical value -9.05 has three significant digits. The numerical value 3.50 has three significant digits. The minimum number of significant digits is three. We round the result to three significant digits to get -4.74×10^0.

Expressions

We now present the algorithm for the evaluation of expressions that involve a mix of addition, subtraction, multiplication and division of measured values.

The Analysis-Synthesis Algorithm for Evaluation of Expressions Involving Addition, Subtraction, Multiplication and Division of Measured Values

1. Analyze:
   a. Analyze the expression into terms using additions and subtractions *outside* brackets.
   b. Analyze each term into factors using multiplications.
   c. Analyze each factor further if needed.
2. Synthesize:
   a. Evaluate the factors. Apply the rounding rules that apply to the particular nature of that factor.[19]

---

[19] As an example, if the factor involves a root, apply the rules for rounding roots to that

b. Multiply the factors to evaluate the terms. To do so
   i. Determine whether the term should be added or subtracted. Use the *Algorithm for Evaluation of Expressions Involving Chain Multiplication/Division of Integers*. From this point on, ignore all signs.
   ii. Determine the *size* of the term. Use the *Algorithm for Evaluation of Expressions Involving Chain Multiplication/Division of Measured Values*.
c. Add and subtract the terms to evaluate the expression. Use the *Algorithm for Evaluation of Expressions Involving Chain Addition/ Subtraction of Measured Values*.

We illustrate the use of the above algorithm using examples.

Example

Evaluate and round according to the Standard Rule.

$$3.8 \times 1.99 + 4.76$$

Solution

$$
\begin{aligned}
3.8 \times 1.99 + 4.76 &= 7.\underset{.}{5}62 + 4.76 \\
&= 12.\overset{.}{3}22 \\
&= 12.\overset{.}{3}
\end{aligned}
$$

Example

Evaluate and round according to the Standard Rule.

$$4.5 - 2.145\,(-3.2)$$

Solution

$$
\begin{aligned}
4.5 - 2.145\,(-3.2) &= 4.5 + 6.\underset{.}{8}64 \\
&= 11.\underset{.}{3}64 \\
&= 11.\overset{.}{4}
\end{aligned}
$$

---

factor.

Example

Evaluate and round according to the Standard Rule.

$$-4.510 \times 3.981 + 9.77 \times 3.11 \times 0.1256$$

Solution

$$
\begin{aligned}
-4.510 \times 3.981 &+ 9.77 \times 3.11 \times 0.1256 \\
&= -17.95431 + 3.81631832 \\
&= -14.13799168 \\
&= -14.14
\end{aligned}
$$

Example

Evaluate and round according to the Standard Rule.

$$1.43\,(3.8) - 7.7\,(14.9) + 21.3\,(16) \quad \text{16 is exact}$$

Solution

$$
\begin{aligned}
1.43\,(3.8) - 7.7\,(14.9) &+ 21.3\,(16) \\
&= 5.434 - 114.73 + 340.8 \\
&= 231.504 \\
&= 230
\end{aligned}
$$

Example

Evaluate and round according to the Standard Rule.

$$-214\,(-4.6)\,(-11.7) - 13.9\,(-4.2)\,(-81)$$

Solution

$$
\begin{aligned}
-214\,(-4.6)\,(-11.7) &- 13.9\,(-4.2)\,(-81) \\
&= -11517.48 - 4728.78 \\
&= -16246.26 \\
&= -16000
\end{aligned}
$$

Example

Evaluate and round according to the Standard Rule.

$$\frac{0.24 \times 0.567}{0.0120} + \frac{-3.51}{2.1 \times 0.817}$$

Solution

$$
\begin{aligned}
\frac{0.24 \times 0.567}{0.0120} + \frac{-3.51}{2.1 \times 0.817} &= 11.34 - 2.045\,812\ldots \\
&= 9.294\,187\ldots \\
&= 9
\end{aligned}
$$

The reader should note that the actual act of rounding happens once at the very end of the calculations. At the end of each step in the algorithm given above for the evaluation of expressions that involve measured values, we use the dot notation to keep track of the place value to which intermediate calculations should be rounded. Actual rounding of intermediate values might result in final answers that are different depending on the particular sequence of steps taken. To see how, consider the following expression

$$457 \times 9.6 \times 23.5$$

To evaluate this expression, one may choose to multiply 457 and 9.6 first and then multiply the result by 23.5, or one may choose to multiply 9.6 and 23.5 first and then multiply the product by 457. Using the Standard Rule and allowing for the rounding of intermediate value, we get

$$
\begin{aligned}
457 \times 9.6 \times 23.5 &= 4400 \times 23.5 \\
&= 100\,000
\end{aligned}
$$

for the former sequence and

$$
\begin{aligned}
457 \times 9.6 \times 23.5 &= 457 \times 230 \\
&= 110\,000
\end{aligned}
$$

for the latter sequence.

9.2.3 The Limited-Significant-Digit Rule

The approximate nature of the Standard Rule and the complexities that arise in the application of the rules tend to make the approach less appealing to the general practitioner. For these reasons, many argue that one might as well round the results of all calculations to a limited number of significant digits. Of the many possibilities, the most common are the three-significant-digit-rule, the four-significant-digit-rule and the five-significant-digit-rule.[20]

The rounding scheme given by the limited-significant-digit rule is quite simple: Perform the calculations exactly and round the final result to a given number of significant digits.

In what follows, we will solve the problems that were posed in the previous subsection on the Standard Rule using the three-significant-digit rule, the four-significant-digit-rule and the five-significant-digit rule. Each solution is followed by a comparison of the result of the calculations with those given in the previous subsection on the Standard Rule.

Example

Evaluate and round according to the three-, four- and five-significant-digit-rule.

$$3.8 \times 1.99 + 4.76$$

Solution

$$3.8 \times 1.99 + 4.76 = 7.562 + 4.76$$
$$= 12.322 \quad \text{Exact}$$

Three-Significant-Digit-Rule result: 12.3
Four-Significant-Digit-Rule result: 12.32
Five-Significant-Digit-Rule result: 12.322
The Standard Rule result: 12.3

---

[20] In most practical applications the level of accuracy given by three, four or five significant digits is sufficient. A measured value with two significant digits is often too rough while measurements with six or more significant digits are much too accurate.

The reason has to do with the relative sizes of place values. In a numerical value such as 27.8149 the size represented by digit 9, i.e., 9 ten thousandths, is much too small compared to the size represented by digit 2, i.e., 2 tens. If 27.8149 is a measure of length in, say, metres, then digit 2 would represent a length of 20 metres while digit 9 would represent 0.0009 metres or 9 tenths of a millimetre. It is uncommon to run into a scenario where an additional length of 9 tenths of a millimetre matters in a length that spans over 20 metres.

Example

Evaluate and round according to the three-, four- and five-significant-digit-rule.

$$4.5 \; - \; 2.145\,(-3.2)$$

Solution

$$4.5 \; - \; 2.145\,(-3.2) \; = \; 4.5 \; + \; 6.864$$
$$= \; 11.364 \quad \text{Exact}$$

Three-Significant-Digit-Rule result: 11.4
Four-Significant-Digit-Rule result: 11.36
Five-Significant-Digit-Rule result: 11.364
The Standard Rule result: 11.4

Example

Evaluate and round according to the three-, four- and five-significant-digit-rule.

$$-4.510 \times 3.981 \; + \; 9.77 \times 3.11 \times 0.1256$$

Solution

$$-4.510 \times 3.981 \; + \; 9.77 \times 3.11 \times 0.1256$$
$$= \; -17.954\,31 \; + \; 3.816\,318\,32$$
$$= \; -14.137\,991\,68 \quad \text{Exact}$$

Three-Significant-Digit-Rule result: -14.1
Four-Significant-Digit-Rule result: -14.14
Five-Significant-Digit-Rule result: -14.138
The Standard Rule result: -14.14

Example

Evaluate and round according to the three-, four- and five-significant-digit-rule.

$$1.43\,(3.8) \;-\; 7.7\,(14.9) \;+\; 21.3\,(16) \quad 16 \text{ is exact}$$

Solution

$$1.43\,(3.8) \;-\; 7.7\,(14.9) \;+\; 21.3\,(16)$$
$$= 5.434 \;-\; 114.73 \;+\; 340.8$$
$$= 231.504 \quad \text{Exact}$$

Three-Significant-Digit-Rule result: 232
Four-Significant-Digit-Rule result: 231.5
Five-Significant-Digit-Rule result: 231.50
The Standard Rule result: 230

Example

Evaluate and round according to the three-, four- and five-significant-digit-rule.

$$-2\underset{.}{1}4\,(-4.6)\,(-1\underset{.}{1}.7) \;-\; 13.9\,(-4.2)\,(-81)$$

Solution

$$-214\,(-4.6)\,(-1\underset{.}{1}.7) \;-\; 13.9\,(-4.2)\,(-81)$$
$$= -11\,517.48 \;-\; 4728.78$$
$$= -16\,246.26 \quad \text{Exact}$$

Three-Significant-Digit-Rule result: $-16\,200$
Four-Significant-Digit-Rule result: $-16\,250$
Five-Significant-Digit-Rule result: $-16\,246$
The Standard Rule result: $-16\,000$

Example

Evaluate and round according to the three-, four- and five-significant-digit-rule.

$$\frac{0.24 \times 0.567}{0.0120} + \frac{-3.51}{2.1 \times 0.817}$$

Solution

$$\frac{0.24 \times 0.567}{0.0120} + \frac{-3.51}{2.1 \times 0.817} = 11.34 - 2.045\,812\ldots$$

$$= 9.294\,187\ldots \quad \text{Exact}$$

Three-Significant-Digit-Rule result: 9.29
Four-Significant-Digit-Rule result: 9.294
Five-Significant-Digit-Rule result: 9.2942
The Standard Rule result: 9

As in the case of the Standard Rule, intermediate values should not be rounded otherwise, depending on the particular sequence of steps, the final result may be different.

In addition to being simpler to apply, the limited-significant-digit-rule frees us from the vexing task of using dots to keep track of the number of significant digits for intermediate values and, instead, allows us to focus on making sense of what is being done as we work our way through the problem without sacrificing much accuracy.

Notwithstanding the above, we emphasize that the rounding schemes given by the Standard Rule and the Limited-Significant-Digit-Rule are approximate. If full accuracy is needed, then the only option is to use the rounding scheme from Interval Math.

Exercise Set 9.2

1. Write each of the following as an interval.

    a. $32.8 < T < 38.7$

    b. $2.5 < m \leq 3.7$

    c. $14\,000 \leq E < 15\,200$

    d. $95.4 \leq A \leq 102.3$

    e. $162 < h < 178$

    f. $52\% < g \leq 87\%$

    g. $21.72 \leq a < 23.80$

    h. $90 \leq v \leq 120$

2. Evaluate each of the following exactly, using the rounding scheme from the Standard Rule and the rounding scheme from the three-, four-, and five-significant-digit rules. All values are measured unless noted otherwise.

a. $318 + 27 + 140$

b. $52 + 33 - 123 + 43$

c. $130 - 240 + 410$

d. $-142 - 210 - 128$

e. $4.25 + 1.86 - 2.1$

f. $3.8 + 4.51 - 5.20$

g. $9.73 + 9.8 - 6.542$

h. $18.25 + 19 - 22.3$

i. $1800 + 1920 - 3110$

j. $4524 + 1830 - 9720$

k. $0.0718 - 0.02347 - 0.0873$

l. $4.5 \times 10^3 + 1.7 \times 10^3$

m. $-1.9 \times 10^{-2} - 9.72 \times 10^{-2}$

n. $7.6 \times 10^2 + 8.8 \times 10^3$

o. $7.4 \times 10^0 + 1.8 \times 10^{-1}$

p. $8.5 \times 10^{-3} + 6.7 \times 10^{-4}$

q. $24\,300 + 44\,000 - 57\,000$

r. $3000 - 2560 + 1020 - 1800$, 3000 is exact

s. $0.026 - 0.00182 + 0.0513$, 0.026 is exact

t. $50\,200 + 32\,000 - 1010 + 21\,000$, 1010 is exact

3. Evaluate each of the following exactly, using the rounding scheme from the Standard Rule and the rounding scheme from the three-, four-, and five-significant-digit rules. All values are measured unless noted otherwise.

a. 4.2×3.71

b. $3.6 \times 2.45 \times 9.8$

c. 0.025×300

d. 1902×4200

e. $0.001\,60 \times 240$

f. 0046×205

g. 32.5×8.4

h. $10.201 \times 42 \times 1200$

i. $\dfrac{62.5}{18}$

j. $\dfrac{9.6}{4.25} \times 3.60$

k. $\dfrac{2400}{30\,000} \times \dfrac{7.69}{1.78 \times 21.9}$

l. $\dfrac{2.681 \times 4.5}{6.23}$

m. $\dfrac{1200}{420 \times 100}$

n. $\dfrac{132\,000}{4200 \times 10.0}$

o. $4.7 \times 10^4 \left(3.2 \times 10^3\right)$

p. $-1.5 \times 10^{-3} \left(6.62 \times 10^5\right)$

q. $-3.7 \times 10^0 \left(2.5 \times 10^{-2}\right)\left(-8.4 \times 10^4\right)$

r. $\dfrac{2.571 \times 10^{-7} \left(1.70 \times 10^3\right)}{-3.6 \times 10^{-1}}$

s. $\dfrac{-4.25 \times 10^{-4}}{5.72 \times 10^3 \left(-1.1 \times 10^{-2}\right)}$

t. $1.6 \times 3.62 \times 1.50$, 1.6 is exact

u. $\frac{9.6}{4.25}$,   9.6 is exact

v. $\frac{9.6}{4.25}$,   4.25 is exact

w. $3.6 \times 60 \times 60$,   Both 60s are exact

x. $\frac{3.6 \times 400}{3.5 \times 1.82}$,   400, 3.5, and 3.6 are all exact

4. Evaluate each of the following exactly, using the rounding scheme from the Standard Rule and the rounding scheme from the three-, four-, and five-significant-digit rules. All values are measured unless noted otherwise.

a. $340 \times 24 + 9.8 \times 1200$

b. $72 \times 130 - 95 \times 210 + 28 \times 150$

c. $42 \times 120 \times 37 - 27 \times 1400 - 1400 \times 2.8$

d. $4.2 \times 6.7 + 18 \times 1.2$

e. $3.0 \times 42.5 + 9.68 \times 2.1 \times 1.04$

f. $1.6 + 12.4 \times 1.72 + 0.094$

g. $\frac{4.17 \times 1.5}{3.61} + 1.9 \times 4.2 - \frac{1.38}{3.6 \times 1.5}$

h. $\frac{4.56}{3.6} + \frac{18}{10.0}$

i. $0.87 \times 3.4 + 4.50 \times 0.652$

j. $14\,200 + 16 \times 2000 - 56\,000$

k. $2450 + 3000 \times 24 \times \frac{1}{650}$

l. $2.57 \times 10^3 \left(1.8 \times 10^2\right) + 3.5 \times 10^2 \left(-4.7 \times 10^4\right)$

m. $-3.2 \times 10^0 \left(2.8 \times 10^{-3}\right) - 8.2 \times 10^{-1} \left(7.84 \times 10^{-1}\right)$

n. $-1.99 \times 10^{-2} \left(-4.76 \times 10^{-4}\right) \left(1.1 \times 10^4\right) + 8.6 \times 10^{-3}$

o. $\frac{1.8 \times 10^2}{-3.22 \times 10^{-1}} - \frac{6.58 \times 10^{-2}}{-1.52 \times 10^{-3}}$

p. $4 \times 9.4 + 3 \times 0.64 - 2 \times 14.5$,   4, 3 and 2 are all exact

q. $-6 \times 3.7 \times 2.1 - (-2 \times 5.5)$,   −6, and −2 are both exact

Part V
Algebraic Systems

In this part of the book we will discuss the origin of algebraic equations and discuss the manner in which such equations are classified and generalized into formulas. Along the way, we will present meaningful and efficient algorithms for simplifying and solving these equations as well as extensions of these algorithms for use when working with formulas.

Chapter 10
Equations

In this chapter we will focus on general aspects of equations that are of importance to us. These include their structure, genesis, analysis, simplification, solutions and notation.

10.1 Structure

Since the main structural components of an equation are expressions, we will begin with the definition of an expression.

An **expression** is a meaningful sequence of numerical values, quantity symbols and operation symbols other than the equality and inequality operation symbols.[1] As an example of an expression we have[2]

$$4n + 2$$

but not

$$2p - 3 = 1 \qquad \text{(Equality operator not allowed)}$$

or

$$5m + \times \div 2 \qquad \text{(Meaningless sequence)}$$

An expression may be as simple as a numerical value such as 14, 0, -127, $\frac{3}{4}$, $\sqrt{2}$, 0.57, -2.1×10^2 and the like, or a quantity symbol such as m, p, T, etc., but not an operation symbol on its own such as $+$, $-$, $\times$, etc. The key is

---

[1]Note that this is an extension of the definition of an expression that was given earlier in the textbook. This extended definition allows for the presence of quantity symbols in addition to numerical values and operation symbols in sequences that qualify as expressions.

[2]For readers who may not be as familiar with notational conventions, the sequence $4n$ in the expression above implies multiplication between the numerical value, 4, and the quantity symbol, n, i.e., $4 \times n$.

that the sequence must evaluate to a value. This means that performing the operations on the values in the sequence results in a value. As an example, the expression

$$4n + 2$$

requires that we multiply 4 by n and then add 2 to the product. While we may not know what n is equal to, we still know that it represents the value of a quantity so that it does make sense to multiply 4 by n and that, once we have done so, add 2 to the product to arrive at a value. It is this feature of an expression that makes it *meaningful* in the sense required in its definition.

An **equation** consists of two expressions that are set equal to each other. As an example, the following equation

$$3m + 4 = 8 + m$$

consists of the two expressions $3m + 4$ and $8 + m$ which are set equal to each other. The equation states that the value of the expression on the left side of the equation,[3] i.e., $3m + 4$, is equal to the value of the expression on the right side of the equation, i.e., $8 + m$.[4] The numerical value that, when used in place of the quantity symbol, m, forces the equality, is known as a solution to the equation $3m + 4 = 8 + m$. As an example, the numerical value 2 is a solution to the equation $3m + 4 = 8 + m$ as substitution of 2 for m on the left and right sides of the equation results in numerical values for the left and right side expressions that are equal:

$$
\begin{aligned}
\text{LS} &= 3m + 4 \\
&= 3 \times 2 + 4 \\
&= 6 + 4 \\
&= 10
\end{aligned}
$$

and

$$
\begin{aligned}
\text{RS} &= 8 + m \\
&= 8 + 2 \\
&= 10
\end{aligned}
$$

---

[3] References to the left and right sides of the equation relate to the left and right sides of the equality sign. In addition, we will use the acronyms LS and RS to refer to the left and right sides of an equation.

[4] Note that the equality is expected to hold for the same value of m throughout the equation, i.e., m cannot assume a value of 4.5 kg on one occurence and a value of 3.2 kg on another occurence.

The numerical value 5, on the other hand, is not a solution to this equation as substitution of 5 for m on the left and right sides of the equation results in numerical values for the left and right side expressions that are *not* equal:

$$
\begin{aligned}
\text{LS} &= 3m + 4 \\
&= 3 \times 5 + 4 \\
&= 15 + 4 \\
&= 19
\end{aligned}
$$

but

$$
\begin{aligned}
\text{RS} &= 8 + m \\
&= 8 + 5 \\
&= 13
\end{aligned}
$$

A solution to an equation that involves multiple unknowns must quote a value for each and every unknown in the equation. As an example, a solution to the equation

$$
3t + 4v = v
$$

is $t = -1$ and $v = 1$ as substitution of -1 for t on the left side of the equation and substitution of 1 for v on both sides of the equation results in numerical values for the left and right side expressions that are equal:

$$
\begin{aligned}
\text{LS} &= 3t + 4v \\
&= 3\,(-1) + 4\,(1) \\
&= -3 + 4 \\
&= 1
\end{aligned}
$$

and

$$
\begin{aligned}
\text{RS} &= v \\
&= 1
\end{aligned}
$$

Exercise Set 10.1

1. What does it mean when we say that a sequence of numerical values, quantity symbols and operation symbols is meaningful?

2. In each case determine whether the sequence of numerical values, quantity symbols and operation symbols qualifies as an expression.

a. $2p + 3$ f. $5n\ -\div\times 2$ k. $t^2 - 1$

b. $m - 2 + 3m$ g. -8 l. $r^2 + 2r - 3 = 0$

c. $2t - 1 = 3$ h. A m. $\frac{3}{5}$

d. $14 - 6E$ i. $+$ n. $L \leq -2$

e. $-5t + 3v$ j. $8 +$ o. $-3(2p + 1)$

3. List the main structural components of each of the following equations.

a. $-t + 7 = 2$ e. $\frac{2}{3}T = \frac{1}{2}T + \frac{3}{5}$

b. $5 - r = 0$ f. $5A + 2 = -3(2A - 1)$

c. $-3n + 2 = 5 + 2n$ g. $2r - 3s = 5$

d. $4(p - 1) = -3$ h. $t^2 + 2t + 3 = 0$

4. In each case determine whether the given values are a solution to the given equation.

a. $3r + 2 = 5$, $r = 1$

b. $-t - 4 = 0$, $t = 3$

c. $2n + 3 = -6 - n$, $n = -3$

d. $7 - 5p = 2p$, $p = 1$

e. $-(r - 1) + 2 = 0$, $r = -1$

f. $3(-2v + 3) = -2 + 7v$, $v = 0$

g. $L^2 - 3L + 2 = 0$, $L = 3$

h. $2m^2 = 6m + 8$, $m = -1$

i. $-\frac{1}{2}E + \frac{2}{3} = \frac{3}{4}E - \frac{1}{2}$, $E = \frac{14}{15}$

j. $\frac{3}{4} + \frac{4}{5}m = \frac{2}{3}m$, $m = \frac{7}{8}$

k. $2p - 3t = -4$, $p = 1, t = 2$

l. $-n + 2m = -3n + m$, $n = -1, m = 1$

m. $t = -v + 3$, $t = 0, v = 3$

10.2   Genesis

In this section we describe how the use of the algebraic technique in the problem-solving process results in the generation of concrete equations and the manner in which a study of these concrete equations results in the generation of abstract equations.

10.2.1   Origin of Concrete Equations

Concrete equations arise from the use of the algebraic technique in the problem-solving process.[5] As an example, consider a problem involving the four values *mass of a molecule of CO_n* where n represents the value of the number of O atoms in a molecule of CO_n, *the mass of a C atom, the atomic mass of O* and *the number of O atoms in a molecule of CO_n*. Such a problem involves four values and, in a given problem involving these four values, any one of them may be unknown giving rise to four variations. Using the algebraic approach, we choose a single line of reasoning to relate the values of the four quantities no matter what the unknown. And of the many options that we have (e.g., *the mass of a molecule of CO_n is equal to the mass of a C atom plus the mass of the O atoms*, or *the mass of a C atom is equal to the mass of a molecule of CO_n minus the mass of the O atoms*, etc.) we choose the simplest. For the present problem, the simplest line of reasoning that relates the values of the four quantities is *the mass of a molecule of CO_n is equal to the mass of a C atom plus the mass of the O atoms*.[6] Using this as the preferred line of reasoning, we now generate a model for each variation.

Variation 1

> Calculate the mass of a molecule of CO_2. The mass of an atom of C is 12.01 amu. The atomic mass of O is 16.00 amu/atom.

An Algebraic Model

> m mass of a molecule of CO_2 (amu)
> $m = 12.01 + 16.00 \times 2$

---

[5]In this textbook we provide a brief summary of the genesis of concrete equations. For a detailed discussion of the topic please see the companion textbook *Semantics and the Syntax of Algebra* by the author.

[6]The simplest line of reasoning is often the one that uses the fewest and simplest of the operations.

Variation 2

The mass of a molecule of XO_2 is 44.01 amu. The atomic mass of O is 16.00 amu/atom. Identify the unknown atom, X, by finding its mass.

An Algebraic Model

m mass of an X atom (amu)
$$44.01 = m + 16.00 \times 2$$

Variation 3

The mass of a molecule of CX_2 is 44.01 amu. The mass of an atom of C is 12.01 amu. Identify the unknown atom, X, by finding its atomic mass.

An Algebraic Model

M atomic mass of X (amu/atom)
$$44.01 = 12.01 + M \times 2$$

Variation 4

The mass of a molecule of CO_n is 44.01 amu. The mass of an atom of C is 12.01 amu. The atomic mass of O is 16.00 amu/atom. How many atoms of O are there in this molecule?

An Algebraic Model

n number of O atoms in a molecule of CO_n (1)
$$44.01 = 12.01 + 16.00 \times n$$

10.2.2   Origin of Abstract Equations

When solving a concrete problem, we use appropriate quantity symbols to represent the unknown values in our equations. This explains the use of such symbols as m for the representation of the value of mass, p for the representation of the value of pressure, and the like, in equations that model concrete problems. However, if we are interested in learning how to manipulate such equations[7] and not the specific applications that gave rise to them, then we use the latter letters of the alphabet, i.e., x, y and z, as symbols for abstract unknown values,[8] often referred to, simply, as *the unknowns* and we refer to such equations as **abstract equations**. We, therefore, write

$$x = 12.01 + 16.00 \times 2$$

in place of

$$m = 12.01 + 16.00 \times 2$$

and we write

$$44.01 = 12.01 + 16.00 \times x$$

in place of

$$44.01 = 12.01 + 16.00 \times n$$

In this book our focus is on the technical aspects of working with equations and, therefore, we will use the generic letters x, y and z to represent the unknown values in our equations.[9]

Exercise Set 10.2

Classify each of the following equations as concrete or abstract.

1. $p \times 1.2 = 3.8 \times 8.314 \times 290$
2. $142 = \frac{1}{2} \times b \times 2$
3. $9x - 2 = 3$
4. $8(4 - 2y) = 0$
5. $4.5 = \frac{1}{2n} \times 13.9$
6. $\frac{2}{3} - \frac{1}{2}z = \frac{3}{4}$
7. $-3(2 - x) = -5 + 3x$
8. $0.13(1 - r)(32.99) = 3.86$
9. $d = 130 + 120 \times 2.5$
10. $-4x - 2 = x$

---

[7] e.g., rearranging such equations, solving such equations, etc.
[8] If more than three unknown values are involved, then we use subscripts on the relevant symbol, e.g., x_1, x_2, x_3, x_4, etc., to represent them with.
[9] For the semantics surrounding equations in applications please see the companion textbook *Semantics and the Syntax of Algebra* by the author.

10.3   Formulation

By *formulation* we mean the manner in which algebraic models are put together. In this section we present two different ways of formulating algebraic
models and discuss the pros and cons of each formulation.

10.3.1   Semantic Formulations

In a **semantic formulation** the order in which various syntactical elements
appear in the model follows the manner in which we normally reason. As an
example, consider the following problem.

Problem

> 87% of those who register in ENG101 pass the course. How many
> people registered in ENG101 this term if 1044 people who regis-
> tered in the course passed the course?

An Algebraic Model

> n the number of people who registered in ENG101 (1)
> $1044 = 87\% \times n$

The order in which the various syntactical elements of the language appear
in this formulation follows the line of reasoning *the number of people who
registered in ENG101 and passed the course is equal to 87% of the number of
people who registered in the course.*[10] It is for this reason that we refer to it
as a semantic formulation of the model.

Note that this formulation is used no matter what the unknown. Here is
an example.

---

[10] Here is the correspondence between the elements in the English statement *the number
of people who registered in ENG101 and passed the course is equal to 87% of the number
of people who registered in the course* and the elements in the mathematical statement
$1044 = 87\% \times n$:

> The number of people who registered in ENG101 and passed the course: 1044
> is equal to: $=$
> 87%: 87%
> of: $\times$
> the number of people who registered in the course: n

Problem

1200 people registered in ENG101. 1044 of them passed the course. What percentage of those who registered in ENG101 passed the course?

An Algebraic Model

r percentage of people who passed the course (1)
$$1044 = r \times 1200$$

The order in which the various syntactical elements of the language appear in this formulation follows the same line of reasoning as the one above, i.e., *the number of people who registered in ENG101 and passed the course is some percentage of the number of people who registered in the course.*

Semantic formulations may be written for both concrete and abstract problems. Here are semantic formulations for the corresponding abstract versions of the word problems above. For the first problem we have

Problem

87% of a number is 1044. Find that number.

An Algebraic Model

x the number sought
$$1044 = 87\% \times x$$

And for the second problem we have

Problem

What percentage of 1200 is 1044?

An Algebraic Model

x the percentage sought
$$1044 = x \times 1200$$

Semantic formulations have the advantage of exposing the logic behind the formulation as the manner in which they are read mimics the manner in which we speak. However, for the purpose of classifying and processing the model, a different formulation is preferred.

10.3.2   Technical Formulations

In a **technical formulation** a different set of conventions are used to order the known and unknown values that appear in an equation. The conventions are[11]

1. If the unknown appears on only one side of the equation, that side appears on the left side of the equation.
2. Within a term, factors that consist entirely of numerical values appear on the left side of factors that contain a mix of numerical values and unknowns and these, in turn, appear on the left side of factors that consist entirely of unknowns.

Following these guidelines, a technical reformulation of the semantic formulation of the concrete and abstract problems posed in the previous subsection are as follows.

Problem

87% of those who register in ENG101 pass the course. How many people registered in ENG101 this term if 1044 people who registered in the course passed the course?

An Algebraic Model

n the number of people who registered in ENG101 (1)

$$1044 = 87\% \times n \quad \text{semantic formulation}$$
$$87\%n = 1044 \quad \text{technical formulation}$$

Problem

1200 people registered in ENG101. 1044 of them passed the course. What percentage of those who registered in ENG101 passed the course?

---

[11]The set of conventions given below will be expanded as the level of complexity of equations increases.

An Algebraic Model

r percentage of people who passed the course (1)

$1044 = r \times 1200$ semantic formulation
$1200r = 1044$ technical formulation

Technical formulations may be written for both concrete and abstract problems. Here are technical formulations for the corresponding abstract versions of the word problems in the previous subsection.

Problem

87% of a number is 1044. Find that number.

An Algebraic Model

x the number sought

$1044 = 87\% \times x$ semantic formulation
$87\%x = 1044$ technical formulation

Problem

What percentage of 1200 is 1044?

An Algebraic Model

x the percentage sought

$1044 = x \times 1200$ semantic formulation
$1200x = 1044$ technical formulation

Technical formulations do not read as well as their semantic counterparts[12] but they are better able to expose structural similarities between equations that model seemingly disparate problems.

---

[12]As an example, the semantic formulation of the model for the last problem above reads as 1044 *is what percentage of* 1200? while its technical counterpart reads as 1200 *times what percentage is* 1044?

Exercise Set 10.3

Reformulate each of the following semantic formulations as a technical formulation.

1. $m \times 1.24 = 19.7 \times 0.77$

2. $4.5 \times 2.7 = 3.2 \times v$

3. $1.22 = \dfrac{0.813}{t}$

4. $4 - x \times 3 = -2$

5. $47.8 = 2.4 \times \dfrac{v^2}{1.4}$

6. $130\,000 = \frac{1}{2} \times x \times 31.8^2$

7. $8.7 = \dfrac{1}{2n} \times 12.9$

8. $(4 - 2x)(-1) = 7$

9. $8.1 = \dfrac{22.4 - v_0}{2.5 - 1.7}$

10. $12.8 = 2.7 \times t + \frac{1}{2} \times 1.13 \times t^2$

11. $1800 = \dfrac{3020 - E_1}{5.7 - 1.8}$

12. $0 = \left(-\frac{2}{3} + x\right)\left(\frac{1}{2}\right)$

13. $155.7 = \frac{1}{2}(b_1 + 7.1)(2.8)$

10.4   Analysis

An equation analyzes into the two expressions that appear on the left and right sides of the equality symbol. Each expression analyzes into terms which are entities within the expression that are being added and/or subtracted. Each term analyzes into factors which are entities within terms that are being multiplied.

Example

Analyze the equation

$$2x + 1 = 3 - 6x$$

Solution

The equation $2x + 1 = 3 - 6x$ analyzes into the two expressions $2x + 1$ and $3 - 6x$.

The expression $2x + 1$ analyzes into the two terms $2x$ and 1. The first term, $2x$, analyzes into the two factors 2 and x. The second term, 1, analyzes into the single factor 1.

The expression $3 - 6x$ analyzes into the two terms 3 and $6x$. The first term, 3, analyzes into the single factor 3. The second term, $6x$, analyzes into the two factors 6 and x.

In some cases further analysis of factors may be needed. Here is an example.

Example

Analyze the equation

$$-4\left(x\ +\ 2\right)\ +\ 3\left(x\ -\ 1\right)\ =\ 2x\ -\ 1$$

Solution

The equation $-4\left(x+2\right)+3\left(x-1\right)=2x-1$ analyzes into the two expressions $-4\left(x+2\right)+3\left(x-1\right)$ and $2x-1$.

The expression $-4\left(x+2\right)+3\left(x-1\right)$ analyzes into the two terms $-4\left(x+2\right)$ and $3\left(x-1\right)$. The first term, $-4\left(x+2\right)$, analyzes into the two factors -4 and $x+2$ with the second factor analyzing further into the two terms x and 2. The second term, $3\left(x-1\right)$, analyzes into the two factors 3 and $x-1$ with the second factor analyzing further into the two terms x and 1.

The expression $2x-1$ analyzes into the two terms $2x$ and 1. The first term, $2x$, analyzes into the two factors 2 and x. The second term, 1, analyzes into the single factor 1.

Example

Analyze the equation

$$-3x^2\ +\ 2x\ -\ 1\ =\ 0$$

Solution

The equation $-3x^2+2x-1=0$ analyzes into the two expressions $-3x^2+2x-1$ and 0.

The expression $-3x^2+2x-1$ analyzes into the three terms $-3x^2$, $2x$ and 1. The first term, $-3x^2$, analyzes into the two factors -3 and x^2 the second of which is an exponent problem. The second term, $2x$, analyzes into the two factors 2 and x. The third term, 1, analyzes into the single factor 1.

The expression 0 analyzes into the single term, 0, which further analyzes into the single factor, 0.

Example

Analyze the equation

$$-3\sqrt{x} + 2 = -18$$

Solution

The equation $-3\sqrt{x} + 2 = -18$ analyzes into the two expressions $-3\sqrt{x} + 2$ and -18.

The expression $-3\sqrt{x} + 2$ analyzes into the two terms $-3\sqrt{x}$ and 2. The first term, $-3\sqrt{x}$, analyzes into the two factors -3 and $\sqrt{x}$ the second of which is a root problem. The second term, 2, analyzes into the single factor, 2.

The expression -18 analyzes into the single term, -18, which further analyzes into the single factor, -18.

Example

Analyze the equation

$$4\log x + 2\log y = 10$$

Solution

The equation $4\log x + 2\log y = 10$ analyzes into the two expressions $4\log x + 2\log y$ and 10.

The expression $4\log x + 2\log y$ analyzes into the two terms $4\log x$ and $2\log y$. The first term, $4\log x$, analyzes into the two factors 4 and $\log x$ the second of which is a log problem. The second term, $2\log y$, analyzes into the two factors 2 and $\log y$ the second of which is a log problem.

The expression 10 analyzes into the single term, 10, which further analyzes into the single factor, 10.

An understanding of the analysis of algebraic equations in the manner described above is essential for the development of efficient techniques to work with equations.

Exercise Set 10.4

Analyze each of the following equations.

1. $3x + 1 = 2$

2. $x^2 - 6x + 8 = 0$

3. $-2x - 3y = 1$

4. $x = -y + 2$

5. $xy = 24$

6. $x^2 + 2xy + y^2 = 0$

7. $z = y(y - 1)$

8. $y = (x - 2)^2$

9. $2x^2 + 3xy + 3y^2 - 6y = 0$

10. $-\frac{1}{2}x = \frac{3}{5}$

11. $\frac{3}{4}x - \frac{2}{3}y = 0$

12. $-(3x - 2x) - 5x = 2(3x + 2)$

13. $2(x - 7) = 1 - 5(x + 4)$

14. $\frac{1}{x} + \frac{1}{y} = \frac{1}{7}$

15. $\frac{2}{1 - x} - \frac{3}{2x + 1} = 1$

16. $2(x + 1) - \log 32 = 3$

17. $2x + 3\log 1 = -8$

18. $x - 2\sqrt{x} + 1 = 0$

19. $2x\sqrt{x} - 3 = 2(\sqrt{x} - 1)$

20. $(x - y)^2 + (x + y)^2 = 1$

10.5   Simplification

Algorithms used in solving equations require that we simplify the expressions on the right and left sides of the given equations, either along the way or towards the end of the process. Simplification rules will be discussed in more detail in the chapters ahead as the need arises. Here, we briefly comment on some of the general features of the activity.

By **simplifying an expression** we mean performing as many calculations as possible. This means working out any factors that can be worked out, multiplying as many factors as possible within the terms, and adding and subtracting as many terms as possible within the expression. As an example, consider the task of simplifying the expression

$$4(-3)x + 3 \times 5 - 2 + (-5)x$$

We begin by analyzing the expression into terms, i.e.,

$$\boxed{4(-3)x} + \boxed{3 \times 5} - \boxed{2} + \boxed{(-5)x}$$

The three factors in the first term, i.e., 4, -3 and x, have been worked out. We can, however, multiply the numerical values 4 and -3 and write $-12x$ in place of $4(-3)x$. As for the second term, both factors 3 and 5 have been worked out. We can multiply them and write 15 in place of 3×5. The third term contains a single factor, 2, which has already been worked out.

As for the last term, we can simplify the sign to get $-5x$. This leads to the following equivalent expression:

$$-12x + 15 - 2 - 5x$$

As always, we associate each addition and each subtraction to the term that follows it. This means that we view the expression above as

$$-12x + 15 - 2 - 5x$$

If the factors that contain the unknown quantity symbols within two or more terms are identical, we refer to the terms as **like terms**. In the expression above, the terms $-12x$ and $5x$ are like terms as the factors that contain the unknown quantity symbols, i.e., x, are identical in both. The terms 15 and 2 are also considered to be like terms since they lack factors that contain unknown quantity symbols.

Like terms can be added and subtracted. In the expression above, we can combine $-12x - 5x$ to get $-17x$. Note that $-12x - 5x$ may be interpreted as a loss of 12 things followed by a further loss of 5 things of the same kind. This adds up to a total loss of 17 things of that kind. We can also combine $+15 - 2$ to get $+13$. This leads to

$$-17x + 13$$

as the simplified form of the expression $-12x + 15 - 2 - 5x$. Note that we cannot combine the two terms $-17x$ and 13 as they are not like terms.[13] The full solution without the intervening explanations is given below.[14]

$$4\,(-3)\,x \; + \; 3 \times 5 \; - \; 2 \; + \; (-5)\,x$$
$$-\,12x \; + \; 15 \; - \; 2 \; - \; 5x$$
$$-\,17x \; + \; 13$$

Expressions that involve rational numbers can be simplified following the same approach. Here's an example:

---

[13]Terms that are not like terms are called **unlike terms**.

[14]Some choose to rearrange the second line in the solution above by grouping the like terms before proceeding to the last line. This would lead to the following solution:

$$4\,(-3)\,x \; + \; 3 \times 5 \; - \; 2 \; + \; (-5)\,x$$
$$-\,12x \; + \; 15 \; - \; 2 \; - \; 5x$$
$$-\,12x \; - \; 5x \; + \; 15 \; - \; 2$$
$$-\,17x \; + \; 13$$

Such a move is unnecessary and we strongly recommend against its adoption. Instead, the reader should train herself or himself to focus on the like terms in the expression on the second line to simplify that line. To simplify the second line in the solution above, we first focus on the like terms that contain x, i.e.,

$$4\,(-3)\,x \; + \; 3 \times 5 \; - \; 2 \; + \; (-5)\,x$$
$$-12x + 15 - 2 - 5x$$

Example

Simplify the expression

$$\frac{3}{4} + \frac{1}{3}x + \frac{2}{5}x + \frac{1}{2}$$

Solution

Analysis of the expression above into terms yields the following:

$$\boxed{\frac{3}{4}} + \boxed{\frac{1}{3}x} + \boxed{\frac{2}{5}x} + \boxed{\frac{1}{2}}$$

The terms $\frac{3}{4}$ and $\frac{1}{2}$ are like terms and so are the terms $\frac{1}{3}x$ and $\frac{2}{5}x$. Style requires that we work with terms that contain quantity symbols first. We, therefore, begin by combining the second and third terms to get

$$\frac{5 + 6}{15}x \ldots$$

We now combine the terms that do not contain quantity symbols, i.e., the first and fourth terms. This leads to

$$\frac{5 + 6}{15}x + \frac{3 + 2}{4}$$

or

$$\frac{11}{15}x + \frac{5}{4}$$

This leads to

$$4(-3)x + 3 \times 5 - 2 + (-5)x$$
$$-12x + 15 - 2 - 5x$$
$$-17x \ldots$$

Next we focus on the like terms that contain no quantity symbols, i.e.,

$$4(-3)x + 3 \times 5 - 2 + (-5)x$$
$$-12x + 15 - 2 - 5x$$
$$-17x \ldots$$

This leads to

$$4(-3)x + 3 \times 5 - 2 + (-5)x$$
$$-12x + 15 - 2 - 5x$$
$$-17x + 13$$

The full solution without the intervening remarks is given below.

$$\frac{3}{4} + \frac{1}{3}x + \frac{2}{5}x + \frac{1}{2}$$

$$\frac{5 + 6}{15}x + \frac{3 + 2}{4}$$

$$\frac{11}{15}x + \frac{5}{4}$$

Presence of negative rationals would require extra care. We illustrate matters using an example.

Example

Simplify the expression

$$-\frac{3}{5} - \frac{1}{4}x + \frac{1}{3} - \frac{1}{6}x$$

Solution

Analysis of the expression above into terms yields the following:

The terms $-\frac{3}{5}$ and $\frac{1}{3}$ are like terms and so are the terms $\frac{1}{4}x$ and $\frac{1}{6}x$. We begin with terms that contain quantity symbols, i.e., the second and fourth terms. This yields

$$-\frac{3 + 2}{12}x \ \ldots$$

The line of reasoning that goes into the above follows the logic introduced in the subsection on integers: We have two losses in a row, resulting in a net loss (represented by the negative sign above) of the *sum* of the sizes of the losses, $\frac{1}{4}$ and $\frac{1}{6}$.

We now combine the terms that do not contain quantity symbols, i.e., the first and third terms. These terms involve a mix of addition and subtraction and in case it is not apparent whether the result will be positive or negative, one can always assume a positive result and change the sign later if needed. This yields the following:[15]

$$-\frac{3 + 2}{12}x + \frac{-9 + 5}{15}$$

or

$$-\frac{5}{12}x + \frac{-4}{15}$$

We now simplify the sign for the second term to arrive at

$$-\frac{5}{12}x - \frac{4}{15}$$

The full solution without the intervening remarks is given below.[16]

$$-\frac{3}{5} - \frac{1}{4}x + \frac{1}{3} - \frac{1}{6}x$$
$$-\frac{3 + 2}{12}x + \frac{-9 + 5}{15}$$
$$-\frac{5}{12}x + \frac{-4}{15}$$
$$-\frac{5}{12}x - \frac{4}{15}$$

---

[15]If the reader can work out the sign of the result based on the sizes of the rational numbers involved, then a shorter solution may be given. However, one must be careful in how a negative result is processed. As an example, if the reader can tell from the relative sizes of $\frac{3}{5}$ and $\frac{1}{3}$ in the expression above that the result will be negative – say, from visualizing the fractions as pie graphs – then one may give the following solution for the simplification of the expression:

$$-\frac{3 + 2}{12}x - \frac{9 - 5}{15}$$

Note the switch in sign of the terms in the numerator of the second fraction as the subtraction that precedes the second fraction affects each term in the numerator.

In case the choice of the sign is in error, the process will end up fixing the sign although this will involve an extra line in the solution.

[16]The full solution using the alternative logic is given below.

$$-\frac{3}{5} - \frac{1}{4}x + \frac{1}{3} - \frac{1}{6}x$$
$$-\frac{3 + 2}{12}x - \frac{9 - 5}{15}$$
$$-\frac{5}{12}x - \frac{4}{15}$$

Our next example contains multiple quantity symbols.

Example

Simplify the expression

$$3x + 2 - 7y + x - 2y + 1$$

Solution

$$3x + 2 - 7y + x - 2y + 1$$
$$4x - 9y + 3$$

We leave it to the reader to work out the logic behind this solution.

The Distributive Property

Consider the expression

$$4(5 + 2)$$

This could be an expression that calculates the cost of buying 4 packages, each of which contains a $5 notebook and a $2 pen.

We can work out the total cost by first adding 5 and 2 to calculate the cost of one package, and then multiplying the result by 4 to compute the cost of 4 packages. Such an approach would generate the following solution

$$4(5 + 2)$$
$$4 \times 7$$
$$28$$

An alternative logic would require that we multiply 4 by 5 to compute the cost of all the notebooks in all the packages, then multiply 4 by 2 to compute the cost of all the pens in all the packages, and then add the results to find the total cost. Such logic can be communicated as follows:

$$4(5 + 2)$$
$$4 \times 5 + 4 \times 2$$
$$20 + 8$$
$$28$$

Both lines of reasoning generate the same answer and in situations where the values involved are all numerical, the first approach is preferred as it is more efficient and is in line with the way we naturally reason.

However, in some cases the first alternative may not be available to us. Such is the case if one of the values within brackets is a quantity symbol. In such cases we can follow the alternative route and multiply the numerical value on the left side of the brackets by each and every term within the brackets. As an example, to simplify the expression

$$4\,(x\,+\,2)$$

we multiply 4 by x and then we multiply 4 by 2 to get

$$4x\,+\,8$$

The logic that permits the move above is called the *distributive property of multiplication over addition and subtraction* often shortened to the **distributive property**.

As we just noted, the distributive property holds over subtraction as well. As an example, the expression

$$6\,(3x\,-\,2)$$

can be simplified as follows:

$$6\,(3x\,-\,2)$$
$$18x\,-\,12$$

In case the numerical value multiplying the brackets is negative, take the sign into account. Here is an example:

$$-\,2\,(3x\,-\,5)$$
$$-\,6x\,+\,10$$

One may view the above as a multiplication of -2 by $+3$ followed by a multiplication of -2 by -5.

Example

Simplify the expression

$$3\,(2x\,+\,1)\,-\,4\,(3\,-\,x)$$

Solution

Analysis of the expression above into terms yields the following:

$$\boxed{3\,(2x\,+\,1)}\,-\,\boxed{4\,(3\,-\,x)}$$

We can apply the distributive property to each term to arrive at the following:

$$3\,(2x\,+\,1)\,-\,4\,(3\,-\,x)$$
$$6x\,+\,3\,-\,12\,+\,4x$$

In simplifying the first term we multiplied $+3$ by $+2$ and then we multiplied $+3$ by $+1$. In simplifying the second term we multiplied -4 by $+3$ and then we multiplied -4 by -1.[17] The rest is simple:

$$3\,(2x\,+\,1)\,-\,4\,(3\,-\,x)$$
$$6x\,+\,3\,-\,12\,+\,4x$$
$$10x\,-\,9$$

In case the expression within brackets can be simplified, we recommend that you do so before applying the distributive property. Here is an example:

Example

Simplify the expression

$$-2\,(3x\,+\,5\,-\,8x\,-\,2)$$

Solution

$$-\,2\,(3x\,+\,5\,-\,8x\,-\,2)$$
$$-\,2\,(-5x\,+\,3)$$
$$10x\,-\,6$$

The alternative, i.e., multiplying -2 into the bracket and then combining the like terms, takes longer and is more involved.[18]

Algorithm for Simplification of Expressions

1. Analyze:

   a. Break the expression into terms.
   b. Break the terms into factors.

---

[17]Note that $3 - x$ is short for $3 - 1x$.
[18]The reader should try this alternative.

2. Synthesize:

   a. Simplify the factors within terms. To do so, work out any operations that involve numerical values. Simplify the expression inside any brackets using this algorithm.

   b. Simplify the terms. To do so, multiply the numerical factors within each term. Process any brackets using the Distributive Property.

   c. Simplify the expression. To do so, combine the like terms within the expression.

To simplify an equation, simplify the expressions on the left and right sides of the equation. Here is an example.

Example

Simplify the equation

$$3x - 2 + 4x = -2(-3x + 1) + 5$$

Solution

$$3x - 2 + 4x = -2(-3x + 1) + 5$$
$$7x - 2 = 6x - 2 + 5$$
$$7x - 2 = 6x + 3$$

Exercise Set 10.5

1. Simplify each of the following expressions.

a. $2x + 3x$

b. $182x + 95x$

c. $15x + x$

d. $y + 9y$

e. $7z - 2z$

f. $142x - 123x$

g. $4x - 12x$

h. $118x - 210x$

i. $6x - x$

j. $z - 6z$

k. $-8x + 3x$

l. $-2x + 7x$

m. $-4y - 6y$

n. $-5y - 2y$

o. $\frac{2}{3}x + \frac{1}{2}x$

p. $\frac{3}{5}x + \frac{5}{6}x$

q. $\frac{4}{5}y + 2y$

r. $4x + \frac{2}{3}x$

s. $\frac{-1}{4}y + \frac{2}{7}y$

t. $\frac{-2}{3}z - \frac{4}{5}z$

u. $\frac{-4}{5}z + \frac{2}{3}z$

v. $\frac{2}{7}x - \frac{3}{4}x$

w. $-\frac{4}{15}x - \frac{3}{10}x$

x. $-\frac{1}{6}x - \frac{3}{4}x$

2. Simplify each of the following expressions.

a. $5x + 2x + 3x$

b. $217x + 82x + 318x$

c. $12x + 3x - 5x$

d. $y - 4y + 2y$

e. $-16y - 7y - 32y + y$

f. $-13z + 42z - 14z$

g. $\frac{5}{6}x + \frac{1}{3}x - \frac{1}{2}x$

h. $\frac{3}{4}x - \frac{1}{6}x - \frac{2}{3}x + \frac{1}{2}x$

i. $\frac{-1}{3}y - \frac{3}{4}y - \frac{2}{5}y$

j. $\frac{-4}{21}z - \frac{1}{14}z + \frac{5}{42}z$

k. $\frac{1}{12}z + \frac{3}{8}z - \frac{5}{16}z$

l. $-\frac{3}{8}y - \frac{1}{6}y - y + \frac{1}{2}y$

m. $\frac{1}{3}x + \frac{2}{3}x - \frac{1}{2}x + \frac{4}{5}x$

n. $\frac{3}{25}x + \frac{4}{15}x - \frac{3}{10}x$

o. $\frac{2}{3}x - x + \frac{1}{2}x$

3. Simplify each of the following expressions.

a. $-(-3)x + (-2)x$

b. $3y - (-4y)$

c. $2z - (-3)z + 2z - 4z$

d. $-(-2)x - 3x + (-3)x$

e. $-4x - (-3)x - (-5)x$

f. $-(-x) + (-3)x - (-2x)$

g. $\frac{2}{3}x - \frac{-2}{5}x$

h. $-\frac{-1}{2}y + \frac{-2}{3}y$

i. $-(-4)x - \frac{-1}{-5}x$

j. $\frac{-2}{-21}x - \frac{3}{-14}x$

k. $-(-1)z + \frac{3}{4}z + 3z$

l. $-\frac{-3}{5}y + \frac{2}{-3}y - \frac{-3}{-4}y$

4. Simplify each of the following expressions.

a. $2(-3)x + (-2)(-4)x$

b. $-3(4)y - (-5)y$

c. $-4z - (-2)(-1)z - 7z$

d. $9x - 2(-1)(-3)x + (-1)x$

e. $-4(-3)(2)y - y$

f. $-2(-4x) + (-3)(-2)x$

g. $\frac{2}{3}x - \frac{-2}{5}x$

h. $-\frac{-1}{2}y + \frac{-2}{-3}y$

5. Simplify each of the following expressions.

a. $4 \times 3x + 2 \times 5x$

b. $-5 \times 2 \times 3x + x$

c. $-8 \times 2x - 5 \times 3x$

d. $4(3)x - 2(3)x$

e. $-(-2)x + 3(-2)(-1)x$

f. $2(-3)(-5)x + 3(-4)x$

g. $\frac{1}{2} \times \frac{2}{3}x + \frac{3}{4} \times \frac{1}{3}x$

h. $-\frac{3}{5} \times \frac{1}{2}x - \frac{1}{4} \times \frac{2}{3}x$

i. $\frac{1}{2}\left(\frac{1}{3}\right)y - \frac{3}{5}\left(-\frac{1}{3}\right)y$

j. $-\frac{3}{5}\left(\frac{-1}{3}\right)x - \left(-\frac{3}{4}\right)\left(\frac{1}{2}\right)x$

6. Simplify each of the following expressions.

a. $-3x + 2y - 4x + y$

b. $-4 + 2x - 3$

c. $2x - (-4)x + 3y - 2y$

d. $-(-4y) + x - (-3)x - 2y$

e. $\frac{1}{-2} + (-3)x - \frac{-2}{3}x - 4$

f. $-2x + (-3)y - 4x + (-y)$

g. $\frac{-2}{3}y + \frac{-1}{-3} - \frac{1}{-2}y + \frac{4}{5}$

h. $2y - \frac{1}{-3} + (-3)y - \frac{1}{-2}y$

i. $\frac{1}{2}x + \frac{2}{3}y - 1 - x$

j. $-\frac{3}{-4}x + \frac{-1}{2}z - \frac{3}{-5}z + \frac{1}{3}x$

k. $\frac{-3}{4}x + \frac{1}{-4}y + \frac{-1}{-5}x + \frac{1}{-4}y$

7. Simplify each of the following expressions.

a. $2(3x + 4)$

b. $-3(-2x - 5) + 3$

c. $4(x + 2) - 3(2x - 4)$

d. $-2(3 - 4x) + 2x$

e. $2y - 3(4y - 2) - 3$

f. $4z - 4(2x + 6)$

g. $-3x + 2(5x - 2)$

h. $\frac{1}{2}\left(x - \frac{2}{3}\right) - \frac{3}{4}\left(x + \frac{1}{9}\right)$

i. $-\frac{3}{5} + 3\left(\frac{1}{5} - \frac{2}{3}x\right)$

j. $\frac{1}{4}z + \frac{1}{5}\left(\frac{5}{6}z - \frac{2}{3}\right) - \frac{1}{4}$

k. $-3(2x + 3y) - 3(5x - y)$

l. $4(-x + y) + 5(x - y)$

m. $\frac{1}{2}(x + 3y) - \frac{2}{3}(-2x + y)$

n. $-\frac{3}{4}\left(\frac{1}{2}x + \frac{2}{3}y\right) - \frac{1}{3}\left(\frac{3}{5}x + \frac{1}{4}y\right)$

8. Simplify each of the following equations.

a. $3x + 2x - 1 = 5 + 2$

b. $-5x + 3 - x = 4x - 3$

c. $-(-3)x + (-2)y = x + (-4)y$

d. $4x - (-2)y - (-3)x = y$

e. $4(-3)x + (-2)(-3)x = 0$

f. $\frac{2}{3}x + \frac{3}{4}x = \frac{-1}{2}x - \frac{2}{5}x$

g. $-\frac{3}{5}z + \frac{1}{2}z = \frac{3}{2}z - \frac{1}{4}z$

h. $-\frac{-1}{2}y - \frac{2}{3}y = \frac{-1}{-3}y + \frac{-3}{4}y$

i. $3(x + 1) - 2x = -3(x - 2)$

j. $-(3x + 1) + 4(1 - x) = 0$

k. $3(x + y) - 2(-x + y) = 0$

l. $-2(3x - 2y) - 3(x + 2y) = 3(-x + 5y)$

m. $\frac{1}{3}(x - 1) + \frac{2}{5}(x + 1) = \frac{3}{4}(-x + 1) - \frac{1}{4}(x - 1)$

n. $\frac{1}{4}\left(\frac{1}{2}x - \frac{2}{3}\right) + \frac{2}{3}x = \frac{-1}{4}\left(-\frac{1}{3}x + \frac{2}{5}\right) - \frac{1}{2}$

10.6   Solutions

By **solving an equation** we mean finding all possible values that, when used in the equation in place of the unknowns, will result in equal values for the expressions on the left and right sides of the equation. We call each of these values a **solution** to the equation.

As an example of this, consider the equation

$$3x + 2 = x + 6$$

The numerical value 2 is a solution to this equation since, if we substitute 2 for x on both sides of the equation,[19] it will result in equal values for the expressions on the left and right sides of the equation: With $x = 2$ the expression on the left side evaluates to

$$\text{LS} = 3x + 2$$
$$= 3 \times 2 + 2$$
$$= 6 + 2$$
$$= 8$$

---

[19]**Substitution** of x by 2 on both sides of the equation implies replacement of x with 2 on both sides of the equation.

and the expression on the right side evaluates to

$$
\begin{aligned}
\text{RS} &= x + 6 \\
&= 2 + 6 \\
&= 8
\end{aligned}
$$

It can be shown that the value 2 is the only value that works in the equation above and, therefore, this equation has a single solution: $x = 2$.

It is possible for an equation to have more than one solution. As an example, both 1 and 3 are solutions to the equation

$$x^2 = 4x - 3$$

With $x = 1$ the expression on the left side evaluates to

$$
\begin{aligned}
\text{LS} &= x^2 \\
&= 1^2 \\
&= 1
\end{aligned}
$$

and the expression on the right side evaluates to

$$
\begin{aligned}
\text{RS} &= 4x - 3 \\
&= 4 \times 1 - 3 \\
&= 4 - 3 \\
&= 1
\end{aligned}
$$

With $x = 3$ the expression on the left side evaluates to

$$
\begin{aligned}
\text{LS} &= x^2 \\
&= 3^2 \\
&= 9
\end{aligned}
$$

and the expression on the right side evaluates to

$$
\begin{aligned}
\text{LS} &= 4x - 3 \\
&= 4 \times 3 - 3 \\
&= 12 - 3 \\
&= 9
\end{aligned}
$$

Both $x = 1$ and $x = 3$ are, therefore, solutions to the equation $x^2 = 4x - 3$. It can be shown that these are the only solutions for the equation above.

An equation may have an infinite number of solutions. As an example, replacement of x with any value in the equation

$$x - 1 = x - 1$$

results in equal values for the expressions on its left and right. Such an equation has an infinite number of solutions. In this case, any real number will work as a solution and, therefore, the solution is $x \in \mathbb{R}$.

It is also possible for an equation not to have any solutions. As an example, the equation

$$\frac{1}{x} = 0$$

has no solutions. Replacement of x by any number other than 0 results in a nonzero value for the expression on the left side while the expression on the right side is 0. In addition, replacement of x by 0 on the left side results in the undefined expression $\frac{1}{0}$ which is not equal to any numbers while the right side does equal to the number 0.

From the above discussion we can conclude that an equation may have no solutions, a limited number of solutions (e.g., only one solution and no more, or only two solutions and no more, etc.), or an infinite number of solutions.

When we speak of solving an equation, we expect that any and all possible solutions to the equation will be listed.

There are two common approaches to solving equations which involve a single unknown that appears only once throughout the equation. One follows a line of reasoning called *natural semantics* while the other follows a different line of reasoning called *standard semantics*. We will discuss both approaches below in the context of the simpler equations that we may encounter. These are equations that involve a single operation and a single unknown with the unknown appearing only once throughout the equation. Tools for dealing with equations that involve multiple operations and unknowns and select cases where the unknowns appears more than once in the equation will be discussed in the chapters ahead.

10.6.1   Natural Semantic Schemes

Natural semantic schemes follow the logic that converts the main operation on one side of the equation to its inverse on the other side of the equation. As an example of how natural semantic schemes work, consider the following problem.

Problem

I bought a pen and a notebook and I paid \$9 in total. Calculate the cost of the pen if the notebook cost \$7.

The following equation models the problem:

$$c + 7 = 9$$

where c represents the value of the cost of the pen.

A common way to solve this problem is to argue that the cost of the pen is equal to the total money spent minus the cost of the notebook. We can express this logic using mathematical language as opposed to English:

$$c = 9 - 7$$

An equation that simplifies to

$$c = \$2$$

The full solution is given below:

$$c + 7 = 9$$
$$c = 9 - 7$$
$$c = \$2$$

We call such a solution a natural semantic solution as the logic used follows the manner in which we naturally reason, i.e., *if the cost of the pen and the notebook is \$9 and the notebook costs \$7 ($c + 7 = 9$), then the cost of the pen is equal to the total money spent, i.e.,* \$9, minus the cost of the notebook, i.e., \$7 ($c = 9 - 7$). Figure 10.6.1 illustrates the steps taken to solve the problem using natural semantics.

Such thought processes can be generalized. Indeed, for the problem above, we can argue that *if two things add up to a total, then one of them is equal to the total minus the other.* This abstract reasoning is illustrated in Figure 10.6.2 below.

Note that the line of reasoning given above is quite general as it applies to any problem whose model involves two pieces whose values add up to a total: It could involve the cost of a pen and a notebook whose values add up to total amount spent. It could be the mass of a C atom and the mass of an O atom that add up to the mass of a CO atom. Such logic always turns the addition of a term on one side of the equation to subtraction of the same term on the other side of the equation.[20] Our first natural semantic

---

[20]Note that subtraction is the inverse operation for addition.

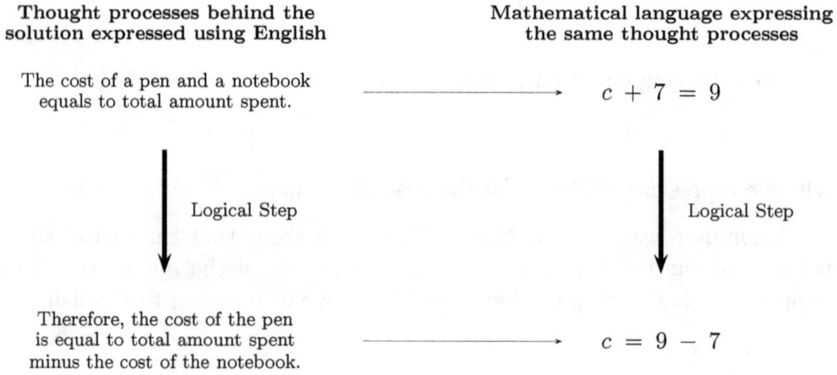

Thought processes behind the solution expressed using English	Mathematical language expressing the same thought processes
The cost of a pen and a notebook equals to total amount spent.	$c + 7 = 9$

Logical Step	Logical Step

| Therefore, the cost of the pen is equal to total amount spent minus the cost of the notebook. | $c = 9 - 7$ |

Figure 10.6.1: Thought processes involved in solving a concrete problem using natural semantic tools

Thought processes behind the solution expressed using English	Mathematical language expressing the same thought processes
The sum of two things is equal to a total.	$x + 7 = 9$

Logical Step	Logical Step

| Therefore, one of them is equal to the total minus the other. | $x = 9 - 7$ |

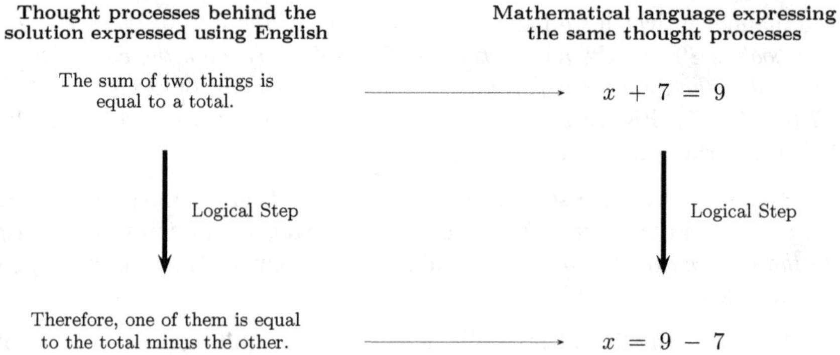

Figure 10.6.2: Thought processes involved in solving an abstract problem using natural semantic tools

powertool, then, states that *if addition of a term is the main operation on one side of an equation, then it can be converted to subtraction of that same term on other side of the equation.* This is illustrated below:

$$c + 7 = 9$$
$$c = 9 - 7$$

The justification for this move is based on the logic that states that *if two things add up to a total, then one of them equals to the total minus the other.*[21]

Problem

After using 3 L of a solution, we are left with 24 L of the solution. What volume of the solution did we start with?

The following equation models the problem:

$$V - 3 = 24$$

where V represents the value of the volume of the solution that we started with.

A common way to solve this problem is to argue that the initial volume of the solution is equal to the sum of the volume that remains and the volume used. We can express this logic using mathematical language as opposed to English:

$$V = 24 + 3$$

---

[21] We warn strongly that the reader should *not* see this move as a *change in sign*, i.e., positive becoming negative. Rather, the move should be seen as a switch from the operation of addition to its inverse operation, subtraction. The erroneous reasoning that sees the move as a switch in sign is not in line with the logic of natural semantics and will later cause confusion when we move beyond equations that involve addition and subtraction. As an example, as we will see shortly, multiplication on one side of the equation turns to division on the other side of the equation. Here there is no apparent switch in the sign of the operand.

Note that a switch in sign is an erroneous logic even if the numerical value appears before the unknown as in

$$7 + c = 9$$

which also requires that we write

$$c = 9 - 7$$

This may leave a strong impression that there a change in sign involved. However, according to our earlier exposition of a positive number such as $+7$ as a shorthand for the longer $0+7$, one may make a mental switch from $+7$ to $0+7$ and then continue with the solution seeing the $+$ as addition. It is this addition by 7 on the left side of the equation that switches to its inverse, subtraction of 7, on the right side of the equation.

An equation that simplifies to

$$V = 27 \, \text{L}$$

The full solution is given below:

$$V - 3 = 24$$
$$V = 24 + 3$$
$$V = 27 \, \text{L}$$

As before, we call such a solution a natural semantic solution as the logic used follows the manner in which we naturally reason. Figure 10.6.3 below illustrates the steps taken to solve the problem using natural semantics.

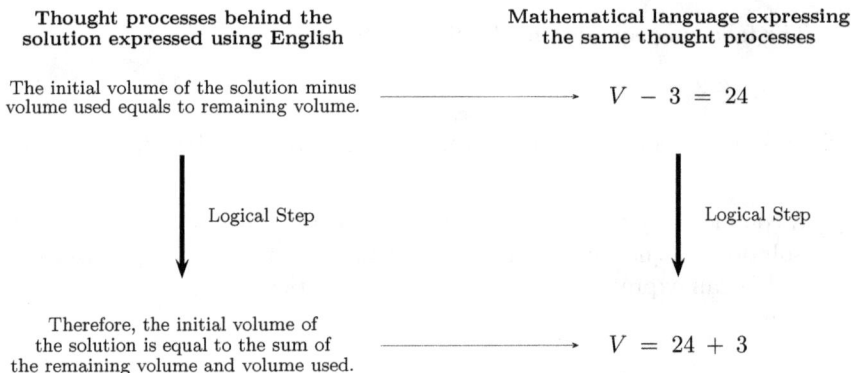

Thought processes behind the solution expressed using English	Mathematical language expressing the same thought processes
The initial volume of the solution minus volume used equals to remaining volume.	$V - 3 = 24$
Logical Step	Logical Step
Therefore, the initial volume of the solution is equal to the sum of the remaining volume and volume used.	$V = 24 + 3$

Figure 10.6.3: Thought processes involved in solving a concrete problem using natural semantic tools

And as before, such thought processes can be generalized. For the problem above, we can argue that *if the total minus a piece is equal to another piece, then the total is equal to the sum of the pieces.* This abstract reasoning is illustrated in Figure 10.6.4 below.

Note that the line of reasoning given above is quite general as it applies to any problem whose model involves two pieces whose values add up to a total: It could involve the initial volume of a solution, volume used and volume that remains. It could be the amount of money made, amount of money spent and amount of money saved. Such logic always turns the subtraction of a term on one side of the equation to addition of the same term on the other side

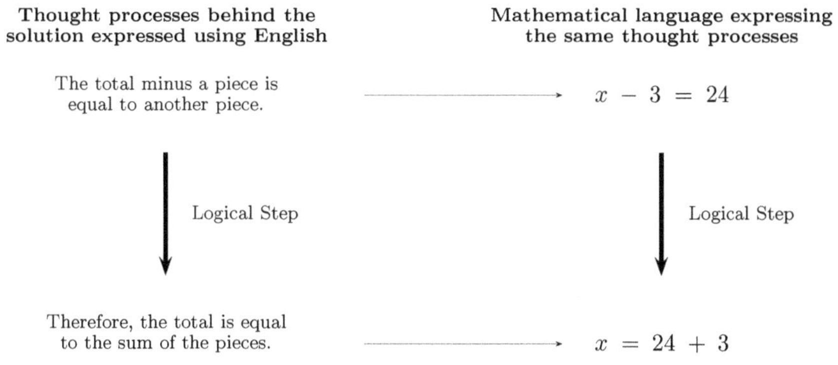

Thought processes behind the solution expressed using English	Mathematical language expressing the same thought processes
The total minus a piece is equal to another piece.	$x - 3 = 24$

Logical Step   Logical Step

| Therefore, the total is equal to the sum of the pieces. | $x = 24 + 3$ |

Figure 10.6.4: Thought processes involved in solving an abstract problem using natural semantic tools

of the equation.[22] Our second natural semantic powertool, then, states that *if subtraction of a term is the main operation on one side of the equation, then it can be converted to addition of that same term on the other side of the equation.* This is illustrated below:

$$V - 3 = 24$$
$$V = 24 + 3$$

The justification for this move is based on the logic that states that *if total minus a piece is equal to another piece, then the total is equal to the sum of the pieces.*[23]

---

[22]Note that addition is the inverse operation for subtraction.

[23]We warn strongly that the reader should not see this move as a *change in sign*, i.e., negative becoming positive. Rather, the move should be seen as a switch from the operation of subtraction to its inverse operation, addition. The erroneous reasoning that sees the move as a switch in sign is not in line with the logic of natural semantics and will later cause confusion when we move beyond equations that involve addition and subtraction. As an example, as we will see shortly, multiplication on one side of the equation turns to division on the other side of the equation. Here there is no switch in the sign of the operand.

Note that a switch in sign is an erroneous logic even if the numerical value appears before the unknown as in

$$-3 + V = 24$$

which also requires that we write

$$V = 24 + 3$$

Problem

4 pens cost \$20. Calculate the cost of a pen.

The following equation models the problem:

$$4c = 20$$

where c represents the value of the cost of a pen.

A common way to solve this problem is to argue that if 4 pens cost \$20, then the cost of one pen is equal to the total cost divided by the number of pens purchased. We can express this logic using mathematical language as opposed to English:

$$c = \frac{20}{4}$$

An equation that simplifies to

$$c = \$5/\text{pen}$$

The full solution is given below:

$$4c = 20$$
$$c = \frac{20}{4}$$
$$c = \$5/\text{pen}$$

As before we refer to such a solution as a natural semantic solution as the logic used follows the manner in which we naturally reason. Figure 10.6.5 below illustrates the steps taken to solve the problem using natural semantics.

And as before, such thought processes can be generalized. For the problem above, we can argue that *if the product of two values is equal to a total, then one of those values is equal to the total divided by the other.* This abstract reasoning is illustrated in Figure 10.6.6 below.

Note that the line of reasoning given above is quite general as it applies to any problem whose model involves the product of two values: This could involve the cost of a number of identical pens, the number of pens, and the

---

This may leave a strong impression that there a change in sign involved. However, according to our earlier exposition of a negative number such as -3 as a shorthand for the longer $0 - 3$, one may make a mental switch from -3 to $0 - 3$ and then continue with the solution seeing the $-$ as subtraction. It is this subtraction of 3 on the left side of the equation that switches to its inverse, addition of 3 on the right side of the equation.

Thought processes behind the solution expressed using English	Mathematical language expressing the same thought processes
4 pens cost \$20.	$4c = 20$

Logical Step

Logical Step

| Therefore, the cost of one pen is equal to the total cost, \$20, divided by the number of pens, 4. | $c = \frac{20}{4}$ |

Figure 10.6.5: Thought processes involved in solving a concrete problem using natural semantic tools

Thought processes behind the solution expressed using English	Mathematical language expressing the same thought processes
The product of two values is equal to a total.	$4x = 20$

Logical Step

| Therefore, one of those values is equal to the total divided by the other. | $x = \frac{20}{4}$ |

Figure 10.6.6: Thought processes involved in solving an abstract problem using natural semantic tools

cost of a pen. It could involve the mass of a molecule of ozone, O_3, the number of atoms of O in a molecule of ozone, and the atomic mass of O. Such logic always turns the multiplication of a factor on one side of the equation into division by the same factor on the other side of the equation.[24,25] Our third natural semantic powertool, then, states that *if multiplication of a factor is the main operation on one side of an equation, then it can be turned to division by the same factor on the other side of the equation.* This is illustrated below:

$$c = \frac{20}{4} \curvearrowright$$

The justification for this move is based on the logic that states that *if the product of two values results in a total, then one of them is equal to the total divided by the other.*

Problem

I am on my way to Montreal. I have covered 140 km so far. This is $\frac{1}{4}$ of the distance from Toronto to Montreal. What is the distance from Toronto to Montreal?

The following equation models the problem:[26]

$$\frac{d}{4} = 140$$

where d represents the distance from Toronto to Montreal.

A common way to solve this problem is to argue that if 140 km is $\frac{1}{4}$ of the distance from Toronto to Montreal, then the distance from Toronto to Montreal is 4 times 140. We can express this logic using mathematical language as opposed to English:

$$d = 140 \times 4$$

An equation that simplifies to

$$d = 560 \text{ km}$$

---

[24] Note that division is the inverse operation for multiplication.

[25] We commented earlier on the manner in which such moves from one side of an equation to the other side should be interpreted as a switch from the main operation on one side of the equation to the inverse of that operation on the other side of the equaiton. Note that there is no *change in sign* involved in the move from multiplication to division.

[26] Technically, the phrase *a quarter of the distance* translates to $\frac{1}{4}x$. However, note that the expression $\frac{1}{4}x$ (read *a quarter of x*) is equivalent to $\frac{x}{4}$ (read *x divided by 4*).

The full solution is given below:

$$\frac{d}{4} = 140$$
$$d = 140 \times 4$$
$$d = 560 \text{ km}$$

As before, we refer to such a solution as a natural semantic solution as the logic used follows the manner in which we naturally reason. Figure 10.6.7 below illustrates the steps taken to solve the problem using natural semantics.

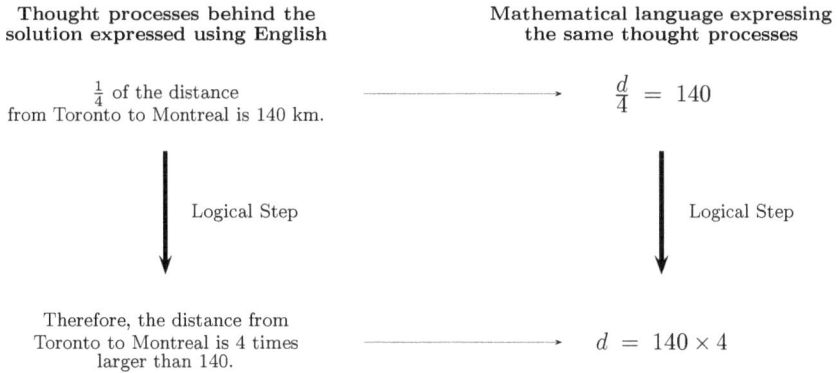

Thought processes behind the solution expressed using English

Mathematical language expressing the same thought processes

$\frac{1}{4}$ of the distance from Toronto to Montreal is 140 km.

$$\frac{d}{4} = 140$$

Logical Step

Logical Step

Therefore, the distance from Toronto to Montreal is 4 times larger than 140.

$$d = 140 \times 4$$

Figure 10.6.7: Thought processes involved in solving a concrete problem using natural semantic tools

And as before, such thought processes can be generalized. We ask the reader to provide such logic and the corresponding equations that express that logic.

Note that the line of reasoning given above is quite general as it applies to any problem whose model involves a piece that results from a fraction of a total: This could involve the distance from Toronto to Montreal and a fraction of that distance. It could involve my savings and the fraction of my income that these savings represent. Such logic always turns the division by a factor on one side of the equation into multiplication by the same factor on the other side of the equation.[27] Our fourth natural semantic powertool,

---

[27]Note that multiplication is the inverse operation for division.

then, states that *if division by a factor is the main operation on one side of an equation, then it can be turned into multiplication by the same factor on the other side of the equation.* This is illustrated below:

$$\frac{d}{4} = 140$$
$$d = 140 \times 4$$

We let the reader provide justification for this move.

What the lines of reasoning above have in common is that they convert the main operation on one side of the equation (be it addition, subtraction, multiplication, etc.) to its inverse on the other side of the equation. Beyond addition, subtraction, multiplication and division, this logic allows us to convert an exponent problem with a missing base, e.g., $x^3 = 8$, to its inverse on the other side of the equation, i.e., $x = \sqrt[3]{8}$, and an exponent problem with a missing exponent, e.g., $2^x = 8$, to its inverse on the other side of the equation, i.e., $x = \log_2 8$. Such logic, which can be applied to any operation and its inverse, maps onto the way we naturally reason and, for this reason we refer to algorithms that utilize it as **natural semantic algorithms.**

Exercise Set 10.6.1

In each case solve the given equation using the natural semantic scheme.

1. $x + 7 = 8$

2. $x + 24 = 9$

3. $x + \frac{1}{2} = \frac{3}{5}$

4. $x + \frac{2}{3} = 3$

5. $16 + x = 42$

6. $2 + x = -1$

7. $\frac{1}{4} + x = -\frac{1}{2}$

8. $\frac{2}{5} + x = \frac{3}{7}$

9. $x - 3 = 7$

10. $x - 18 = 5$

11. $x - 1 = -6$

12. $x - \frac{3}{5} = \frac{1}{4}$

13. $-5 + x = 2$

14. $-8 + x = -1$

15. $-\frac{3}{5} + x = \frac{2}{7}$

16. $\frac{-1}{4} + x = -\frac{1}{3}$

17. $4x = 28$

18. $3x = -6$

19. $-5x = 12$

20. $-2x = 22$

21. $-7x = -15$

22. $\frac{3}{4}x = \frac{1}{2}$

23. $-\frac{1}{3}x = \frac{2}{5}$

24. $\frac{4}{7}x = -\frac{1}{2}$

25. $\frac{-3}{8}x = -\frac{1}{4}$

26. $\frac{x}{4} = \frac{3}{5}$

27. $\frac{x}{3} = 2$

28. $\frac{x}{-2} = \frac{3}{7}$

29. $\frac{x}{-4} = -3$

30. $\frac{x}{-8} = \frac{3}{5}$

31. $x^2 = 25$

32. $x^2 = 7$

33. $x^2 = 0$

34. $x^2 = -9$

35. $x^2 = \frac{1}{4}$

36. $x^2 = \frac{4}{9}$

37. $x^3 = 1$

38. $x^3 = -8$

39. $x^3 = 2$

40. $x^3 = -64$

41. $x^3 = -\frac{8}{27}$

42. $x^4 = 16$

43. $x^4 = -16$

44. $x^5 = 32$

45. $x^5 = -32$

46. $x^8 = 1$

47. $x^{100} = 1$

48. $x^{100} = -1$

49. $x^{101} = 1$

50. $x^{101} = -1$

51. $2^x = 8$

52. $2^x = 1$

53. $3^x = 81$

54. $3^x = -3$

55. $3^x = \frac{1}{3}$

56. $4^x = \frac{1}{16}$

57. $10^x = 10\,000$

58. $2^x = 3$

59. $10^x = 4$

10.6.2 Standard Semantic Schemes

An alternative to the natural semantic scheme uses the argument that *the same thing can be done to both sides of an equation*. Algorithms that utilize this kind of logic are called **standard semantic algorithms**. The logic is based on our understanding that, if two things are equal, we can do the same thing to both and end up with values that are equal. As an analogy, consider a balance with weights in the left and right pans of the balance.[28] If the weights balance, then we can remove equal weights from both pans and the remining weights would still balance or we can double the weights in the pans and still have them balance.

Example

Solve the equation

$$x + 4 = 7$$

using standard semantics.

Solution

$$x + 4 = 7$$
$$x + 4 - 4 = 7 - 4$$
$$x = 3$$

---

[28]These correspond to the left and right sides of an equation.

In order to solve the problem above, we need to isolate the unknown, x. To do so, we need to get rid of addition by 4 on the left side of the equation. Since, based on standard semantics, we can do the same thing to both sides of an equation, we can get rid of addition by 4 on the left side of the equation by subtracting 4 from both sides of the equation.[29] This leads to the solution above.

Standard semantics does make sense but the logic that it uses is not in line with the logic that we naturally use when we solve such problems. Consider, as an example, the case when the equation above models a problem where one buys a pen and a notebook and pays $7 in total with the cost of the notebook being $4, i.e.,

$$c + 4 = 7$$

where c represents the value of the cost of the pen. To calculate the cost of the pen, natural semantics would require that we reason as follows: *If the cost of the pen and the notebook equals to total amount spent, then the cost of the pen is equal to total amount spent minus the cost of the notebook.* This is in line with the way we normally reason. An approach that uses standard semantics, however, would require that we reason as follows: *If the cost of the pen and the notebook equals to total amount spent, then, to find the cost of the pen, subtract the cost of the notebook from both sides of the equation!* Who reasons like this? And if one chooses to use a certain logic when one solves the problem in real life, but switches to a different logic when solving the problem using algebra, then it will be unlikely that she or he will ever use algebra to solve problems in real life as the logic used is foreign.

Example

Solve the equation

$$x - 2 = 14$$

using standard semantics.

Solution

$$x - 2 = 14$$
$$x - 2 + 2 = 14 + 2$$
$$x = 16$$

In order to solve the problem above, we need to isolate the unknown, x. To do so, we need to get rid of subtraction of 2 on the left side of the equation.

---

[29]Note that subtraction is the inverse operation for addition.

Since, based on standard semantics, we can do the same thing to both sides of an equation, we can get rid of subtraction of 2 on the left side of the equation by adding 2 to both sides of the equation.[30] This leads to the solution above.

We let the reader provide the argument behind the solution to the next two examples.

Example

Solve the equation

$$4x = 20$$

using standard semantics.

Solution

$$4x = 20$$
$$\frac{4x}{4} = \frac{20}{4}$$
$$x = 5$$

Example

Solve the equation

$$\frac{x}{3} = 7$$

using standard semantics.

Solution

$$\frac{x}{3} = 7$$
$$3 \times \frac{x}{3} = 3 \times 7$$
$$x = 21$$

The logic can be used to solve problems that involve more advanced operations. Here is an example.

---

[30] Note that addition is the inverse operation for subtraction.

Example

Solve the equation

$$x^3 = 64$$

using standard semantics.

Solution

$$x^3 = 64$$
$$\sqrt[3]{x^3} = \sqrt[3]{64}$$
$$x = 4$$

To solve for x in the problem above, we apply the inverse operation for an exponent problem with a missing base, i.e., the root operation, to both sides of the equation.

Example

Solve the equation

$$2^x = 8$$

using standard semantics.

Solution

$$2^x = 8$$
$$\log_2 2^x = \log_2 8$$
$$x = 3$$

To solve for x in the problem above, we apply the inverse operation for an exponent problem with a missing exponent, i.e., the logarithm operation, to both sides of the equation.

Exercise Set 10.6.2

Solve the equations in Exercise Set 10.6.1 using the standard semantic scheme.

As noted earlier, standard semantics is inefficient and the logic that it uses is not in line with the manner in which we normally reason. In this textbook we will not discuss this approach further.[31]

10.7   Notation

In this section we will discuss the use of notation in theoretical and applied discussions.

10.7.1   Theoretical Equations

The discussion so far has focused on the use of the formal notation in working with basic equations. As discussed earlier in the textbook, the formal notation is exact and descriptive, two characteristics that are required features of notations for use in conducting theoretical discussions. Equations that arise from theoretical considerations, therefore, use the formal notation to express the relationships between the quantities involved and one is expected to continue to use the formal notation in solving or otherwise rearranging such equations. This allows us to continue to make sense of what is happening as we go through one line to the next in our solution process.

10.7.2   Applied Equations

Equations that arise from applications of theory, however, often involve measured numerical values which are, by convention, written using base-ten notations such as the decimal notation or scientific notation. Since equations that model applied problems involve measured values which are often approximate, the models themselves represent approximations and, therefore, one may, and indeed one is advised to, use the relevant base-ten notation throughout the solution.[32]

---

[31]The reader interested in learning more about the standard semantic scheme and further comparison of natural semantics vs. standard semantics in the context of more complex problems is invited to read Appendix P.

[32]Unless the given equation begins with decimals, one should not use the notation in the solution process. However, if the given equation involves even a single decimal number, then one may, and indeed should, use the decimal notation throughout.

Exercise Set 10.7

In each case solve the given equation using the natural semantic scheme.

1. $x + 0.2 = 15.8$

2. $x + 1.75 = 2.99$

3. $18.2 + x = -16.7$

4. $x - 3.7 = 5.6$

5. $x - 12 = 17.9$

6. $x - 8.8 = -6.4$

7. $-6.4 + x = 9.6$

8. $-8.76 + x = -1.97$

9. $0.7x = 12$

10. $7.9x = -6.2$

11. $-8.7x = -4.51$

12. $\frac{x}{0.65} = 4.7$

13. $\frac{x}{-4.8} = 2.1$

14. $\frac{x}{-8.1} = -7.5$

15. $x^2 = 0.25$

16. $x^2 = 0.4$

17. $x^2 = 18.6$

18. $x^2 = -9.1$

19. $x^3 = 0.82$

20. $x^3 = -7.7$

21. $x^3 = 6.1$

22. $x^3 = -6.1$

23. $x^4 = 0.0001$

24. $x^4 = 12.8$

25. $x^5 = -99.7$

26. $x^8 = -42.7$

27. $x^{2.3} = 45.8$

28. $x^{3.8} = 423$

29. $10^x = 42.8$

30. $10^x = -15.8$

31. $2^x = 0.68$

32. $2^x = 3.4$

33. $3^x = 72.6$

34. $3^x = -82.9$

35. $7.2^x = 182$

36. $2.6^x = 0.81$

Chapter 11
Equations in One Unknown

In this chapter we will discuss the genesis of equations in one unknown and present an algorithm for solving such equations. Initially we will focus on equations in which the unknown appears only once throughout the equation. Following this, we will discuss select cases where the unknown appears more than once in the equation.

11.1   Genesis

We have shown how algebraic equations arise during the modelling stage in the problem-solving process. Equations in one unknown, in particular, arise from the modelling of problems in which the values of all quantities in the model other than the one represented by the unknown are known. We have already seen many such models. Here's another.

Problem

You wish to cover a distance of 420 km today. This morning you drove at a speed of 120 km/h for 2.5 h. How fast will you have to drive to cover the rest of the distance in 3.5 h?

An Algebraic Model

v speed to cover the rest of the distance (km/h)

$$420 = 120 \times 2.5 + v \times 3.5 \qquad \text{Semantic formulation}$$
$$120 \times 2.5 + 3.5v = 420 \qquad \text{Technical formulation}$$

A technical formulation of the abstract version of this equation is

$$120 \times 2.5 + 3.5x = 420$$

This is an example of an equation that contains a single unknown. In the case of the model above, the unknown appears as a factor within a term and, therefore, relevant operations on the symbol are multiplication and addition. The model may, of course, involve other operations or combination of operations. Examples of such models are

$$\frac{3.00 \times 2 + 4x}{2 + 4} = 3.40$$

$$0.13\,(1 - x)\,(25.99) = 3.04$$

$$\frac{360\,000}{x} = \frac{320\,000}{280}$$

$$\frac{1}{2^x} \times 4.7 = 1.3$$

In the next section we will present an efficient, meaningful algorithm for solving such equations.

11.2 Solutions

We will begin this section with an algorithm for solving equations in which the unknown appears only once throughout the equation.

Solution Technique for Equations Involving a Single Unknown with the Unknown Appearing Once Throughout the Equation

1. Switch sides or otherwise bring the unknown to the left side of the equation.[1]
2. Solve for the term that contains the unknown.
3. Solve for the factor that contains the unknown.
4. Repeat Steps 1, 2 and 3 until they cannot be applied anymore.
5. Solve for the argument of the main operation that contains the unknown.
6. Repeat Steps 1, 2, 3, 4 and 5 until the problem is solved.

The steps in the algorithm above may be carried out using either the natural semantic scheme or the standard semantic scheme. As noted earlier, the natural semantic scheme is both more efficient and more in line with the manner in which we normally reason. We will, therefore, focus on the use of the natural semantic scheme in carrying out the steps in the algorithm above.

---

[1] This is a common convention that we adhere to in this textbook.

The natural semantic scheme may be carried out with or without intermediate simplifications. The former is commonly used when working with equations in which there is a single unknown. The latter is commonly used when working with formulas. In this section we will focus on the use of the natural semantic scheme with intermediate simplifications but we will also discuss the natural semantic scheme without intermediate simplification to set the stage later when we come to discuss the manner in which one works with formulas.

In the natural semantic scheme with intermediate simplifications the steps in the algorithm above are carried out by converting the main operation for an argument (term, factor, etc.) on one side of the equation to the inverse of that operation for the same argument on the other side of the equation. In addition, the steps are alternated with simplification of the equations as they arise with a simplification step following the last step of the algorithm. We will illustrate the ideas using examples.

Example

Solve the equation

$$4x + 3 = 11$$

Solution

Step 1 in the algorithm above does not apply as the unknown is already on the left side of the equation.

The equation is already simplified as the factors have been worked out and the terms are unlike terms so they cannot be combined.

Step 2 in the algorithm above requires that we solve for the term that contains the unknown.[2] As shown below, this is the term $4x$.

$$\boxed{4x} + \boxed{3} = \boxed{11}$$

In order to solve for this term we need to get rid of addition by 3. Using natural semantics, we can achieve this by converting addition of 3 on the left side of the equation to subtraction of 3 on the right side of the equation, i.e.,

$$4x = 11 - 3$$

---

[2]This means isolating the term that contains the unknown on the left side of the equation.

Our next step is to simplify the equation that we have arrived at. The expression on the left side of this equation cannot be simplified further but the expression on the right side can be simplified by carrying out the subtraction $11 - 3$. This yields

$$4x = 8$$

Step 3 in the algorithm above requires that we solve for the factor that contains the unknown.[3] The expression on the left side of this equation has two factors: 4 and x. We, therefore, need to solve for factor x. In order to solve for this factor we need to get rid of multiplication by 4. Using natural semantics, we can achieve this by converting multiplication by 4 on the left side of the equation[4] to division by 4 on the right side of the equation, i.e.,

$$x = \frac{8}{4}$$

We now simplify the equation above to arrive at

$$x = 2$$

The full solution without the intervening comments is given below:

$$4x + 3 = 11$$
$$4x = 11 - 3$$
$$4x = 8$$
$$x = \frac{8}{4}$$
$$x = 2$$

The equation has now been solved. The solution to the given equation is $x = 2$.

---

[3] This means isolating the factor that contains the unknown on the left side of the equation.

[4] Note that the multiplication we are speaking of is the one that precedes 4, i.e., we see the expression $4x$ as

$$1 \times 4 \times x$$

which shows that 4 is being multiplied with and we see the expression $\frac{-7x}{3}$ as

$$\frac{1 \times (-7) \times x}{3}$$

which shows that -7 is being multiplied with.

In an expression that involves chain multiplication/division, we can always place a leading $1\times$ before the first factor in that expression.

We can test this solution by substituting 2 for x in the equation above and check to see whether this substitution results in equal values for the expressions on the left and right sides of the equation. As shown below, with $x = 2$ the expressions on the left and right sides of the equation simplify to 11:

$$\begin{aligned} \text{LS} &= 4x + 3 \\ &= 4 \times 2 + 3 \\ &= 8 + 3 \\ &= 11 \end{aligned}$$

and

$$\text{RS} = 11$$

Example

Solve the equation

$$8 - 5x = 3$$

Solution

Step 1 in the algorithm above does not apply as the unknown is already on the left side of the equation.

The equation is already simplified as the factors within the terms have been worked out and the terms are unlike terms so that they cannot be combined.

Step 2 in the algorithm above requires that we solve for the term that contains the unknown. This is the term $5x$. In order to solve for this term we need to get rid of addition by 8.[5] Using natural semantics, we can achieve this by converting addition of 8 on the

---

[5] In light of our earlier discussions on the equivalence of 8 and $0 + 8$, the reader should see the expression $8 - 5x$ as

$$0 + 8 - 5x$$

which shows that 8 is being added and see the expression $-3 + x$ as

$$0 - 3 + x$$

which shows that 3 is being subtracted.

In an expression that involves chain addition/subtraction, we can always place a leading $0 +$ or $0 -$, as the case may be, before the first term in that expression.

left side of the equation to subtraction of 8 on the right side of the equation, i.e.,

$$-5x = 3 - 8$$

Our next step is to simplify the equation that we have arrived at. The expression on the left side of this equation cannot be simplified further but the one on the right side can be simplified by carrying out the subtraction $3 - 8$. This yields

$$-5x = -5$$

Step 3 in the algorithm above requires that we solve for the factor that contains the unknown. There are two factors on the left side of this equation: -5 and x. In order to solve for the factor x, we need to get rid of multiplication by -5. Using natural semantics, we can achieve this by converting multiplication by -5 on the left side of the equation to division by -5 on the right side of the equation, i.e.,

$$x = \frac{-5}{-5}$$

We now simplify the equation above to arrive at

$$x = 1$$

The full solution without the intervening comments is given below:

$$8 - 5x = 3$$
$$-5x = 3 - 8$$
$$-5x = -5$$
$$x = \frac{-5}{-5}$$
$$x = 1$$

We leave it to the reader to check to see whether this solution is correct.

Example

Solve the equation

$$-2 - 3x = 4$$

Solution

Step 1 in the algorithm above does not apply as the unknown is already on the left side of the equation.

The equation is already simplified as the factors within the terms have been worked out and the terms are unlike terms so that they cannot be combined.

Step 2 in the algorithm above requires that we solve for the term that contains the unknown. This is the term $3x$. In order to solve for this term we need to get rid of subtraction by 2. Using natural semantics, we can achieve this by converting subtraction of 2 on the left side of the equation to addition of 2 on the right side of the equation, i.e.,

$$-3x \ = \ 4 \ + \ 2$$

Our next step is to simplify the equation that we have arrived at. The expression on the left side of this equation cannot be simplified further but the one on the right side can be simplified by carrying out the addition $4 + 2$. This yields

$$-3x \ = \ 6$$

Step 3 in the algorithm above requires that we solve for the factor that contains the unknown. On the left side of the equation there are two factors: -3 and x. In order to solve for factor x, we need to get rid of multiplication by -3. Using natural semantics, we can achieve this by converting multiplication by -3 on the left side of the equation to division by -3 on the right side of the equation, i.e.,

$$x \ = \ \frac{6}{-3}$$

We now simplify the equation above to arrive at

$$x \ = \ -2$$

The full solution without the intervening comments is given below:

$$-2 \ - \ 3x \ = \ 4$$
$$-3x \ = \ 4 \ + \ 2$$
$$-3x \ = \ 6$$
$$x \ = \ \frac{6}{-3}$$
$$x \ = \ -2$$

We leave it to the reader to check to see whether this solution is correct.

Exercise Set 11.2A

Solve each of the following equations using natural semantics with intermediate simplifications.

1. $-x = 14$	12. $8x = 1$	23. $x - 4 = 2$
2. $-x = 5$	13. $3x = -2$	24. $3 - x = 7$
3. $-x = -2$	14. $6x = -1$	25. $-2 + 4x = 6$
4. $-x = -7$	15. $-3x = 8$	26. $3x - 5 = 1$
5. $-x = 0$	16. $-7x = 2$	27. $-2x - 6 = -14$
6. $4x = 24$	17. $-5x = -4$	28. $4 = 3 + 2x$
7. $-5x = 20$	18. $-3x = -3$	29. $5 = -x + 3$
8. $-2x = 18$	19. $-4x = 0$	30. $-7 + 8x = 0$
9. $-3x = -12$	20. $x + 2 = 18$	31. $9 - 2x = 2$
10. $-9x = -18$	21. $5 + x = -3$	32. $6x + 1 = -3$
11. $3x = 2$	22. $x + 1 = 0$	33. $-8x - 6 = -7$

The numerical values in an equation may be rational of course. In such cases it is assumed that the discussion is theoretical and, therefore, one may *not* convert the rational numbers to decimals; rather, it is expected that the practitioner keeps the discussion formal by restricting herself or himself to the use of the formal notation.[6] The steps in solving such equations are identical to those presented earlier. We will illustrate the ideas using examples.

Example

Solve the equation

$$\frac{2}{3} + \frac{1}{2}x = \frac{11}{15}$$

Solution

Step 1 in the algorithm above does not apply as the unknown is already on the left side of the equation.

---

[6] Recall that such conversions often lead to repeating digits that are difficult to work with and, if roudned, result in approximate solutions which, in theoretical contexts, are unacceptable.

The equation is already simplified as the factors within the terms have been worked out and the terms are unlike terms so that they cannot be combined.

Step 2 in the algorithm above requires that we solve for the term that contains the unknown. This is the term $\frac{1}{2}x$. In order to solve for this term we need to get rid of addition by $\frac{2}{3}$. Using natural semantics, we can achieve this by converting addition of $\frac{2}{3}$ on the left side of the equation to subtraction of $\frac{2}{3}$ on the right side of the equation, i.e.,

$$\frac{1}{2}x = \frac{11}{15} - \frac{2}{3}$$

Our next step is to simplify the equation that we have arrived at. The expression on the left side of this equation cannot be simplified further but the one on the right side can be simplified by carrying out the subtraction $\frac{11}{15} - \frac{2}{3}$. This yields

$$\frac{1}{2}x = \frac{11 - 10}{15}$$
$$\frac{1}{2}x = \frac{1}{15}$$

Step 3 in the algorithm above requires that we solve for the factor that contains the unknown. On the left side of the equation there are two factors: $\frac{1}{2}$ and x. In order to solve for the factor x, we need to get rid of multiplication by $\frac{1}{2}$. Using natural semantics, we can achieve this by converting multiplication by $\frac{1}{2}$ on the left side of the equation to division by $\frac{1}{2}$ on the right side of the equation, i.e.,[7]

$$x = \frac{2}{1} \times \frac{1}{15}$$

We now simplify the equation above to arrive at

$$x = \frac{2}{15}$$

---

[7]In the solution given above, we have turned multiplication by $\frac{1}{2}$ on the left side of the equation to multiplication by $\frac{2}{1}$ on the right side of the equation. This move can be arrived at through two line of reasoning: The first sees the move as a shortcut that turns division by a fraction to multiplication by its reciprocal. The second sees the expression on the left as multiplication of x by 1 and division of x by 2. This is because the expression $\frac{1}{2}x$ (i.e., $\frac{1}{2}$ of x) may be seen as $\frac{1x}{2}$ (i.e., x divided by 2) showing tht x is being multiplied by 1 and divided by 2. In applying the natural semantic scheme, we convert multiplication by 1 on the left to division by 1 on the right, and we convert division by 2 on the left, to multiplication by 2 on the right.

The full solution without the intervening comments is given below:

$$\frac{2}{3} + \frac{1}{2}x = \frac{11}{15}$$

$$\frac{1}{2}x = \frac{11}{15} - \frac{2}{3}$$

$$\frac{1}{2}x = \frac{11 - 10}{15}$$

$$\frac{1}{2}x = \frac{1}{15}$$

$$x = \frac{2}{1} \times \frac{1}{15}$$

$$x = \frac{2}{15}$$

We leave it to the reader to check to see whether this solution is correct.

Example

Solve the equation

$$-\frac{3}{4} - \frac{2}{3}x = \frac{7}{12}$$

Solution

$$-\frac{3}{4} - \frac{2}{3}x = \frac{7}{12}$$

$$-\frac{2}{3}x = \frac{7}{12} + \frac{3}{4}$$

$$-\frac{2}{3}x = \frac{7 + 9}{12}$$

$$-\frac{2}{3}x = \frac{16}{12}$$

$$-\frac{2}{3}x = \frac{4}{3}$$

$$x = -\frac{3}{2} \times \frac{4}{3}$$

$$x = -\frac{2}{1}$$

$$x = -2$$

We let the reader provide the logic behind each move in the solution above and check to see whether the solution arrived at is correct.

Example

Solve the equation

$$\frac{3}{4}x = 5$$

Solution

We provide the full solution below. We ask the reader to supply justification for each step.

$$\frac{3}{4}x = 5$$

$$x = \frac{4}{3} \times 5$$

$$x = \frac{20}{3}$$

Exercise Set 11.2B

Solve each of the following equations using natural semantics with intermediate simplifications.

1. $\frac{2}{3}x = 24$

2. $\frac{1}{3}x = 2$

3. $\frac{3}{5}x = -1$

4. $\frac{3}{7}x = 21$

5. $\frac{4}{5}x = 0$

6. $-\frac{2}{5}x = 8$

7. $-\frac{3}{8}x = -5$

8. $\frac{5}{6}x = \frac{1}{4}$

9. $\frac{2}{3}x = \frac{6}{5}$

10. $\frac{-3}{4}x = \frac{-9}{8}$

11. $\frac{2}{-5}x = -\frac{1}{5}$

12. $3x = \frac{1}{2}$

13. $-2x = \frac{4}{5}$

14. $-x = \frac{5}{6}$

15. $9x = -\frac{3}{4}$

16. $\frac{3}{5}x + \frac{1}{2} = \frac{1}{3}$

17. $\frac{4}{7}x + \frac{3}{4} = \frac{2}{5}$

18. $-\frac{1}{3}x - \frac{2}{9} = -\frac{5}{6}$

19. $\frac{3}{5} + \frac{1}{8}x = \frac{1}{2}$

20. $-\frac{4}{9} + 2x = \frac{1}{3}$

21. $-4x - \frac{2}{5} = \frac{1}{4}$

22. $-x + 1 = \frac{3}{4}$

23. $2x - 3 = \frac{4}{5}$

24. $3 - 5x = \frac{9}{2}$

25. $-2 + \frac{2}{3}x = \frac{1}{4}$

26. $-\frac{5}{6}x - \frac{3}{4} = 1$

27. $\frac{8}{3}x - \frac{8}{3} = 2$

28. $\frac{1}{3} = -2 + \frac{4}{7}x$

29. $-3 = -\frac{7}{5}x + 2$

30. $-\frac{7}{8} = \frac{3}{5}x - \frac{1}{4}$

Our next example involves brackets:

Example

Solve the equation

$$4\,(x\,+\,3)\,=\,20$$

Solution

We skip Step 1 as the unknown is already on the left.

According to our earlier discussion on simplification of equations, simplification of the expression on the left side would require that we multiply 4 into the bracket. However, such a move, in addition to other shortcomings, is not in line with natural semantics. We will explain the reasons shortly but for now we will not apply the distributive law to simplify the expression on the left.

We skip Step 2 as the term that contains the unknown is already solved for.

We move on to Step 3 and solve for the factor that contains the unknown. The expression on the left of the equation analyzes into two factors: 4 and $x+3$. To solve for the factor that contains the unknown, i.e., $x+3$, we convert multiplication by 4 on the left side to division by 4 on the right side. This yields

$$x\,+\,3\,=\,\frac{20}{4}$$

Simplifying the expression on the right yields

$$x\,+\,3\,=\,5$$

The opening up of the brackets around the factor $x+3$ makes it possible to apply Step 2 again. We solve for the term that contains the unknown:

$$x\,=\,5\,-\,3$$

Simplification of the expression on the right side of the equation yields

$$x\,=\,2$$

The full solution without the intervening explanations is given below.

$$4\,(x\,+\,3)\ =\ 20$$
$$x\,+\,3\ =\ \frac{20}{4}$$
$$x\,+\,3\ =\ 5$$
$$x\ =\ 5\,-\,3$$
$$x\ =\ 2$$

Let us now look at the alternative approach that would fully simplify the expression on the left side of the equation at the outset by multiplying 4 into the bracket, i.e.,

$$4\,(x\,+\,3)\ =\ 20$$
$$4x\,+\,12\ =\ 20$$

We now return to Step 2 and solve for the term that contains the unknown:

$$4x\ =\ 20\,-\,12$$

Simplifying the expression on the right yields

$$4x\ =\ 8$$

We now move on to Step 3 and solve for the factor that contains the unknown:

$$x\ =\ \frac{8}{4}$$

Simplifying the expression on the right yields

$$x\ =\ 2$$

The full solution without the intervening explanations is given below.

$$4\,(x\,+\,3)\ =\ 20$$
$$4x\,+\,12\ =\ 20$$
$$4x\ =\ 20\,-\,12$$
$$4x\ =\ 8$$
$$x\ =\ \frac{8}{4}$$
$$x\ =\ 2$$

The two solutions are written side by side below:

$$4\,(x\,+\,3)\;=\;20 \qquad\qquad 4\,(x\,+\,3)\;=\;20$$
$$x\,+\,3\;=\;\frac{20}{4} \qquad\qquad 4x\,+\,12\;=\;20$$
$$x\,+\,3\;=\;5 \qquad\qquad 4x\;=\;20\,-\,12$$
$$x\;=\;5\,-\,3 \qquad\qquad 4x\;=\;8$$
$$x\;=\;2 \qquad\qquad x\;=\;\frac{8}{4}$$
$$\qquad\qquad\qquad\qquad\qquad x\;=\;2$$

Let us put some context behind the equation above. Suppose you buy 4 packages each of which contains a pen and a \$3 notebook and that you pay \$20 in total. We wish to calculate the cost of a pen. The following equation models the problem:

$$4\,(c\,+\,3)\;=\;20$$

where c represents the value of the cost of a pen in dollars.

There are two ways to work out the cost of a pen: The first proceeds by finding the cost of one package through the division of the total amount of money spent, i.e., 20, by the number of packages, i.e., 4. This yields

$$c\,+\,3\;=\;\frac{20}{4}$$

Simplifying the right side yields the cost of one package as \$5, i.e.,

$$c\,+\,3\;=\;5$$

We now calculate the cost of a pen by subtracting the cost of a notebook from the cost of a package. This yields

$$c\;=\;5\,-\,3$$

which simplifies to

$$c\;=\;2$$

The line of reasoning above represents the most common logic used to solve the given problem and corresponds to the solution given on the left in the side-by-side presentation of the solutions above.

An alternative logic proceeds as follows: We begin by working out the cost of all the pens and all the notebooks in all the packages and set the sum of these costs to total amount of money spent, i.e.,

$$4c\,+\,12\;=\;20$$

Next we subtract the cost of the notebooks from total amount spent to get the cost of the pens. This yields

$$4c = 20 - 12$$

Simplifying the right side yields the cost of the pens:

$$4c = 8$$

We can now work out the cost of a pen through the division of the cost of all the pens, i.e., 8, by the number of pens, i.e., 4. This yields

$$c = \frac{8}{4}$$

which simplifies to

$$c = 2$$

as the cost of a pen.

This logic, which corresponds to the solution given on the right in the side-by-side presentation of the solutions above, works. However, it is not the logic that is commonly used to solve such problems and is, therefore, not the preferred approach to solve such problems.[8] The first solution uses algebra as a language to express our natural thought processes and is, in this sense, superior to the second solution which uses a more obscure line of reasoning.[9]

---

[8]It is also less efficient: Note that the logic used on the left in the side-by-side presentation requires a single division followed by a subtraction whereas the logic used on the right in the side-by-side presentation requires two multiplications (which would grow in number if the package contained more items), a subtraction and a division. Furthermore, the multiplications usually generate larger numerical values that are more time-consuming to work with. And finally, when working with formulas, the alternative logic results in multiple appearance of the quantity symbol that multiplies the bracket which is highly undesirable as it makes it difficult to relate the values of the various quantities involved in the formula. See the chapter on formulas for more on this.

[9]Readers who prefer to use the alternative logic may, of course, choose to do so but it is likely that the preference is rooted in familiarity. We recommend such readers to familiarize themselves with the recommended logic before making a decision as to the one they prefer to use.

Exercise Set 11.2C

Solve each of the following equations using natural semantics with intermediate simplifications.

1. $4(x + 3) = 20$

2. $3(x + 5) = 24$

3. $2(8 + x) = 18$

4. $3(2 + x) = 6$

5. $5(x - 4) = 10$

6. $7(x - 2) = -7$

7. $32 = 8(5 - y)$

8. $18 = 3(2 - y)$

9. $-2(y - 5) = 8$

10. $-3(4 + y) = -21$

11. $-2(5 - y) = -6$

12. $-(2y + 3) = -3$

13. $-(4 - 5z) = 6$

14. $5(3z - 1) = -5$

15. $-2(5 - 3z) = 8$

16. $3(4z + 2) = 7$

17. $-2(-z + 1) = 5$

18. $\frac{1}{2}(2x + 4) = 3$

19. $\frac{2}{3}\left(\frac{3}{4}x - \frac{1}{2}\right) = \frac{1}{4}$

20. $\frac{3}{4}\left(\frac{5}{3}y - 1\right) = \frac{1}{3}$

21. $-\frac{2}{3}\left(z - \frac{3}{4}\right) = \frac{1}{5}$

22. $4(x + 2) - 1 = 19$

23. $3(2x + 1) + 4 = 1$

24. $-(3x - 1) = 0$

25. $5\left[2(x - 1) + 3\right] = 45$

26. $-3\left[5 - 3(x + 4)\right] = 12$

27. $2\left[3 - (2x + 1)\right] = 1$

28. $-2\left[5(x - 3) + 4\right] = 3$

We now move on to the case when the unknown appears as an argument of division.

Example

Solve the equation

$$\frac{x}{4} = 2$$

Solution

Steps 1, 2 and 3 do not apply. We will leave it to the reader to explain why. We move on to Step 5 and solve for the argument of the main operation on the left that contains the unknown, x. The main operation on the left is division (represented by the horizontal line) and its argument that contains the unknown, x, is its numerator. To solve for x, we turn division by 4 on the left side to multiplication by 4 on the right side. This yields

$$x = 4 \times 2$$

which simplifies to

$$x = 8$$

The full solution without the intervening explanations is given below.

$$\frac{x}{4} = 2$$
$$x = 4 \times 2$$
$$x = 8$$

Example

Solve the equation

$$\frac{x + 3}{4} = 2$$

Solution

Steps 1, 2 and 3 do not apply. We will leave it to the reader to explain why. We move on to Step 5 and solve for the argument of the main operation on the left that contains the unknown, x. The main operation on the left is division (represented by the horizontal line) and its argument that contains the unknown, x, is its numerator, $x + 3$. To solve for $x + 3$, we turn division by 4 on the left side to multiplication by 4 on the right side. This yields

$$x + 3 = 4 \times 2$$

which simplifies to

$$x + 3 = 8$$

We now return to Step 2 and solve for the term that contains the unknown, i.e.,

$$x = 8 - 3$$

This simplifies to

$$x = 5$$

The full solution without the intervening explanations is given below.

$$\frac{x + 3}{4} = 2$$
$$x + 3 = 4 \times 2$$
$$x + 3 = 8$$
$$x = 8 - 3$$
$$x = 5$$

Example

Solve the equation

$$\frac{12}{x} = 3$$

Solution

Steps 1, 2 and 3 do not apply. We will leave it to the reader to explain why. We move on to Step 5 and solve for the argument of the main operation on the left that contains the unknown, x. The main operation on the left is division (represented by the horizontal line) and its argument that contains the unknown, x, is its denominator. To solve for x, we turn division by x on the left side to multiplication by x on the right side and to isolate x we turn multiplication by 3 on the right to division by 3 on the left. This yields

$$\frac{12}{3} = x$$

We now return to Step 1 and switch sides to bring the unknown to the left.

$$x = \frac{12}{3}$$

This simplifies to

$$x = 4$$

The full solution without the intervening comments in given below:

$$\frac{12}{x} = 3$$

$$\frac{12}{3} = x \qquad \text{This step is often not written}$$

$$x = \frac{12}{3}$$

$$x = 4$$

Example

Solve the equation

$$\frac{12}{x + 1} = 3$$

Solution

We will provide the full solution below. We ask the reader to justify the steps.

$$\frac{12}{x + 1} = 3$$

$$\frac{12}{3} = x + 1 \qquad \text{This step is often not written}$$

$$x + 1 = \frac{12}{3}$$

$$x + 1 = 4$$

$$x = 4 - 1$$

$$x = 3$$

Exercise Set 11.2D

Solve each of the following equations using natural semantics with intermediate simplifications.

1. $\frac{x}{5} = 2$

2. $\frac{x}{2} = -3$

3. $\frac{x}{-3} = 4$

4. $\frac{y}{-5} = -1$

5. $\frac{3}{x} = -4$

6. $\frac{2}{y} = 5$

7. $\frac{-5}{z} = \frac{2}{3}$

8. $\frac{4x}{5} = \frac{1}{6}$

9. $\frac{-x}{2} = -3$

10. $\frac{3y}{-2} = 0$

11. $\frac{2}{-3y} = 0$

12. $\frac{4}{5x} = -2$

13. $\frac{5}{2z} = \frac{1}{4}$

14. $\frac{3}{-x} = -\frac{1}{3}$

15. $\frac{x + 2}{2} = -3$

16. $\frac{2x - 1}{3} = 5$

17. $\frac{1 - y}{2} = \frac{1}{2}$

18. $\frac{3}{x - 3} = 4$

19. $\frac{-1}{2 - 3y} = -5$

20. $\frac{4}{3 - 2z} = \frac{3}{5}$

21. $\frac{1}{2 - 5z} = -1$

Higher order operations may be present as in the following example.

Example

Solve the equation

$$-4x^3 + 2 = -2$$

Solution

We skip Step 1 as the unknown is already on the left. Furthermore,
the expression is already simplified. We, therefore, move to Step
2 and solve for the term that contains the unknown:

$$-4x^3 = -2 - 2$$

Simplifying the expression on the right yields

$$-4x^3 = -4$$

We now move on to Step 3 and solve for the factor that contains
the unknown. The expression on the left analyzes into two factors.
These are -4 and x^3. We solve for the factor that contains the
unknown by converting multiplication by -4 on the left side of

the equation to division by -4 on the right side of the equation, i.e.,

$$x^3 = \frac{-4}{-4}$$

Simplifying the expression on the right yields

$$x^3 = 1$$

Steps 2 and 3 no longer apply. We move on to Step 5. Since the main operation on the left is exponentiation with a missing base, we convert the operation to its inverse, a root, on its right. This yields

$$x = \sqrt[3]{1}$$

Simplifying the expression on the right yields

$$x = 1$$

The full solution without the intervening explanations is given below.

$$-4x^3 + 2 = -2$$
$$-4x^3 = -2 - 2$$
$$-4x^3 = -4$$
$$x^3 = \frac{-4}{-4}$$
$$x^3 = 1$$
$$x = \sqrt[3]{1}$$
$$x = 1$$

Example

Solve the equation

$$\frac{12}{x^2 + 1} = 3$$

Solution

We will provide the full solution below. We ask the reader to justify the steps.

$$\frac{12}{x^2 + 1} = 6$$

$$\frac{12}{6} = x^2 + 1 \qquad \text{This step is often not written}$$

$$x^2 + 1 = \frac{12}{6}$$

$$x^2 + 1 = 2$$

$$x^2 = 2 - 1$$

$$x^2 = 1$$

$$x = \pm\sqrt{1}$$

$$x = \pm 1$$

Here is an example involving a missing exponent.

Example

Solve the equation

$$3^{2x} - 1 = 80$$

Solution

We will provide the full solution below. We ask the reader to justify the steps.

$$3^{2x} - 1 = 80$$

$$3^{2x} = 80 + 1$$

$$3^{2x} = 81$$

$$2x = \log_3 81$$

$$2x = 4$$

$$x = \frac{4}{2}$$

$$x = 2$$

Exercise Set 11.2E

Solve each of the following equations using natural semantics with intermediate simplifications.

1. $x^2 = 16$

2. $x^2 = 25$

3. $x^2 = 11$

4. $x^2 = -5$

5. $y^3 = 125$

6. $y^3 = 1$

7. $y^3 = -64$

8. $z^2 + 3 = 4$

9. $4z^3 + 1 = -3$

10. $2z^3 - 2 = -4$

11. $4\left(x^3 + 1\right) = 36$

12. $-\left(5 - x^4\right) = 76$

13. $-\left(2x^4 - 1\right) = 1$

14. $4^x = 64$

15. $3^x = 9$

16. $4^x = \frac{1}{16}$

17. $3^y = \frac{1}{81}$

18. $5^y + 2 = 27$

19. $-3 + 2^y = -2$

20. $4\left(5^z + 1\right) = 4$

21. $-3\left(2 - 4^z\right) = 6$

22. $\log x = 3$

23. $\log x = -1$

24. $\log_2 x = 5$

25. $\log_4 y = -2$

26. $4\log_3 y = 0$

27. $-\log_2 z = -3$

28. $\log z + 3 = 5$

29. $\log\left(x + 3\right) = 1$

30. $-2\log\left(y + 1\right) = -2$

31. $-\log\left(z - 2\right) = -1$

32. $\frac{\log x}{4} = 1$

33. $\log\left(x - 52\right) = 2$

When the unknown appears more than once but as identical factors within multiple terms, regrouping becomes necessary. This requires that we open up any brackets that contain the unknown using the distributive property and combine the resulting like terms.

Example

Solve the equation

$$4\left(3x + 2\right) - 2x = 28$$

Solution

Since the unknown appears in multiple terms, we simplify the terms and combine like terms before we proceed. The second term is already simplified but the first term can be simplified by multiplying 4 into the bracket. This yields

$$12x + 8 - 2x = 28$$

Simplifying the left side yields

$$10x + 8 = 28$$

We now isolate the term that contains the unknown.

$$10x = 28 - 8$$

Simplifying the right side yields

$$10x = 20$$

We now isolate the factor that contains the unknown.

$$x = \frac{20}{10}$$

Simplifying the right side yields

$$x = 2$$

The full solution without the intervening explanations is given below.

$$4\,(3x + 2) - 2x = 28$$
$$12x + 8 - 2x = 28$$
$$10x + 8 = 28$$
$$10x = 28 - 8$$
$$10x = 20$$
$$x = \frac{20}{10}$$
$$x = 2$$

Example

Solve the equation

$$-3\,(-5 + 4x) + 2\,(3x - 1) = 19$$

Solution

The full solution is given below. We ask the reader to justify the steps.

$$-3\left(-5+4x\right)+2\left(3x-1\right)=19$$
$$15-12x+6x-2=19$$
$$-6x+13=19$$
$$-6x=19-13$$
$$-6x=6$$
$$x=\frac{6}{-6}$$
$$x=-1$$

If the unknown appears on both sides of the equation, we solve for the terms that contain the unknown[10] and then combine them. Here is an example.

Example

Solve the equation

$$4\left(x-5\right)=-\left(x+2\right)$$

Solution

$$4\left(x-5\right)=-\left(x+2\right)$$
$$4x-20=-x-2$$
$$4x+x=-2+20$$
$$5x=18$$
$$x=\frac{18}{5}$$

---

[10]This requires that we isolate the terms that contain the unknown on the left side of the equation.

Exercise Set 11.2F

Solve each of the following equations using natural semantics with intermediate simplifications.

1. $2(x + 1) + 3(x - 2) = 6$
2. $5(x + 2) - 2(x + 5) = -3$
3. $-3(x + 4) + 2x = -12$
4. $-(x - 1) - 4x = -14$
5. $5(x - 3) = 2(x + 3)$
6. $4(5 - 2x) = -8x + 20$
7. $3(2x + 5) = x - 3$

8. $-(x + 4) = 3(x - 1)$
9. $4x = -2(x + 6)$
10. $3(2x - 1) = 6x$
11. $8(2x + 1) - 3(x - 2) = -x$
12. $4(x^3 + 3) - 2x^3 = 28$
13. $3x^2 - 2(x^2 + 5) = -9$
14. $9(x^2 - 1) + 3(x^2 - 1) = -12$

In the natural semantic scheme without intermediate simplifications the steps in the algorithm for solving equations given earlier are carried out by converting the main operation for an argument (term, factor, etc.) on one side of the equation to the inverse of that operation for the same argument on the other side of the equation. However, no simplifications are performed in between the steps. Rather, the simplifications are performed after the equation is solved for the unknown. We will illustrate the ideas using examples.

Example

Solve the equation

$$4x + 3 = 11$$

Solution

Step 1 does not apply as the unknown is already on the left side of the equation.

Step 2 requires that we solve for the term that contains the unknown.

$$4x = 11 - 3$$

Rather than simplifying the right side, we move on to Step 3 and solve for the factor that contains the unknown.

$$x = \frac{11 - 3}{4}$$

The equation is now solved for x. We now simplify the expression on the right side.

$$x = \frac{8}{4}$$
$$x = 2$$

The full solution without the intervening comments is given below:

$$4x + 3 = 11$$
$$4x = 11 - 3$$
$$x = \frac{11 - 3}{4}$$
$$x = \frac{8}{4}$$
$$x = 2$$

Here is another example.

Example

Solve the equation

$$8 - 5x = 3$$

Solution

Step 1 does not apply as the unknown is already on the left side of the equation.

Step 2 requires that we solve for the term that contains the unknown.

$$-5x = 3 - 8$$

We move on to Step 3 and solve for the factor that contains the unknown.

$$x = \frac{3 - 8}{-5}$$

The equation has been solved for x. We now proceed to simplify the expression on the right side.

$$x = \frac{-5}{-5}$$
$$x = 1$$

The full solution without the intervening explanations is given below.

$$8 - 5x = 3$$
$$-5x = 3 - 8$$
$$x = \frac{3 - 8}{-5}$$
$$x = \frac{-5}{-5}$$
$$x = 1$$

Example

Solve the equation

$$-2 - 3x = 4$$

Solution

We provide the full solution below. We ask the reader to supply justification for each step.

$$-2 - 3x = 4$$
$$-3x = 4 + 2$$
$$x = \frac{4 + 2}{-3}$$
$$x = \frac{6}{-3}$$
$$x = -2$$

Exercise Set 11.2G

Solve each of the following equations using natural semantics without intermediate simplifications.

1. $-2 + 4x = 6$
2. $3x - 5 = 1$
3. $-2x - 6 = -14$

4. $4 = 3 + 2x$
5. $5 = -x + 3$
6. $-7 + 8x = 0$

7. $9 - 2x = 2$
8. $6x + 1 = -3$
9. $-8x - 6 = -7$

We present a few examples with rational numbers.

Example

Solve the equation

$$\frac{2}{3} + \frac{1}{2}x = \frac{11}{15}$$

Solution

Step 1 does not apply as the unknown is already on the left side of the equation.

Step 2 requires that we solve for the term that contains the unknown. This is the term $\frac{1}{2}x$.

$$\frac{1}{2}x = \frac{11}{15} - \frac{2}{3}$$

Step 3 requires that we solve for the factor that contains the unknown.

$$x = \frac{2}{1}\left(\frac{11}{15} - \frac{2}{3}\right)$$

The equation has now been solved for x. We now simplify the expression on the right side to arrive at

$$x = \frac{2}{1} \times \frac{11 - 10}{15}$$

$$x = \frac{2}{1} \times \frac{1}{15}$$

$$x = \frac{2}{15}$$

The full solution without the intervening comments is given below:

$$\frac{2}{3} + \frac{1}{2}x = \frac{11}{15}$$

$$\frac{1}{2}x = \frac{11}{15} - \frac{2}{3}$$

$$x = \frac{2}{1}\left(\frac{11}{15} - \frac{2}{3}\right)$$

$$x = \frac{2}{1} \times \frac{11 - 10}{15}$$

$$x = \frac{2}{1} \times \frac{1}{15}$$

$$x = \frac{2}{15}$$

Example

Solve the equation

$$-\frac{3}{4} - \frac{2}{3}x = \frac{7}{12}$$

Solution

We will present the full solution below. We let the reader provide the logic behind each move.

$$-\frac{3}{4} - \frac{2}{3}x = \frac{7}{12}$$

$$-\frac{2}{3}x = \frac{7}{12} + \frac{3}{4}$$

$$x = -\frac{3}{2}\left(\frac{7}{12} + \frac{3}{4}\right)$$

$$x = -\frac{3}{2} \times \frac{7 + 9}{12}$$

$$x = -\frac{3}{2} \times \frac{16}{12}$$

$$x = -\frac{2}{1}$$

$$x = -2$$

Exercise Set 11.2H

Solve each of the following equations using natural semantics without intermediate simplifications.

1. $\frac{3}{5}x + \frac{1}{2} = \frac{1}{3}$

2. $\frac{4}{7}x + \frac{3}{4} = \frac{2}{5}$

3. $-\frac{1}{3}x - \frac{2}{9} = -\frac{5}{6}$

4. $\frac{1}{8} + \frac{3}{5}x = \frac{1}{2}$

5. $-\frac{4}{9} + 2x = \frac{1}{3}$

6. $-4x - \frac{2}{5} = -\frac{1}{4}$

7. $-x + 1 = \frac{3}{4}$

8. $2x - 3 = \frac{4}{5}$

9. $3 - 5x = \frac{9}{2}$

10. $-2 + \frac{2}{3}x = \frac{1}{4}$

11. $-\frac{5}{6}x - \frac{3}{4} = 1$

12. $\frac{8}{3}x - \frac{8}{3} = 2$

13. $\frac{1}{3} = -2 + \frac{4}{7}x$

14. $-3 = -\frac{7}{5}x + 2$

Our next example involves brackets:

Example

Solve the equation

$$4\left(x + 3\right) = 20$$

Solution

We skip Step 1 as the unknown is already on the left.

We skip Step 2 as the term that contains the unknown is already solved for.

We move on to Step 3 and solve for the factor that contains the unknown.

$$x + 3 = \frac{20}{4}$$

The opening up of the brackets around the factor $x + 3$ makes it possible to apply Step 2 again. We solve for the term that contains the unknown:

$$x = \frac{20}{4} - 3$$

The unknown has now been solved for. We now move on to the simplification stage and simplify the expression on the right side of the equation to get

$$x = 5 - 3$$
$$x = 2$$

The full solution without the intervening explanations is given below.

$$4(x + 3) = 20$$
$$x + 3 = \frac{20}{4}$$
$$x = \frac{20}{4} - 3$$
$$x = 5 - 3$$
$$x = 2$$

Exercise Set 11.2I

Solve each of the following equations using natural semantics without intermediate simplifications.

1. $4(x + 3) = 20$

2. $3(x + 5) = 24$

3. $2(8 + x) = 18$

4. $3(2 + x) = 6$

5. $5(x - 4) = 10$

6. $7(x - 2) = -7$

7. $32 = 8(5 - y)$

8. $18 = 3(2 - y)$

9. $-2(y - 5) = 8$

10. $-3(4 + y) = -21$

11. $-2(5 - y) = -6$

12. $-(2y + 3) = -3$

13. $-(4 - 5z) = 6$

14. $5(3z - 1) = -5$

15. $-2(5 - 3z) = 8$

16. $3(4z + 2) = 7$

17. $-2(-z + 1) = 5$

18. $\frac{1}{2}(2x + 4) = 3$

19. $\frac{2}{3}\left(\frac{3}{4}x - \frac{1}{2}\right) = \frac{1}{3}$

20. $\frac{3}{4}\left(\frac{5}{3}y - 1\right) = \frac{1}{3}$

21. $-\frac{2}{3}\left(z - \frac{3}{4}\right) = \frac{1}{5}$

22. $4(x + 2) - 1 = 19$

23. $3(2x + 1) + 4 = 1$

24. $-(3x - 1) = 0$

25. $5\left[2\left(x - 1\right) + 3\right] = 45$ 27. $2\left[3 - \left(2x + 1\right)\right] = 1$

26. $-3\left[5 - 3\left(x + 4\right)\right] = 12$ 28. $-2\left[5\left(x - 3\right) + 4\right] = 3$

We now move on to the case when the unknown appears as an argument of division.

Example

Solve the equation

$$\frac{x + 3}{4} = 2$$

Solution

Steps 1, 2 amd 3 do not apply. We move on to Step 5 and solve for the argument of the main operation on the left that contains the unknown x. The main operation on the left is division (represented by the horizontal line) and its argument that contains the unknown is its numerator, $x + 3$. To solve for $x + 3$, we turn division by 4 on the left side to multiplication by 4 on the right side. This yields

$$x + 3 = 4 \times 2$$

We now return to Step 2 and solve for the term that contains the unknown, i.e.,

$$x = 4 \times 2 - 3$$

The equation has now been solved for x. We move on to simplify the expression on the right side.

$$x = 8 - 3$$
$$x = 5$$

The full solution without the intervening explanations is given below.

$$\frac{x + 3}{4} = 2$$
$$x + 3 = 4 \times 2$$
$$x = 4 \times 2 - 3$$
$$x = 8 - 3$$
$$x = 5$$

Example

Solve the equation

$$\frac{12}{x + 1} = 3$$

Solution

We will provide the full solution below. We ask the reader to justify the steps.

$$\frac{12}{x + 1} = 3$$

$$\frac{12}{3} = x + 1 \qquad \text{This step is often not written}$$

$$x + 1 = \frac{12}{3}$$

$$x = \frac{12}{3} - 1$$

$$x = 4 - 1$$

$$x = 3$$

Exercise Set 11.2J

Solve each of the following equations using natural semantics without intermediate simplifications.

1. $\dfrac{x + 2}{2} = -3$

2. $\dfrac{2x - 1}{3} = 5$

3. $\dfrac{1 - y}{2} = \dfrac{1}{2}$

4. $\dfrac{3}{x - 3} = 4$

5. $\dfrac{-1}{2 - 3y} = -5$

6. $\dfrac{4}{3 - 2z} = \dfrac{3}{5}$

7. $\dfrac{1}{2 - 5z} = -1$

8. $\dfrac{3z + 1}{2 - 5z} = -1$

9. $\dfrac{4 - z}{z + 5} = 2$

We present a few examples involving higher order operations.

Example

Solve the equation

$$-4x^3 + 2 = -2$$

Solution

We skip Step 1 as the unknown is already on the left. Furthermore, the expression is already simplified. We, therefore, move to Step 2 and solve for the term that contains the unknown:

$$-4x^3 = -2 - 2$$

We now move on to Step 3 and solve for the factor that contains the unknown.

$$x^3 = \frac{-2 - 2}{-4}$$

Steps 2 and 3 no longer apply. We move on to Step 5. Since the main operation on the left is exponentiation with a missing base, we convert the operation to its inverse, a root, on its right. This yields

$$x = \sqrt[3]{\frac{-2 - 2}{-4}}$$

Simplifying the expression on the right yields

$$x = \sqrt[3]{\frac{-4}{-4}}$$
$$x = \sqrt[3]{1}$$
$$x = 1$$

The full solution without the intervening explanations is given below.

$$-4x^3 + 2 = -2$$
$$-4x^3 = -2 - 2$$
$$x^3 = \frac{-2 - 2}{-4}$$
$$x = \sqrt[3]{\frac{-2 - 2}{-4}}$$
$$x = \sqrt[3]{\frac{-4}{-4}}$$
$$x = \sqrt[3]{1}$$
$$x = 1$$

Example

Solve the equation

$$\frac{12}{x^2 + 1} = 3$$

Solution

We will provide the full solution below. We ask the reader to justify the steps.

$$\frac{12}{x^2 + 1} = 6$$

$$\frac{12}{6} = x^2 + 1 \qquad \text{This step is often not written}$$

$$x^2 + 1 = \frac{12}{6}$$

$$x^2 = \frac{12}{6} - 1$$

$$x = \pm\sqrt{\frac{12}{6} - 1}$$

$$x = \pm\sqrt{2 - 1}$$

$$x = \pm\sqrt{1}$$

$$x = \pm 1$$

Here is an example involving a missing exponent.

Example

Solve the equation

$$3^{2x} - 1 = 80$$

Solution

$$3^{2x} - 1 = 80$$

$$3^{2x} = 80 + 1$$

$$2x = \log_3 (80 + 1)$$

$$x = \frac{1}{2} \log_3 (80 + 1) \quad \text{or} \quad x = \frac{\log_3 (80 + 1)}{2}$$

$$x = \frac{1}{2} \log_3 81 \quad \text{or} \quad x = \frac{\log_3 81}{2}$$

$$x = \frac{1}{2} \times 4 \quad \text{or} \quad x = \frac{4}{2}$$

$$x = 2$$

We ask the reader to justify the steps in the solution above.

Exercise Set 11.2K

Solve each of the following equations using natural semantics without intermediate simplifications.

1. $z^2 + 3 = 4$

2. $4z^3 + 1 = -3$

3. $2z^3 - 2 = -4$

4. $4\left(x^3 + 1\right) = 36$

5. $-\left(5 - x^4\right) = 76$

6. $-\left(2x^4 - 1\right) = 1$

7. $5^y + 2 = 27$

8. $-3 + 2^y = -2$

9. $4\left(5^z + 1\right) = 4$

10. $-3\left(2 - 4^z\right) = 6$

11. $\log z + 3 = 5$

12. $\log (x + 3) = 2$

13. $-2 \log y + 1 = -1$

14. $-\log (z - 2) = -1$

15. $\dfrac{\log x}{2} = 1$

16. $\dfrac{4}{\log (x - 1)} = 2$

We end this section with a few examples where the unknown appears more than once.

Example

Solve the equation

$$4(3x + 2) - 2x = 28$$

Solution

Since the unknown appears in multiple terms, we begin by simplifying the terms and combining like terms. The second term is already simplified but the first term can be simplified by multiplying 4 into the bracket. This yields

$$4 \times 3x + 4 \times 2 - 2x = 28$$

We now group the terms that contain the unknown into a single term using the distributive property in reverse.[11]

$$(4 \times 3 - 2)\,x + 4 \times 2 = 28$$

We now isolate the term that contains the unknown.

$$(4 \times 3 - 2)\,x = 28 - 4 \times 2$$

We now isolate the factor that contains the unknown.

$$x = \frac{28 - 4 \times 2}{4 \times 3 - 2}$$

The equation has now been solved for the unknown. Simplifying the right side yields

$$x = \frac{28 - 8}{12 - 2}$$
$$x = \frac{20}{10}$$
$$x = 2$$

The full solution without the intervening explanations is given below.

$$4\,(3x + 2) - 2x = 28$$
$$4 \times 3x + 4 \times 2 - 2x = 28$$
$$(4 \times 3 - 2)\,x + 4 \times 2 = 28$$
$$(4 \times 3 - 2)\,x = 28 - 4 \times 2$$
$$x = \frac{28 - 4 \times 2}{4 \times 3 - 2}$$
$$x = \frac{28 - 8}{12 - 2}$$
$$10x = \frac{20}{10}$$
$$x = 2$$

---

[11]This step is a key step as it turns the multiple occurrences of the unknown into a single occurrence.

Example

Solve the equation

$$-3\left(-5 + 4x\right) + 2\left(3x - 1\right) = 19$$

Solution

We will provide the full solution below. We ask the reader to justify the steps.

$$-3\left(-5 + 4x\right) + 2\left(3x - 1\right) = 19$$
$$-3\left(-5\right) - 3\left(4\right)x + 2\left(3\right)x - 2\left(1\right) = 19$$
$$\left[-3\left(4\right) + 2\left(3\right)\right]x - 3\left(-5\right) - 2\left(1\right) = 19$$
$$\left[-3\left(4\right) + 2\left(3\right)\right]x = 19 + 3\left(-5\right) + 2\left(1\right)$$
$$x = \frac{19 + 3\left(-5\right) + 2\left(1\right)}{-3\left(4\right) + 2\left(3\right)}$$
$$x = \frac{19 - 15 + 2}{-12 + 6}$$
$$x = \frac{6}{-6}$$
$$x = -1$$

We present one more example.

Example

Solve the equation

$$4\left(x - 5\right) = -\left(x + 2\right)$$

Solution

$$4\,(x\,-\,5)\,=\,-\,(x\,+\,2)$$
$$4x\,-\,4\times5\,=\,-x\,-\,2$$
$$4x\,+\,x\,=\,-2\,+\,4\times5$$
$$(4\,+\,1)\,x\,=\,-2\,+\,4\times5$$
$$x\,=\,\frac{-2\,+\,4\times5}{4\,+\,1}$$
$$x\,=\,\frac{-2\,+\,20}{5}$$
$$x\,=\,\frac{18}{5}$$

Exercise Set 11.2L

Solve each of the following equations using natural semantics without intermediate simplifications.

1. $2\,(x\,+\,1)\,+\,3\,(x\,-\,2)\,=\,6$
2. $5\,(x\,+\,2)\,-\,2\,(x\,+\,5)\,=\,-3$
3. $-3\,(x\,+\,4)\,+\,2x\,=\,-12$
4. $-\,(x\,-\,1)\,-\,4x\,=\,-14$
5. $5\,(x\,-\,3)\,=\,2\,(x\,+\,3)$
6. $4\,(5\,-\,2x)\,=\,-8x\,+\,20$
7. $3\,(2x\,+\,5)\,=\,x\,-\,3$
8. $-\,(x\,+\,4)\,=\,3\,(x\,-\,1)$
9. $4x\,=\,-2\,(x\,+\,6)$
10. $3\,(2x\,-\,1)\,=\,6x$
11. $8\,(2x\,+\,1)\,-\,3\,(x\,-\,2)\,=\,-x$
12. $4\,(x^3\,+\,3)\,-\,2x^3\,=\,28$
13. $3x^2\,-\,2\,(x^2\,+\,5)\,=\,-9$
14. $9\,(x^2\,-\,1)\,+\,3\,(x^2\,-\,1)\,=\,-12$

Chapter 12
Equations in Two Unknowns

Equations in two unknowns are statements that describe the relationship between the values of the quantities that the two unknowns represent.[1] In this chapter, after a discussion on the sources of such equations, we will present a systematic approach for classifying the many different types of equations that we encounter when we model the solution to common word problems as equations. As we will see, the classification scheme is based on the type of relationship between the values of the quantities that the two unknowns represent, starting with the simplest of the relationships and moving on to more and more complex ones.

12.1　Genesis

It is often the case that we wish to know how changes in the value of one quantity, called the **independent variable**, force changes in the value of another quantity, called the **dependent variable**. To work out such relationships, we must, of course, make sure that the values of any other quantities, i.e., the values of quantities other than that of the independent and dependent variables, remain constant otherwise we will not be able to associate observed changes in the value of the dependent variable to changes in the value of the independent variable.

As an example of this, suppose we wish to establish the manner in which changes in the value of the temperature of helium (the independent variable) in a container affects the value of its pressure (the dependent variable). However, the value of the pressure of the helium is not related only to the value of its temperature. We can change the value of the pressure of the helium by changing the value of the volume of the container within which it is confined. We can also change the value of the pressure of the helium by changing the value of the amount of helium in the container. If we were to allow simultane-

---

[1]We will clarify what we mean by the word *relationship* in the passages ahead.

ous changes in the values of temperature, volume and amount of the helium in the container, then we would not be able to tie in the changes in the value of the pressure of the helium to changes in the value of its temperature as some of the change in the value of the pressure may come from changes in the values of the volume of the container and the amount of helium in the container. To tie in the changes in the value of the pressure of the helium to changes in the value of the temperature of the helium, we must make sure that the values of the volume and amount of helium in the container remain constant, i.e., that their values remain unchanged. To keep the value of the volume the same, we might choose to place the helium inside a steel tank (as opposed to, say, a balloon) and to keep the value of the amount the same, we might choose to seal the tank so that no helium enters the tank and no helium leaves the tank. Only then can we relate changes in the value of the temperature of the helium, which might be brought about through heating or cooling of the tank, to changes in the value of its pressure.

The relationship between the values of the independent variable and the dependent variable is often expressed as an equation. As an example, through experiments we can show that, with the values of the volume and amount of helium being constant, the equation that relates the value of the pressure of the helium in pascals to the value of its temperature in kelvins in the example given above has the following form:

$$p = kT$$

where k is a constant, i.e., a known number in $\mathbb{R}$. In this equation the value of the pressure of the helium, p, is related to the value of its temperature, T, through multiplication by a constant. Such a relationship is observed in many disparate systems. As an example, if we travel at constant speed, we can relate the value of distance covered to the value of travel time through the equation

$$d = kt$$

where k is constant[2] and d and t represent the values of distance covered in kilometres and travel time in hours. In this equation the value of distance covered, d, is related to the value of travel time through multiplication by a constant. While the contexts are very different, the equations $p = kT$ and $d = kt$ have the same form. This common form is usually expressed as an abstract equation. For the problems posed above, the abstract form of the equation is

$$y = kx, \qquad k \in \mathbb{R}$$

where the side note $k \in \mathbb{R}$ states that k is a constant and x and y represent the values of the two abstract quantities involved. Abstract forms are helpful

---

[2]In this case k represents the constant speed of the car.

because any conclusions made based on the study of an abstract form applies equally to its concrete instances. As an example, any conclusions made based on the study of the abstract form $y = kx$ applies to its concrete counterparts $p = kT$, $d = kt$, and other pairs of quantities whose values are related through multiplication by a constant.

For more on the analysis of concrete forms of equations we refer the reader to the companion textbook *Semantics and the Syntax of Algebra* by the author. In this textbook we focus on the abstract formulations of these equations and conclusions that may be drawn from a study of these abstract formulations.

12.2   Classification

In this section of the textbook we will discuss some of the common forms of equations that we encounter in basic applications of math. The many forms are divided into two categories which we label in this textbook as *single-operation forms* and *multiple-operations forms*.

12.2.1   Single-Operation Forms

By a **single-operation form** we mean an equation where the dependent variable, y, is isolated on the left side of the equation and a single operation applies to the independent unknown, x, on the right side of the equation.

An example of a single-operation form is the equation $y = 3x$. Here y is isolated on the left side of the equation. Furthermore, on the right side, x is being multiplied by, as is seen in the more explicit $y = 3 \times x$. A second example of a single-operation form is the equation $y = x^k$. Here y is isolated on the left side of the equation and, on the right side, x is being raised to the power of k.

The coverage begins with equations that involve operations that are associated with the simplest types of problems that we, as humans, solve followed by increasingly more complex types of problems that we, as humans, encounter. This order maps onto the operations of addition, subtraction, multiplication, division, power functions, root functions, exponential functions and logarithmic functions.[3]

---

[3]The list continues but this is the extent to which we cover single-operation forms in this textbook.

12.2.1.1   Direct Superposition

The simplest type of relationship that we observe between the values of two quantities is **direct superposition**.

Definition

The dependent variable, y, is said to be directly superpositional to the independent variable, x, if an increase or decrease in x by a certain amount forces a corresponding increase or decrease in y by the same amount.[4]

Model

The model for relationships of type direct superposition comes in two flavours: The **formal form** and the **conservation form**.[5]

Formal Form

A relationship of type *direct superposition* between the values of two quantities can be modelled as

$$y = k + x, \qquad k \in \mathbb{R}$$

where the abstract quantity symbols x and y represent the values of the two quantities and the side condition, $k \in \mathbb{R}$, states that, for a given concrete problem of this type, k must be a constant, i.e., a known number in $\mathbb{R}$. We refer to this formulation as the formal form of the model for direct superposition and we refer to k as the **constant of superpositionality**.

The reason such an equation models a relationship of type direct superposition between y and x is that, as shown below, x is being added on the

---

[4]Two quantities are said to have a **direct relationship** in the sense of superposition if they both increase in value or both decrease in value. Note that a direct relationship in the sense of superposition imposes no requirements on the equality of the sizes of increase or decrease in the values of the quantities involved. If the relationship is of type *direct superposition*, however, then the sizes of increase or decrease in the values of the quantities involved are required to be the same.

[5]Both forms are suitable when working with problems whose solutions involve a single superposition step. For chain superposition problems, however, it is possible to extend the formal form but not the conservation form. In single-superposition applications, the version that is most suitable to the semantics of the problem is adopted. For a more detailed discussion of the use of the different forms of the equation for direct superposition please see the companion textbook *Semantics and the Syntax of Algebra* by the author.

right side of the equation.

$$y = k + \overset{\curvearrowright}{x}$$

If we now increase x by, say, 2 units, then we will be adding 2 more units to k on the right side of the equation. This will result in an increase of 2 units in y as well. Equivalently, if we decrease x by, say, 2 units, then we will be adding 2 fewer units to k on the right side of the equation. This will result in a decrease of 2 units in y as well.

The argument given above is quite general: Increasing x by any amount forces an increase in y by the same amount and decreasing x by any amount forces a decrease in y by the same amount. Such dependence is exactly what a relationship of type direct superposition is about.

Conservation Form

The fact that, in a relationship of type *direct superposition*, an increase or decrease in the value of one of the quantities by a certain amount is matched by a corresponding increase or decrease in the value of the other quantity by the same amount implies that the values of the two quantities involved rise and fall by equal amounts. If this is so, then the difference between the values of the two quantities involved remains the same, i.e, the difference between the values of the two quantities involved is constant. We can use mathematical language to express this view of a relationship of type *direct superposition* as

$$y - x = k, \quad k \in \mathbb{R}$$

Note that in this formulation the value of the expression on the left side of the equation is constant. In a given problem, this value, i.e., $y - x$, never increases or decreases and is, therefore, in some sense *conserved*. For this reason we refer to the formulation $y - x = k$ as the conservation form of the equation $y = k + x$.

The conservation form of the model, i.e., $y - x = k$, can be shown to be the same as the formal form of the model, i.e., $y = k + x$, through conversion of addition of x on the right side of the equation $y = k + x$ to subtraction of x on the its left, i.e.,

$$y = k + x$$
$$y - x = k$$

Symmetry

Relationships of type direct superposition are symmetric: If y is directly superpositional to x, then x is directly superpositional to y.[6] This can be seen through a rearrangement of the equation $y = k + x$ as shown below:

$$y = k + x$$
$$k + x = y$$
$$x = -k + y$$

Note that y is being added on the right side of the equation. Furthermore, since k is in $\mathbb{R}$, then so is $-k$. Therefore x is directly superpositional to y.

Change in Values of Variables

The fact that, in a relationship of type *direct superposition*, an increase or decrease in the value of one quantity by a certain amount forces a corresponding increase or decrease in the value of the other quantity by an equal amount implies that any change in x is matched by an equal change in y. In mathematical notation, we write this as[7]

$$\Delta y = \Delta x$$

We can show this to be the case as follows: By definition

$$\Delta y = y_2 - y_1$$

where y_1 represents the initial value of y and y_2 represents the final value of y. Since $y = k + x$, there is an x value, x_1, that corresponds to y_1 through $y_1 = k + x_1$ and there is an x value, x_2, that corresponds to y_2 through $y_2 = k + x_2$. Substituting $k + x_1$ for y_1 and $k + x_2$ for y_2 in the equation above yields

$$\Delta y = k + x_2 - (k + x_1)$$

Simplifying the expression on the right side yields

$$\Delta y = k + x_2 - k - x_1$$
$$\Delta y = x_2 - x_1$$
$$\Delta y = \Delta x$$

This proves that for a relationship of type direct superposition, i.e., $y = k + x$, $k \in \mathbb{R}$, $\Delta y = \Delta x$.

---

[6]This means that, if it is the case that an increase or decrease in x by a certain amount forces a corresponding increase or decrease in y by the same amount, then an increase or decrease in y by some amount forces a corresponding increase or decrease in x by the same amount. Therefore, whichever variable is labelled as independent, the relationship between the two variables will be of type *direct superposition*.

[7]The Greek capital letter Δ stands for the phrase *change in the value of* so that Δx reads as *change in the value of* x and Δy reads as *change in the value of* y.

Graph

Provided units of equal size are used in the graph of an equation of type direct superposition, the graph will be a straight line going up at $45°$ relative to the horizontal.[8] The line has an x-intercept of $(-k, 0)$, a y-intercept of $(0, k)$ and a slope of 1. The x-intercept can be found by setting $y = 0$ in the equation $y = k + x$. This yields

$$\begin{aligned} y &= k + x \\ 0 &= k + x \\ k + x &= 0 \\ x &= 0 - k \\ x &= -k \end{aligned}$$

showing that the x-intercept is $(-k, 0)$. For the y-intercept we set $x = 0$ in the equation $y = k + x$. This yields

$$\begin{aligned} y &= k + x \\ y &= k + 0 \\ y &= k \end{aligned}$$

showing that the y-intercept is $(0, k)$. To calculate the slope, we use the formula

$$m = \frac{y_2 - y_1}{x_2 - x_1}$$

with P_1 as the x-intercept, i.e., $(-k, 0)$, and P_2 as the y-intercept, i.e., $(0, k)$.[9] This yields

$$\begin{aligned} m &= \frac{y_2 - y_1}{x_2 - x_1} \\ m &= \frac{k - 0}{0 - (-k)} \\ m &= \frac{k}{0 + k} \\ m &= \frac{k}{k} \\ m &= 1 \end{aligned}$$

Alternatively, we can use the formula

$$m = \frac{\Delta y}{\Delta x}$$

---

[8] For a detailed discussion of fundamental concepts that relate to graphs please see Appendix O.

[9] If the intercepts are the same, i.e., they are both $(0, 0)$, then a second point will be needed. Such a point can be found by setting x equal to any nonzero value and finding the corresponding y through $y = k + x$.

along with the fact that, as shown above, for a relationship of type *direct superposition*, $\Delta y = \Delta x$, to get

$$m = \frac{\Delta y}{\Delta x}$$
$$m = \frac{\Delta x}{\Delta x}$$
$$m = 1$$

The graph of the line $y = k + x$ is shown in Figure 12.2.1 below.

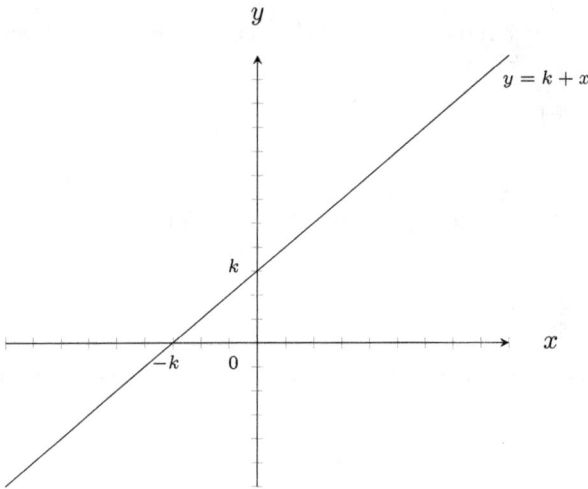

Figure 12.2.1: Graph of the equation for relationships of type direct superposition

Some of the more important features of the line $y = k + x$ are shown in Figure 12.2.2 below.[10]

In $y = k + x$, $k \in \mathbb{R}$, the effect of k on the graph of the equation is to raise or lower the line. The higher the value of k, the higher the line. Figure 12.2.3 shows the graphs of the five equations $y = -6 + x$, $y = -3 + x$, $y = x$ (which is the same as $y = 0 + x$), $y = 3 + x$ and $y = 6 + x$. Note that the angle between the line and the horizontal remains at 45° in all cases.

---

[10]The most important facts to keep in mind about a relationship of type *direct superposition* are its form, i.e., $y = k + x$, $k \in \mathbb{R}$, the fact that change in y must equal to change in x, i.e., $\Delta y = \Delta x$, and that its graph is a straight line with a slope of 1.

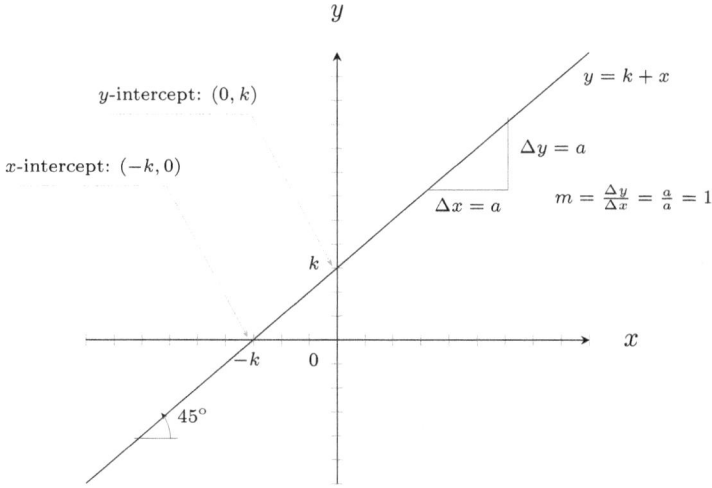

Figure 12.2.2: Features of the graph of the equation
for relationships of type direct superposition

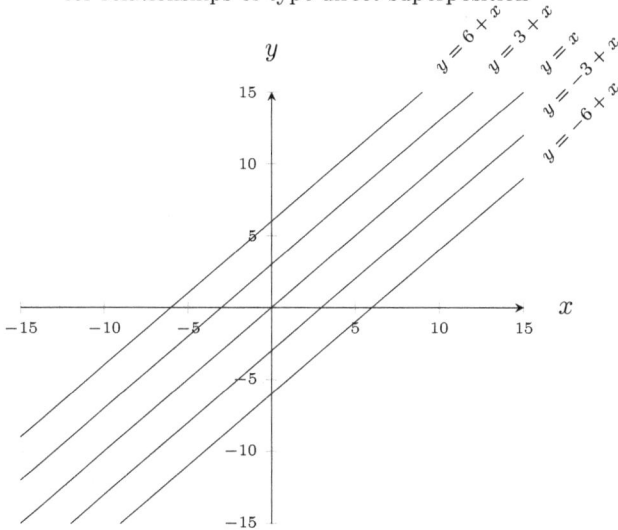

Figure 12.2.3: Effect of k in $y = k + x$, $k \in \mathbb{R}$, on the
graph of the equation for relationships of type *direct
superposition*

Applications

We now present an abstract and a concrete example of a problem of type
direct superposition.

Example

Given the equation

$$y = 4 + x$$

1. Name the type of the relationship between x and y and support that claim.
2. Write the equation in conservation form.
3. State the relationship between change in x and change in y.
4. Describe the shape of the graph of the equation.
5. Find the x- and y-intercepts of the line.
6. Graph the line.
7. Find the slope of the line.
8. What is the angle between the line and the horizontal?

Solution

1. The relationship between x and y is direct superposition. The equation $y = 4 + x$ is similar in form to $y = k + x$, $k \in \mathbb{R}$, with $k = 4$. Since we have shown that in an equation of the form $y = k + x$, $k \in \mathbb{R}$ the variables x and y are related through direct superposition, we can state the same claim for the particular case $y = 4 + x$.

2. The conservation form of the equation $y = 4 + x$ is $y - x = 4$. This tells us that the difference between y and x (in that order) is always 4, i.e., y is always 4 units larger than x.

3. $\Delta y = \Delta x$

4. The graph of the equation $y = 4 + x$ is a straight line.

5. We can use our earlier findings that the x- and y-intercepts of the graph of the equation $y = k + x$, $k \in \mathbb{R}$ are $(-k, 0)$ and $(0, k)$ to conclude that the x- and y-intercepts of the equation $y = 4 + x$ are $(-4, 0)$ and $(0, 4)$.

   Alternatively, we can find the x-intercept by setting $y = 0$, and the y-intercept by setting $x = 0$, in the equation $y = 4 + x$. For the x-intercept, this yields

$$
\begin{aligned}
y &= 4 + x \\
0 &= 4 + x \\
4 + x &= 0 \\
x &= 0 - 4 \\
x &= -4
\end{aligned}
$$

which shows that the x-intercept is $(-4, 0)$. For the y-intercept we have:

$$y = 4 + x$$
$$y = 4 + 0$$
$$y = 4$$

which shows that the y-intercept is $(0, 4)$.

6. Since the graph of the equation $y = 4 + x$ is a line, we need two points on the line to graph the line. Here we use the x- and y-intercepts as our points with coordinates $(-4, 0)$ and $(0, 4)$. The graph of the equation $y = 4 + x$ is shown in Figure 12.2.4 below.

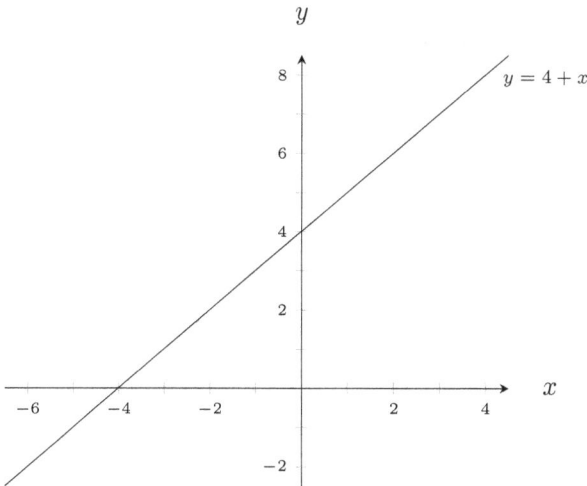

Figure 12.2.4: Graph of the equation $y = 4 + x$

7. Since the equation is of type direct superposition, the slope of the line is 1. Alternatively, we can compute the slope using the equation $m = \frac{y_2 - y_1}{x_2 - x_1}$ with P_1 as the x-intercept, i.e., $(-4, 0)$, and P_2 as the y-intercept, i.e., $(0, 4)$. This yields

$$m = \frac{y_2 - y_1}{x_2 - x_1}$$
$$m = \frac{4 - 0}{0 - (-4)}$$
$$m = \frac{4}{0 + 4}$$
$$m = \frac{4}{4}$$
$$m = 1$$

8. The line makes an angle of 45 ° with the horizontal. Note that for this to show itself graphically, equal unit sizes should be chosen for x- and y-axes.

We now present a concrete example of a problem of type *direct superposition*.[11]

Example

The equation relating the value of the temperature in kelvins, T, and the value of the temperature in degrees Celsius, t, is[12]

T temperature (K)
t temperature (°C)
T_0 temperature at which water freezes under standard
        conditions (K)

$$T = T_0 + t$$

The measured value of T_0, a constant, is 273.15 K.

What information do you extract from this equation?

---

[11] For more concrete examples of problems of type *direct superposition* please see the companion textbook *Semantics and the Syntax of Algebra* by the author.

[12] In physics it is shown that -273.15 °C represents the value of the coldest possible temperature. This means that the model should be accompanied by the statement $t \geq -273.15$. This simply means that the relationship between T and t is as described by the model *so long as* $t \geq -273.15$. The equation does not make sense for $t < -273.15$ as this is physically impossible.

Side conditions such as the one given by the statement $t \geq -273.15$ above, are referred to as the **domain** of the equation which is the set of permissible values for the independent variable.

The statement on the domain of an equation is provided as part of the model using the following format:

$$T = T_0 + t, \qquad t \in \mathbb{R}, \, t \geq -T_0$$

When the statement of the domain is missing, it is understood that the domain is the largest set of numbers for which the equation makes sense. As an example, the missing statement of the domain in the model

$$y = \sqrt{x}$$

implies that the domain is the set of nonnegative real numbers, i.e.,

$$y = \sqrt{x}, \qquad x \in \mathbb{R}, \, x \geq 0$$

Other than in a few isolated cases, in this textbook we will not discuss the domain in any great detail.

Solution

From the form of the equation we can conclude that T is directly superpositional to t. The equation $T = T_0 + t$ is similar in form to $y = k + x$, $k \in \mathbb{R}$ with $k = T_0$. Since we have shown that in an equation of the form $y = k + x$, $k \in \mathbb{R}$ the variables x and y are related through direct superposition, we can state the same claim for the variables t and T in the particular case $T = T_0 + t$.

The conservation form of the equation $T = T_0 + t$ is $T - t = T_0$. This tells us that the difference between T and t (in that order) is always T_0, i.e., the value of the temperature in kelvins, T, is always T_0 or 273.15 units larger than the value of the temperature in degrees Celsius, t.

Since T is directly superpositional to t, $\Delta T = \Delta t$. This tells us that the temperature in kelvins and the temperature in degrees Celsius change by equal amounts. In particular, a change in the value of temperature by 1 K corresponds to a change in the value of temperature by 1 °C. This implies that the size of the unit on the kelvin scale is equal to the size of the unit on the degree Celsius scale.

Exercise Set 12.2.1.1

1. What does it mean when we say that *the relationship between two quantities is direct in the sense of superposition?*

2. What does it mean when we say that *the relationship between two quantities is direct superposition?*

3. What is the formal model for a relationship of type direct superposition?

4. What is the conservation model for a relationship of type direct superposition?

5. What does it mean when we say that *relationships of type direct superposition are symmetric?*

6. How does the change in the value of one quantity relate to change in the value of another quantity if the two quantities are related through direct superposition?

7. What is the shape of the graph of the model for a relationship of type direct superposition?

8. What is the slope of the line that represents the graph of a relationship of type direct superposition?

9. What angle does the line that represents the graph of a relationship of type direct superposition make with the horizontal?

10. What is the effect of k in $y = k + x$, $k \in \mathbb{R}$, on the line that represents the graph of the equation?

11. Which of the following equations are models of relationships of type direct superposition?

a. $y = 2 + x$ k. $-x + y = \sqrt{3}$

b. $y = -6 + x$ l. $y = x - 1$

c. $y = x$ m. $y = -x + 4$

d. $y = 2 - x$ n. $y = x + \frac{1}{3}$

e. $y = \frac{1}{3} + x$ o. $y - 3 = x$

f. $y + x = 4$ p. $x = -4 + y$

g. $y - x = -3$ q. $x = 2 - y$

h. $y = 4 + 2x$ r. $-2x = 5 + y$

i. $y = -5x$ s. $2y = 5 + 2x$

j. $x - y = 0$ t. $3y = -3 + 4x$

12. Each of the following is a model for a direct superposition relationship between y and x. In each case rewrite the equation in formal and conservation forms.

a. $y = 7 + x$ g. $x - y = -1$

b. $y = -2 + x$ h. $y + 4 - x = 0$

c. $y = x$ i. $y - x - 2 = 0$

d. $x = 3 + y$ j. $x - y = 0$

e. $y = x + \frac{2}{5}$ k. $-y + x = \sqrt{2}$

f. $y - x = 4$ l. $-x + 1 = -y$

13. Rewrite each of the equations in Question 12 above in formal form, find the x- and y-intercepts of the line, calculate the slope of the line using the formula $m = \frac{y_2 - y_1}{x_2 - x_1}$ and graph the line.

12.2.1.2   Inverse Superposition

The next simplest type of relationship that we observe between the values of two quantities beyond direct superposition is **inverse superposition**.

Definition

The dependent variable, y, is said to be inversely superpositional to the independent variable, x, if an increase or decrease in x by a certain amount forces a corresponding decrease or increase in y by the same amount.[13]

Model

The model for relationships of type inverse superposition comes in two flavours: The formal form and the conservation form.[14]

Formal Form

A relationship of type *inverse superposition* between the values of two quantities can be modelled as

$$y = k - x, \qquad k \in \mathbb{R}$$

where the abstract quantity symbols x and y represent the values of the two quantities and the side condition, $k \in \mathbb{R}$, states that, for a given concrete problem of this type, k must be a constant, i.e., a known number in $\mathbb{R}$. We refer to this formulation as the formal form of the model for inverse superposition and we refer to k as the constant of superpositionality.

The reason such an equation models a relationship of type inverse superposition between y and x is that, as shown below, x is being subtracted on the right side of the equation.

$$y = k - \overset{\curvearrowright}{x}$$

If we now increase x by, say, 2 units, then we will be subtracting 2 more units from k on the right side of the equation. This will result in a decrease of 2 units in y. Equivalently, if we decrease x by, say, 2 units, then we will be

---

[13]Two quantities are said to have an **inverse relationship** in the sense of superposition if one increases in value while the other decreases in value or one decreases in value while the other increases in value. Note that an inverse relationship imposes no requirements on the equality of the sizes of increase and decrease in the values of the quantities involved. If the relationship is of type *inverse superposition*, however, then the sizes of increase and decrease in the values of the quantities involved are required to be the same.

[14]Both forms are suitable when working with problems whose solutions involve a single superposition step. For chain superposition problems, however, it is possible to extend the formal form but not the conservation form. In single-superposition applications, the version that is most suitable to the semantics of the problem is adopted. For a more detailed discussion of the use of the different forms of the equation for inverse superposition please see the companion textbook *Semantics and the Syntax of Algebra* by the author.

subtracting 2 fewer units from k on the right side of the equation. This will result in an increase of 2 units in y.

The argument given above is quite general: Increasing x by any amount forces a decrease in y by the same amount and decreasing x by any amount forces an increase in y by the same amount. Such dependence is exactly what a relationship of type inverse superposition is about.

Conservation Form

The fact that, in a relationship of type *inverse superposition*, an increase or decrease in the value of one of the quantities by a certain amount is matched by a corresponding decrease or increase in the value of the other quantity by the same amount implies that any increase in the value of one of the quantities is offset by the decrease in the value of the other quantity and that any decrease in the value of one of the quantities is offset by the increase in the value of the other quantity. If this is so, then the sum of the values of the two quantities involved remains the same, i.e, the sum of the values of the two quantities involved is constant. We can use mathematical language to express this view of a relationship of type *inverse superposition* as

$$y + x = k, \quad k \in \mathbb{R}$$

Note that in this formulation the value of the expression on the left side of the equation is constant. In a given problem, this value, i.e., $y + x$, never increases or decreases and is, therefore, in some sense *conserved*. For this reason we refer to the formulation $y + x = k$ as the conservation form of the equation $y = k - x$.

The conservation form of the model, i.e., $y + x = k$, can be shown to be the same as the formal form of the model, i.e., $y = k - x$, through conversion of subtraction of x on the right side of the equation $y = k - x$ to addition of x on its left, i.e.,

$$y = k - x$$
$$y + x = k$$

Symmetry

Relationships of type inverse superposition are symmetric: If y is inversely superpositional to x, then x is inversely superpositional to y.[15] This can be

---

[15]This means that, if it is the case that an increase or decrease in x by a certain amount forces a corresponding decrease or increase in y by the same amount, then an increase or decrease in y by some amount forces a corresponding decrease or increase in x by the same amount. Therefore, whichever variable is labelled as independent, the relationship between the two variables will be of type *inverse superposition*.

seen through a rearrangement of the equation $y = k - x$ as shown below:

$$y = k - x$$
$$x = k - y$$

Note that y is being subtracted on the right side of the equation. Since k is constant, x is inversely superpositional to y.

Change in Values of Variables

The fact that, in a relationship of type *inverse superposition*, an increase or decrease in the value of one quantity by a certain amount forces a corresponding decrease or increase in the value of the other quantity by an equal amount implies that change in x is equal to the negative of the change in y. In mathematical notation, we write this as

$$\Delta y = -\Delta x$$

We can show this to be the case as follows: By definition

$$\Delta y = y_2 - y_1$$

where y_1 represents the initial value of y and y_2 represents the final value of y. Since $y = k - x$, there is an x value, x_1, that corresponds to y_1 through $y_1 = k - x_1$ and there is an x value, x_2, that corresponds to y_2 through $y_2 = k - x_2$. Substituting $k - x_1$ for y_1 and $k - x_2$ for y_2 in the equation above yields

$$\Delta y = k - x_2 - (k - x_1)$$

Simplifying the expression on the right side yields

$$\Delta y = k - x_2 - k + x_1$$
$$\Delta y = -x_2 + x_1$$
$$\Delta y = -(x_2 - x_1)$$
$$\Delta y = -\Delta x$$

This proves that for a relationship of type inverse superposition, i.e., $y = k - x$, $k \in \mathbb{R}$, $\Delta y = -\Delta x$.

Graph

Provided units of equal size are used in the graph of an equation of type inverse superposition, the graph will be a straight line going down at $45°$ relative to the horizontal.[16] The line has an x-intercept of $(k, 0)$, a y-intercept of $(0, k)$ and a slope of -1. The x-intercept can be found by setting $y = 0$ in the equation $y = k - x$. This yields

$$
\begin{aligned}
y &= k - x \\
0 &= k - x \\
x &= k - 0 \\
x &= k
\end{aligned}
$$

showing that the x-intercept is $(k, 0)$. For the y-intercept we set $x = 0$ in the equation $y = k - x$. This yields

$$
\begin{aligned}
y &= k - x \\
y &= k - 0 \\
y &= k
\end{aligned}
$$

showing that the y-intercept is $(0, k)$. To calculate the slope, we use the formula

$$
m = \frac{y_2 - y_1}{x_2 - x_1}
$$

with P_1 as the x-intercept, i.e., $(k, 0)$, and P_2 as the y-intercept, i.e., $(0, k)$. This yields

$$
\begin{aligned}
m &= \frac{y_2 - y_1}{x_2 - x_1} \\
m &= \frac{k - 0}{0 - k} \\
m &= \frac{k}{-k} \\
m &= -1
\end{aligned}
$$

Alternatively, we can use the formula

$$
m = \frac{\Delta y}{\Delta x}
$$

along with the fact that, as shown above, for a relationship of type *inverse superposition*, $\Delta y = -\Delta x$ to get

$$
\begin{aligned}
m &= \frac{\Delta y}{\Delta x} \\
m &= \frac{-\Delta x}{\Delta x} \\
m &= -1
\end{aligned}
$$

The graph of the line $y = k - x$ is shown in Figure 12.2.5 below.

---

[16] For a detailed discussion of fundamental concepts that relate to graphs please see Appendix O.

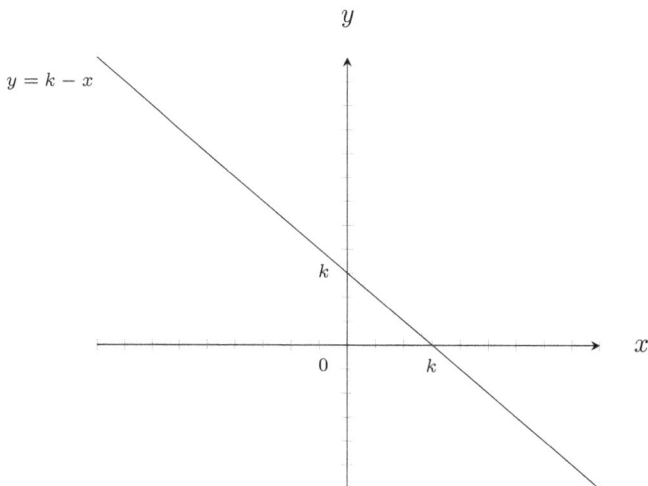

Figure 12.2.5: Graph of the equation for relationships of type inverse superposition

Some of the more important features of the line $y = k - x$ are shown in Figure 12.2.6 below.[17]

In $y = k - x$, $k \in \mathbb{R}$, the effect of k on the graph of the equation is to raise or lower the line. The higher the value of k, the higher the line. Figure 12.2.7 shows the graphs of the five equations $y = -6 - x$, $y = -3 - x$, $y = -x$ (which is the same as $y = 0 - x$), $y = 3 - x$ and $y = 6 - x$. Note that the angle between the line and the horizontal remains at $-45°$ in all cases.

---

[17]The most important facts to keep in mind about a relationship of type *inverse superposition* are its form, i.e., $y = k - x$, $k \in \mathbb{R}$, the fact that change in y must equal to the negative of change in x, i.e., $\Delta y = -\Delta x$, and that its graph is a straight line with a slope of -1.

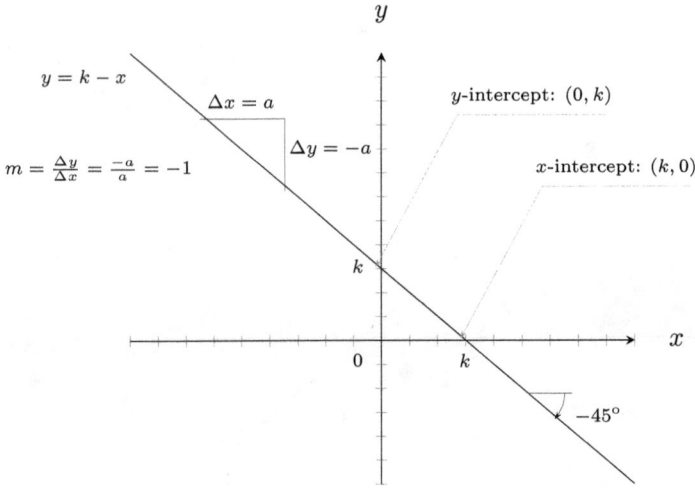

Figure 12.2.6: Features of the graph of the equation for relationships of type inverse superposition

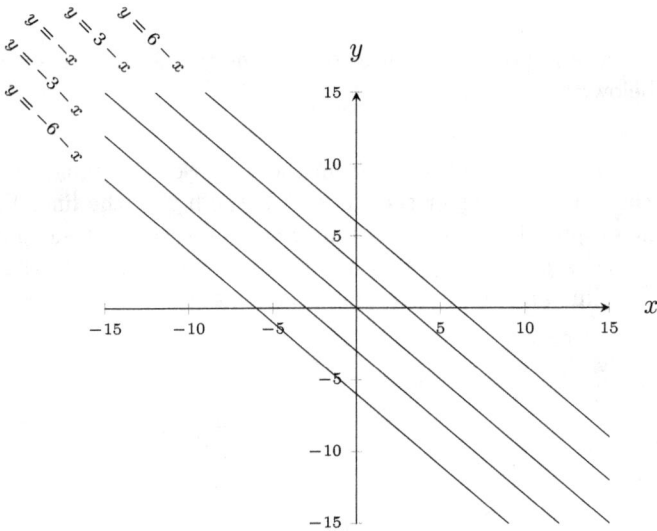

Figure 12.2.7: Effect of k in $y = k - x$, $k \in \mathbb{R}$, on the graph of the equation for relationships of type *inverse superposition*

Applications

We now present an abstract and a concrete example of a problem of type *inverse superposition.*

Example

Given the equation

$$y = 4 - x$$

1. Name the type of the relationship between x and y and support that claim.
2. Write the equation in conservation form.
3. State the relationship between change in x and change in y.
4. Describe the shape of the graph of the equation.
5. Find the x- and y-intercepts of the line.
6. Graph the line.
7. Find the slope of the line.
8. What is the angle between the line and the horizontal?

Solution

1. The relationship between x and y is inverse superposition. The equation $y = 4 - x$ is similar in form to $y = k - x$, $k \in \mathbb{R}$, with $k = 4$. Since we have shown that in an equation of the form $y = k - x$, $k \in \mathbb{R}$ the variables x and y are related through inverse superposition, we can state the same claim for the particular case $y = 4 - x$.

2. The conservation form of the equation $y = 4 - x$ is $y + x = 4$. This tells us that the sum of y and x is always 4.

3. $\Delta y = -\Delta x$

4. The graph of the equation $y = 4 - x$ is a straight line.

5. We can use our earlier findings that the x- and y-intercepts of the graph of the equation $y = k - x$, $k \in \mathbb{R}$ are $(k, 0)$ and $(0, k)$ to conclude that the x- and y-intercepts of the equation $y = 4 - x$ are $(4, 0)$ and $(0, 4)$.

Alternatively, we can find the x-intercept by setting $y = 0$, and the y-intercept by setting $x = 0$, in the equation $y = 4 - x$. For the x-intercept, this yields

$$
\begin{aligned}
y &= 4 - x \\
0 &= 4 - x \\
x &= 4 - 0 \\
x &= 4
\end{aligned}
$$

which shows that the x-intercept is $(4, 0)$. For the y-intercept we have:

$$
\begin{aligned}
y &= 4 - x \\
y &= 4 - 0 \\
y &= 4
\end{aligned}
$$

which shows that the y-intercept is $(0, 4)$.

6. Since the graph of the equation $y = 4 - x$ is a line, we need two points on the line to graph the line. Here we use the x- and y-intercepts as our points with coordinates $(4, 0)$ and $(0, 4)$. The graph of the equation $y = 4 - x$ is shown in Figure 12.2.8 below.

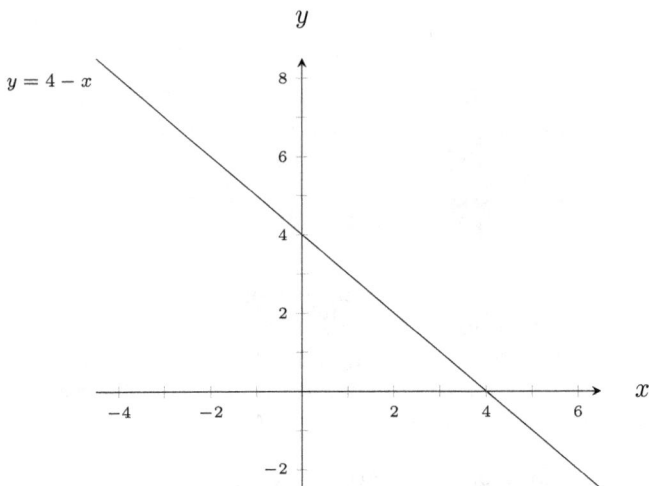

Figure 12.2.8: Graph of the equation $y = 4 - x$

7. Since the equation is of type inverse superposition, the slope of the line is -1. Alternatively, we can compute the slope using the equation $m = \frac{y_2 - y_1}{x_2 - x_1}$ with P_1 as the x-intercept, i.e., $(4, 0)$, and P_2 as the y-

intercept, i.e., $(0, 4)$. This yields

$$m = \frac{y_2 - y_1}{x_2 - x_1}$$
$$m = \frac{4 - 0}{0 - 4}$$
$$m = \frac{4}{-4}$$
$$m = -1$$

8. The line makes an angle of -45 ° with the horizontal. Note that for this to show itself graphically, equal unit sizes should be chosen for x- and y-axes.

We now present a concrete example of a problem of type *inverse superposition*.[18]

Example

The following equation relates the value of distance remaining to the value of distance from Toronto to Montreal and distance covered.

d distance from Toronto to Montreal (km)
d_c distance covered (km)
d_r distance remaining (km)

$$d_r = d - d_c$$

The measured value of distance from Toronto to Montreal, d, is 504 km.

What information do you extract from this equation?

Solution

From the form of the equation we can conclude that d_r is inversely superpositional to d_c. The equation $d_r = d - d_c$ is similar in form to $y = k - x, k \in \mathbb{R}$ with $k = d$, a constant that represents the value of the distance from Toronto to Montreal. Since we have shown that in an equation of the form $y = k - x, k \in \mathbb{R}$ the variables x and y are related through inverse superposition, we can state

---

[18]For more concrete examples of problems of type *inverse superposition* please see the companion textbook *Semantics and the Syntax of Algebra* by the author.

the same claim for the variables d_c and d_r in the particular case $d_r = d - d_c$.

The conservation form of the equation $d_r = d - d_c$ is $d_r + d_c = d$. This tells us that the sum of d_r and d_c is equal to the constant, d. This makes sense: The sum of the distance remaining and distance covered is fixed and is equal to the distance from Toronto to Montreal.

Since d_r is inversely superpositional to d_c, $\Delta d_r = -\Delta d_c$. As one would expect, this tells us that any increase in the value of distance covered by a certain amount will result in a reduction in the value of distance remaining by the same amount.

Exercise Set 12.2.1.2

1. What does it mean when we say that *the relationship between two quantities is inverse in the sense of superposition?*

2. What does it mean when we say that *the relationship between two quantities is inverse superposition?*

3. What is the formal model for a relationship of type inverse superposition?

4. What is the conservation model for a relationship of type inverse superposition?

5. What does it mean when we say that *relationships of type inverse superposition are symmetric?*

6. How does the change in the value of one quantity relate to change in the value of another quantity if the two quantities are related through inverse superposition?

7. What is the shape of the graph of the model for a relationship of type inverse superposition?

8. What is the slope of the line that represents the graph of a relationship of type inverse superposition?

9. What angle does the line that represents the graph of a relationship of type inverse superposition make with the horizontal?

10. What is the effect of k in $y = k - x$, $k \in \mathbb{R}$, on the line that represents the graph of the equation?

11. Which of the following equations are models of relationships of type inverse superposition?

a. $y = 8 - x$

b. $y = -3 - x$

c. $y = -x$

d. $y = 5 + x$

e. $y = \frac{2}{5} - x$

f. $y - x = 10$

g. $y + x = -5$

h. $y = 6 - 3x$

i. $y = 9x$

j. $x + y = 0$

k. $-x - y = \sqrt{2}$

l. $y = -x - 1$

m. $y = x + 5$

n. $y = -x - \frac{3}{4}$

o. $y - 3 = -x$

p. $x = -7 - y$

q. $-x = 1 - y$

r. $-3x = 8 - y$

s. $-5y = -1 + 5x$

t. $-2y = -3 - 2x$

12. Each of the following is a model for an inverse superposition relationship between y and x. In each case rewrite the equation in formal and conservation forms.

a. $y = 7 - x$

b. $y = -2 - x$

c. $y = -x$

d. $x = 3 - y$

e. $y = -x + \frac{2}{5}$

f. $y + x = 4$

g. $x + y = -1$

h. $y + 4 + x = 0$

i. $y + x - 2 = 0$

j. $x + y = 0$

k. $-y - x = \sqrt{2}$

l. $x + 1 = -y$

13. Rewrite each of the equations in Question 12 above in formal form, find the x- and y-intercepts of the line, calculate the slope of the line using the formula $m = \frac{y_2 - y_1}{x_2 - x_1}$ and graph the line.

12.2.1.3   Direct Proportion

The next simplest type of relationship that we observe between the values of two quantities beyond direct superposition and inverse superposition is **direct proportion**.

Definition

The dependent variable, y, is said to be directly proportional to the independent variable, x, if scaling x by a certain factor[19] forces a corresponding

---

[19]The phrase *scaling the value of a quantity by a certain factor* means multiplying the value of the quantity by that factor.

scaling of y by the same factor.[20,21]

Model

The model for relationships of type direct proportion comes in two flavours: The formal form and the conservation form.[22]

Formal Form

A relationship of type *direct proportion* between the values of two quantities can be modelled as

$$y = kx, \qquad k \in \mathbb{R}, \ k \neq 0$$

where the abstract quantity symbols x and y represent the values of the two quantities and the side conditions, $k \in \mathbb{R}$, $k \neq 0$, state that, for a given concrete problem of this type, k must be a nonzero constant, i.e., a known nonzero number in $\mathbb{R}$.[23] We refer to this formulation as the formal form of the model for direct proportion and we refer to k as the **constant of proportionality**.

The reason such an equation models a relationship of type direct proportion between y and x is that, as shown below, x is being multiplied by on the right side of the equation.

$$y = k \times \overset{\curvearrowright}{x}$$

---

[20] Two quantities are said to have a direct relationship in the sense of proportion if scaling the value of one quantity results in scaling the value of the other quantity with the values of the two quantities both increasing or both decreasing in size. Note that a direct relationship in the sense of proportion imposes no requirements on the equality of the sizes of the factors by which the values of the two quantities are scaled. If the relationship is of type *direct proportion*, however, then the factors by which the values of the two quantities are scaled are required to be the same.

[21] Note that this definition implies that multiplying x by any number results in the multiplication of y by the same number and dividing x by any number results in the division of y by the same number.

[22] Both forms are suitable when working with problems whose solutions involve a single proportion step. For chain proportion problems, however, it is possible to extend the formal form but not the conservation form. In single-proportion applications, the version that is most suitable to the semantics of the problem is adopted. For a more detailed discussion of the use of the different forms of the equation for direct proportion please see the companion textbook *Semantics and the Syntax of Algebra* by the author.

[23] If $k = 0$, one may still argue that scaling x by any factor forces y to scale by the same factor and choose to refer to $y = 0$ as an equation describing a relationship of type *direct proportion* between y and x. In this textbook we choose not to do so as we consider the equation $y = 0$ to imply that y is 0 regardless of what x is and that, therefore, changes in x do not affect y.

If we now scale x by, say, a factor of 2,[24] then we will be multiplying k on the right side of the equation by a value that is twice its original value. This will result in a scaling of y by a factor of 2 as well. And if we scale x by, say, a factor of $\frac{1}{2}$,[25] then we will be multiplying k on the right side of the equation by a value that is half its original value. This will result in a scaling of y by a factor of $\frac{1}{2}$ as well.

The argument given above is quite general: Scaling x by any factor forces a scaling of y by the same factor. Such dependence is exactly what a relationship of type direct proportion is about.

Conservation Form

The fact that, in a relationship of type *direct proportion*, scaling the value of one of the quantities by a certain factor is matched by a corresponding scaling of the value of the other quantity by the same factor implies that the values of the two quantities involved get multiplied by the same number or that they get divided by the same number (both change to twice their original values, both change to half their original values, etc.). If this is so, then the quotient of the values of the two quantities involved remains the same, i.e, the quotient of the values of the two quantities involved is constant. We can use mathematical language to express this view of a relationship of type *direct proportion* as

$$\frac{y}{x} = k, \quad k \in \mathbb{R}, k \neq 0$$

Note that in this formulation the value of the expression on the left side of the equation is constant. In a given problem, this value, i.e., $\frac{y}{x}$, never increases or decreases and is, therefore, in some sense *conserved*. For this reason we refer to the formulation $\frac{y}{x} = k$ as the conservation form of the equation $y = kx$.

The conservation form of the model, i.e., $\frac{y}{x} = k$, can be shown to be the same as the formal form of the model, i.e., $y = kx$, through conversion of multiplication by x on the right side of the equation $y = kx$ to division by x on the its left, i.e.,

$$y = kx$$
$$\frac{y}{x} = k$$

---

[24] This means changing the value of x to twice its original value.
[25] This means changing the value of x to half its original value.

Symmetry

Relationships of type direct proportion are symmetric: If y is directly proportional to x, then x is directly proportional to y.[26] This can be seen through a rearrangement of the equation $y = kx$ as shown below:

$$y = kx$$
$$kx = y$$
$$x = \frac{1}{k}y$$

Note that y is being multiplied on the right side of the equation. Furthermore, since k is a nonzero number in $\mathbb{R}$, then so is $\frac{1}{k}$. Therefore, x is directly proportional to y.

Change in Values of Variables

The fact that, in a relationship of type *direct proportion*, scaling the value of one of the quantities by a certain factor forces the scaling of the value of the other quantity by the same factor implies that any change in x is matched by a change in y that scales the change in x by a factor that is constant. In fact, for the equation $y = kx$, $k \neq 0$, change in y is related to change in x through the equation

$$\Delta y = k\Delta x$$

We can show this to be the case as follows: By definition

$$\Delta y = y_2 - y_1$$

where y_1 represents the initial value of y and y_2 represents the final value of y. Since $y = kx$, there is an x value, x_1, that corresponds to y_1 through $y_1 = kx_1$ and there is an x value, x_2, that corresponds to y_2 through $y_2 = kx_2$. Substituting kx_1 for y_1 and kx_2 for y_2 in the equation above yields

$$\Delta y = kx_2 - kx_1$$

Using the distributive property is reverse on the right side yields

$$\Delta y = k(x_2 - x_1)$$
$$\Delta y = k\Delta x$$

This proves that for a relationship of type direct proportion, i.e., $y = kx$, $k \in \mathbb{R}$, $k \neq 0$, $\Delta y = k\Delta x$.

---

[26]This means that, if it is the case that scaling x by a certain factor forces a corresponding scaling of y by the same factor, then scaling y by some factor forces a corresponding scaling of x by the same factor. Therefore, whichever variable is labelled as independent, the relationship between the two variables will be of type *direct proportion*.

Graph

The graph of an equation of type direct proportion is a nonhorizontal, non-vertical straight line through the origin.[27] The x- and y-intercepts are both $(0,0)$. The x-intercept can be found by setting $y = 0$ in $y = kx$. This yields

$$y = kx$$
$$0 = kx$$
$$kx = 0$$
$$x = \frac{0}{k}$$
$$x = 0$$

showing that the x-intercept is $(0,0)$. For the y-intercept we set $x = 0$ in the equation $y = kx$. This yields

$$y = kx$$
$$y = k \times 0$$
$$y = 0$$

showing that the y-intercept is $(0,0)$. To calculate the slope, we use the formula

$$m = \frac{y_2 - y_1}{x_2 - x_1}$$

We can use the x- or y-intercept, i.e., $(0,0)$, as P_1 but we need another point as P_2. We can set $x = 1$ in $y = kx$ to get

$$y = kx$$
$$y = k \times 1$$
$$y = k$$

which shows that the point $(1, k)$ is on the line. We set P_1 as the x- or y-intercept, i.e., $(0,0)$, and P_2 as the point $(1, k)$. This yields

$$m = \frac{y_2 - y_1}{x_2 - x_1}$$
$$m = \frac{k - 0}{1 - 0}$$
$$m = \frac{k}{1}$$
$$m = k$$

---

[27]For a detailed discussion of fundamental concepts that relate to graphs please see Appendix O.

Alternatively, we can use the formula

$$m = \frac{\Delta y}{\Delta x}$$

along with the fact that, as shown above, for a relationship of type *direct proportion*, $\Delta y = k\Delta x$, to get

$$m = \frac{\Delta y}{\Delta x}$$
$$m = \frac{k\Delta x}{\Delta x}$$
$$m = k$$

The graph of the line $y = kx$ is shown in Figure 12.2.9 below.

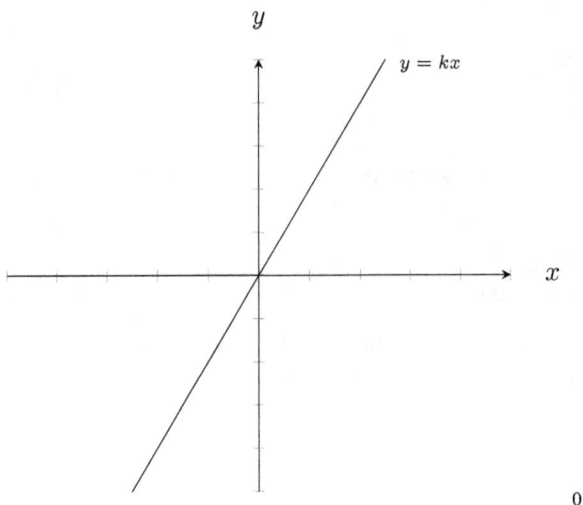

Figure 12.2.9: Graph of the equation for relationships of type direct proportion

Some of the more important features of the line $y = kx$ are shown in Figure 12.2.10 below.[28]

---

[28]The most important facts to keep in mind about a relationship of type *direct proportion* are its form, i.e., $y = kx$, the fact that change in y must equal to k times change in x, i.e., $\Delta y = k\Delta x$, and that its graph is a straight line through the origin.

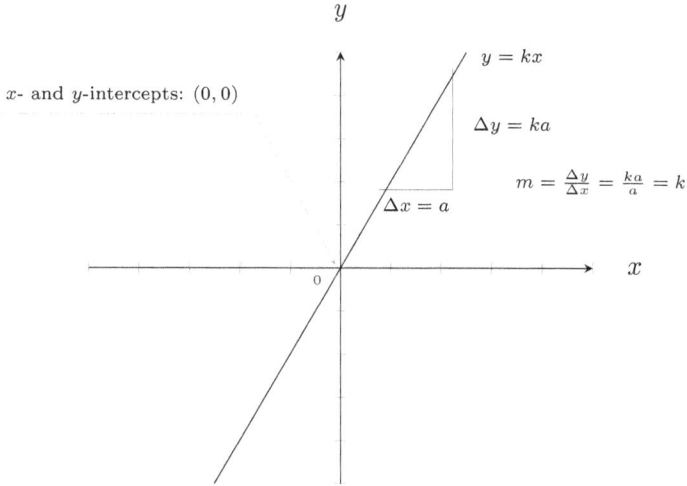

Figure 12.2.10: Features of the graph of the equation for relationships of type direct proportion

In $y = kx$, $k \in \mathbb{R}$, $k \neq 0$, the effect of k on the graph of the equation is to rotate the line. The higher the size of k, the steeper line. Figure 12.2.11 shows the graphs of three equations with positive values of k: $\frac{1}{2}$, 1 and 2, with corresponding equations $y = \frac{1}{2}x$, $y = x$, and $y = 2x$. Note that as k increases, the line rotates counterclockwise about the origin. For $0 < k < 1$, the angle that the line makes with the horizontal is less than 45°. When $k = 1$, the angle between the line and the horizontal becomes 45°.[29] For $k > 1$ the angle between the line and the horizontal exceeds 45° and approaches 90° as k becomes larger and larger.

The discussion for negative values of k is left as an exercise.

---

[29]This means that the relationship between y and x in $y = x$ is both direct superposition and direct proportion.

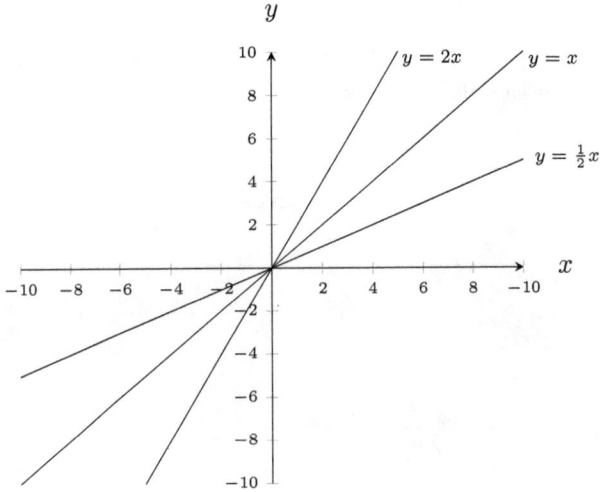

Figure 12.2.11: Effect of positive values of k in $y = kx$, $k \in \mathbb{R}$, on the graph of the equation for relationships of type *direct proportion*

Applications

We now present an abstract and a concrete example of a problem of type *direct proportion*.

Example

Given the equation

$$y = \frac{1}{2}x$$

1. Name the type of the relationship between x and y and support that claim.
2. Write the equation in conservation form.
3. State the relationship between change in x and change in y.
4. Describe the shape of the graph of the equation.
5. Find the x- and y-intercepts of the line.
6. Graph the line.
7. Find the slope of the line.
8. Describe the size of the angle between the line and the horizontal relative to an angle of $45°$.

Solution

1. The relationship between x and y is direct proportion. The equation $y = \frac{1}{2}x$ is similar in form to $y = kx$, $k \in \mathbb{R}$, $k \neq 0$, with $k = \frac{1}{2}$. Since we have shown that in an equation of the form $y = kx$, $k \in \mathbb{R}$, $k \neq 0$ the variables x and y are related through direct proportion, we can state the same claim for the particular case $y = \frac{1}{2}x$.

2. The conservation form of the equation $y = \frac{1}{2}x$ is $\frac{y}{x} = \frac{1}{2}$. This tells us that the quotient of y and x (in that order) is always $\frac{1}{2}$, i.e., y is always $\frac{1}{2}$ the size of x.

3. $\Delta y = \frac{1}{2}\Delta x$

4. The graph of the equation $y = \frac{1}{2}x$ is a straight line.

5. We can use our earlier findings that the x- and y-intercepts of the graph of the equation $y = kx$, $k \in \mathbb{R}$, $k \neq 0$ are both $(0,0)$ to conclude that the x- and y-intercepts of the equation $y = \frac{1}{2}x$ are both $(0,0)$.

   Alternatively, we can find the x-intercept by setting $y = 0$, and the y-intercept by setting $x = 0$, in the equation $y = \frac{1}{2}x$. For the x-intercept, this yields

$$y = \frac{1}{2}x$$
$$0 = \frac{1}{2}x$$
$$\frac{1}{2}x = 0$$
$$x = \frac{2}{1} \times 0$$
$$x = 0$$

   which shows that the x-intercept is $(0,0)$. For the y-intercept we have:

$$y = \frac{1}{2}x$$
$$y = \frac{1}{2} \times 0$$
$$y = 0$$

   which shows that the y-intercept is $(0,0)$.

6. Since the graph of the equation $y = \frac{1}{2}x$ is a line, we need two points on the line to graph the line. Here we use the x- or y-intercept, i.e., $(0,0)$,

as our first point. To find a second point, we set x equal to 2 and find the corresponding y using the equation $y = \frac{1}{2}x$.[30] For $x = 2$, we have

$$y = \frac{1}{2} \times 2$$
$$y = 1$$

This yields the point $(2, 1)$ which we will use as our second point. The graph of the equation $y = \frac{1}{2}x$ is shown in Figure 12.2.12 below.

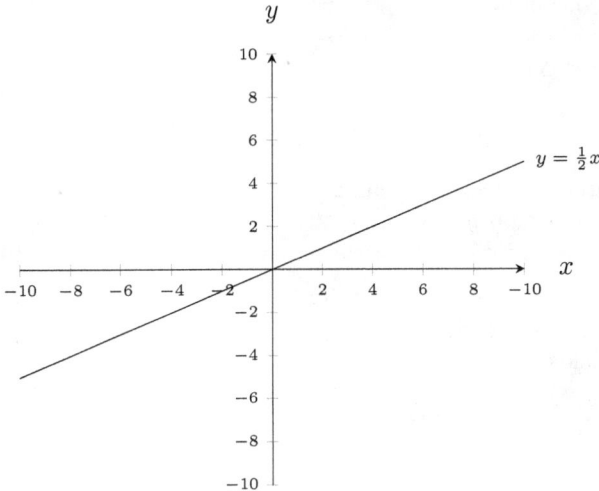

Figure 12.2.12: Graph of the equation $y = \frac{1}{2}x$

7. Since the slope of the line corresponding to the equation for direct proportion, $y = kx$, $k \in \mathbb{R}$, $k \neq 0$ is k, the slope of the line corresponding to the equation $y = \frac{1}{2}x$ is $\frac{1}{2}$. Alternatively, we can compute the slope using the equation $m = \frac{y_2 - y_1}{x_2 - x_1}$ with P_1 as the x- or y-intercept, i.e., $(0, 0)$, and P_2 as the point $(2, 1)$. This yields

$$m = \frac{y_2 - y_1}{x_2 - x_1}$$
$$m = \frac{1 - 0}{2 - 0}$$
$$m = \frac{1}{2}$$

---

[30] Any value other than 0 for x would work.

8. Since the slope of the line is between 0 and 1, the angle between the line and the horizontal is between $0°$ than $45°$.[31]

We now present a concrete example of a problem of type *direct proportion*.[32]

Example

A car is travelling at a speed of 120 km/h. For such a motion, distance covered in kilometres and travel time in hours are related through the equation

d distance covered (km)
t travel time (h)
$d = 120t$

What information do you extract from this equation?

Solution

From the form of the equation we can conclude that distance covered, d, is directly proportional to travel time t. The equation $d = 120t$ is similar in form to $y = kx$, $k \in \mathbb{R}$, $k \neq 0$ with $k = 120$, the speed of the car. Since we have shown that in an equation of the form $y = kx$, $k \in \mathbb{R}$, $k \neq 0$ the variables x and y are related through direct proportion, we can state the same claim for the variables d and t in the particular case $d = 120t$.

The conservation form of the equation $d = 120t$ is $\frac{d}{t} = 120$. This tells us that the quotient of d and t (in that order) is always 120 km/h, i.e., the speed of the car.

Since d is directly proportional to t, $\Delta d = 120\Delta t$. This tells us that changes in d are 120 times larger than changes in t. As an example, a change in travel time of 1 h (an additional hour of driving) corresponds to a change in distance covered of 120 km (an additional coverage of 120 km).

---

[31] Note that for the angle to show correctly, equal unit sizes should be chosen for x- and y-axes.

[32] For more concrete examples of problems of type *direct proportion* please see the companion textbook *Semantics and the Syntax of Algebra* by the author.

Exercise Set 12.2.1.3

1. What does it mean when we say that *the relationship between two quantities is direct in the sense of proportion?*

2. What does it mean when we say that *the relationship between two quantities is direct proportion?*

3. What is the formal model for a relationship of type direct proportion?

4. What is the conservation model for a relationship of type direct proportion?

5. What does it mean when we say that *relationships of type direct proportion are symmetric?*

6. How does the change in the value of one quantity relate to change in the value of another quantity if the two quantities are related through direct proportion?

7. What is the shape of the graph of the model for a relationship of type direct proportion?

8. What is the slope of the line that represents the graph of a relationship of type direct proportion?

9. What angle does the line that represents the graph of a relationship of type direct proportion make with the horizontal?

10. What is the effect of k in $y = kx$, $k \in \mathbb{R}$, $k < 0$, on the line that represents the graph of the equation? Illustrate using an example.

11. Which of the following equations are models of relationships of type direct proportion?

a. $y = 2x$

b. $y = -6x$

c. $y = x$

d. $y = \frac{2}{5}x$

e. $y = -\frac{1}{3}x$

f. $\frac{y}{x} = 7$

g. $\frac{y}{x} = -\frac{3}{4}$

h. $yx = -3$

i. $xy = \frac{1}{5}$

j. $\frac{x}{y} = -1$

k. $\frac{x}{y} = 4$

l. $y = \frac{x}{8}$

m. $y = -\frac{x}{2}$

n. $y = \frac{3}{x}$

o. $y = -\frac{1}{x}$

p. $x = -4y$

q. $x = \frac{1}{6}y$

r. $-2x = 5y$

s. $3y = 6x$

t. $2y = -8 + 5x$

u. $3x - 1 = y$

v. $-2x - 3y = 0$

w. $7x + 2y = 4$

12. Each of the following is a model for a direct proportion relationship between y and x. In each case rewrite the equation in formal and conservation forms.

a. $y = 7x$

e. $y = \frac{2}{5}x$

i. $y - x = 0$

b. $y = -2x$

f. $\frac{y}{x} = 4$

j. $x - y = 0$

c. $y = x$

g. $\frac{x}{y} = -1$

k. $-\frac{y}{x} = \sqrt{2}$

d. $x = 3y$

h. $4y - x = 0$

l. $x = -y$

13. Rewrite each of the equations in Question 7 above in formal form, find the x- and y-intercepts of the line, calculate the slope of the line using the formula $m = \frac{y_2 - y_1}{x_2 - x_1}$ and graph the line.

12.2.1.4   Inverse Proportion

The next simplest type of relationship that we observe between the values of two quantities beyond direct superposition, inverse superposition and direct proportion is **inverse proportion**.

Definition

The dependent variable, y, is said to be inversely proportional to the independent variable, x, if scaling x by a certain factor forces a corresponding scaling of y by the inverse of the same factor.[33,34]

Model

The model for relationships of type inverse proportion comes in two flavours: The formal form and the conservation form.[35]

---

[33]Two quantities are said to have an inverse relationship in the sense of proportion if scaling the value of one quantity results in scaling the value of the other quantity with one of the values increasing in size and the other value decreasing in size. Note that an inverse relationship in the sense of proportion imposes no requirements on the relationship between the sizes of the factors by which the values of the two quantities are scaled. If the relationship is of type *inverse proportion*, however, then the factors by which the values of the two quantities are scaled are required to be inverses of each other.

[34]Note that this definition implies that multiplying x by any number results in the division of y by the same number and dividing x by any number results in the multiplication of y by the same number.

[35]Both forms are suitable when working with problems whose solutions involve a single proportion step. For chain proportion problems, however, it is possible to extend the formal form but not the conservation form. In single-proportion applications, the version

Formal Form

A relationship of type *inverse proportion* between the values of two quantities can be modelled as

$$y = \frac{k}{x}, \qquad k \in \mathbb{R}, \, k \neq 0$$

where the abstract quantity symbols x and y represent the values of the two quantities and the side conditions, $k \in \mathbb{R}$, $k \neq 0$, state that, for a given concrete problem of this type, k must be a nonzero constant, i.e., a known nonzero number in $\mathbb{R}$.[36] We refer to this formulation as the formal form of the model for inverse proportion and we refer to k as the **constant of proportionality**.

The reason such an equation models a relationship of type inverse proportion between y and x is that, as shown below, x is being divided by on the right side of the equation.

$$y = \frac{k}{x} \curvearrowright$$

If we now scale x by, say, a factor of 2,[37] then we will be dividing k on the right side of the equation by a value that is twice its original value. This will result in a scaling of y by a factor of $\frac{1}{2}$. And if we scale x by, say, a factor of $\frac{1}{2}$,[38] then we will be dividing k on the right side of the equation by a value that is half its original value. This will result in a scaling of y by a factor of 2.

The argument given above is quite general: Scaling x by any factor forces a scaling of y by the inverse of the same factor. Such dependence is exactly what a relationship of type inverse proportion is about.

Conservation Form

The fact that, in a relationship of type *inverse proportion*, scaling the value of one of the quantities by a certain factor is matched by a corresponding

---

that is most suitable to the semantics of the problem is adopted. For a more detailed discussion of the use of the different forms of the equation for inverse proportion please see the companion textbook *Semantics and the Syntax of Algebra* by the author.

[36] If $k = 0$, one may still argue that scaling x by any factor forces y to scale by the inverse of the same factor and choose to refer to the lines $y = 0$ and $x = 0$ as equations describing a relationship of type *inverse proportion* between y and x. In this textbook we choose not to do so as we consider the equation $y = 0$ to imply that y is 0 regardless of what x is and that, therefore, changes in x do not affect y. Similarly, we consider the equation $x = 0$ to imply that x is 0 regardless of what y is and that, therefore, changes in y do not affect x.

[37] This means changing the value of x to twice its original value.

[38] This means changing the value of x to half its original value.

scaling of the value of the other quantity by the inverse of the same factor implies that if the value of one of the quantities involved is multiplied by some number, then the value of the other quantity gets divided by that number and if the value of one of the quantities involved is divided by some number, then the value of the other quantity gets multiplied by the same number. If this is so, then the product of the values of the two quantities involved remains the same, i.e, the product of the values of the two quantities involved is constant. We can use mathematical language to express this view of a relationship of type *inverse proportion* as

$$yx = k, \quad k \in \mathbb{R}, k \neq 0$$

Note that in this formulation the value of the expression on the left side of the equation is constant. In a given problem, this value, i.e., yx, never increases or decreases and is, therefore, in some sense *conserved*. For this reason we refer to the formulation $yx = k$ as the conservation form of the equation $y = \frac{k}{x}$.

The conservation form of the model, i.e., $xy = k$, can be shown to be the same as the formal form of the model, i.e., $y = \frac{k}{x}$, through conversion of division by x on the right side of the equation $y = \frac{k}{x}$ to multiplication by x on the its left, i.e.,

$$y = \frac{k}{x}$$
$$yx = k$$

Symmetry

Relationships of type inverse proportion are symmetric: If y is inversely proportional to x, then x is inversely proportional to y.[39] This can be seen through a rearrangement of the equation $y = \frac{k}{x}$ as shown below:

$$y = \frac{k}{x}$$
$$x = \frac{k}{y}$$

Note that y is being divided by on the right side of the equation. Furthermore, k is a nonzero number in $\mathbb{R}$. Therefore, x is inversely proportional to y.

---

[39]This means that, if it is the case that scaling x by a certain factor forces a corresponding scaling of y by the inverse of the same factor, then scaling y by some factor forces a corresponding scaling of x by the inverse of the same factor. Therefore, whichever variable is labelled as independent, the relationship between the two variables will be of type *inverse proportion*.

Change in Values of Variables

The fact that, in a relationship of type *inverse proportion*, scaling the value of one quantity by a certain factor forces a scaling of the value of the other quantity by the inverse of the same factor implies that change in y is directly proportional to change in $\frac{1}{x}$. In mathematical notation, we write this as

$$\Delta y = k\Delta\frac{1}{x}$$

We can show this to be the case as follows. By definition

$$\Delta y = y_2 - y_1$$

where y_1 represents the initial value of y and y_2 represents the final value of y. Since $y = \frac{k}{x}$, there is an x value, x_1, that corresponds to y_1 through $y_1 = \frac{k}{x_1}$ and there is an x value, x_2, that corresponds to y_2 through $y_2 = \frac{k}{x_2}$. Substituting $\frac{k}{x_1}$ for y_1 and $\frac{k}{x_2}$ for y_2 in the equation above yields

$$\Delta y = \frac{k}{x_2} - \frac{k}{x_1}$$

Using the distributive property in reverse on the right side yields

$$\Delta y = k\left(\frac{1}{x_2} - \frac{1}{x_1}\right)$$
$$\Delta y = k\Delta\frac{1}{x}$$

This proves that for a relationship of type inverse proportion, i.e., $y = \frac{k}{x}$, $k \in \mathbb{R}$, $k \neq 0$, $\Delta y = k\Delta\frac{1}{x}$.[40]

---

[40]There is no simple relationship between Δy and Δx. Expanding $\Delta\frac{k}{x}$ as shown below

$$\Delta y = \Delta\frac{k}{x}$$
$$\Delta y = \frac{k}{x_2} - \frac{k}{x_1}$$
$$\Delta y = \frac{kx_1 - kx_2}{x_1x_2}$$
$$\Delta y = \frac{-k(x_2 - x_1)}{x_1x_2}$$
$$\Delta y = -\frac{k}{x_1x_2}\Delta x$$

shows that change in y, i.e., Δy, does not depend just on change in x, i.e., Δx, but also the actual values of x_1 and x_2 both of which can vary.

Graph

The graph of an equation of type inverse proportion consists of two curves with a discontinuity at $x = 0$.[41] There are no intercepts as the expressions that correspond to $x = 0$ or $y = 0$ are both undefined. As an example, for the y-intercept we have $x = 0$ which yields

$$y = \frac{k}{x}$$
$$y = \frac{k}{0}$$

Since $k \neq 0$, the expression $\frac{k}{x}$ is undefined[42] showing that y-intercept is undefined, i.e., there is no y-intercept.

We can plot the graph of $y = \frac{k}{x}$ by plotting points. The graph of $y = \frac{k}{x}$, $k \in \mathbb{R}$, $k \neq 0$ is shown in Figure 12.2.13 below.

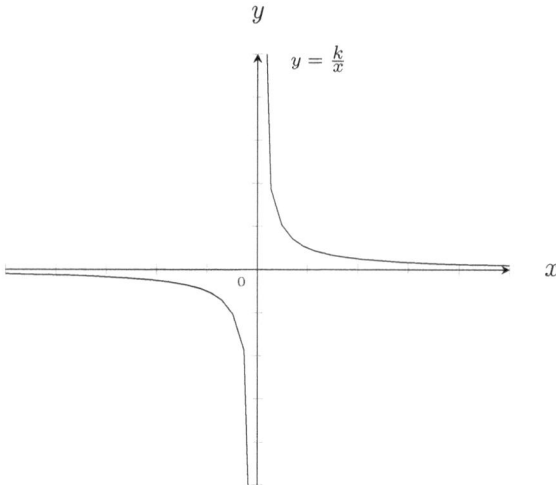

Figure 12.2.13: Graph of the equation for relationships of type inverse proportion

Some of the more important features of the graph of $y = \frac{k}{x}$ are shown in

---

[41]For a detailed discussion of fundamental concepts that relate to graphs please see Appendix O.

[42]If $k = 0$, for $x \neq 0$ we will have $y = 0$. This is the equation for the x-axis without the origin and for $x = 0$ we will have $y = \frac{0}{0}$ which is indeterminate. This means that y can equal to any number which maps onto the y-axis. Therefore, the overall graph is the union of the two axes which means that any point on the x-axis is an x-intercept and any point on the y-axis is a y-intercept.

Figure 12.2.14 below.[43]

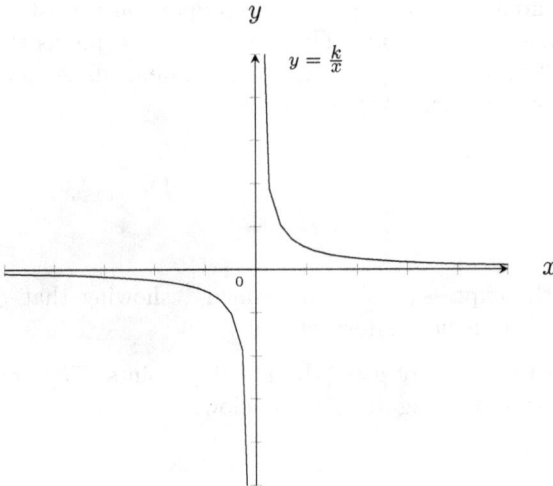

Figure 12.2.14: Features of the graph of the equation for relationships of type inverse proportion

In $y = \frac{k}{x}$, $k \in \mathbb{R}$, $k \neq 0$, the effect of k on the graph of the equation is to move the two branches of the graph toward and away from the origin. The higher the size of k, the farther away the graph of the equation is from the origin. Figure 12.2.15 shows the graphs of three equations with positive values of k: $\frac{1}{2}$, 1 and 2, with corresponding equations $y = \frac{1}{2x}$, $y = \frac{1}{x}$, and $y = \frac{2}{x}$. Note that as k increases, the graph moves away from the origin. Furthermore, when k is 0 the graph maps onto the two axes.

---

[43]The most important facts to keep in mind about a relationship of type *inverse propor-tion* are its form, i.e., $y = \frac{k}{x}$, the fact that change in y must equal to k times change in $\frac{1}{x}$, i.e., $\Delta y = k\Delta\frac{1}{x}$, that there are no intercepts, and that its graph is a double curve with a discontinuity at $x = 0$.

The discussion for negative values of k is left as an exercise.

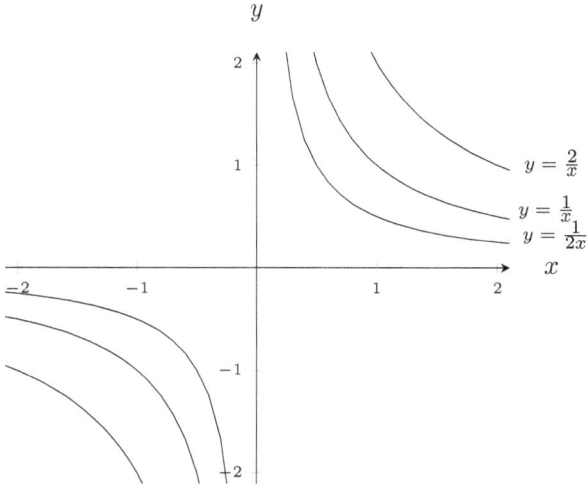

Figure 12.2.15: Effect of positive values of k in $y = \frac{k}{x}$, $k \in \mathbb{R}$, on the graph of the equation for relationships of type *inverse proportion*

Applications

We now present an abstract and a concrete example of a problem of type *inverse proportion*.

Example

Given the equation

$$y = \frac{3}{x}$$

1. Name the type of the relationship between x and y and support that claim.
2. Write the equation in conservation form.
3. State the relationship between change in y and change in $\frac{1}{x}$.
4. Find the x- and y-intercepts of the line.
5. Graph the line.

Solution

1. The relationship between x and y is inverse proportion. The equation $y = \frac{3}{x}$ is similar in form to $y = \frac{k}{x}$, $k \in \mathbb{R}$, $k \neq 0$, with $k = 3$. Since we have shown that in an equation of the form $y = \frac{k}{x}$, $k \in \mathbb{R}$, $k \neq 0$, the variables x and y are related through inverse proportion, we can state the same claim for the particular case $y = \frac{3}{x}$.

2. The conservation form of the equation $y = \frac{3}{x}$ is $xy = 3$. This tells us that the product of x and y is always 3, i.e., y is always 3 times the size of the inverse of x.

3. $\Delta y = 3\Delta\frac{1}{x}$

4. Since $k \neq 0$ there are no x- or y-intercepts as substitution of either x or y by 0 results in the undefined expression $\frac{3}{0}$. As an example, if we set $x = 0$ in $y = \frac{3}{x}$ we would arrive at $y = \frac{3}{0}$. Since $\frac{3}{0}$ is undefined then so is y. This implies that there is no y for $x = 0$, i.e., there is no y-intercept. We will leave it to the reader to show that the equation $y = \frac{3}{x}$ does not have an x-intercept either.

5. Since the graph of the equation $y = \frac{3}{x}$ is a curve, we will need many points to graph the curve.[44] Setting x equal to $\frac{1}{4}, \frac{1}{2}, 1, 2$, and 4 on the positive side and $-\frac{1}{4}, -\frac{1}{2}, -1, -2$ and -4 on the negative side generates the points $\left(\frac{1}{4}, 12\right)$, $\left(\frac{1}{2}, 6\right)$, $(1, 3)$, $\left(2, \frac{3}{2}\right)$, and $\left(4, \frac{3}{4}\right)$ on the positive side and $\left(-\frac{1}{4}, -12\right)$, $\left(-\frac{1}{2}, -6\right)$, $(-1, -3)$, $\left(-2, -\frac{3}{2}\right)$, and $\left(-4, -\frac{3}{4}\right)$ on the negative side.[45] The graph of the equation $y = \frac{3}{x}$ is shown in Figure 12.2.16 below.

We now present a concrete example of a problem of type *inverse proportion*.[46]

Example

The speed of travel and travel time for a trip from Toronto to Montreal are related through the equation

v travel speed (km/h)
t travel time (h)

$$v = \frac{504}{t}$$

---

[44]This is where an a priori knowledge of the general shape of the graph can be useful as it allows one to use fewer points to graph the curve.

[45]We ask the reader to work out the y values that correspond to each selected x value.

[46]For more concrete examples of problems of type *inverse proportion* please see the companion textbook *Semantics and the Syntax of Algebra* by the author.

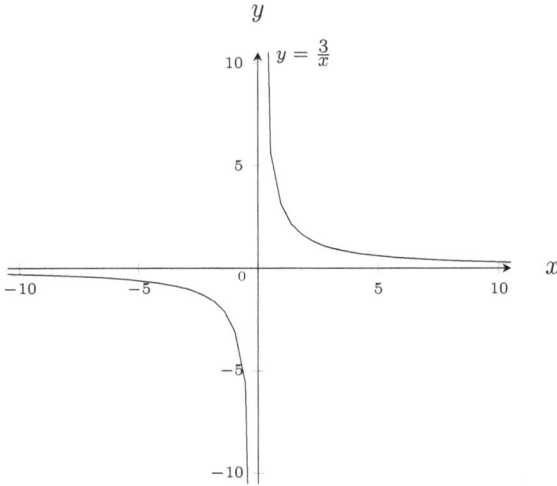

Figure 12.2.16: Graph of the equation $y = \frac{3}{x}$

where the constant 504 represents the distance from Toronto to Montreal in kilometres.

What information do you extract from this equation?

Solution

From the form of the equation we can conclude that travel speed, v, is inversely proportional to travel time, t. The equation $v = \frac{504}{t}$ is similar in form to $y = \frac{k}{x}$, $k \in \mathbb{R}$, $k \neq 0$ with $k = 504$, a constant that represents the distance from Toronto to Montreal. Since we have shown that in an equation of the form $y = \frac{k}{x}$, $k \in \mathbb{R}$, $k \neq 0$ the variables x and y are related through inverse proportion, we can state the same claim for the variables v and t in the particular case $v = \frac{504}{t}$. This implies that, for a fixed distance, driving twice faster results in half the travel time or that driving at half the speed would double travel time and the like.

The conservation form of the equation $v = \frac{504}{t}$ is $vt = 504$. This tells us that the product of speed, v, and travel time, t is always 504 km, i.e., the distance from Toronto to Montreal.

Since v is inversely proportional to t, $\Delta v = 504\Delta\frac{1}{t}$.

Exercise Set 12.2.1.4

1. What does it mean when we say that *the relationship between two quantities is inverse in the sense of proportion?*

2. What does it mean when we say that *the relationship between two quantities is inverse proportion?*

3. What is the formal model for a relationship of type inverse proportion?

4. What is the conservation model for a relationship of type inverse proportion?

5. What does it mean when we say that *relationships of type inverse proportion are symmetric?*

6. How does the change in the value of one quantity relate to change in the value of another quantity if the two quantities are related through inverse proportion?

7. What is the shape of the graph of the model for a relationship of type inverse proportion?

8. What is the effect of k in $y = \frac{k}{x}$, $k \in \mathbb{R}$, $k < 0$, on the graph of the equation? Illustrate using an example.

9. Which of the following equations are models of relationships of type inverse proportion?

a. $y = \frac{2}{x}$

b. $y = \frac{-6}{x}$

c. $y = \frac{1}{x}$

d. $y = \frac{2}{5x}$

e. $y = -\frac{1}{3x}$

f. $yx = 7$

g. $yx = -\frac{3}{4}$

h. $yx = -3$

i. $\frac{y}{x} = \frac{1}{5}$

j. $\frac{x}{y} = -1$

k. $\frac{x}{y} = 4$

l. $y = \frac{8}{x}$

m. $y = -\frac{2}{x}$

n. $y = 3x$

o. $y = -x$

p. $x = \frac{-2}{y}$

q. $x = \frac{1}{6}y$

r. $-2x = 5y$

s. $3y = 6x$

t. $y = -8 + \frac{5}{x}$

u. $3x - 1 = y$

v. $-2x - \frac{3}{y} = 0$

w. $7x + 2y = 4$

10. Each of the following is a model for an inverse proportion relationship between y and x. In each case rewrite the equation in formal and conservation forms.

a. $y = \frac{7}{x}$

b. $y = \frac{-2}{x}$

c. $y = \frac{1}{x}$

d. $x = \frac{3}{y}$

e. $y = \frac{2}{5x}$

f. $yx = 4$

g. $yx = -1$

h. $4y - \frac{1}{x} = 0$

i. $\frac{1}{y} - x = 0$

j. $x - \frac{1}{y} = 0$

k. $xy = \sqrt{2}$

l. $x = -\frac{1}{y}$

11. Rewrite each of the equations in Question 10 above in formal form, find the x- and y-intercepts of the equation and graph the equation.

12.2.1.5   Advanced Relationships

The four basic relationships that we have described so far are those whose formal representations involve addition, subtraction, multiplication and division of a constant, k, to, from or by one unknown, x, to generate a value for the other unknown, y. The two quantities may of course be related through other operations performed on x. As an example, the relationship between x and y may be such that, to obtain y, we may have to raise x to the power of k or raise k to the power of x, etc. In trigonometry we may relate y to $\sin x$, $\cos x$, $\tan x$, etc. and there are many other functions that arise from the study of relationships between the values of two quantities.

Further to the above, for every operation introduced there is an inverse operation.[47] As an example, the inverse operation for raising to a power is to take the corresponding root.

We will not delve deeply into a study of the more advanced operations in this textbook but, to provide the flavour, we will provide a brief catalogue of operations that involve exponents.

---

[47]We have seen this already in our coverage of the basic four arithmetic operations with addition and subtraction as well as multiplication and division being inverse operations.

The Direct Power Relationship

The formal form of the model is

$$y = x^k, \qquad k \in \mathbb{N}, k > 1$$

The conservation form of the model is

$$\log_x y = k, \qquad k \in \mathbb{N}, k > 1$$

The x- and y-intercepts are both $(0,0)$.

The graph of an equation for a direct power relationship looks differently depending on whether k is even or odd. The graphs of equations of type *direct power* for even and odd values of k are given below. The larger the value of k, the steeper the curve.

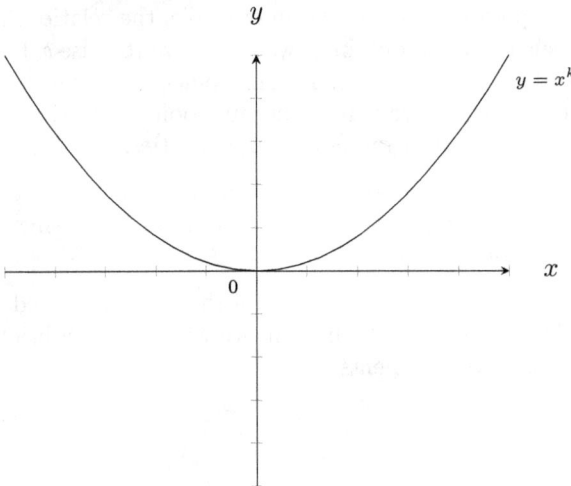

Figure 12.2.17: Graph of the equation for relationships of type direct power for even values of the exponent

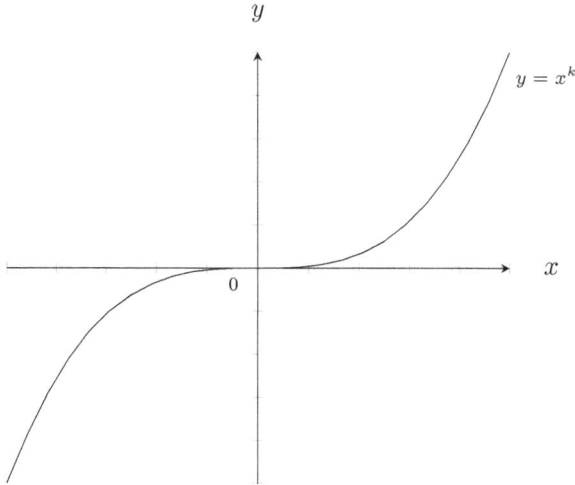

Figure 12.2.18: Graph of the equation for relationships of type direct power for odd values of the exponent

The Inverse Power Relationship

The formal form of the model is[48]

$$y = \sqrt[k]{x}, \qquad k \in \mathbb{N}, k > 1$$

The conservation form of the model is

$$\log_x y = \frac{1}{k}, \qquad k \in \mathbb{N}, k > 1$$

The x- and y-intercepts are both $(0, 0)$.

The graph of an equation for an inverse power relationship looks differently depending on whether k is even or odd. The graph of an equation for a direct power relationship looks differently depending on whether k is even or odd. The graphs of equations of type *inverse power* for even and odd values of k are given below. The larger the value of k, the flatter the curve.

---

[48] An alternative formulation is $y = x^{\frac{1}{k}}$.

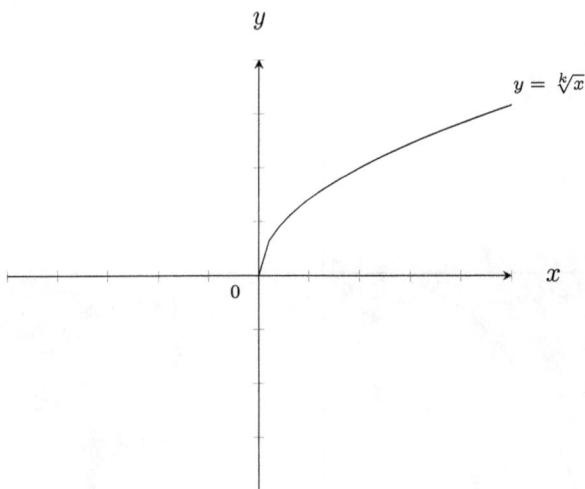

Figure 12.2.19: Graph of the equation for relationships of type inverse power for even values of the constant

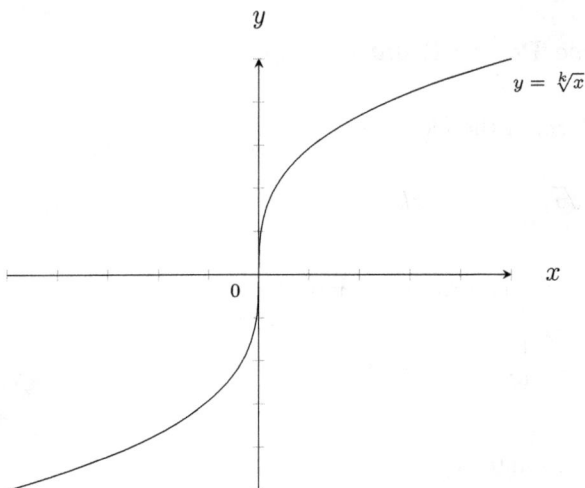

Figure 12.2.20: Graph of the equation for relationships of type inverse power for odd values of the constant

Direct Exponential Relationship

The formal form of the model is

$$y = k^x, \qquad k \in \mathbb{R},\, k > 0$$

The conservation form of the model is

$$\frac{\log y}{x} = \log k, \qquad k \in \mathbb{R},\, k > 0$$

There is no x-intercept. The y-intercept is $(0, 1)$.

The graph of an equation for a direct exponential relationship looks like the following.

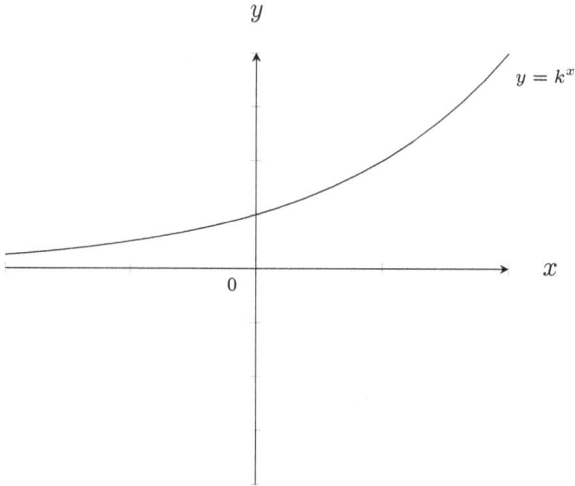

Figure 12.2.21: Graph of the equation for relation-ships of type direct exponential

The larger the value of k, the steeper the curve.

The Inverse Exponential Relationship

The formal form of the model is

$$y = \log_k x, \qquad k \in \mathbb{R},\, k > 0$$

The conservation form of the model is

$$\frac{\log x}{y} = \log k, \qquad k \in \mathbb{R}, \ k > 0$$

The x-intercept is $(1, 0)$. There is no y-intercept.

The graph of an equation for an inverse power relationship looks like the following:

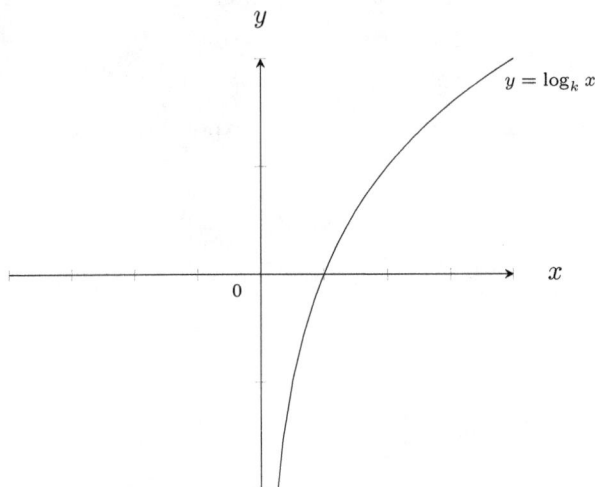

Figure 12.2.22: Graph of the equation for relationships of type inverse exponential

The larger the value of k, the flatter the curve.

Exercise Set 12.2.1.5

1. Write the formal and conservation models of the equations for relationships of type

   a. direct power relationship
   b. inverse power relationship
   c. direct exponential relationship
   d. inverse exponential relationship

2. Classify each of the following equations as models for relationships of type direct power, inverse power, direct exponential or inverse exponential.

a. $y = 3^x$ c. $y = \sqrt[5]{x}$

b. $y = x^3$ d. $y = \log_4 x$

3. For each part in Question 2 find the x- and y-intercepts and graph the equation.

12.2.2   Multiple-Operations Forms

Multiple-operations forms are forms in which multiple operations apply to syntactic entities that contain x. As an example of such forms, consider the equation

$$y = 2 + 3x$$

The operations that involve x in this equation are $\times$, the operation that is directly applied to x, and $+$, which is applied to the term $3x$ which contains x.

Multiple-operation equations come in a variety of forms. One can limit oneself to the operations of addition, subtraction and multiplication to generate such equations as $y = -1 + 2x$, equations that we refer to as linear equations. Limiting oneself to the operations of addition, subtraction, multiplication and division generates such equations as $y = \frac{2x-1}{3x+2}$, equations that we refer to as rational equations. We can limit ourselves to the operations addition, subtraction, multiplication and exponentiation to generate equations such as $y = x^2 - 3x - 4$, equations that we refer to as polynomials.

Our coverage of multiple-operations forms will focus on linear equations.

12.2.2.1   Linear Equations

Linear equations are equations whose graphs are straight lines. The formal form of such equations is

$$y = mx + b, \qquad m, b \in \mathbb{R}$$

where m and b are constants.[49]

---

[49]The form given above is sufficient for our purposes but the reader should note that the formal form given above is not able to represent lines whose graphs appear as vertical lines. The most general form of a linear equation is

$$Ax + By + C = 0, \qquad A, B, C \in \mathbb{R}$$

Depending on any restrictions, this equations can be rearranged to many different forms. As an example, for nonvertical lines we can rearrange the equation above to $y = mx + b$. For equations that have distinct x- and y-intercepts the equation can be rearranged to $\frac{x}{a} + \frac{y}{b} = 1$ with $(a, 0)$ being the x-intercept and $(0, b)$ being the y-intercept and there are others.

Examples of linear equations are $y = 3x + 7$, $y = -2x + 1$, $y = 2x$ (which can be seen as $y = 2x + 0)^{50}$, $y = \frac{1}{2}x - 2$ and the like.

Linear equations are symmetric. If y is a linear function of x, then the two can be related through $y = mx + b$. Solving this equation for x yields

$$y = mx + b$$
$$mx + b = y$$
$$mx = y - b$$
$$x = \frac{1}{m}(y - b)$$
$$x = \frac{1}{m}y - \frac{b}{m}$$
$$x = \frac{1}{m}y + \left(-\frac{b}{m}\right)$$

If m is a nonzero number in $\mathbb{R}$, then so is $\frac{1}{m}$ and since m is a nonzero number in $\mathbb{R}$ and b is in $\mathbb{R}$, so is $-\frac{b}{m}$. Therefore, x is a linear function of y.

If m is 0, then the inverse becomes undefined so that the graph of the inverse is a vertical line – still linear in form.

Changes in y are directly proportional to changes in x with m as the constant of proportionality, i.e.,

$$\Delta y = m\Delta x$$

We can show this to be the case as follows:

$$\Delta y = y_2 - y_1$$
$$\Delta y = mx_2 + b - (mx_1 + b)$$
$$\Delta y = mx_2 + b - mx_1 - b$$
$$\Delta y = mx_2 - mx_1$$
$$\Delta y = m(x_2 - x_1)$$
$$\Delta y = m\Delta x$$

The graph of a linear equation is a straight line[51]. Since a straight line is determined by two points, we can graph the line by finding two points on the line.

As an example, consider the linear equation

$$y = 2x + 5$$

---

[50] Note that the equation $y = 2x$ represents a relationship of type direct proportion between y and x. The single-operation forms *direct superposition*, *inverse superposition* and *direct proportion* are all special cases of linear relationships.

[51] Hence, the label.

We set x equal to -2 and 2 and for each, we find the corresponding y using the equation $y = 2x + 5$. This yields the points $(-2, 1)$ and $(2, 9)$. The graph of the line is shown in Figure 12.2.23 below.

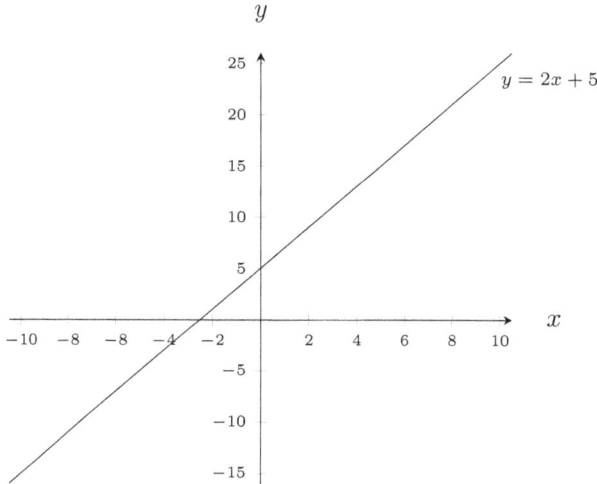

Figure 12.2.23: Graph of the equation $y = 2x + 5$

Exercise Set 12.2.2.1

1. Write the formal model for a linear relationship.

2. What does it mean when we say that *linear relationships are symmetric?*

3. How does the change in the value of one quantity relate to change in the value of another quantity if the relationship between the two quantities is linear?

4. What is the shape of the graph of the model for a linear relationship?

5. What is the slope of the line that represents the graph of a linear relationship?

6. Classify each of the following equations as linear or nonlinear.

   a. $y = 2x + 1$ d. $y = \frac{2}{5}x - \frac{1}{4}$ g. $y = 3x$

   b. $y = -6x - 5$ e. $y = -x + 3$ h. $y = -3x$

   c. $y = x + 3$ f. $y = x + 1$ i. $y = 4$

j. $y = x^2$

o. $y = -\dfrac{1}{x+1}$

t. $2y = -8 + 5x$

k. $y = \sqrt{x} + 1$

p. $x = -4y$

u. $3x - 1 = y$

l. $y = \dfrac{x}{10} + 3$

q. $x = \dfrac{1}{6}y$

m. $y = -\dfrac{x}{2}$

r. $-2x = 5y^2$

v. $y = \log x$

n. $y = \dfrac{3}{x}$

s. $y = x^2 + 3x + 2$

w. $x^2 + 2xy - y^2 = 0$

7. Each of the following is a model for a linear relationship between y and x. In each case rewrite the equation in formal form.

a. $y = 7x + 3$

b. $y = -2x - 1$

c. $y = x$

d. $x = 3y$

e. $y = \dfrac{3}{4}x$

f. $\dfrac{y}{x} = 4$

g. $\dfrac{y}{2x} = -1$

h. $4y - x = 0$

i. $x = 3y - 7$

j. $x + y = 0$

k. $-\dfrac{y}{x} = -6$

l. $x = -y$

m. $y = 2(x - 2)$

n. $3(y - 1) = -2(5 - x)$

o. $-(x - y) = 2 - x$

8. Rewrite each of the equations in Question 7 above in formal form, find the x- and y-intercepts and the slope of the line and graph the line.

12.2.2.2   Nonlinear Equations

Nonlinear equations are equations whose graphs are curves. Generally, equations in which operations beyond addition, subtraction and multiplication are used are nonlinear. Examples of nonlinear equations are $y = x^2 + x + 1$, $y = \dfrac{1}{x+1}$, $y = x^3$,[52] $y = \dfrac{1}{x} - 2$ and the like.

Nonlinear equations that involve multiple operations will not be discussed in this textbook.

---

[52] Note that the equation $y = x^3$ represents a relationship of type direct power between y and x. Single-operation forms beyond *direct superposition*, *inverse superposition* and *direct proportion* are all special cases of nonlinear relationships.

Chapter 13
Formulas

13.1   Genesis

Formulas are statements about relationships between the values of the quantities that appear in their expressions. They arise as a result of definitions or experimentations.

Formulas from Definitions

Defined formulas are formulas that arise as a result of defined relationships between the values of quantities involved in that formula.

As an example of a defined formula, consider the definition of the unit *kilometre*. By definition, 1 km is equal to 1000 m. This relationship is not subject to experimental verification as it is defined, i.e., we all agree that we can refer to a length of exactly 1000 m as 1 km. Numerical values that arise as a result of definitions are seen as being exact.

With the defined relationship between 1 km and 1000 m we can generate a formula that converts the unit of distance from km to m. This formula is given below.

$$d_{km} \qquad \text{distance (km)}$$
$$d_{m} \qquad \text{distance (m)}$$
$$d_{m} = 1000 d_{km}$$

Note that defined numerical values appear as actual numbers in defined formulas.

Defined formulas arise in the study of sciences as well. Such formulas arise from the definition of certain quantities in terms of those of other quantities. As an example, in physics we define speed as the rate of distance covered per unit time. This definition immediately leads to the formula

v speed (km/h)
d distance covered (km)
t travel time (h)

$$v = \frac{d}{t}$$

Formulas from Experimental Data

Formulas may also arise from the manner in which experimental data show the relationship between the values of various quantities involved in the formula. As an example of this, experimental data show that the pressure of helium in a balloon, in pascals, is directly proportional to its temperature in kelvins and its amount in moles, and is inversely proportional to its volume in cubic metres. Such relationships can be expressed as a single formula:

p pressure (Pa)
V volume (m^3)
n amount (mol)
T temperature (K)
R universal gas constant (J/(mol·K))

$$pV = nRT$$

As a second example, consider the manner in which a radioisotope decays. Experimental data show that after a fixed length of time[1], called *half-life*, half the mass of the radioisotope decays. This means that, after 1 half-life we will have $\frac{1}{2} \times m_0$ (read *half of m_0*) left, that after 2 half-lives we will have $\frac{1}{2} \times \frac{1}{2} \times m_0$ (read *half of half of m_0*) left, and so on. This behaviour leads to the following formula for the calculation of mass of radioisotope left after n half-lives.

m_0 initial mass of radioisotope (g)
m mass of radioisotope left (g)
n number of half-lives passed (1)

$$m = \frac{1}{2^n} m_0$$

The exponent on 2 keeps track of the number of times the factor $\frac{1}{2}$ is repeated.

---

[1] The particular value of this time is dependent on the type of the radioisotope and can range from a fraction of a second to thousands of years.

13.2   Relating the Values of Quantities in Formulas

The discussion that follows is limited to the case where the relevant quantity symbols appear only once throughout the formula. This includes formulas in which the quantity symbols, although appearing multiple times, can be factored. For the most part we will limit ourselves to linear relationships.[2] Nonlinear relationships are limited to those that involve division, exponents, roots and logarithms.

We will begin with single-operation relationships before we discuss multiple-operations relationships.

13.2.1   Single-Operation Relationships

13.2.1.1   Relationships of Type Superposition

Superposition describes the relationship between the values of quantities that are represented by the terms in a formula. In other words, in a formula, any term is superpositional to any other term.

In addition, if the terms appear on opposite sides of the formula, as in formal formulations given earlier for superposition relationships, then the relationship is of type direct superposition provided the terms are both being added or both being subtracted otherwise the relationship is of type inverse superposition.

If the terms appear on the same side of the formula, as in conservation formulations given earlier for superposition relationships, then the relationship is of type direct superposition provided one term is added while the other is subtracted otherwise the relationship is of type inverse superposition.

We will illustrate the points above using examples.

Problem

Consider the formula

$$d = d_0 + vt$$

This formula describes the distance between an object and an observer for motion along a straight line away or toward the observer. In this formula, d_0 represents the initial distance between the object and the observer, v represents the speed of the object

---

[2]This includes superposition, direct proportion and a combination of these.

relative to the observer, t represents the duration of motion, and d represents the final distance between the object and the observer. The formula states that the final distance between the object and the observer, d, is equal to the sum of the initial distance between the two, d_0, and the distance that the object covers due to its state of motion, vt. This latter quantity can be calculated by multiplying the speed of the object and travel time.[3]

1. The formula above analyzes into three terms. These are d, d_0 and vt. Describe the relationship between the various terms in this formula.

2. Suppose d_0 increases by 40 units. How should vt change to keep d the same?

Solution

1. The relationship between d and d_0 is direct superposition. This is because the terms appear on opposite sides of the formula and are being added.[4]

   The relationship between d and vt is also direct superposition. This is because the terms appear on opposite sides of the formula and are being added.

   The relationship between d_0 and vt is inverse superposition. This is because the terms appear on the same side of the formula and they are both being added.

2. To keep d the same, we can decrease vt by 40 units to compensate for the increase of 40 units in d_0. Since the relationship between d and d_0 as well as d and vt is direct superposition, an increase in d_0 by 40 units tends to increase d by 40 units and a decrease in vt by 40 units tends to decrease d by 40 units. This result in a net change of 0 in d.

---

[3] As an example, if the object is moving at a speed of 120 km/h for 2 h, it will cover a distance of 120×2 or 240 km.

For more on the semantics behind formulas please see the companion textbook *Semantics and the Syntax of Algebra* by the author.

[4] As with equations, we view the expressions on the left and right sides a formula that involve multiple terms as starting with a leading 0+. As an example, we see the formula

$$d = d_0 + vt$$

as

$$0 + d = 0 + d_0 + vt$$

This shows that the terms d and d_0 are being added as the operation on their left side is a +.

Problem

By definition

$$\Delta x = x_2 - x_1$$

where Δx represents the change in the value of x, and x_1 and x_2 represent the initial and final values of x respectively.

1. The formula above analyzes into three terms. These are Δx, x_2 and x_1. Describe the relationship between the various terms in this formula.

2. Suppose x_2 increases by 6 units. How should x_1 change to keep Δx the same?

3. Suppose x_1 increases by 8 units. How should x_2 change so that Δx increases by 3 units?

4. Suppose Δx decreases by 2 units. How should x_2 change so that x_1 decreases by 5 units?

5. Suppose x_1 decreases by 10 units. How should Δx change so that x_2 remains the same?

Solution

1. The relationship between Δx and x_2 is direct superposition. This is because the terms appear on opposite sides of the equation and are being added.
The relationship between Δx and x_1 is inverse superposition. This is because the terms appear on opposite sides of the formula with one being added and the other being subtracted.
The relationship between x_1 and x_2 is direct superposition. This is because the terms appear on the same side of the formula with one being added and the other being subtracted.

2. Since the relationship between x_2 and Δx is direct superposition, an increase of 6 units in x_2 would tend to increase Δx by 6 units as well. We can compensate for this gain by increasing the value of x_1 by 6 units. Since x_1 is inversely superpositional to Δx, an increase of 6 units in x_1 forces a decrease of 6 units in Δx. This cancels the increase in Δx due to the increase in x_2 and keeps Δx the same.

3. Since the relationship between Δx and x_1 is inverse superposition, an increase of 8 units in x_1 tends to decrease Δx by 8 units. To increase Δx by 3 units, we need to increase x_2 by 11 units. Since the relationship between x_2 and Δx is direct superposition, an increase of 11 units in

x_2 forces an increase of 11 units in Δx. The net effect of increasing x_1 by 8 units and increasing x_2 by 11 units is to increase Δx by 3 units.

4. Since the relationship between Δx and x_1 is inverse superposition, a decrease of 2 units in Δx tends to increase x_1 by 2 units. To decrease x_1 by 5 units, we need to decrease x_2 by 7 units. Since the relationship between x_2 and x_1 is direct superposition, a decrease of 7 units in x_2 forces a decrease of 7 units in x_1. The net effect of increasing x_1 by 2 units and decreasing it by 7 units is to decrease it by 5 units.

5. Since the relationship between x_1 and x_2 is direct superposition, a decrease of 10 units in x_1 tends to decrease x_2 by 10 units. To keep x_2 the same, we need to increase Δx by 10 units. Since the relationship between x_2 and Δx is direct superposition, an increase of 10 units in Δx forces an increase of 10 units in x_2. The net effect of decreasing x_2 by 10 units and increasing x_2 by 10 units is to leave its value unchanged.

Exercise Set 13.2.1.1

1. Consider the formula for the mechanical energy of an object moving within a gravitational field.

$$E_m = E_k + E_p$$

In this formula E_m represents the object's mechanical energy, E_k represents the object's kinetic energy, and E_p represents the object's potential energy with all energies measured in joules.
It is known that movement of an object within a gravitational field is conserved. This means that, for such motion, E_m is constant. If so, what happens to the potential energy of an object moving within a gravitational field if its kinetic energy increases by 12 000 J?

2. A garden has the shape of a square with a side length of s and 4 adjacent identical circular sections of radius r. The area of this garden is given by the formula

$$A = s^2 + 4\pi r^2$$

where A represents the area of the garden.

a. What happens to the area of the garden if the area of the square decreases by 0.75 m^2 and the area of the circular sections increases by 0.5 m^2?

b. The area of the square increases by 4.2 m^2. How would you change the area of the circular sections so that, in the balance, the area of the garden increases by 3.5 m^2?

13.2.1.2   Relationships of Type Proportion

A proportion equation is an equation in which the expressions on the left and right side of the equation analyze into one term each. In addition, if the main operation on either side is division, then the dividend and divisor analyze into one term each. The following are examples of proportion formulas.

$$pV = nRT$$
$$t = (1 + r_t)(1 - r_d)p$$
$$v = \frac{d}{t}$$
$$\frac{N_1}{V_1} = \frac{N_2}{V_2}$$

Proportion describes the relationship between the values of quantities that are represented by the factors in a proportion formula. In other words, in a proportion formula, any factor is proportional to any other factor.[5]

In addition, if the factors appear on opposite sides of the equation, as in formal formulations given earlier for proportion relationships, then the relationship is of type direct proportion provided the factors are both being multiplied by or both being divided by otherwise the relationship is of type inverse proportion.

If the factors appear on the same side of the equation, as in conservation formulations given earlier for proportion problems, then the relationship is of type direct proportion provided one factor is multiplied by while the other is divided by otherwise the relationship is of type inverse proportion.

We will illustrate the points above using examples.

Problem

Consider the formula

$$\frac{p_1}{n_1} = \frac{p_2}{n_2}$$

This formula describes the relationship between the pressure and amount of a gas inside a container of fixed volume and temperature.[6]

---

[5]By the word factor here we mean, not only factors in the official sense, but factors that appear as divideneds and divisors as well. As an example, in the formula $ab = \frac{cd}{ef}$, the list of factors in this sense includes a, b, c, d, e, and f.

[6]In this formula, p_1 and n_1 represent the initial values of pressure and amount and p_2

1. Explain why this formula is a proportion formula.

2. Relate the values of the various variables to each other.

3. Suppose n_1 doubles while n_2 remains the same. How would one change p_2 to keep p_1 the same?

4. Suppose n_1 doubles and p_2 triples. How would one change n_2 to change p_1 by a factor of 12?

Solution

1. The formula is a proportion formula because the expressions on the left and right sides of the formula analyze into one term each, i.e., $\frac{p_1}{n_1}$ and $\frac{p_2}{n_2}$.

2. The relationship between p_1 and n_1 is direct proportion. This is because the factors appear on the same side of the formula with one factor being multiplied by (i.e., p_1) and the other being divided by (i.e., n_1).[7]

   The relationship between p_1 and p_2 is also direct proportion. This is because the factors appear on opposite sides of the formula and are both being multiplied by.

   The relationship between p_1 and n_2 is inverse proportion. This is because the factors appear on opposite sides of the formula with one factor being multiplied by and the other being divided by.

   The relationship between n_1 and p_2 is inverse proportion. This is because the factors appear on opposite sides of the formula with one factor being divided by while the other is being multiplied by.

   The relationship between n_1 and n_2 is direct proportion. This is because the factors appear on opposite sides of the formula with both factors being divided by.

   The relationship between p_2 and n_2 is direct proportion. This is because the factors appear on the same side of the formula with one factor being multiplied by and the other being divided by.

---

and n_2 represent the final values of pressure and amount with pressure measured in pascals and amount measured in moles. For more on the semantics behind formulas please see the companion textbook *Semantics and the Syntax of Algebra* by the author.

[7] We can always view a proportion formula as being preceded by a $1\times$ so that we see

$$\frac{p_1}{n_1}$$

as

$$\frac{1 \times p_1}{n_1}$$

which shows that p_1 is being multiplied by.

3. Since the relationship between n_1 and p_1 is direct proportion, an increase in n_1 by a factor of 2 would tend to increase p_1 by a factor of 2 well. Since n_2 does not change, it has no effect on p_1. We can compensate for the doubling of p_1 by halving the value of p_2. Since p_2 is directly proportional to p_1, a halving of p_2 forces a halving of p_1. This cancels the doubling of p_1 due to the doubling of n_1 and keeps p_1 the same.

4. Since the relationship between n_1 and p_1 is direct proportion, an increase in n_1 by a factor of 2 tends to increase p_1 by a factor of 2 as well. Furthermore, since the relationship between p_2 and p_1 is direct proportion, an increase in p_2 by a factor of 3 forces an increase in p_1 by a factor of 3 as well.[8] To increase p_1 by a factor of 12, we need to decrease n_2 by a factor of $\frac{1}{2}$. Since the relationship between p_1 and n_2 is inverse proportion, a decrease in n_2 by a factor of $\frac{1}{2}$ forces an increase in p_1 by a factor of 2. The net effect of increasing p_1 by factor of 2, a factor of 3 and a factor of 2 is to increase p_1 by a factor of 12.

Problem

Experiments show that the acceleration of an object is related to the force acting on it through the formula

$$F = ma$$

where F represents the force acting on the object in N, m represents the mass of the object in kg, and a represents the object's acceleration in m/s^2.

1. Explain why this formula is a proportion formula.

2. Relate the values of the various variables to each other.

3. Suppose the mass of the object remains the same while the force acting on it doubles. How would this affect the acceleration of the object?

4. Suppose the acceleration of the object doubles while its mass decreases by a factor of $\frac{1}{4}$. What can be said about the change in the force acting on the object?

---

[8]The net effect of the changes in n_1 and p_2 is to increase p_1 by a factor of 6.

Solution

1. The expressions on the left and right side of the formula above analyze into one term each. Therefore, the formula is a proportion formula.

2. The relationship between F and m is direct proportion. This is because the factors appear on opposite sides of the formula and are both being multiplied by.
   The relationship between F and a is also direct proportion. This is because the factors appear on opposite sides of the formula and are both being multiplied by.
   The relationship between m and a is inverse proportion. This is because the factors appear on the same side of the formula and are both being multiplied by.

3. Since the relationship between F and a is direct proportion, an increase in F by a factor of 2 forces an increase in a by a factor of 2. Since m does not change, it will have no effect on the change in a.

4. Since the relationship between F and a is direct proportion, an increase in a by a factor of 2 forces an increase in F by a factor of 2 as well. Since the relationship between F and m is direct proportion, a decrease in m by a factor of $\frac{1}{4}$ forces a decrease in F by a factor of $\frac{1}{4}$. The combined effect of changes in a and m force a change in F by a factor of $2 \times \frac{1}{4}$ or $\frac{1}{2}$, i.e., F halves.

Exercise Set 13.2.1.2

1. Consider the formula for the area of a triangle.

$$A \;=\; \frac{1}{2}bh$$

where A represents the area of the triangle, b represents its base and h represents its height.

   a. Explain why this formula is a proportion formula.
   b. Relate the values of the various variables to each other.
   c. Suppose the base of a triangle doubles while its height remains the same. How would this affect the area of the triangle?
   d. Suppose the base of a triangle remains the same while its height halves. How would this affect the area of the triangle?
   e. Suppose the base of a triangle doubles while its height halves. How would this affect the area of the triangle?
   f. Suppose the base of a triangle doubles. How would you change its height so that the area of the triangle grows by a factor of 10?

2. The formula for the kinetic energy of an object of mass m in kg moving at a speed v in m/s is given by

$$E = \frac{1}{2}mv^2$$

where E represents the kinetic energy of the object in J.

a. Explain why this formula is a proportion formula.
b. Relate the values of the various variables to each other.
c. Suppose the mass of an object doubles while its speed remains the same. How would this affect the kinetic energy of the object?
d. Suppose the mass of an object remains the same while its speed triples. How would this affect the kinetic energy of the object?
e. Suppose the mass of an object doubles while its speed halves. How would this affect the kinetic energy of the object?
f. Suppose the kinetic energy of an object doubles. How would you change its speed so that m grows by a factor of 16?

3. The formula for the volume of a rectangular solid with length l, width w, and height h is given by the formula

$$V = lwh$$

where V represents the volume of the object.

a. Explain why this formula is a proportion formula.
b. Relate the values of the various variables to each other.
c. Suppose the length of a rectangular solid doubles while its width and height remain the same. How would this change its volume?
d. Suppose the length of a rectangular solid remains the same while its width doubles and its height triples. How would these changes affect the volume of the rectangular solid?
e. Suppose the width of a rectangular solid decreases by a factor of $\frac{1}{2}$ while its height increases by a factor of 3. How would you change the length so that the volume remains unchanged?

4. The following formula describes the behaviour of a gas inside a container.

$$pV = nRT$$

where p represents the pressure of the gas in Pa, V represents the volume of the container in m³, n represent the amount of gas in mol and T represents the temperature of the gas in K. R is a constant called the *universal gas constant* and its value is approximately 8.314 J/(mol·K).

a. Explain why this formula is a proportion formula.

b. Relate the values of the various variables to each other.

c. Suppose the temperature of the gas doubles but that there is no change in the amount of the gas or the volume of the container. How would this affect the pressure of the gas?

d. Through heating, the temperature of the gas triples. How would you change the volume of the container to keep the pressure the same? The amount of helium does not change.

e. The temperature of helium inside a steel cylinder drops by a factor of $\frac{1}{2}$. How would you change the amount of helium in the tank to force the pressure to double?

13.2.1.3 Advanced Relationships

Relationships of type superposition and proportion deal with scenarios where the variable of interest is being added to, subtracted from, multiplied by or divided by. There are, of course, other operations that a variable may be subject to. Examples of these are raising the variable to an exponent, taking the root of the variable, raising a value to the power of the variable, taking the logarithm of the variable and so on. In this section we will briefly discuss such relationships.

Power Relationships

y is said to be a power function of x if y is related to a power of x.

Example

The area of a square is given by the formula

$$A = s^2$$

where A represents the area of the square and s represents its side length.

In this formula A is a power function of s.

Root Relationships

y is said to be a root function of x if y is related to a root of x.

Example

The side length of a cube is given by the formula

$$s = \sqrt[3]{V}$$

where s represents the side length of the cube and V represents its volume.

In this formula s is a root function of V.

Exponential Relationships

y is said to be an exponential function of x if y is related to a value to the power of x.

Example

The following formula can be used to calculate the number of a type of bacteria whose number doubles every hour, starting with 1 bacteria.

$$n = 2^t$$

In this formula n represents the number of bacteria and t represents time passed in h.

In this formula n is an exponential function of t.

Logarithmic Relationships

y is said to be a logarithmic function of x if y is related to the logarithm of x.

Example

The following formula can be used to calculate the duration of investment for a given amount of interest earned on \$1 invested in an account that pays interest at a given annual rate, with the interest compounded annually.

$$n = \log_{1+r} a$$

In this formula, n represents the number of terms[9], r represents the rate of compound interest per term, and a represents the amount of money in the account after n terms.

In this formula n is a logarithmic function of a.

There are, of course, other types of single operation relationships between the values of two quantities. These other types of relationships are not covered in this textbook.

13.2.2   Multiple-Operations Relationships

The discussion on multiple-operations relationships is divided into a discussion on linear relationships and another on nonlinear relationships.

13.2.2.1   Linear Relationships

We have seen that, in a formula, the terms that are being added and subtracted are superpositional to each other. We have also shown that, in a proportion formula, the factors that are being multiplied by and divided by are proportional to each other.

A linear relationship is a mix of superposition and direct proportion. Linear relationships, therefore, have the form of

$$y = a \pm bx$$

where the $\pm$ notation states that either $+$ or $-$ is acceptable.

As an example of a linear relationship, consider the following problem.

Problem

The following formula can be used to calculate the cost of ordering a number of pens and notebooks.

$$c = c_p n_p + c_n n_n$$

where c represents the cost of the order, c_p represents the cost of a pen, n_p represents the number of pens ordered, c_n represents the cost of a notebook, and n_n represents the number of notebooks ordered.

---

[9]This relates to the duration of investment. In the problem above, this duration is measured in years.

In this formula, the three terms c, $c_p n_p$ and $c_n n_n$ are superpositional to each other. This makes sense as the terms represent the cost of the order, the cost of the pens, and the cost of the notebooks.

The relationship between c and $c_p n_p$ is direct superposition. This is expected: If the cost of the pens increases by, say, $120, then the cost of the order will also increase by $120 and if the cost of the pens decreases by, say, $34.50, then the cost of the order will also decrease by $34.50.

The relationship between c and $c_n n_n$ is also direct superposition. This is expected as well for the same reasons as those given for the relationship between c and $c_p n_p$ above.

The relationship between $c_p n_p$ and $c_n n_n$ is inverse superposition. For a fixed cost of order, an increase in $c_p n_p$ will have to be compensated for by a decrease in $c_n n_n$. As an example, if the cost of the order should remain the same, an increase in the cost of the pens by $210 would require a decrease in the cost of the notebooks by $210.

Note that in the statements above we have said that the cost of the order, c, is directly superpositional to the cost of the pens, $c_p n_p$. This latter quantity, i.e., the cost of the pens or $c_p n_p$, is different from c_p, which is the cost of a pen, and n_p, which is the number of pens ordered. The fact that c is directly superpositional to $c_p n_p$ does not imply that the same relationship exists between c and c_p, or c and n_p. Note that an increase in the number of pens by a certain amount does not raise the cost of the order by the same amount. As an example, if a pen costs $2, then increasing the number of pens by 4 does not increase the cost of the order by $4 but by $8. Nor is c directly proportional to c_p as doubling the cost of a pen doubles the cost of the pens, not the cost of the order.

The relationship between c and c_p as well as that between c and n_p are both linear. The same is true about the relationship between c and c_n as well as c and n_n. We can prove the latter by rewriting

$$c = c_p n_p + c_n n_n$$

as

$$c = a + b n_n$$

where the term $c_p n_p$ and the factor c_n have been turned into the constants a and b.[10] This maps onto the form of the linear formula, $y = a \pm bx$.

---

[10]Constants may represent any syntactic component of an expression that makes sense,

Problem

The formula for distance between an observer and an object moving along a straight line away or towards the observer at constant speed is given by the formula

$$d = d_0 + vt$$

where d represents distance between the observer and the object, d_0 represents the initial distance between the observer and the object, v represents the speed of the object, and t represents the duration of motion.[11]

In this formula the relationships between d and v as well as that between d and t are both linear. This can be shown to be the case for the relationship between d and v by writing the formula

$$d = d_0 + vt$$

as

$$d = a + bv$$

---

e.g., a term, the sum of two terms, a factor within a term, a combination of factors within a term, etc. However, constants may not replace a mix of syntactic elements of different kinds. As an example, a constant may not replace a term and a factor within another term. This explains why, in the formula $c = c_p n_p + c_n n_n$, we may replace $c_p n_p$ (a term) with a constant or replace c_n (a factor within a term) with another constant but that we may *not* replace $c_p n_p + c_n$ with a single constant as this would require combining unlike elements: A term $(c_p n_p)$ and a factor within another term (c_n).

When trying to work out your constants, begin by looking for a combination of one or more terms that may be turned into a constant. Following this, look within each term to see whether a combination of one or more factors may be turned into a constant. As an example, to relate the variables d and v in the formula

$$h = h_0 + vt + \frac{1}{2}gt^2$$

we begin by turning $h_0 + \frac{1}{2}gt^2$ into a constant. This leads to

$$h = a + vt$$

Next, we look within the term vt and in there we turn the factor t into a constant. This leads to

$$h = a + bv$$

which has the form of the linear equation $y = a \pm bx$.

[11] For some time now we have dropped the use of the phrase *value of* in such statements as *v represents the value of the speed of the object* and, instead, have used the shortened version *v represents the speed of the object*. This convention is quite common in scientific discourse and we will follow this convention from now on but we hope that the use of the full phrase for so long has trained the reader to understand that a quantity symbol refers to *the value of* a quantity, not the quantity itself.

The importance of linear relationships is in the fact that, in such relationships, changes in the values of the two quantities involved are directly proportional to each other.[12]

Exercise Set 13.2.2.1

1. Which of the variables on the right side of each formula below is linearly related to the variable on the left side of the formula?

   a. $d_0 = d - vt$

   b. $m = \frac{1}{2^n} m_0$

   c. $e = \frac{1}{2} v^2 + gh$

   d. $m_{CO_2} = m_C + 2m_O$

13.2.2.2   Nonlinear Relationships

Linear relationships include direct superposition, inverse superposition, direct proportion and a mix of these. Any other relationship is a nonlinear relationship. Examples of single-operation nonlinear relationships are inverse proportion, the power relationship, the root relationship, etc. Multiple-operations relationships may also be nonlinear. Examples of such relationships are the relationship between m and n in the formula for radioisotope decay, $m = \frac{1}{2^n} m_0$, the relationship between A and r_o in the formula for the surface area of a thin washer, $A = \pi \left(r_o - r_i \right)^2$, and the like.

We will not discuss such relationships further in this textbook.

13.3   Rearranging Formulas

The discussion that follows is limited to rearranging formulas where the variable being solved for appears once throughout the formula. This includes formulas in which the variable, although appearing multiple times, can be factored. Nonlinearity is limited to division, exponents, roots and logarithms.

We will begin with an algorithm for rearranging formulas for a variable where the variable appears only once throughout the formula.[13,14]

---

[12]Therefore, as long as changes in the values of the variables are concerned, the behaviour is similar to that of direct proportion relationship.

[13]This algorithm is identical to the one given earlier for solving equations involving a single unknown with the unknown appearing only once throughout the equation. See Chapter 11.

[14]Strictly speaking, only the last two steps are required. The steps prior to the last two

Algorithm for Solving for a Variable in a Formula with the Variable Appearing Once Throughout the Formula

1. Switch sides, if necessary, or otherwise bring the variable to the top, left side of the formula.[15]
2. Solve for the term that contains the variable.
3. Solve for the factor that contains the variable.
4. Repeat Steps 1, 2 and 3 until they cannot be applied anymore.
5. Solve for the argument of the main operation that contains the variable.
6. Repeat Steps 1, 2, 3, 4 and 5 until the problem is solved.

The steps in the algorithm above may be carried out using either the natural semantic scheme or the standard semantic scheme. As noted earlier, the natural semantic scheme is both more efficient and more in line with the manner in which we normally reason. We will, therefore, focus on the use of the natural semantic scheme in carrying out the steps in the algorithm above.

With formulas it is often the case that intermediate simplifications are not possible. This is due to the fact that, apart from exact numerical values, the values of quantities in formulas are represented as variables not numerical values. While one can always simplify the expression 2×3 and write 6 in its place, we cannot combine the product pV into a single variable.[16] For the most part, the process of rearrangement proceeds without intermediate simplifications.

The reader should also keep in mind that not all the steps in the algorithm above are always taken when rearranging a formula for a given variable: The task may end after Step 2 or it may be the case that Step 2 is not applicable in which case we move on to Step 3 and so on.

Here is an example where the rearrangement ends after Steps 1 and 2.

---

spell out the details when the main operation is addition, subtraction, multiplication or division.

[15]This is a common convention that we adhere to in this textbook.

[16]It is of course possible to define a quantity symbol, say b, to represent the product pV. However, such definitions are often difficult to remember and they mask the presence of p and V, the variables that we truly care about. This latter point is important and in many cases it is preferable not to simplify even numerical expressions in formulas as the nonsimplified forms are often more informative. As an example, the formula for the total length of the sides of 2 identical equilateral triangles is $l = 2 \times 3 \times s$ where l represents the total length of the sides of the triangles and s is the length of a side. Here we may wish to keep 2 and 3 as they are because they are reminders of the fact that we have 2 triangles and that each triangle has 3 sides. This is not immediately apparent from the simplified form $l = 6s$.

Example

Solve the following formula for v_0.

$$v = v_0 + at$$

Solution

Switch sides to bring the variable of interest, v_0, to the top, left side of the formula.

$$v_0 + at = v$$

The expression on the left side of the formula analyzes into two terms: v_0 and at. We solve for the term that contains the variable of interest, v_0, by converting addition of at on the left side to subtraction of at on the right side:

$$v_0 = v - at$$

The full solution without the intervening explanations is given below.

$$v = v_0 + at$$
$$v_0 + at = v$$
$$v_0 = v - at$$

Here is an example where the task of rearranging the formula ends after Steps 1, 2 and 3.

Example

Solve the following formula for t.

$$d = d_0 + vt$$

Solution

Following Step 1, we switch sides to bring the variable of interest, t, to the top, left side of the formula.

$$d_0 + vt = d$$

The expression on the left side of the formula above analyzes into two terms: d_0 and vt. Following Step 2, we solve for the term that contains the variable of interest, t, by converting addition of d_0 on the left side to subtraction of d_0 on the right side:

$$vt = d - d_0$$

The single-term expression on the left side of the formula analyzes into two factors: v and t. We solve for the factor that contains the unknown by converting multiplication by v on the left side to division by v on the right side:[17]

$$t = \frac{d - d_0}{v} \quad \text{or} \quad t = \frac{1}{v}(d - d_0)$$

The full solution without the intervening explanations is given below.

$$d = d_0 + vt$$
$$d_0 + vt = d$$
$$vt = d - d_0$$
$$t = \frac{d - d_0}{v} \quad \text{or} \quad t = \frac{1}{v}(d - d_0)$$

Here is a problem where Steps 1 and 2 are not applicable and the problem is solved after Step 3.

Example

Solve the following formula for m_1.

$$m_1 v_1 = m_2 v_2$$

Solution

Step 1 does not need to be taken as the variable of interest, m_1, is already on the top, left side of the formula.

Step 2 does not need to be taken as the term that contains the variable of interest, m_1, is already solved for as it is isolated on the left side of the formula.

---

[17]The two alternatives below have their roots in the equivalence between statements such as $\frac{x}{2}$ (x *divided by* 2) and $\frac{1}{2}x$ (*half of* x). Both forms are used frequently by those who are fluent in the art of rearranging formulas and the reader would do well to keep the equivalency between the two forms in mind.

We move on to Step 3. The term $m_1 v_1$ contains two factors: m_1 and v_1. We solve for the factor that contains the variable of interest, m_1, by turning multiplication by v_1 on the left to division by v_1 on the right.[18]

$$m_1 = \frac{v_2}{v_1} m_2$$

The full solution without the intervening explanations is given below.

$$m_1 v_1 = m_2 v_2$$

$$m_1 = \frac{v_2}{v_1} m_2$$

Here is a problem whose solution reaches Step 4.

Example

Solve

$$E - E_0 = P(t - t_0)$$

for t_0.

Solution

Following Step 1, we switch sides to bring the variable of interest, t_0, to the top, left side of the formula.

$$P(t - t_0) = E - E_0$$

Step 2 does not need to be taken as the term that contains the variable of interest t_0, is already solved for as it is isolated on the left side of the formula.

The term on the left side of the formula consists of two factors: P and $t - t_0$. Following Step 3, we solve for the factor that contains the variable of interest, t_0, by converting multiplication by P on the left side to division by P on the right side.

$$t - t_0 = \frac{1}{P}(E - E_0)$$

---

[18]Note that alternative expressions such as $\frac{m_2 v_2}{v_1}$, $\frac{1}{v_1} m_2 v_2$ and the like would also be acceptable. The one we have opted for is the one that is often semantically more relevant as it states that m_1 and m_2 are directly proportional to each other and are related through a factor of $\frac{v_2}{v_1}$.

We now return to Step 2 and solve for the term that contains the variable of interest, t_0.[19]

$$-t_0 = -t + \frac{1}{P}(E - E_0)$$

The single-term expression on the left side of this formula analyzes into two factors -1 and t_0. Following Step 3, we solve for the factor that contains the variable of interest, t_0.

$$t_0 = t - \frac{1}{P}(E - E_0)$$

The full solution without the intervening explanations is given below.

$$E - E_0 = P(t - t_0)$$
$$P(t - t_0) = E - E_0$$
$$t - t_0 = \frac{1}{P}(E - E_0)$$
$$-t_0 = -t + \frac{1}{P}(E - E_0)$$
$$t_0 = t - \frac{1}{P}(E - E_0)$$

Here is an example where Steps 2, 3 and 4 do not need to be performed and the problem ends after we take Step 5.

Example

The following formula relates the area of a square to its side length.

$$A = s^2$$

Solve this formula for s.

---

[19]With an eye on the next step, we choose an order for the terms t and $\frac{1}{P}(E - E_0)$ that is more semantically relevant. To explain the difference in semantics, consider the expressions

$$a - a_o$$

and

$$-a_0 + a$$

The two expressions are syntactically equivalent but, interpreting a as *amount of money I have* and a_o as *amount of money I owe*, then the first expression (i.e., the difference between what I have and what I owe) is more meaningful than the second expression (i.e., the negative of what I owe plus what I have).

Solution

The full solution is given below. We ask the reader to justify the steps.

$$A = s^2$$
$$s^2 = A$$
$$s = \pm\sqrt{A}$$
$$s = \sqrt{A}$$

Note that we have dropped the negative sign as a square's side length is nonnegative.

We present more examples involving operations beyond addition, subtraction, multiplication and division.

Example

Solve the formula

$$m = \frac{1}{2^n} m_0$$

for n.

Solution

The full solution is given below. We ask the reader to justify the steps.

$$m = \frac{1}{2^n} m_0$$
$$2^n = \frac{m_0}{m}$$
$$n = \log_2 \frac{m_0}{m}$$

Example

Solve the formula

$$d = d_0 + \frac{1}{2}at^2$$

for t.

Solution

The full solution is given below. We ask the reader to justify the steps.

$$d = d_0 + \frac{1}{2}at^2$$

$$d_0 + \frac{1}{2}at^2 = d$$

$$\frac{1}{2}at^2 = d - d_0$$

$$t^2 = \frac{2}{a}(d - d_0)$$

$$t = \pm\sqrt{\frac{2}{a}(d - d_0)}$$

Our last example involves a formula where the variable of interest, although appearing multiple times, can be factored.[20]

Example

Solve the formula

$$l = l_0 + \alpha(t - t_0)l_0$$

for l_0.

Solution

Following Step 1, we switch sides to bring the variable of interest, l_0, to the top, left side of the formula.

$$l_0 + \alpha(t - t_0)l_0 = l$$

The variable of interest, l_0, appears in multiple terms but as identical factors. We can, therefore, factor l_0 to get.[21]

$$\left[1 + \alpha(t - t_0)\right]l_0 = l$$

The single-term expression on the left side analyzes into two factors: $1 + \alpha(t - t_0)$ and l_0. Following Step 3, we solve for the factor

---

[20] **Factoring** converts a multi-term expression into a single-term expression by turning the multi-term expression into a product of factors.
[21] Note that this is the distributive property in reverse.

that contains the variable of interest, l_0, by converting multiplication by $1 + \alpha\,(t - t_0)$ on the left to division by $1 + \alpha\,(t - t_0)$ on the right.

$$l_0 = \frac{l}{1 + \alpha\,(t - t_0)}$$

The full solution without the intervening explanations is given below.

$$l = l_0 + \alpha\,(t - t_0)\,l_0$$
$$l_0 + \alpha\,(t - t_0)\,l_0 = l$$
$$\left[1 + \alpha\,(t - t_0)\right]l_0 = l$$
$$l_0 = \frac{l}{1 + \alpha\,(t - t_0)}$$

Exercise Set 13.3

1. Solve each of the following formulas for the indicated variable.

a. $g = \dfrac{GM}{R^2}$

    i. G ii. M iii. R

b. $a = \dfrac{v}{t}$

    i. v ii. t

c. $m_1 v_1 = m_2 v_2$

    i. m_1 ii. v_1 iii. m_2 iv. v_2

d. $T = \dfrac{1}{f}$

    i. f

e. $d = \frac{1}{2}at^2$

    i. a ii. t

f. $F = ma$

    i. m ii. a

g. $a = \dfrac{v^2}{r}$

    i. v ii. r

h. $F = \dfrac{Gm_1m_2}{d^2}$

    i. G ii. m_1 iii. m_2 iv. d

i. $E_k = \frac{1}{2}mv^2$

    i. m ii. v

j. $d = \dfrac{v^2}{2g}$

    i. v ii. g

k. $P = \dfrac{W}{t}$

    i. W ii. t

l. $V = lwh$

    i. l ii. w iii. h

m. $\nu = f\lambda$

    i. f ii. λ

n. $V = IR$

    i. I ii. R

o. $\dfrac{V_0}{V_i} = \dfrac{N_0}{N_i}$

    i. V_0 ii. V_i iii. N_0 iv. N_i

p. $M = \dfrac{-p}{s}$

    i. p ii. s

q. $\lambda = \dfrac{h}{mv}$

    i. h ii. m iii. v

r. $E = mc^2$

    i. m ii. c

s. $m = \frac{1}{2^n} m_0$

    i. n ii. m_0

2. Solve each of the following formulas for the indicated variable.

  a. $a = \dfrac{v - v_0}{t - t_0}$

    i. v ii. v_0 iii. t iv. t_0

  b. $F = \dfrac{mv - m_0 v_0}{t - t_0}$

    i. m ii. v iii. m_0 iv. v_0 v. t vi. t_0

  c. $d = v_0 t + \frac{1}{2} a t^2$

    i. v_0 ii. a

  d. $d = d_0 + v_0 t + \frac{1}{2} a t^2$

    i. d_0 ii. v_0 iii. a

  e. $p = 2(l + w)$

    i. l ii. w

  f. $A = \frac{1}{2}(b_1 + b_2) h$

    i. b_1 ii. b_2 iii. h

  g. $l_2 - l_1 = \alpha l_1 (T_2 - T_1)$

    i. l_2 ii. l_1 iii. α iv. T_2 v. T_1

  h. $Q = mC(T_2 - T_1)$

     i. m ii. C iii. T_2 iv. T_1

i. $p = \dfrac{sf}{s - f}$

     i. s ii. f

j. $x_2 - x_1 = \dfrac{S}{a}\lambda$

     i. x_2 ii. x_1 iii. S iv. a v. λ

k. $E_2 - E_1 = hf$

     i. E_2 ii. E_1 iii. h iv. f

Appendix A
Division by 0

Before we present the complexities that arise when one attempts to divide a number by 0, let us refresh the reader's memory on some division facts.

Consider the problem of the division of 14 by 2. The answer to this problem is 7, a fact that we can write as

$$\frac{14}{2} = 7$$

We can check the validity of this calculation by multiplying 2 by 7 to arrive at 14, i.e.,

$$2 \times 7 = 14$$

This means that we can think of the problem

$$\frac{14}{2} = ?$$

as

$$2 \times ? = 14$$

i.e., I multiplied 2 by a number and I got 14. What is that number? The answer, of course, is 7.

A.1 The Undefined Case

Consider now the case when we divide a nonzero number by 0 as in $\frac{4}{0}$. The question before us is

$$\frac{4}{0} = ?$$

This is equivalent to the following question:

$$0 \times ? \,=\, 4$$

i.e., I multiplied 0 by a number and I got 4. What is that number?

A moment of reflection will convince the reader that there is no such number and, therefore, I could not have done what I claim that I did. The division of a nonzero number by 0, such as the division $\frac{4}{0}$, is, therefore, meaningless. Formally we call such an expression **undefined** which means that we do not attach any meaning whatsoever to such an expression.

As strange as it may sound, situations do arise in practice when we encounter expressions of this kind. When we do encounter them, it means that the problem under consideration is impossible, i.e., it has no solutions. Here is a simple example of a problem that leads to such a conclusion.

Problem I am going through Toronto, driving east at a speed of 100 km/h. My friend is 100 km behind me, following me at a speed of 100 km/h. At what time will my friend catch up with me?

It is somewhat obvious that the answer is that my friend will never catch up with me, i.e., there is no time at which my friend will have caught up with me.[1] We could, of course, write an equation to model this problem using the equations of physics. However, when we attempt to solve such an equation, we run into the undefined case $\frac{1}{0}$ indicating that there is no answer to this problem. It would, in fact, be very strange if such an equation had a specific answer such as 2 or 5 as this would imply that my friend would catch up with me after 2 hours or 5 hours. Something has to break down along the way and something *does* break down. Looking at the specifics, an equation that models this problem is[2]

$$100(t \,-\, 1) \,=\, 100t$$

with the expression of the left side of the equality representing my friend's location after t hours and the expression on the right side of the equality representing my location after t hours. If my friend can catch up with me at some point, then we will be at the same location, which is what the equality in the equation above suggests.

---

[1] Note that *there is not time at which* ... does not mean *at time* 0 which might be the time I was going through Toronto.

[2] In this formulation distances are measured relative to Toronto and the 0 on the time scale is set to the time I was going through Toronto. One could, of course, set other references as the 0 for distance covered and elapsed time but the solution to any such equation would be the same: undefined.

When we attempt to solve this equation we arrive at

$$100(t - 1) = 100t$$
$$t - 1 = \frac{100}{100}t$$
$$t - 1 = t$$
$$t - t = 1$$
$$0t = 1$$
$$t = \frac{1}{0}$$

The expression $\frac{1}{0}$ is undefined. As such, there is no number that can solve the original equation, i.e., my friend will never catch up with me.

A.2   The Indeterminate Case

A different scenario results when we consider the division of 0 by 0. Let us write this as $\frac{0}{0}$. The question before us is

$$\frac{0}{0} = ?$$

or, its equivalent counterpart

$$0 \times ? = 0$$

i.e., I multiplied 0 by a number and I got 0. What is that number? The logical answer to such a question, of course, is that you could have used any number. Formally we call such an expression **indeterminate** and we understand that this means that any number could work as an answer.

As with the case where the expression is undefined, there are scenarios under which we encounter the expression $\frac{0}{0}$. In such situations, it is understood that the value of the quantity that we seek can equal to any number. Here is a problem whose solution is indeterminate.

Problem My friend and I left Toronto at the same time, driving east at a speed of 100 km/h. At what time will we be together?

It is somewhat obvious that the answer is that my friend and I will always be together, i.e., we will be together at all times. As with the case of the previous problem, we could write an equation to model this problem using the equations of physics. However, when we attempt to solve such an equation, we run into the indeterminate case $\frac{0}{0}$ indicating that any number could be the answer to this problem. It would, in fact, be very strange if such an equation had a specific answer such as 2 or 5 as this would imply that my friend and I will be together only 2 hours into the drive (but not before or after 2 hours as 2 is the only answer that we arrived at) or only 5 hours into the drive (but not before or after 5 hours as 5 is the only answer that we arrived at). Looking at the specifics, an equation that models this problem is

$$100t = 100t$$

with the expression of the left side of the equality representing my friend's location after t hours and the expression on the right side of the equality representing my location after t hours. If my friend and I can be together at some point, then we will be at the same location, which is what the equality in the equation above suggests.

When we attempt to solve this equation we arrive at

$$100t = 100t$$
$$100t - 100t = 0$$
$$0t = 0$$
$$t = \frac{0}{0}$$

The expression $\frac{0}{0}$ is indeterminate. As such, it can equal to any number, which implies that my friend and I will be together at all times, or in other words, my friend and I will always be together.

If you feel uncomfortable with the solutions to the two problems that we have posed in this appendix, it is perhaps in the deeply seated assumption that a math problem should always have a unique solution to it. *This is not so.* You should understand that there are problems for which there are no solutions. And of those that do have solutions, the solution may be unique, or there could be two possible solutions and no more, three possible solutions and no more, four possible solutions and no more, and so on. Or it could be the case that a problem may have an infinite number of solutions. What makes math so exceptional, is that it can help us determine which of these possibilities is true for a given problem including problems that have no solutions to them or those for which there is an infinite number of solutions.

Appendix B
On the Existence of Irrationals

To prove that irrational numbers exist, we must show that there exist quantities whose numerical representation cannot be expressed as a rational number. A standard proof is to show the existence of $\sqrt{2}$ and to show that it is not rational.

B.1 Existence of $\sqrt{2}$

To show the existence of $\sqrt{2}$, we will use the Pythagorean Theorem to find the length of the hypotenuse of a right-angle triangle whose short sides are each 1 unit long.

According to **the Pythagorean Theorem**, in a right-angle triangle, such as the one shown in Figure B.1.1 below, the square of the length of the hypotenuse is equal to the sum of the squares of the lengths of the short sides, i.e.,

$$c^2 = a^2 + b^2, \qquad a, b, c \geq 0$$

The side condition a, b, $c \geq 0$ relates to the fact that, by definition, length cannot be negative.

We will not prove the Pythagorean Theorem here. The interested reader can refer to any standard textbook on geometry for a proof of this theorem.

Applying the Pythagorean Theorem to a right-angle triangle whose short

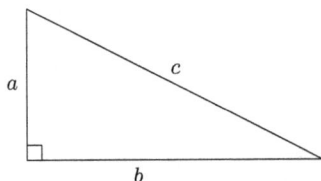

Figure B.1.1: A right-angle triangle of arbitrary side lengths. According to the Pythagorean Theorem, $c^2 = a^2 + b^2$, $a, b, c \geq 0$.

sides are each 1 unit long will yield

$$c^2 = 1^2 + 1^2, \quad c \geq 0$$
$$c^2 = 1 + 1, \quad c \geq 0$$
$$c^2 = 2, \quad c \geq 0$$
$$c = \pm\sqrt{2}, \quad c \geq 0$$
$$c = \sqrt{2}$$

See Figure B.1.2 below.

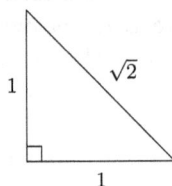

Figure B.1.2: A right-angle triangle with short sides each 1 unit long. According to the Pythagorean Theorem, the length of the hypotenuse is $\sqrt{2}$.

Therefore, there *are* quantities whose numerical value is $\sqrt{2}$. This proves the existence of $\sqrt{2}$.

B.2   Irrationality of $\sqrt{2}$

As noted above, for the triangle in Figure B.1.2

$$c^2 = 2, \qquad c \geq 0 \tag{B.2.1}$$

If all numbers are rational, then so must be the length of the hypotenuse of any triangle and, therefore, we must be able to find nonzero integers p and q such that

$$c = \frac{p}{q}, \qquad p, q \in \mathbb{I}; \ q \neq 0; \ \frac{p}{q} \geq 0$$

Using this, we can write Equation B.2.1 as

$$\left(\frac{p}{q}\right)^2 = 2, \qquad p, q \in \mathbb{I}; \ q \neq 0; \ \frac{p}{q} \geq 0 \tag{B.2.2}$$

The side conditions in Equation B.2.2 can be simplified as follows: First we note that $q \neq 0$ as noted. However, it is also the case that $p \neq 0$ as otherwise $\frac{p}{q}$ would equal 0 which would imply that c is 0 which is clearly not the case for the triangle in Figure B.1.2. Therefore, the requirement $\frac{p}{q} \geq 0$ becomes $\frac{p}{q} > 0$. This means either p and q are both positive or that they are both negative. We can drop the second case as the negative signs would cancel out. We, therefore, arrive at

$$\left(\frac{p}{q}\right)^2 = 2, \qquad p, q \in \mathbb{N} \tag{B.2.3}$$

We can re-write Equation B.2.3 as

$$\frac{p^2}{q^2} = 2, \qquad p, q \in \mathbb{N}$$

which is in turn equivalent to

$$p^2 = 2q^2, \qquad p, q \in \mathbb{N} \tag{B.2.4}$$

This shows that p^2 is even (as it has a factor of 2 in it) which in turn implies that p is even.[1] But if p is even, then there must be a natural number, m,

---

[1] If p is odd, then there is a natural number, s, such that $p = 2s + 1$. Therefore

$$p^2 = (2s + 1)^2$$
$$p^2 = 4s^2 + 4s + 1$$
$$p^2 = 2\left(2s^2 + 2s\right) + 1$$

or

$$p^2 = 2t + 1$$

where $t = 2s^2 + 2s$ is a natural number. This shows that p^2 is odd.

such that

$$p = 2m, \qquad p, m \in \mathbb{N}$$

Replacing p in Equation B.2.4 with $2m$ we get

$$(2m)^2 = 2q^2, \qquad m, q \in \mathbb{N}$$

Carrying out the exponent and switching sides yields

$$2q^2 = 4m^2, \qquad q, m \in \mathbb{N}$$

This leads to

$$q^2 = \frac{4}{2}m^2, \qquad q, m \in \mathbb{N}$$

or

$$q^2 = 2m^2, \qquad q, m \in \mathbb{N}$$

which means that q^2 is even and that, therefore, q is even. Since q is even, there must be a natural number n such that

$$q = 2n, \qquad q, n \in N$$

Replacing p and q in Equation B.2.3 with $2m$ and $2n$, we arrive at

$$\left(\frac{2m}{2n}\right)^2 = 2, \qquad m, n \in \mathbb{N}$$

or, after reduction

$$\left(\frac{m}{n}\right)^2 = 2, \qquad m, n \in \mathbb{N}$$

But this brings us back to where we started (See Equation B.2.3). By going through the same argument, we can arrive at the conclusion that m and n are both even and can, therefore, be reduced using 2, arriving at another fraction which can in turn be reduced using 2. We can, therefore, reduce the fraction indefinitely using 2. This is impossible as there are no fractions that can be reduced indefinitely.[2] Therefore, there are no integers p and q such that $\sqrt{2} = \frac{p}{q}$. This proves the irrationality of $\sqrt{2}$.

We have shown that $\sqrt{2}$ exists and is not rational. Irrationals do exist.

---

[2] This is true as to every pair of natural numbers we can assign a finite greatest common divisor.

Appendix C
On the Equivalence of the Division and Fraction Interpretations of the Horizontal Line

The horizontal line can be interpreted as both a division and the result of that division. As an example, the notation $\frac{4}{6}$ can be interpreted as both the size of a piece arrived at as a result of the division of four things into six parts and as four-sixths of one. This equivalence is shown in Figure C.0.1.

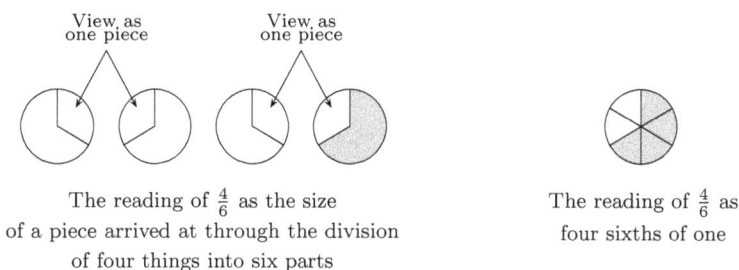

The reading of $\frac{4}{6}$ as the size
of a piece arrived at through the division
of four things into six parts

The reading of $\frac{4}{6}$ as
four sixths of one

Figure C.0.1: Equivalence of the result of the division of four things into six parts and four sixths of one

The two interpretations result in the same overall size. While the cuts are different, the shaded parts in Figure C.0.1 have the same overall size. Note that the interpretation of $\frac{4}{6}$ as a division requires that we divide 4 things (four pies, four pizzas, etc.) into 6 parts whereas the interpretation of $\frac{4}{6}$ as a fraction is understood to mean *four sixths of* 1, i.e., divide 1 (one pie, one pizza, etc.) into 6 parts and take 4 parts.

The reduced form of $\frac{4}{6}$, i.e., $\frac{2}{3}$, can also be interpreted as a division and a fraction. As a reduced division, $\frac{2}{3}$ is to be seen as the division of two things

577

(two pies, two pizzas, etc.) into three parts whereas as a reduced fraction, it
is seen as *two thirds of* 1, i.e., divide 1 (one pie, one pizza, etc.) into 3 parts
and take 2 parts. The two interpretations are illustrated in Figure C.0.2.

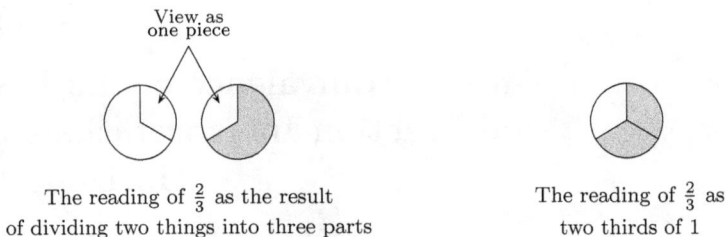

The reading of $\frac{2}{3}$ as the result   The reading of $\frac{2}{3}$ as
of dividing two things into three parts   two thirds of 1

Figure C.0.2: Equivalence of the result of the division of two
things into three parts and two thirds of 1

Comparison of the shaded areas in Figures C.0.1 and C.0.2 shows that the
interpretations of $\frac{4}{6}$ as a division and a fraction and their reduced counterparts
$\frac{2}{3}$ as a division and a fraction are all the same.

Appendix D
Notes on Algorithms for Basic Arithmetic Operations on Whole Numbers

In this appendix we refresh the reader's memory on the finer details of algorithms for addition, subtraction, multiplication and division of whole numbers.[1,2]

D.1   The Subtraction Algorithm

In this section we comment on subtractions involving borrowing when there are zeros present.

Consider the subtraction $3007 - 249$ written vertically below.

$$
\begin{array}{rrrr}
3 & 0 & 0 & 7 \\
- & 2 & 4 & 9 \\
\hline
\end{array}
$$

Since we can not subtract 9 from 7, we need to borrow. However, there are no *tens* or *hundreds* and, therefore, we borrow from *thousands*. Borrow-

---

[1] We assume that the reader already has this background and intend to present a brief review here.

[2] The discussion in this appendix focuses on the use of general algorithms in working out the addition, subtraction, multiplication and division of whole numbers. In practice, such algorithms are used as a last resort. As an example, if one seeks to work out the subtraction $1002 - 998$, one would likely work one's way up from 998 to 1002 or work one's way down from 1002 to 998 as opposed to using the general subtraction algorithm. Such alternatives are useful and they should be employed whenever there is a fit. However, such alternative are often of limited use and suffer from extreme inefficiency if used outside the limits within which they work well.

ing a *thousand* and breaking it into 10 *hundreds*[3] gives us

```
     2  10
     3̸  0̸  0  7
  −     2  4  9
  _____
```

Next we borrow a *hundred* and break it into 10 *tens*. This gives us

```
        9
     2  1̸0̸ 10
     3̸  0̸  0̸  7
  −     2  4  9
  _____
```

And finally we borrow a *ten* and break it into 10 *ones* to arrive at

```
        9   9
     2  1̸0̸ 1̸0̸ 17
     3̸  0̸  0̸  7̸
  −     2  4  9
  _____
```

We subtract 9 from 17 to get 8:

```
        9   9
     2  1̸0̸ 1̸0̸ 17
     3̸  0̸  0̸  7̸
  −     2  4  9
  _____
                8
```

Moving back column by column and performing the subtractions as we encounter them results in

```
        9   9
     2  1̸0̸ 1̸0̸ 17
     3̸  0̸  0̸  7̸
  −     2  4  9
  _____
     2  7   5   8
```

We can now write

$$3007 - 249 = 2758$$

---

[3] We are using the structural feature of the place value scheme discussed earlier to relate the size of the place value *thousands* to that of the place value *hundreds*. Given that a given place value is 10 times larger than the one on its right, a *thousand* breaks into 10 *hundreds*.

D.2   The Multiplication Algorithm

We illustrate matters using 539×46.[4]

Step 1.
Set up the multiplication as shown.

$$\begin{array}{r} 5\ 3\ 9 \\ \times\ \ 4\ 6 \\ \hline \end{array}$$

Step 2.
To find how many ones we have,
multiply 6 by 9 to get 54.
54 breaks into 4 ones and 5 tens
Write 4 in the ones column and carry
5 to the tens column as shown.

$$\begin{array}{r} 5 \\ 5\ 3\ 9 \\ \times\ \ 4\ 6 \\ \hline 4 \end{array}$$

Step 3.
To find how many tens we have,
multiply 6 by 3 to get 18.
Add the 5 tens that were carried
over to get a total of 23 tens.
23 tens breaks into 3 tens and
2 hundreds. Write 3 in the tens
column and carry 2 to the hundreds
column as shown.

$$\begin{array}{r} 2\ 5 \\ 5\ 3\ 9 \\ \times\ \ 4\ 6 \\ \hline 3\ 4 \end{array}$$

Step 4.
To find how many hundreds we have,
multiply 6 by 5 to get 30. Add the
2 hundreds that were carried over to
get a total of 32 hundreds. 32
hundreds breaks into 2 hundreds and
3 thousands so that we can write 32
as shown.

$$\begin{array}{r} 2\ 5 \\ 5\ 3\ 9 \\ \times\ \ 4\ 6 \\ \hline 3\ 2\ 3\ 4 \end{array}$$

---

[4]The multiplication algorithm given above is based on the **distributive property of multuplication over addition**, i.e.,

$$a\,(b\ +\ c)\ =\ ab\ +\ ac$$

The distributive property itself is based on the familiar reasoning that states that, in order to double the value of something we can double the values of its parts and, in order to triple the value of something, we can triple the values of its parts and so on. We can write the product 539×46 as $539\,(40 + 6)$ which, according to the distributive property, can be evaluated as $539 \times 40 + 539 \times 6$.

Having multiplied digit 6 in 46 by the digits in 539 in the manner shown above, we clear the digits that were carried and move on to multiply digit 4 in 46 by the digits in 539 in a similar manner. However, the resulting digits are shifted back over one place value as digit 4 in 46 is in the tens place and, therefore, multiplication of 4 by 9 will tell us how many *tens* we have and multiplication of 4 by 3 will tell us how many *hundreds* we have and so on.

Step 5. Multiply 4 by 9 to get 36. Write 6 down and carry 3 as shown.	3 5 3 9 × 4 6 3 2 3 4 6
Step 6. Multiply 4 by 3 to get 12. Add the 3 that was carried over to get 15. Write 5 down and carry 1 as shown.	1 3 5 3 9 × 4 6 3 2 3 4 5 6
Step 7. Multiply 4 by 5 to get 20. Add the 1 that was carried over to get 21. Write 21 as shown.	1 3 5 3 9 × 4 6 3 2 3 4 2 1 5 6
Step 8. Add the numbers below the line to arrive at the result.	1 3 5 3 9 × 4 6 3 2 3 4 + 2 1 5 6 2 4 7 9 4

We can now write

$$539 \times 46 = 24\,794$$

D.3 The Division Algorithm

We illustrate matters using $\frac{364}{7}$.

Step 1.
Set up the division as shown.

$$7 \overline{\smash{)}\ 3\ 6\ 4}$$

Step 2.
Since 3 is too small for 7, we
start with 36. Ask: How many 7s
go into 36? The answer is 5.
Write 5 as shown.

$$\begin{array}{r} 5 \\ 7 \overline{\smash{)}\ 3\ 6\ 4} \end{array}$$

Step 3.
Multiply 5 by 7 to get 35.
Write 35 as shown.

$$\begin{array}{r} 5 \\ 7 \overline{\smash{)}\ 3\ 6\ 4} \\ 3\ 5 \end{array}$$

Step 4.
Subtract 35 from 36 to get 1.
Write 1 as shown.

$$\begin{array}{r} 5 \\ 7 \overline{\smash{)}\ 3\ 6\ 4} \\ -\ 3\ 5 \\ \hline 1 \end{array}$$

Step 5.
Bring 4 down as shown.

$$\begin{array}{r} 5 \\ 7 \overline{\smash{)}\ 3\ 6\ 4} \\ -\ 3\ 5 \\ \hline 1\ 4 \end{array}$$

Step 6.
Ask: How many 7s go into 14?
Answer: 2. Write 2 as shown.

$$\begin{array}{r} 5\ 2 \\ 7 \overline{\smash{)}\ 3\ 6\ 4} \\ -\ 3\ 5 \\ \hline 1\ 4 \end{array}$$

Step 7.
Multiply 2 by 7 to get 14.
Write 14 as shown.

$$\begin{array}{r} 5\,2 \\ 7\,\overline{)\,3\,6\,4} \\ -\,3\,5 \\ \hline 1\,4 \\ 1\,4 \end{array}$$

Step 8.
Subtract 14 from 14 to
get 0. Since there are no
more digits to pull down
we stop.

$$\begin{array}{r} 5\,2 \\ 7\,\overline{)\,3\,6\,4} \\ -\,3\,5 \\ \hline 1\,4 \\ -\,1\,4 \\ \hline 0 \end{array}$$

remainder

We can now write

$$\frac{364}{7} = 52$$

If, *after subtraction*, there are no more digits left to pull down, the division is finished and you can stop.

A common mistake is to 'miss a 0' in the quotient. As an example, the quotient $4872 \div 12$ yields 406 as the answer. However, frequently enough we see students write 46 in place of 406. A more difficult case is when the 0 is on the right end of the quotient. An example of this is the quotient $4500 \div 2$ which yields 2250 as the answer. Frequently we see students write 225 in place of 2250. To make sure you do not fall into such 'traps', keep the following in mind: *Each time you pull a digit down as in Step 5 in the example above, you must answer questions like the one in Step 6 above.* Sometimes the answer may be 0.

Finally, keep in mind that the **remainder** may not be 0. As an example, when we divide 7 by 2 we arrive at a quotient of 3 and a remainder of 1. We write this as follows.

$$\frac{7}{2} = 3\,\text{R}1$$

Appendix E
On the Superiority of the Horizontal Line Over Alternative Notations for Division

The horizontal line has many advantages over other notations used to denote division and it is because of these advantages that the formal notation has adopted its use. Examples of such use can be seen in the setup of such theoretical formulas as $E_k = \frac{1}{2}mv^2$, $m = \frac{1}{2\pi}m_0$, $A = \frac{1}{2}bh$, $\frac{p_1}{T_1} = \frac{p_2}{T_2}$, and the like. In this appendix we point to some of the technical advantages of using the horizonal line as the symbol for division.[1]

The first advantage of horizontal line as the symbol for division is in its natural ability to group its operands without the need to use brackets. The reason this is desirable is that, officially, brackets are used to denote multiplication and the use of brackets in both multiplication and division would make the analysis of terms into factors difficult.[2] As an example of how other notations often require the use of brackets in the set up of a division problem, consider the problem of finding the quotient of the sum of 6 and 8 and the sum of 4 and 3. Using the horizontal line to denote division, we can write this as

$$\frac{6 + 8}{4 + 3}$$

The use of $\div$ requires that we use brackets:

$$(6 + 8) \div (4 + 3)$$

Without the brackets, the expression would become

$$6 + 8 \div 2 + 3$$

---

[1]For semantic advantages of the use of the horizontal line please see the companion textbook *Semantics and the Syntax of Algebra* by the author.

[2]Recall that factors are entities within terms that are *multiplied* not those that are divided. By limiting the use of brackets to denote multiplication, we can tell at a glance that each set of outer brackets constitutes a factor within the term in which it resides.

which has a different meaning than $(6+8) \div (4+3)$: The expression $(6+8) \div$ $(4+3)$ analyzes into one term containing one factor. That factor contains a division and requires that we add 6 and 8 to get 14, add 4 and 3 to get 7, and then divide 14 by 7 to get 2 while the expression $6+8 \div 2+3$ analyzes into three terms: 6, $8 \div 2$ which evaluates to 4, and 3 resulting in $6+4+3$ which adds up to 13. This may be the correct expression for some other problem but it does not model the problem that we are trying to solve here.

The $\div$ notation also affects readability. The expression

$$\frac{\frac{1}{3} + \frac{2}{5}}{\frac{3}{4} - \frac{1}{2}}$$

clearly tells us to add $\frac{1}{3}$ and $\frac{2}{5}$, and divide the result by the difference between $\frac{3}{4}$ and $\frac{1}{2}$. Such quick analysis is not possible in formulations that use alternative notations for division:

$$(1 \div 3 + 2 \div 5) \div (3 \div 4 - 1 \div 2)$$

Not only is this expression more difficult to comprehend, it also uses brackets, interfering with the notation used for multiplication.

Further to this interference with the notation used to represent multiplication, overuse of brackets in this manner has a severe effect on readability.

The use of the horizontal line to denote division also aids the task of evaluating expressions. As an example of how it does so, consider the expression

$$\frac{12}{2} \times 3$$

Here we understand that, starting with 12, we must divide by 2 and multiply by 3 and, what's more, the two tasks of dividing by 2 and multiplying by 3 can be done in any order. We could write this expression as $12 \times \frac{3}{2}$ or $\frac{12 \times 3}{2}$ and they would still evaluate to the same value. What we should *not* do is multiply 2 by 3 and divide 12 by the result of that product. The notation is strong enough to make such a move clearly unacceptable. The $\div$ notation is not so clear. In the equivalent formulation of the above expression using the $\div$ notation, i.e.,

$$12 \div 2 \times 3$$

it is a common mistake to multiply 2 by 3 and then divide 12 by the result of that product. This may be due to the link that we make between the ranking of the operations in terms of ease of use and the order in which they should be carried out. The $\div$ notation impedes our ability to make sense of such expressions while the horizontal line facilitates it.

The visual cues that enforce the equality of $\frac{12}{2} \times 3$, $12 \times \frac{3}{2}$ and $\frac{12\times3}{2}$ also help the user remember the equivalence, a skill that is important in working with algebraic expressions and equations as well and one that is difficult to master using the $\div$ notation.

The use of the $\div$ notation also adversely affects our ability to simplify algebraic expressions and solve algebraic equations efficiently and its use hinders the task of classifying algebraic equations.

Of interest is also the manner in which expressions such as $4 \div 3 \times 6$ can be evaluated. The $\div$ notation makes it difficult to see that we could multiply 4 by 6 and *then* divide the result by 3. The horizontal line, on the other hand, makes it easier for us to see that this can be done if the reader keeps in mind that, as we noted above, $\frac{4}{3} \times 6$ is equivalent to $\frac{4 \times 6}{3}$.[3]

Alternative division notations such as / or : suffer from the same problems as $\div$.

---

[3]Such equivalencies have semantic significance as well. See the companion textbook *Semantics and the Syntax of Algebra* by the author for the semantics of the phrase *fraction of*.

Appendix F
On the Superiority of the Analysis-Synthesis Technique over BEDMAS

In this appendix we will discuss some of the technical advantages of the analysis-synthesis technique over BEDMAS. For semantic advantages of the analysis-synthesis technique over BEDMAS please see the companion textbook *Semantics and the Syntax of Algebra* by the author.

To begin, BEDMAS seems arbitrary. Why is it BEDMAS and not MAS-BED or some other sequencing of these letters? One may follow our lead and argue that the order reflects the increasing level of complexity of operations but if this was the case, then shouldn't BEDMAS have been written backwards?[1] In addition, even the reverse sequencing of the operations seems out of order as addition is simpler than subtraction, implying that the sequence should have been BEDMSA and not BEDMAS.[2]

BEDMAS is rigid. It leaves a strong impression that the particular sequence is the only sequence that works whereas, in reality, any of BEDMSA, BEMDAS or BEMDSA would work just as well as BEDMAS.

BEDMAS can be misleading. In BEDMAS, the occurrence of letter A before letter S implies that addition should be done before subtraction.[3] However, precisely what gets added can become a source of confusion. To explain how, consider the expression

$$18 - 4 + 7$$

the requirement to add first leaves the impression that one must add 4 and 7 and then subtract the sum from 18. This is not so. To properly apply

---

[1] The order corresponds to the steps in the synthesis stage of the analysis-synthesis technique.

[2] Interestingly enough, BEDMSA works just as well as BEDMAS.

[3] This interpretation works but is not required.

BEDMAS one must adopt the view that we have been promoting throughout this textbook: That each + and each − refers to the term that follows it. This is shown below.

$$18 \;-\; 4 \;+\; 7$$

If we follow BEDMAS, we would have to add 7 but the 7 should be added to 18, not 4. Once the sum of 18 and 7 is computed, we can subtract 4 from it. Such an understanding is not communicated by the sequence of the letters in BEDMAS and the common interpretation is usually the incorrect one given earlier.

In BEDMAS the BED part is unnecessary. Beyond the MAS part any operation may have priority and it is notation that specifies which operation is to be considered as the main operation beyond MAS.

BEDMAS is limited. It speaks of additions, subtractions, multiplications, divisions and exponents. Where is R for roots, L for logarithms, etc.?

BEDMAS is inefficient. As an example of why this is so, consider the task of evaluating the following expression:

$$-3\,(-4)\,(2) \;-\; 2\,(-5)\,(-1) \;-\; 4\,(-3)$$

The use of the analysis-synthesis technique leads to the following solution:

$$
\begin{aligned}
-3\,(-4)\,(2) \;-\; &2\,(-5)\,(-1) \;-\; 4\,(-3) \\
&= 24 \;-\; 10 \;+\; 12 \\
&= 26
\end{aligned}
$$

The use of BEDMAS leads to the following solution:

$$
\begin{aligned}
-3\,(-4)\,(2) \;-\; &2\,(-5)\,(-1) \;-\; 4\,(-3) \\
&= 12\,(2) \;-\; 2\,(-5)\,(-1) \;-\; 4\,(-3) \\
&= 24 \;-\; 2\,(-5)\,(-1) \;-\; 4\,(-3) \\
&= 24 \;-\; (-10)\,(-1) \;-\; 4\,(-3) \\
&= 24 \;-\; 10 \;-\; 4\,(-3) \\
&= 24 \;-\; 10 \;-\; (-12) \\
&= 24 \;-\; 10 \;+\; 12 \\
&= 14 \;+\; 12 \\
&= 26
\end{aligned}
$$

Second line in the solution using the analysis-synthesis technique

Third line in the solution using the analysis-synthesis technique

The approach using BEDMAS also suffers from the need to constantly reason one's way through the thicket: We would have to start by arguing that, since the product of two negative numbers is positive, then $-3\,(-4)$ evaluates to

12. Next we would multiply 12 by 2 to get 24. We now move on to the next multiplication,[4] i.e., $2(-5)$. We now argue that, since the product of a positive number and a negative number is negative, $2(-5)$ evaluates to -10. We would have to continue in this fashion until all the terms are worked out. This leads us to the 6th line in the solution above that uses BEDMAS. Here, now, we would have to argue that subtraction of a negative number is the same as adding the absolute value of that number. Following these simplifications, we still need to repeatedly apply the *Algorithm for Evaluation of Addition and Subtraction of Integers* to arrive at the final result.

But perhaps the biggest shortcoming of BEDMAS is in its inability to allow the user to extract meaning from the activity.[5]

---

[4] For no apparent reason other than the fact that the rule says so.

[5] For the semantic advantages of the analysis-synthesis technique over BEDMAS please see the companion textbook *Semantics and the Syntax of Algebra* by the author.

A Proof for the Divisibility Test by 3

The divisibility test for 3 provided in this text is repeated below.

A Divisibility Test for 3 A given number is divisible by 3 if the sum of its digits is divisible by 3 otherwise it is not divisible by 3.

To prove this theorem, we write the given number as

$$d_n \cdots d_3 d_2 d_1 d_0$$

As an example, in 4308, d_3 is 4, d_2 is 3, d_1 is 0 and d_0 is 8. Rewriting this in expanded form, we arrive at

$$d_n \times 10^n + \cdots + d_3 \times 10^3 + d_2 \times 10^2 + d_1 \times 10^1 + d_0 \times 10^0$$

We will now rewrite the expression above as

$$d_n \left(10^n - 1 + 1\right) + \cdots + d_3 \left(10^3 - 1 + 1\right) + d_2 \left(10^2 - 1 + 1\right)$$
$$+ d_1 \left(10^1 - 1 + 1\right) + d_0$$

The reason behind this reformulation will become apparent soon. The above expression can, in turn, be written as

$$d_n \left(10^n - 1\right) + d_n + \cdots + d_3 \left(10^3 - 1\right) + d_3 + d_2 \left(10^2 - 1\right) + d_2$$
$$+ d_1 \left(10^1 - 1\right) + d_1 + d_0$$

Regrouping the terms, we arrive at

$$d_n \left(10^n - 1\right) + \cdots + d_3 \left(10^3 - 1\right) + d_2 \left(10^2 - 1\right) + d_1 \left(10^1 - 1\right)$$
$$+ d_n + \cdots + d_3 + d_2 + d_1 + d_0$$

Note that each expression within brackets consist of a sequence of 9s. As an example, $10^3 - 1$ is equal to $1000 - 1$ which evaluates to 999. This means that each term that contains brackets is divisible by 3. Whether or not the whole expression is divisible by three, then, depends on whether or not the sum

$$d_n + \cdots + d_3 + d_2 + d_1 + d_0$$

is divisible by 3.

Appendix H
On the Fractional Part of a Mixed Number

In the section on rational numbers we noted that the fractional part of a mixed number must be proper. This requirement implies that the number $3\frac{3}{5}$ may be referred to as a mixed number as its fractional part, $\frac{3}{5}$ is proper. However, although the number $2\frac{8}{5}$ is meaningful, it may not be called a mixed number. Let us see why.

In Figure H.0.1 we show different views of the same quantity.

In part (a) of this figure, we view the quantity as 3 whole pies and an extra $\frac{3}{5}$, i.e., $3\frac{3}{5}$. This is a mixed number. Note that the fractional part is proper. This means that the extra bits and pieces do not add up to a whole pie, i.e., we have formed the maximum number of whole pies that we can.

In part (b) of Figure H.0.1 we break a whole pie into 5 parts (as many pieces as the rightmost pie is cut into). This leads to the view that we have 2 whole pies and $\frac{8}{5}$, i.e., $2\frac{8}{5}$.

In part (c) of Figure H.0.1 we break another whole pie into 5 parts. This leads to the view that we have 1 whole pie and $\frac{13}{5}$, i.e., $1\frac{13}{5}$.

In part (d) of Figure H.0.1 all whole pies are cut into 5 pieces each, leading to the view that we have $\frac{18}{5}$. In this view, there are no whole pies left. This is what we call an improper fraction.

The views in parts (a) and (d) are special. The first forms as many whole pies as it possibly can whereas the last breaks all the pies into fifths. These two views are important in applications as well and it is for this reason that we name them, calling the first a mixed number and the second an improper fraction. The other views, those in parts (b) and (c), while they appear from time to time along the way as we work with fractions, are not as important and are, therefore, not named.

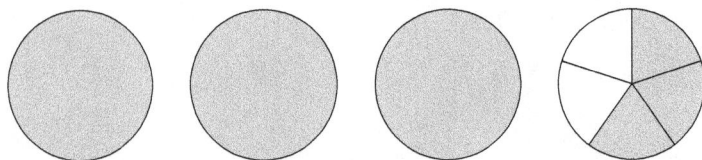

(a) The quantity viewed as $3\frac{3}{5}$ (a mixed number)

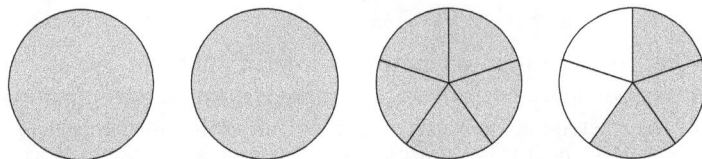

(b) The quantity viewed as $2\frac{8}{5}$

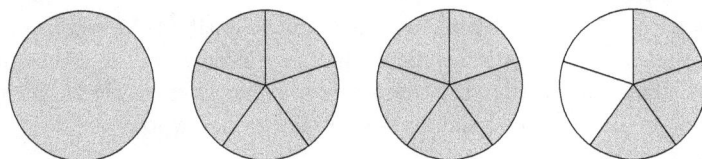

(c) The quantity viewed as $1\frac{13}{5}$

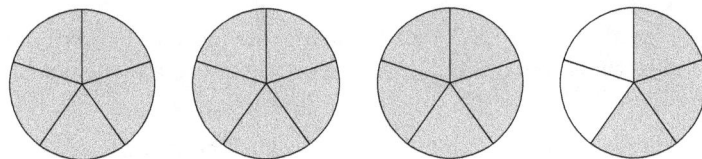

(d) The quantity viewed as $\frac{18}{5}$ (an improper fraction)

Figure H.0.1: Different views of the same quantity

Appendix I
Conversion of Unlike Fractions to Their Equivalent, Like Fractions

In this appendix we will show how to convert a set of unlike fractions to their equivalent, like fractions. The process involves two stages. The first stage involves the finding of a common denominator, preferably the smallest possible common denominator called the **least common denominator** or **LCD**. A common denominator ensures that our fractions will be like fractions. The second stage involves the changing of the numerators to ensure equivalency with the original fractions. We will explain the process using examples.

I.1   The Least Common Denominator

Consider the unlike fractions below:

$$\frac{3}{4}, \frac{2}{5}, \frac{1}{2}$$

By *a* common denominator of these fractions we mean a number that is divisible by the denominators of these fractions, i.e., 4, 5 and 2. There are many such common denominators for a given set of fractions. As an example, for the given fractions above we can come up with 20 and 40 as two possible common denominators but there are others: 60, 80, 100, 120, etc. are all common denominators for these fractions. The smallest number in the list, 20, is referred to as the least common denominator.

One way to find a common denominator for a given set of fractions is to multiply their denominators. In the case of the fractions above, this would yield 40.[1] At times, this approach might yield the least common denominator.

---

[1]This may not seem that bad but often the multiplication of denominators generates much larger common denominators. As an example, the least common denominator for the fractions $\frac{5}{12}$, $\frac{1}{18}$ and $\frac{4}{9}$ is 36 while multiplying the denominators generates the number 1944.

Often, however, it won't. And it is always advantageous to find the *least* common denominator as its use generates smaller intermediate values and requires fewer reductions of the results of additions and subtractions of the fractions.

We will now introduce an efficient algorithm for finding the least common denominator of a given set of fractions. We will illustrate the idea using the fractions listed above as an example.

Example

Find the least common denominator of

$$\frac{3}{4}, \frac{2}{5}, \frac{1}{2}$$

Solution

To begin, list the denominators in tabular form as shown below:

2	5	4

Next, start with the first prime number, 2, and ask yourself the following question: Are any of the numbers in the table above divisible by 2? In this case the answer is yes. In fact, both 2 and 4 in the table are divisible by 2. Write 2 down on the left side of the table as shown below:

	2	5	4
2			

Divide by 2 the numbers in the table that are divisible by 2 and record the results as shown below:

	2	5	4
2			
	1	5	2

We return to the first prime number, 2, and repeat the question: Are any of the new numbers in the table (i.e., those in the second row of the table) divisible by 2? The answer, again, is yes. Write 2 on the left side of the table again as shown below:

2	2	5	4
2	1	5	2

Divide by 2 the numbers in the second row of the table that are divisible by 2 and record the results in the table as shown below:

2	2	5	4
2	1	5	2
	1	5	1

We return to the first prime number 2 again and we ask the question: Are any of the new numbers in the table (i.e., those in the third row) divisible by 2? This time the answer is no. We move on to the next prime number, 3, and ask: Are any of the numbers in the third row of the table divisible by 3? The answer is no. We move on to the next prime number, 5, and ask: Are any of the numbers in the third row of the table divisible by 5? The answer is yes. Write 5 down as shown below:

2	2	5	4
2	1	5	2
5	1	5	1

Divide by 5 any of the new numbers in the table (those in the third row of the table) that are divisible by 5 and write the result down as shown below:

2	2	5	4
2	1	5	2
5	1	5	1
	1	1	1

This tabular process stops when all the numbers along the bottom row of the table turn into 1s. At this point, we can multiply the primes listed on the left side of the table to get the least common denominator for the unlike fractions on our list. For the present example, this yields

$$2 \times 2 \times 5 = 20$$

as expected.

The process described above generates the least common denominator by factoring out the numbers that are common to the denominators of the unlike fractions, 2, 5 and 4 and counting these commonalities only once. As an example, the first row ensures that, if a 2 is common to both 2 and 4, that the 2 is counted only once.

We present another example, this time showing the complete table only.

Example

Find the least common denominator of the following fractions:

$$\frac{3}{4}, \frac{1}{6}, \frac{5}{14}$$

Solution

2	4	6	14
2	2	3	7
3	1	3	7
7	1	1	7
	1	1	1

The least common denominator is

$$2 \times 2 \times 3 \times 7 = 84$$

While this number may look large, it is the *smallest* common denominator that we can find (the smallest number that is divisible by the denominators 4, 6 and 14). Multiplying the denominators would result in 336 which is significantly larger.

Exercise Set I.1

Find the least common denominator of each of the following sets of fractions.

1. $\frac{2}{5}, \frac{5}{6}, \frac{1}{3}$

2. $\frac{1}{4}, \frac{2}{5}, \frac{2}{3}$

3. $\frac{3}{8}, \frac{1}{4}, \frac{1}{2}$

4. $\frac{5}{12}, \frac{1}{8}, \frac{1}{6}$

5. $\frac{1}{12}, \frac{2}{3}, \frac{1}{5}, \frac{5}{6}$

6. $\frac{1}{15}, \frac{5}{12}, \frac{5}{8}$

7. $\frac{1}{17}, \frac{5}{34}, \frac{3}{2}$

8. $\frac{9}{14}, \frac{2}{21}, \frac{3}{7}$

9. $\frac{3}{16}, \frac{1}{4}, \frac{2}{3}$

10. $\frac{4}{25}, \frac{1}{15}, \frac{2}{3}$

11. $\frac{7}{10}, \frac{1}{2}, \frac{3}{4}, \frac{1}{8}$

12. $\frac{1}{18}, \frac{3}{15}, \frac{1}{12}$

13. $\frac{11}{7}, \frac{1}{5}, \frac{1}{3}$

14. $\frac{2}{22}, \frac{2}{33}, \frac{15}{11}$

15. $\frac{3}{19}, \frac{1}{2}, \frac{4}{38}, \frac{2}{3}$

I.2   Changing the Numerators

Having found the least common denominator of a set of fractions, our next step is to modify the numerators of the fractions in the set to ensure that the equivalent fractions have the same size as those in the set. Let us return to our first example in the previous section and see how this can be done.

Recall that our original aim was to convert the unlike fractions

$$\frac{3}{4}, \frac{2}{5}, \frac{1}{2}$$

to their equivalent, like fractions. Finding the least common denominator solves part of this quest: It provides us with a common denominator for the fractions listed. For the fractions in the list above, the least common denominator is 20.

Consider now the first fraction in the list, $\frac{3}{4}$. We wish to convert this into an equivalent fraction with a denominator of 20. The two fractions are shown in Figure I.2.1 below.

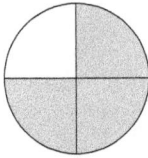

(a) Graph of $\frac{3}{4}$: We have 3 shaded pieces out of a total of 4 pieces

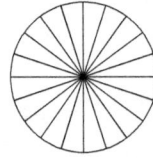

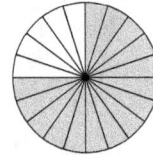

(b) Graph of $\frac{}{20}$: We have as yet an unknown number of shaded pieces out of a total of 20 pieces

Figure I.2.1: The fractions $\frac{3}{4}$ and $\frac{}{20}$

As shown in Figure I.2.2 below, for the fractions $\frac{3}{4}$ and $\frac{}{20}$ to have the same size, the sizes of the shaded sections of the two pie graphs must be the same.

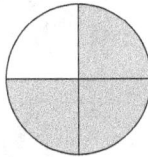

(a) Graph of $\frac{3}{4}$: We have 3 shaded pieces out of a total of 4 pieces

(b) Graph of $\frac{15}{20}$: We have 15 shaded pieces out of a total of 20 pieces

Figure I.2.2: The fractions $\frac{3}{4}$ and $\frac{15}{20}$

As Figure I.2.2 shows, in going from $\frac{3}{4}$ to $\frac{}{20}$, the total number of pieces has grown 5 times, from 4 to 20. But the total number of shaded pieces has also grown 5 times, from 3 to 15. For the fractions $\frac{3}{4}$ and $\frac{}{20}$ to be equivalent, then, the numerator of $\frac{}{20}$ must be 15.

So, to find the numerator of $\frac{}{20}$, we need to ask *what did I multiply 4 by to get 20?* Once we find the answer, 5, we multiply the numerator, 3, by 5 as well to get 15.[2] We arrive at the fraction $\frac{15}{20}$.

To answer the question *what did I multiply 4 by to get* 20? we can divide 20 by 4. This generates 5 right away.[3] We then multiply the numerator, 3, by 5 to get the missing numerator. The following diagram illustrates the idea:

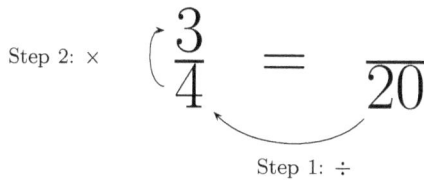

$$\text{Step 2: } \times \quad \frac{3}{4} \;=\; \frac{}{20}$$

$$\text{Step 1: } \div$$

Figure I.2.3: Steps for working out the numerator of the fraction with the least common denominator: Divide 20 by 4 to get 5. Multiply 5 by 3 to get the missing numerator, i.e., 15.

For the second fraction in the list, $\frac{2}{5}$, we divide 20 by 5 to get 4. This means that we multiplied the denominator of $\frac{2}{5}$ by 4 to get 20. Multiply the numerator, 2, by 4 as well to get $\frac{8}{20}$.

The algorithm that we recommend, then, divides the least common denominator by each of the denominators in the fractions in the list and multiplies the corresponding numerators by the results of those divisions. For the last fraction in the list, $\frac{1}{2}$, divide 20 by 2 to get 10. Multiply the numerator, 1, by 10 as well to get 10. The equivalent, like fractions that we seek for

$$\frac{3}{4}, \frac{2}{5}, \frac{1}{2}$$

are

$$\frac{15}{20}, \frac{8}{20}, \frac{10}{20}$$

Each fraction in the list above is equivalent in size to the corresponding one in the earlier list. The new fractions are, however, like fractions.

---

[2]Note that this is the opposite of the reduction of fractions, an activity called **building fractions**.

[3]The use of trial and error is not recommended: The least common denominator of $\frac{3}{14}$, $\frac{1}{12}$ and $\frac{5}{9}$ is 252. To answer the question *what did I multiply 14 by to get 252?*, it is easier to divide 252 by 14 rather than repeatedly guess at the number that multiplies 14 to generate 252.

For the second set of fractions

$$\frac{3}{4}, \frac{1}{6}, \frac{5}{14}$$

we found the least common denominator to be 84. We write:

$$\overline{84}, \overline{84}, \overline{84}$$

We now divide 84 by each denominator in the list and multiply the result by the corresponding numerator. This yields the following:

$$\frac{63}{84}, \frac{14}{84}, \frac{30}{84}$$

The unlike fractions

$$\frac{3}{4}, \frac{1}{6}, \frac{5}{14}$$

have been converted to their equivalent, like fractions

$$\frac{63}{84}, \frac{14}{84}, \frac{30}{84}$$

Having converted our unlike fractions to equivalent, like fractions, we can now easily compare them, add them and subtract them.

Exercise Set I.2

For each question in Exercise Set I.1, convert the fractions in the given list to their equivalent, like fractions using the least common denominator.

Appendix J
An Alternative Algorithm for Ordering, Addition and Subtraction of Rational Numbers

In the body of the text we explained that one way to order, add and subtract rational numbers is to convert any mixed numbers and integers to improper fractions at the outset. An alternative approach is to convert any improper fractions to mixed numbers and integers before we order, add or subtract them. In this appendix we will show how this alternative algorithm works through examples.

Example

Arrange

$$\frac{11}{4}, \ 1\frac{4}{5}, \ \frac{13}{3}$$

in increasing order.

Solution

We begin by listing these fractions vertically.

$$\frac{11}{4}$$

$$1\frac{4}{5}$$

$$\frac{13}{3}$$

Next we convert improper fractions to mixed numbers and integers.

$$2\frac{3}{4}$$

$$1\frac{4}{5}$$

$$4\frac{1}{3}$$

Since in a mixed number the fractional part is less than 1, we can order these mixed numbers by ordering their whole parts. This leads to

$$2\frac{3}{4} \qquad 2$$

$$1\frac{4}{5} \qquad 1$$

$$4\frac{1}{3} \qquad 3$$

The original fractions can be ordered accordingly:

$$\frac{11}{4} \qquad 2$$

$$1\frac{4}{5} \qquad 1$$

$$\frac{13}{3} \qquad 3$$

This leads to the following order:

$$1\frac{4}{5}, \frac{11}{4}, \frac{13}{3}$$

In case the whole parts are the same, we can order the mixed numbers by ordering their fractional parts. Here is an example.

Example

Arrange

$$2\frac{3}{4}, \frac{9}{4}, 2\frac{1}{2}$$

in increasing order.

Solution

We begin by listing these fractions vertically.

$$2\frac{3}{4}$$

$$\frac{9}{4}$$

$$2\frac{1}{2}$$

Next we convert improper fractions to mixed numbers and integers.

$$2\frac{3}{4}$$

$$2\frac{1}{4}$$

$$2\frac{1}{2}$$

Since the whole parts are the same, we order the fractions by comparing their fractional parts. To do so, we convert the fractional parts to their equivalent, like fractions using the least common denominator. This leads to

$$2\frac{3}{4}$$

$$2\frac{1}{4}$$

$$2\frac{2}{4}$$

This leads to the following order.

$$2\frac{3}{4} \qquad 3$$

$$2\frac{1}{4} \qquad 1$$

$$2\frac{2}{4} \qquad 2$$

The original fractions can be ordered accordingly:

$$2\frac{3}{4} \qquad 3$$

$$\frac{9}{4} \qquad 1$$

$$2\frac{1}{2} \qquad 2$$

This leads to the following order:

$$\frac{9}{4},\, 2\frac{1}{2},\, 2\frac{3}{4}$$

Alternative Algorithm for Ordering Rational Numbers

To order a set of rational numbers

1. Convert all improper fractions to mixed numbers and integers.
2. If the whole parts of the mixed numbers are different, order the mixed numbers according to their whole parts[1] otherwise order the mixed numbers by comparing their fractional parts.
3. Order the original numbers in the list accordingly.

The same approach can be used when adding and subtracting fractions. Instead of converting mixed numbers and integers to improper fractions at the outset, we can convert all improper fractions to mixed numbers and integers before we add and subtract them. When adding or subtracting expressions that involve mixed numbers, we can add or subtract the whole parts and fractional parts separately.

Example

$$15\frac{1}{3} + 18 = 33\frac{1}{3}$$

In this case it is much easier to simply add the whole parts rather than convert both arguments to improper fractions, add the improper fractions (by finding the least common denominator, replacing the fractions that figure in the sum with their equivalent, like ones, and adding the numerators of these equivalent, like fractions), and finally rewrite the answer in mixed form.

Example

$$24 + \frac{9}{7} = 24 + 1\frac{2}{7}$$
$$= 25\frac{2}{7}$$

---

[1] Proper fractions have a 0 as their whole part and whole numbers and integers have a 0 as their fractional part.

Here again it is easier to convert the improper fraction, $\frac{9}{7}$, to a mixed number and then add the whole parts. The alternative would require that we convert 24 to an improper fraction, add the improper fractions (by finding the least common denominator, replacing the fractions that figure in the sum with their equivalent, like ones, and adding the numerators of these equivalent, like fractions), and finally rewrite the answer in mixed form.

If the fractional parts add up to an improper fraction, we need to take an extra step to write the result in mixed number format.

Example

$$
\begin{aligned}
32\frac{3}{4} + 21\frac{1}{3} &= 53\frac{9 + 4}{12} \\
&= 53\frac{13}{12} \\
&= 53 + \frac{13}{12} \\
&= 53 + 1\frac{1}{12} \\
&= 54\frac{1}{12}
\end{aligned}
$$

Here is an example involving subtraction.

Example

$$
\begin{aligned}
12\frac{3}{4} - 5\frac{1}{2} &= 12\frac{3}{4} - 5\frac{2}{4} \\
&= 7\frac{1}{4}
\end{aligned}
$$

The only problem with such an approach to subtraction of fractions is that we may end up with negative values when we subtract the fractional or the whole parts.

Example

$$9\frac{1}{2} - 6\frac{3}{4} = 9\frac{2}{4} - 6\frac{3}{4}$$
$$= 3\frac{2-3}{4}$$
$$= 3\frac{-1}{4}$$

The result $3\frac{-1}{4}$ is short for the longer $3 - \frac{1}{4}$.[2]

$$9\frac{1}{2} - 6\frac{3}{4} = 9\frac{2}{4} - 6\frac{3}{4}$$
$$= 3\frac{2-3}{4}$$
$$= 3\frac{-1}{4}$$
$$= 3 - \frac{1}{4}$$

To speed up such subtractions, we can *borrow* a 1 from 3 and break it down into $\frac{4}{4}$.[3],[4]

$$9\frac{1}{2} - 6\frac{3}{4} = 9\frac{2}{4} - 6\frac{3}{4}$$
$$= 3\frac{2-3}{4}$$
$$= 3\frac{-1}{4}$$
$$= 3 - \frac{1}{4}$$
$$= 2\frac{4}{4} - \frac{1}{4}$$
$$= 2\frac{3}{4}$$

In general, one must exercise care when applying the algorithm introduced in this appendix to the subtraction of rational numbers, especially when the first operand is negative.

---

[2]It is, in fact, short for the longer $3 + \frac{-1}{4}$ which simplifies to $3 - \frac{1}{4}$.

[3]The reason 4 is chosen is because it is the same as the denominator in $\frac{1}{4}$, making the two fractions *like* fractions which eases the task of subtraction.

[4]Some choose to revert back to the original problem and borrow right from the start. This is undesirable as it means that the work done before one runs into a negative result on the subtraction of the fractional parts has been wasted.

Example

$$-2\frac{1}{3} - 4\frac{2}{5} = -2\frac{5}{15} - 4\frac{6}{15}$$
$$= -6\frac{-5 - 6}{15}$$
$$= -6\frac{11}{15}$$

The expression above involves a loss followed by another loss. The result is, therefore, negative as it represents the total *loss*. Furthermore, the whole parts and fractional parts of the operands get added as we seek *total* loss.

Of all the cases that one can set up, the one where the difference between the whole parts is negative while the difference between the fractional parts is positive can be most misleading as the notation fails to convey what needs to be done.

$$5\frac{3}{4} - 7\frac{2}{3} = -2\frac{9 - 8}{12}$$
$$= -2\frac{1}{12}$$

The result above, i.e., $-2\frac{1}{12}$, is short for $-2-\frac{1}{12}$ while semantics would require that the problem above should have simplified to $-2+\frac{1}{12}$.

We present a couple of examples involving whole numbers.

Example

$$8\frac{4}{5} - 2 = 6\frac{4}{5}$$

We subtract 2 from 8 to get 6. There is nothing to subtract from $\frac{4}{5}$.

Example

$$20 - 8\frac{4}{5} = 19\frac{5}{5} - 8\frac{4}{5}$$
$$= 11\frac{1}{5}$$

While the mixed number notation may be a useful notation for use in basic applications of math, at higher levels where negative numbers enter the game and multiple operations are involved, we often find that the need to go back and forth between the mixed and improper formats and the need to closely follow what is being done to make sure we add and subtract the whole and the fractional parts correctly, undermines any advantage that the mixed number format may offer. For this reason, the use of the improper format is preferred at higher level of study of math.

Ideally, you should know both alternatives and decide on the spot which of the two is most efficient for the addition and/or subtraction of the numbers in a given expression, or whether you need to switch back and forth between the two approaches.

Appendix K
0 as Exponent

Our aim in this appendix is to show that any number other than 0 raised to the power of 0 equals 1 and that 0^0 is indeterminate. These assertions have their roots in the reduction of fractions.

Consider the expression

$$\frac{4^5}{4^2}$$

One way to simplify such an expression is to rewrite the numerator and the denominator in expanded form and then reduce the resulting factors. This is shown below:

$$\frac{4 \times 4 \times 4 \times 4 \times 4}{4 \times 4}$$

Since the numerator and the denominator contain multiplication only, we can cancel out two 4s in the numerator with two 4s in the denominator. This yields

$$4 \times 4 \times 4$$

or 4^3 as the result. While this approach works, it is tedious and, if the exponents are large, can become impractical. As a shortcut to this longer approach we note that, since two 4s in the numerator can be cancelled out with two 4s in the denominator, we will end up with

$$4^{5-2}$$

or 4^3 as expected.

For us, then, *subtraction of exponents has its root in the reduction of fractions.*[1] Note also that the first operand in the subtraction refers to the

---

[1] The reader should train herself or himself to relate a subtraction of exponents to an act of reduction of fractions as shown above.

number of times 4 appears in the numerator and that the second operand in the subtraction refers to the number of times 4 appears in the denominator.

Consider now the case

$$\frac{4^5}{4^5}$$

If we were to use the shortcut above to evaluate this expression, we would arrive at

$$4^{5-5}$$

or 4^0. Unlike the previous example where the subtraction of exponents yielded a positive value which we could interpret, it is not quite apparent what this 0 exponent could mean. Let us then return to the longer approach of expanding the expressions in the numerator and the denominator of $\frac{4^5}{4^5}$ and see where it will lead us.

$$\frac{4 \times 4 \times 4 \times 4 \times 4}{4 \times 4 \times 4 \times 4 \times 4}$$

As it is apparent, all the 4s cancel out leaving us with 1 on both the numerator and the denominator. Since division of 1 by 1 equals 1, we have

$$\frac{4^5}{4^5} = 1$$

A little reflection now shows why an exponent of 0 sohuld lead to 1. An exponent of 0 can arise from subtraction of equal quantities. As discussed above, the first operand refers to the number of times 4 appears in the numerator and the second operand refers to the number of times 4 appears in the denominator. Since these numbers are equal, we have an equal number of 4s in the numerator and the denominator. This means that they all cancel out, leaving us with 1 as the answer.

The only exception to this rule is when the base itself is 0. Relating an expression of the form 0^0 to subtraction of equal quantities in the exponent (e.g., 0^{2-2} or 0^{6-6} and the like), we find that the root of the expression is in fractions with 0 on top and botom (e.g., $\frac{0^2}{0^2}$ or $\frac{0^6}{0^6}$ and the like). Simplification of the numerator and denominator of such fractions leads to the expression $\frac{0}{0}$ which, as shown in Appendix A, is indeterminate.

Appendix L
Negative Exponents

Our aim in this appendix is to show that a negative exponent on a nonzero base implies repeated division by that base with the size of the exponent indicating the number of times the division needs to be performed. In addition, when the base is 0, the expression is undefined. The assertions above have their roots in the reduction of fractions.

Consider the expression

$$\frac{4^5}{4^2}$$

One way to simplify such an expression is to rewrite the numerator and the denominator in expanded form and then reduce the resulting factors. This is shown below:

$$\frac{4 \times 4 \times 4 \times 4 \times 4}{4 \times 4}$$

Since the numerator and the denominator contain multiplication only, we can cancel out two 4s in the numerator with two 4s in the denominator. This yields

$$4 \times 4 \times 4$$

or 4^3 as the result. While this approach works, it is tedious and, if the exponents are large, can become impractical. As a shortcut to this longer approach we note that, since two 4s in the numerator can be cancelled out with two 4s in the denominator, we will end up with

$$4^{5-2}$$

or 4^3 as expected.

For us, then, *subtraction of exponents has its root in the reduction of fractions.*[1] Note also that the first operand in the subtraction refers to the number of times 4 appears in the numerator and that the second operand in the subtraction refers to the number of times 4 appears in the denominator.

Consider now the case

$$\frac{4^2}{4^5}$$

If we were to use the shortcut above to evaluate this expression, we would arrive at

$$4^{2-5}$$

or 4^{-3}. Unlike the previous example where the subtraction of exponents yielded a positive value which we could interpret, it is not quite apparent what this negative exponent could mean. Let us then return to the longer approach of expanding the expressions in the numerator and the denominator of $\frac{4^2}{4^5}$ and see where it will lead us.

$$\frac{4 \times 4}{4 \times 4 \times 4 \times 4 \times 4}$$

As it is apparent, two 4s in the numerator reduce with two 4s in the denominator leaving us with three 4s in the denominator, i.e.,

$$\frac{1}{4 \times 4 \times 4}$$

Since division of a product implies repeated division,[2] the expression above is equivalent to repeated division by 4.

---

[1] The reader should train herself or himself to relate a subtraction of exponents to an act of reduction of fractions as shown above.

[2] If you don't know this already, then you should: Division by a product of factors can be performed by dividing by each factor. As an example

$$\frac{210}{2 \times 5 \times 3}$$

can be performed by either (a) working out the denominator by performing the multiplications to get 30 and then divide 210 by 30 or (b) dividing 210 by 2, the result of the division by 5 and the result of this division by 3. The divisions can be performed in any order.

This is similar to the case of the subtraction of a sum which can be turned into repeated subtraction of the terms in the sum. As an example, if I had $62, bought a pen for $3.20, a notebook for $20.99 and an eraser for $3.25, I can calculate the amount of money that I have left by (a) adding the expenses to find total amount spent, which in this case evaluates to $27.44 and subtract this from the $62 that I started with or (b) subtracting the expenses one at a time: Subtract $3.20 from $62, then subtract $20.99 from the result, and then subtract $3.25 from the result of this subtraction. The expenses can be subtracted in any order.

We now understand two things: First, the exponent -3 that we came across has its root in some subtraction, such as $2 - 5$ in the example above. Such subtractions must have a larger number being subtracted from a smaller number. This larger number points to the number of times 4 appears in the denominator (here, 5 times) and the smaller number points to the number of times 4 appears in the numerator (here, 2 times) which results in the presence of 4s in the denominator after cancellation. This, in turn, implies the need for division. In short, a negative exponent implies repeated division by the base imvolved. We can write

$$4^{-3} = \frac{1}{4^3}$$

The only exception to this rule is when the base itself is 0. Relating an expression of the form 0^{-3} to $\frac{1}{0^3}$ and noting that 0^3 is equal to 0, we see at once that we are dealing with division of a nonzero number by 0 which is an undefined expression.[3] Accordingly, we do not associate meaning to expressions of the form 0^e where e is a negative integer and we call such expressions *undefined*.[4]

---

[3]See Appendix A on matters relating to division by 0.

[4]We have commented elsewhere and do so again here that undefined expressions *do* appear in practice when one attempts to solve a problem that has no solutions.

Appendix M
Working with Exponents

In this appendix we will explain the manner in which one can simplify algebraic expressions that involve exponents.

Consider the expression

$$x^3 x^5$$

The expanded form of this expression is

$$\overbrace{x\ x\ x}^{x^3}\ \overbrace{x\ x\ x\ x\ x}^{x^5}$$

As the reader can see, the expanded form of $x^3 x^5$ turns into a multiplication of 8 identical factors, x.

$$\overbrace{x\ x\ x}^{x^3}\ \overbrace{x\ x\ x\ x\ x}^{x^5}\ =\ \overbrace{x\ x\ x\ x\ x\ x\ x\ x}^{x^8}$$

This is because x^3 involves the multiplication of 3 identical factors, x, and x^5 involves the multiplication of 5 of the same factors, x, so that multiplication of x^3 and x^5 generates a chain multiplication of $3 + 5$ or 8 identical factors, x. This implies the following:

$$\begin{aligned} x^3 x^5 &= x^{3+5} \\ &= x^8 \end{aligned}$$

i.e., when expressions involving exponents on the same base are multiplied together, the exponents add up. The general statement is written as

$$x^m x^n = x^{m+n}$$

It can be shown that this identity works, not just with whole numbers as exponents, but other number types as well. As an example

$$x^7x^{-2} = x^{7-2}$$
$$= x^5$$

To explain why we can subtract the exponents as shown above, we convert the second factor, x^{-2} to $\frac{1}{x^2}$ to get[1]

$$x^7x^{-2} = x^7 \times \frac{1}{x^2}$$

Carrying out the multiplication yields

$$x^7x^{-2} = x^7 \times \frac{1}{x^2}$$
$$= \frac{x^7}{x^2}$$
$$= x^{7-2}$$

Justification for the last step is given in Appendix L.

Examples

Simplify each of the following expressions.

1. x^5x^8 3. $x^{-7}x^3$ 5. $z^2z^{-3}z^0$

2. y^8y^{-1} 4. $x^{-2}x^{-3}$ 6. $z^{-5}z^{-1}z^{-2}$

Solutions

1.   3.

$$x^5x^8 = x^{5+8}$$
$$= x^{13}$$

2.

$$y^8y^{-1} = y^{8-1}$$
$$= y^7$$

$$x^{-7}x^3 = x^{-7+3}$$
$$= x^{-4}$$
$$= \frac{1}{x^4}$$

---

[1]See Appendix L for an explanation on the equivalence of x^{-n} and $\frac{1}{x^n}$.

4.

$$x^{-2}x^{-3} = x^{-2-3}$$
$$= x^{-5}$$
$$= \frac{1}{x^5}$$

6.

$$z^{-5}z^{-1}z^{-2} = z^{-5-1-2}$$
$$= z^{-8}$$
$$= \frac{1}{z^8}$$

5.

$$z^2 z^{-3} z^0 = z^{2-3+0}$$
$$= z^{-1}$$
$$= \frac{1}{z^1}$$
$$= \frac{1}{z}$$

Consider now the expression

$$\frac{x^5}{x^3}$$

Rewriting this expression in expanded form yields

$$\frac{\overbrace{x\ x\ x\ x\ x}^{x^5}}{\underbrace{x\ x\ x}_{x^3}}$$

Since this is an expression involving chain multiplication/division, we can reduce factors in the numerator with factors in the denominator. In fact, we can reduce 3 factors in the numerator with 3 factors in the denominator, leaving us with $5 - 3$ or 2 factors in the numerator:

$$\frac{x\ x\ x\ x\ x}{x\ x\ x} = x\ x$$

As noted above, a shortcut for this reduction is to subtract the exponents to find out how many extra factors, x, we are left with after reduction:

$$\frac{x^5}{x^3} = x^{5-3}$$
$$= x^2$$

The observation above can be generalized:

$$\frac{x^m}{x^n} = x^{m-n}$$

The identity above holds, not just for whole numbers as exponents but also other type of numbers as well. As an example,

$$\frac{x^3}{x^5} = x^{3-5}$$
$$= x^{-2}$$

The negative exponent implies that, after reduction, there are extra factors, x, in the denominator, in fact, 2 of them, i.e.,

$$\frac{x^3}{x^5} = x^{3-5}$$
$$= x^{-2}$$
$$= \frac{1}{x^2}$$

As a second example

$$\frac{x^3}{x^{-5}} = x^{3+5}$$
$$= x^8$$

Note that

$$\frac{1}{x^{-5}} = \frac{1}{\frac{1}{x^5}}$$
$$= \frac{\frac{1}{1}}{\frac{1}{x^5}}$$
$$= \frac{1 \times x^5}{1 \times 1}$$
$$= \frac{x^5}{1}$$
$$= x^5$$

Examples

Simplify each of the following expressions.

1. $\dfrac{x^8}{x^2}$　　　　2. $\dfrac{z^6}{z^{-2}}$　　　　3. $\dfrac{x^{-1}}{x^3}$　　　　4. $\dfrac{y^{-6}}{y^{-3}}$

Solutions

1.

$$\frac{x^8}{x^2} = x^{8-2}$$
$$= x^6$$

2.

$$\frac{z^6}{z^{-2}} = z^{6+2}$$
$$= z^8$$

3.

$$\frac{x^{-1}}{x^3} = x^{-1-3}$$
$$= x^{-4}$$
$$= \frac{1}{x^4}$$

4.

$$\frac{y^{-6}}{y^{-3}} = y^{-6-3}$$
$$= y^{-9}$$
$$= \frac{1}{y^9}$$

One can combine the identities above to simplify expressions involving a chain multiplication/division of the same base raised to different exponents. As an example

$$\frac{x^{-4}x^6}{x^2 x^{-5}} = x^{-4+6-2+5}$$
$$= x^5$$

Examples

Simplify each of the following expressions.

1. $\dfrac{x^9 x^3}{x^8}$　　　　2. $\dfrac{y^6}{y^{-5}y^3}$　　　　3. $\dfrac{x^{-2}x^{-3}}{x^{-7}}$　　　　4. $\dfrac{z^{-6}z}{z^0}$

Solutions

1.

$$\frac{x^9 x^3}{x^8} = x^{9+3-8}$$

$$= x^4$$

2.

$$\frac{y^6}{y^{-5}y^3} = y^{6+5-3}$$

$$= y^8$$

3.

$$\frac{x^{-2}x^{-3}}{x^{-7}} = x^{-2-3+7}$$

$$= x^2$$

4.

$$\frac{z^{-6}z}{z^0} = z^{-6+1-0}$$

$$= z^{-5}$$

$$= \frac{1}{z^5}$$

Consider now the expression

$$\left(x^3\right)^4$$

Expanding the expression above yields

$$\left(x^3\right)^4 = (x\,x\,x)\,(x\,x\,x)\,(x\,x\,x)\,(x\,x\,x)$$

We have 4 groups of x^3 which yields a chain multiplication of 3×4 or 12 factors, x:

$$\left(x^3\right)^4 = (x\,x\,x)\,(x\,x\,x)\,(x\,x\,x)\,(x\,x\,x)$$

$$= x\,x\,x\,x\,x\,x\,x\,x\,x\,x\,x\,x$$

$$= x^{12}$$

A shortcut for the argument above is

$$\left(x^3\right)^4 = x^{3\times 4}$$

$$= x^{12}$$

The argument above can be generalized:

$$\left(x^m\right)^n = x^{mn}$$

This identity works for all exponents and not just whole numbers.

Examples

Simplify each of the following expressions.

1. $\left(x^4\right)^5$ 2. $\left(x^7\right)^{-2}$ 3. $\left(x^{-3}\right)^4$ 4. $\left(z^{-2}\right)^{-3}$

Solutions

1.

$$\left(x^4\right)^5 = x^{4\times 5}$$
$$= x^{20}$$

2.

$$\left(x^7\right)^{-2} = x^{7(-2)}$$
$$= x^{-14}$$
$$= \frac{1}{x^{14}}$$

3.

$$\left(x^{-3}\right)^4 = x^{-3\times 4}$$
$$= x^{-12}$$
$$= \frac{1}{x^{12}}$$

4.

$$\left(z^{-2}\right)^{-3} = z^{-2(-3)}$$
$$= z^6$$

Note that the expression $(x^m)^n$ is different from the expression x^{m^n}. In the case of $(x^m)^n$, it is the expression within brackets, i.e., x^m that is repeatedly multiplied by itself whereas in x^{m^n}, it is the exponent m that is repeatedly multiplied by itself. As an example

$$\left(x^4\right)^5 = x^{4\times 5}$$
$$= x^{20}$$

whereas

$$x^{4^5} = x^{4\times 4\times 4\times 4\times 4}$$
$$= x^{1024}$$

Before we end this appendix, we would like to note that the identities given above hold provided the expressions are defined. As an example, the expression x^{-2} is undefined if x is 0.

Exercise Set M

1. Simplify each of the following expressions.

   a. $x^6 x^7$ c. $x^0 x^{-1}$ e. $x^{-4} x^{-12}$ g. $x^2 x^0 x^{-1}$

   b. $x^{19} x^{-3}$ d. $y^{-5} y^5$ f. $x^{-5} x^6 x^0$ h. $z^{-3} z^{-3} z^{-3}$

2. Simplify each of the following expressions.

   a. $\dfrac{x^8}{x^2}$ b. $\dfrac{x^{-17}}{x^{-12}}$ c. $\dfrac{x^2}{x^0}$ d. $\dfrac{y^{-3}}{y^{-5}}$

3. Simplify each of the following expressions.

   a. $\dfrac{x^5 x^4}{x^7}$ c. $\dfrac{x^3 x^{-1}}{x^6}$ e. $\dfrac{z^0}{z^3 z^7}$ g. $\dfrac{x^2}{x^0 x^{-3}}$

   b. $\dfrac{x^{-10}}{x^8 x^5}$ d. $\dfrac{y^{-3} y^{-2}}{y^{-5} y^{-1}}$ f. $\dfrac{x^{-4} x^{-7}}{x^{-6}}$ h. $\dfrac{z^{-1} z^0}{z^{-2} z^5 z^2}$

4. Simplify each of the following expressions.

   a. $\left(x^3\right)^6$ c. $\left(x^{-7}\right)^3$ e. $\left(x^0\right)^{-2}$ g. $\left(x^2\right)^8$

   b. $\left(x^6\right)^3$ d. $\left(x^{-1}\right)^5$ f. $\left(x^5\right)^5$ h. $\left(x^{-7}\right)^2$

5. Simplify each of the following expressions.

   a. x^{3^4} b. x^{2^5} c. x^{6^1} d. $x^{(-3)^2}$

Appendix N
Percentage

N.1   Introduction

Percentages are expressions that attempt to show how the size of one quantity compares to 100 units of another quantity. As an example, when we state that 63% *of the eligible voters voted*, we aim to convey the fact that, out of every 100 eligible voters, 63 voted. In this example, the two quantities involved are *the number of eligible voters* and *the number of eligible voters who voted*.

It is understood that percent notation uses division to compare the values of the two quantities involved, e.g., the statement 63% *of the eligible voters voted* is interpreted as $\frac{63}{100}$ *(read sixty-three hundredths) of the eligible voters voted*.

In the expression 63%, we refer to the numerical part, i.e., 63, as the **rate** and we read the % sign as **percent**.

In this appendix we will discuss the manner in which one can convert notation between percentage and formal and base-ten notations.

N.2   Conversion to and from the Formal Notation

We will begin with the conversion from percent notation to formal notation. Following this, we will present the manner in which formal notation can be converted to percent notation.

N.2.1   Conversion from Percent Notation to Formal Notation

We will use examples to describe the manner in which percent notation can be converted to formal notation.

Example

Convert 34% to formal notation.

Solution

Since we want to write the result in formal notation and the rate is a whole number, we use the fraction notation to process the division by 100. If it is possible to reduce the resulting fraction, we will do so.

$$34\% = \frac{34}{100}$$
$$= \frac{17}{50}$$

Example

Convert 2% to formal notation.

Solution

$$2\% = \frac{2}{100}$$
$$= \frac{1}{50}$$

If the rate is a whole number larger than or equal to 100, we can place the rightmost two digits over 100 and write the rest on the whole side. This is faster than placing the full rate over 100 and converting the result to a mixed number. The following example illustrates the idea.

Example

Convert 345% to formal notation.

Solution

$$345\% = 3\frac{45}{100}$$
$$= 3\frac{9}{20}$$

When the rate is a rational number, division by 100 can be processed using either the $\div$ notation or the horizontal line. We recommend the latter.

Example

Convert $\frac{3}{4}\%$ to formal notation.

Solution

Alternative 1 – Use the $\div$ notation

$$\frac{3}{4}\% = \frac{3}{4} \div 100$$
$$= \frac{3}{4} \div \frac{100}{1}$$
$$= \frac{3}{4} \times \frac{1}{100}$$
$$= \frac{3}{400}$$

Alternative 2 – Use the horizontal line (Recommended)

$$\frac{3}{4}\% = \frac{\frac{3}{4}}{100}$$
$$= \frac{\frac{3}{4}}{\frac{100}{1}}$$
$$= \frac{3 \times 1}{4 \times 100}$$
$$= \frac{3}{400}$$

Example

Convert $2\frac{1}{5}\%$ to formal notation.

Solution

Alternative 1 – Use the $\div$ notation

$$
\begin{aligned}
2\frac{1}{5}\% &= 2\frac{1}{5} \div 100 \\
&= \frac{11}{5} \div \frac{100}{1} \\
&= \frac{11}{5} \times \frac{1}{100} \\
&= \frac{11}{500}
\end{aligned}
$$

Alternative 2 – Use the horizontal line (Recommended)

$$
\begin{aligned}
2\frac{1}{5}\% &= \frac{2\frac{1}{5}}{100} \\
&= \frac{\frac{11}{5}}{\frac{100}{1}} \\
&= \frac{11 \times 1}{5 \times 100} \\
&= \frac{11}{500}
\end{aligned}
$$

N.2.2   Conversion from Formal Notation to Percent Notation

To convert from formal notation to percent notation, we multiply the given value by 100%. This will not change the size of the value that we start with as 100% is equal to[1] 1 and multiplication of a value by 1 does not change the value. We can also think of multiplication by 100% as multiplication by 100 followed by division by 100 (the latter maps onto our interpretation of the

---

[1]

$$
\begin{aligned}
100\% &= \frac{100}{100} \\
&= 1
\end{aligned}
$$

% sign) and the combined effect of the two will leave the size of the value that we start with unchanged. The second action, i.e., division by 100, is not actually carried out but is left as %. This means that, while we view 100% as a single entity to begin with, we continue by splitting the 100 from the % sign, carry the multiplication by 100 and leave the % (division by 100) untouched.

Example

Convert $\frac{3}{5}$ to percent notation.

Solution

$$\begin{aligned}
\frac{3}{5} &= \frac{3}{5} \times 100\% \\
&= \frac{3}{5} \times \frac{100}{1}\% \\
&= \frac{60}{1}\% \\
&= 60\%
\end{aligned}$$

Conversion of mixed numbers to percentage can be done by converting the mixed number into an improper fraction before we multiply the fraction by 100%. However, an easier approach is to convert the fractional part of the mixed number to percent as in the previous example, and then add 100 times the whole part to the result.

Example

Convert $5\frac{4}{15}$ to percent notation.

Solution

$$\frac{4}{15} = \frac{4}{15} \times 100\%$$

$$= \frac{4}{15} \times \frac{100}{1}\%$$

$$= \frac{80}{3}\%$$

$$= 26\frac{2}{3}\%$$

Since $\frac{4}{15}$ converts to $26\frac{2}{3}\%$, we can conclude that $5\frac{4}{15}$ converts to $526\frac{2}{3}\%$.

Exercise Set N.2

1. Convert each of the following to formal notation.

   a. 57%   d. 8%   g. $27\frac{1}{8}\%$ j. $\frac{5}{6}\%$

   b. 20%   e. 1%   h. $2\frac{3}{4}\%$ k. $\frac{3}{8}\%$

   c. 32%   f. 99%   i. $35\frac{2}{5}\%$ l. $\frac{9}{10}\%$

2. Convert each of the following to percent notation.

   a. $\frac{57}{100}$ d. $\frac{2}{25}$ g. $\frac{217}{800}$ j. $\frac{1}{120}$

   b. $\frac{1}{5}$ e. $\frac{1}{100}$ h. $\frac{11}{400}$ k. $\frac{3}{800}$

   c. $\frac{8}{25}$ f. $\frac{99}{100}$ i. $\frac{177}{500}$ l. $\frac{9}{1000}$

N.3   Conversion to and From Base-Ten Notations

In this section of the text we will discuss how percent notation can be converted to decimal notation and vice versa.

N.3.1   Conversion from Percent Notation to Base-Ten Notations

To convert from percent notation to decimal notation, we divide the rate by 100 and drop the % sign.

Example

Convert 48% to decimal notation.

Solution

$$48\% = 48 \div 100$$
$$= 0.48$$

Example

Convert 30% to decimal notation.

Solution

$$30\% = 30 \div 100$$
$$= 0.3$$

Note that in the example above we dropped the right-end 0.

Example

Convert 2% to decimal notation.

Solution

$$2\% = 2 \div 100$$
$$= 0.02$$

Example

Convert 235% to decimal notation.

Solution

$$235\% = 235 \div 100$$
$$= 2.35$$

Example

Convert 6.9% to decimal notation.

Solution

$$6.9\% = 6.9 \div 100$$
$$= 0.069$$

Example

Convert 0.7% to decimal notation.

Solution

$$0.7\% = 0.7 \div 100$$
$$= 0.007$$

N.3.2   Conversion from Base-Ten Notations to Percent Notation

To convert from decimal notation to percent notation, we multiply the given value by 100%. As discussed earlier, this will not change the size of the value that we start with as 100% is equivalent to 1 and multiplication by 1 does not change the size of a number. We can also think of multiplication by 100% as multiplication by 100 followed by division by 100 (the latter maps onto our interpretation of the % sign) and the combined effect of the two will leave the size of the value that we start with unchanged. The second action, i.e., the division by 100, is not actually carried out but is left as %. This means that, while we view 100% as a single entity, we continue by splitting the 100 from the % sign, carry the multiplication by 100 and leave the % (division by 100) untouched.

Example

Convert 0.32 to percent notation.

Solution

$$0.32 = 0.32 \times 100\%$$
$$= 32\%$$

Example

Convert 0.2 to percent notation.

Solution

$$0.2 = 0.2 \times 100\%$$
$$= 20\%$$

Example

Convert 0.07 to percent notation.

Solution

$$0.07 = 0.07 \times 100\%$$
$$= 7\%$$

Example

Convert 0.054 to percent notation.

Solution

$$0.054 = 0.054 \times 100\%$$
$$= 5.4\%$$

Example

Convert 3.2 to percent notation.

Solution

$$3.2 = 3.2 \times 100\%$$
$$= 320\%$$

Exercise Set N.3

1. Convert each of the following to decimal notation.

a. 57%	e. 1%	i. 14.2%	m. 350.9%
b. 20%	f. 99%	j. 9.6%	n. 0.02%
c. 32%	g. 45.7%	k. 2.4%	o. 0.5%
d. 8%	h. 23.9%	l. 120.5%	p. 0.004%

2. Convert each of the following to percent notation.

a. 0.57	f. 0.99	k. 0.009	p. 0.024
b. 0.2	g. 0.27125	l. 0.457	q. 1.205
c. 0.32	h. 0.0275	m. 0.239	r. 3.509
d. 0.08	i. 0.354	n. 0.142	s. 0.0002
e. 0.01	j. 0.00375	o. 0.096	t. 0.000 04

Appendix O
Graphing in Two Dimensions

Graphs in two dimensions provide a snapshot of the relationship between the values of two quantities. Their strength lies in their ability to allow the user to quickly relate the value of one quantity to that of another as well as understand at a glance the manner in which change in the value of one quantity forces change in the value of the other quantity. On the downside, graphs are never exact and, if exact values are needed, one will have to resort to the use of the underlying equations.[1]

In this appendix we will discuss the fundamentals of graphing in two dimensions. We will begin with a discussion of the Cartesian coordinate system, the most common coordinate system used for graphing equations, and move on to discuss the manner in which graphs are generated with our focus on graphing by plotting points. The appendix ends with a presentation of some of the key features of graphs: The intercepts and the slope.

O.1   The Cartesian Coordinate System

The Cartesian coordinate system is the most basic and most widely used coordinate system.[2] In two dimensions, the Cartesian coordinate system consists of two number lines, one horizontal, the other vertical, intersecting at the points marked as 0 on the two lines, with the same scale used on both lines.[3] In abstract settings, the horizontal axis is referred to as the x-**axis**

---

[1] The approximate nature of graphs implies that we can use base-ten notations to display the values of the two quantities along the axes. This is another use of base-ten notations in math.

[2] The Cartesian coordinate system is named after the French mathematician and philosopher, Rene Descartes (1596-1650).

[3] In practical settings where the sizes of the values on the horizontal axis are very different from the sizes of the values on the vertical axis, we often drop this last requirement and use different scales on the two axes. For more on the practical applications of coordinate systems please see the companion textbook *Semantics and the Syntax of Algebra* by the author.

and the vertical axis is called the *y*-**axis**.[4] The Cartesian coordinate system is shown in Figure O.1.1.

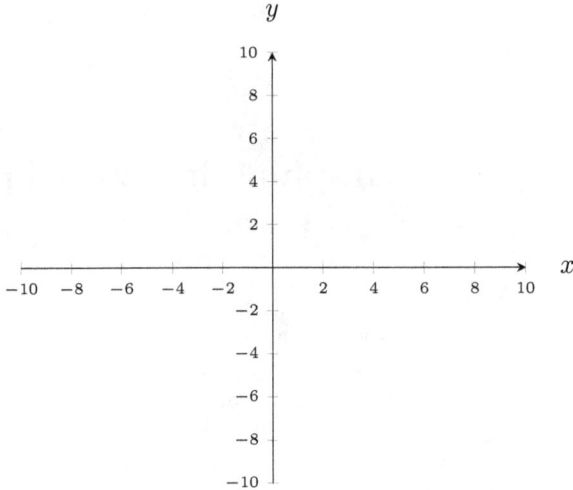

Figure O.1.1: The Cartesian coordinate system with a step size of 2 on both axes

As for step size, any value that can be skip-counted easily is acceptable. Examples of acceptable step sizes are 1, 2, 5, 10, 20, etc., but not 3, 7, 11, and the like.

O.2 Graphs as Relationships

Suppose the relationship between x and y is such that when x is 3, y is 6. We write this relationship as the ordered pair $(3, 6)$. An **ordered pair** consists of two values within parentheses separated by a comma. The first value relates to the independent variable, x, and the second value relates to the dependent variable, y.

We can also represent the relationship between the values assigned to x and y above as a point in the plane of a Cartesian coordinate system. This is shown in Figure O.2.2 below.

---

[4]In applications the abstract quantity symbols, x and y, are replaced by the relevant quantity symbols. As an example, in a problem where the horizontal axis represents travel time and the vertical axis represents distance covered, the horizontal axis is referred to as the t-axis and the vertical axis is referred to as the d-axis.

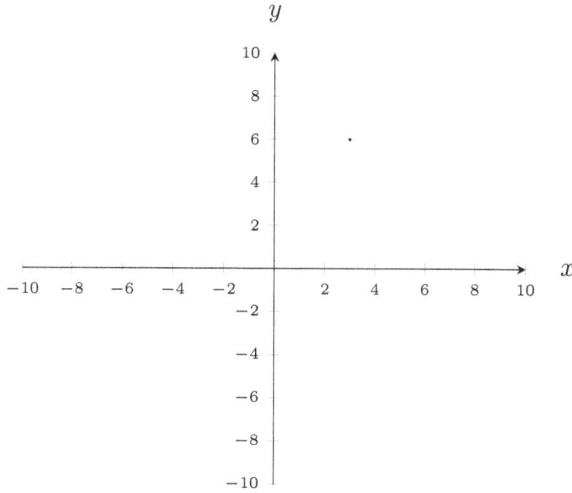

Figure O.2.2: Representation of the association between x and y values as a point in the plane of a Cartesian coordinate system

Note the manner in which the point in the plane associates the value of 3 for x to the value of 6 for y: If you draw an imaginary line from the point straight down to the x-axis you will cut the x axis at 3 and if you draw an imaginary line from the point to the y-axis you will cut the y-axis at 6. The imaginary lines are shown as dashed lines in Figure O.2.3 below.

To find the ordered pair that is associated with a point, then, we draw imaginary lines from the point onto the x- and y-axes and read the values where the imaginary lines cut the two axes. As shown in the example above, the ordered pair associated with the point in Figure O.2.3 is $(3, 6)$.

We can also map an ordered pair to its associated point in the plane by finding the x and y values listed by the ordered pair on the x- and y-axes, draw an imaginary vertical line from the x value on the x-axis and an imaginary horizontal line from the y value on the y-axis to find their point of intersection. As an example, to find the point in the plane associated with the ordered pair $(3, 6)$, we find the value 3 on the x-axis and draw an imaginary vertical line at 3 on the x-axis, and we find the value 6 on the y-axis and draw an imaginary horizontal line at 6 on the y-axis. The point at which these imaginary lines intersect is the point that we associate with the ordered pair.

We refer to the values listed by the ordered pair as the coordinates of the

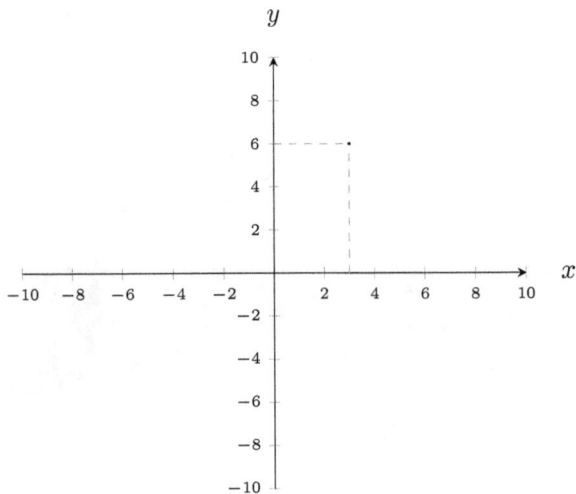

Figure O.2.3: An illustration of the manner in which a point relates x and y values

point. As an example, we refer to the values 3 and 6 listed by the ordered pair $(3, 6)$ as the coordinates of the point in Figure O.2.2. The first value in the ordered pair is referred to as the x-**coordinate** of the associated point. As an example, we refer to 3 as the x-coordinate of the point in Figure O.2.2. Similarly, the second value in the ordered pair is referred to as the y-**coordinate** of the associated point. As an example, we refer to 6 as the y-coordinate of the point in the Figure O.2.2.

Multiple points can be used to represent multiple relationships between x and y values. As an example, suppose that x and y are related such that when x is 3, y is 6, when x is 0, y is -1.5 and when x is -2, y is 4. These relationships can be expressed as the ordered pairs $(3, 6)$, $(0, -1.5)$, and $(-2, 4)$ and displayed as the three points in Figure O.2.4 below.

An ordered pair, of course, represents a single relationship between x and y values. Such representations become inefficient as the number of specific relationships between x and y values increases. Furthermore, the notation cannot keep up with the demand for the representation of an infinite number of such relationships.

As an example, suppose y is always twice larger than x. We can list a few specific relationships between x and y values by listing the ordered pairs $(3, 6)$, $(1.5, 3)$, $(0, 0)$, $(-2, -4)$, etc. but what about $(4, 8)$, $(1.4, 2.8)$ and the infinite other possibilities where y is twice larger than x?

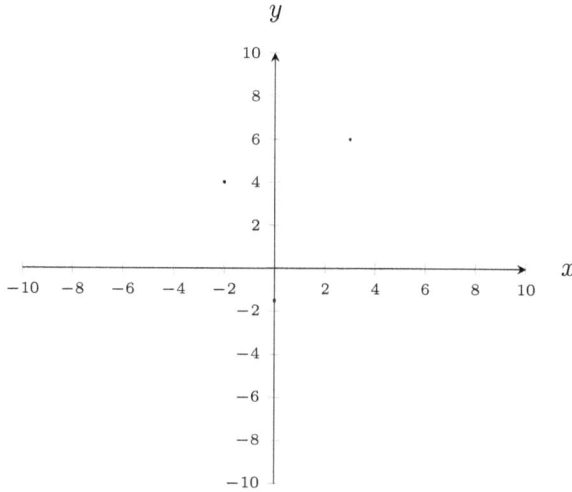

Figure O.2.4: Representation of multiple relationships between x and y values

In such situations we express the relationship as an equation. As an
example, the equation

$$y = 2x, \qquad x \in \mathbb{R}$$

expresses the fact that y is always twice larger than x. This side condition,
$x \in \mathbb{R}$, states that x can be any real number, e.g., -3, 0, $\frac{1}{2}$, $\sqrt{3}$, etc.[5]

We can now pick any value from the set of real numbers for x and use

---

[5]Such statements as $x \in \mathbb{R}$ in the model above are called the *domain* of the equation. The **domain** of an equation is a statement of permissible values for the independent
variable. Note that the model

$$y = 2x, \qquad x \in \mathbb{R}$$

is different from the model

$$y = 2x, \qquad x \in \mathbb{N}$$

While the first allows x to be $\frac{1}{2}$, the second does not as the domain of the second equation
is $\mathbb{N}$, i.e., x can only assume the values 1, 2, 3, 4, and so on.

Every model includes a statement for the domain of the equation. By agreement, if the
domain is not given, it is assumed to be the largest set for which the equation makes sense.
As an example, limiting ourselves to $\mathbb{R}$ as the largest set of numbers known to us, the model

$$y = 2x$$

is seen to be the same as

$$y = 2x, \qquad x \in \mathbb{R}$$

the given equation to find the corresponding value of y. As an example, if we pick the value 3 for x, then y will be 6. This selection generates the ordered pair $(3, 6)$. And if we pick the value -2 for x, then y will be -4. This seclection generates the ordered pair $(-2, -4)$. It is obvious that we can generate an infinite number of such relationships and express each as an ordered pair. Figure O.2.5 shows the graph of ten such ordered pairs for the equation $y = 2x$.

For the relationship $y = 2x$, the points appear to fall on a straight line. This is in fact the case: If we were going to add more points to the graph of the equation $y = 2x$, we would arrive at the graph in Figure O.2.6 below.

---

and not

$$y = 2x, \qquad x \in \mathbb{N}$$

As a second example, the equation

$$y = \frac{1}{x}$$

is seen to be the same as

$$y = \frac{1}{x}, \qquad x \in \mathbb{R}, \ x \neq 0$$

Note that x cannot be 0 otherwise we run into the undefined expression $\frac{1}{0}$.

In applications, situations in which the domain is in some way restricted are common. As an example, the relationship between the value of temperature in degrees Celsius, t, and the value of temperature in kelvins, T, is given by the equation

$$t = T - T_0, \qquad T \in \mathbb{R}, \ T \geq 0$$

where T_0 is a constant. It is shown in physics that T is nonnegative, i.e., there is a coldest possible temperature which occurs as 0 K. This corresponds to a temperature of -273.15 °C. For more on the domain of models that arise from a study of concrete problems please see the companion textbook *Semantics and the Syntax of Algebra* by the author.

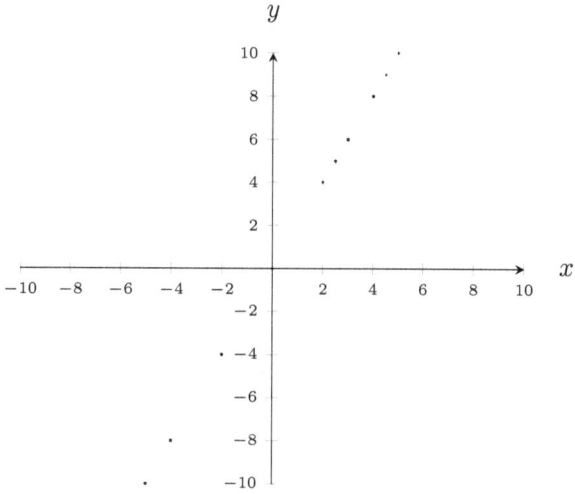

Figure O.2.5: Graph of ten relationships between x and y values for the model $y = 2x$

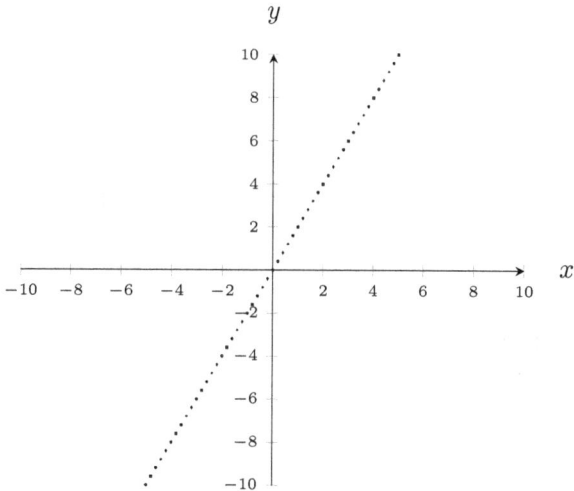

Figure O.2.6: Graph of fifty relationships between x and y values for the model $y = 2x$

As more and more points are added to the graph above, the graph begins to appear more and more as a straight line. We, therefore, draw a straight line through these points and call this line the graph of the equation $y = 2x$. Such a graph displays *all* ordered pairs where the second value is twice the first.[6] This is shown in Figure O.2.7 below.

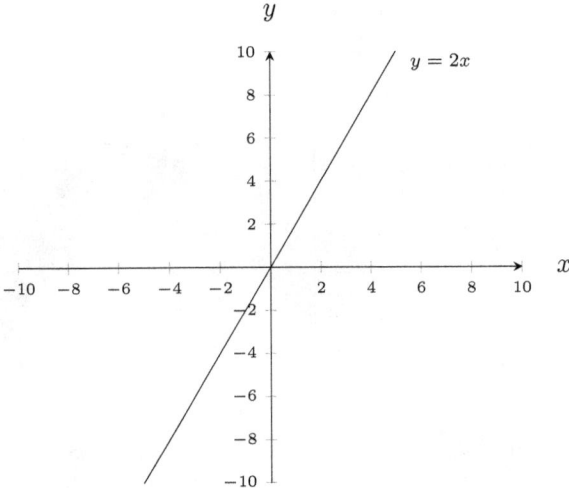

Figure O.2.7: Graph of the equation $y = 2x$

As a matter of style, we write the equation for the graph next to the graph.[7]

Each point on the line in the graph above relates to a single ordered pair where the second value is twice the first. As an example, consider the point on the line in Figure O.2.8 below.

To find the x-coordinate of this point we draw an imaginary line from the point straight down to the x-axis. The value on the x-axis at which this imaginary line meets the x-axis is the x-coordinate of the point. As shown in Figure O.2.9 below, the x-coordinate of the point in Figure O.2.8 is 4.

To find the y-coordinate of this point we draw an imaginary line from the point across to the y-axis. The value on the y-axis at which this imaginary

---

[6]The graph is assumed to continue on both sides indefinitely.

[7]As stated earlier, graphs, no matter how accurately plotted, are inherently approximate. As an example, a point in the plane may appear to have an x-coordinate of 3. However, we can never be sure that this value is exact: It could be 3.001 as an example. Equations, on the other hand, are able to work with exact values. While graphs are able to pack a lot of information in one visual image (e.g., related values of the quantities involved, the manner in which change in the value of one quantity forces change in the value of the other quantity, etc.), all such information is approximate. Inclusion of the actual equation ensures that, if needed, the user can switch to the use of exact values.

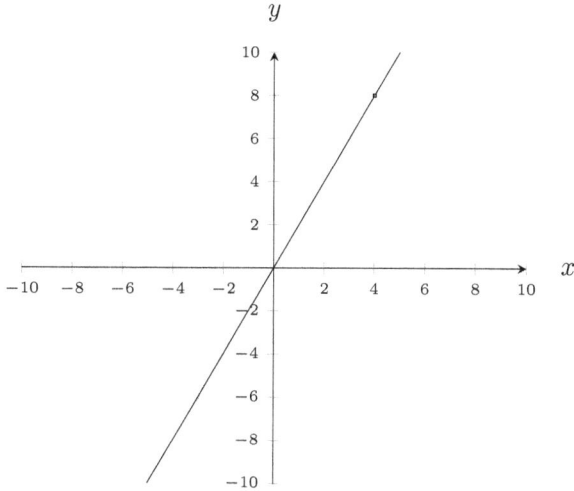

Figure O.2.8: A point on the graph of the line $y = 2x$

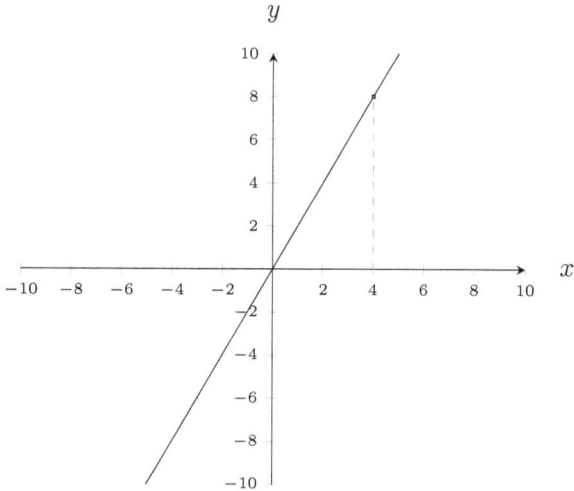

Figure O.2.9: Reading the x-coordinate of a point on a line in the plane of a Cartesian coordinate system

line meets the y-axis is the y-coordinate of the point. As shown in Figure O.2.10 below, the y-coordinate of the point in Figure O.2.8 is 8.

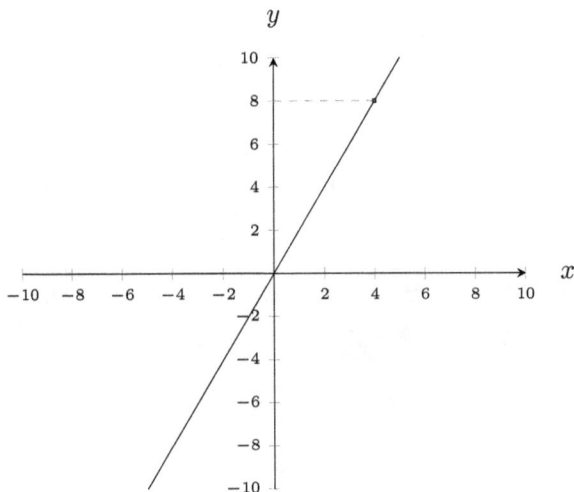

Figure O.2.10: Reading the y-coordinate of a point on a line in the plane of a Cartesian coordinate system

This associates the ordered pair $(4, 8)$ to the point in Figure O.2.8. Both representations state that, when x is 4, y is 8 with the ordered pair $(4, 8)$ doing so exactly and the graph, approximately.

Small circles are used to represent restricted domains. An example of such circles is given in Figure O.2.11 below.

The graph above is a segment of the line $y = 2x$. The segment runs from -4 to 3 on the x-axis. The presence of the filled dot on the left indicates that x can equal to -4. The presence of the empty dot on the right indicates that x cannot equal to 3, i.e., the domain is $x \in \mathbb{R}$, $-4 \leq x < 3$.

The discussion above leads to the following steps for graphing equations by plotting points.

Steps for Graphing Equations by Plotting Points

1. Choose a scale and a range for the x values and draw a Cartesian coordinate system.
2. Generate a table of values for x and y. Select a number of values for x and write them in the x column. In case the domain is restricted, include the values that correspond to the end points of the domain in the x column.

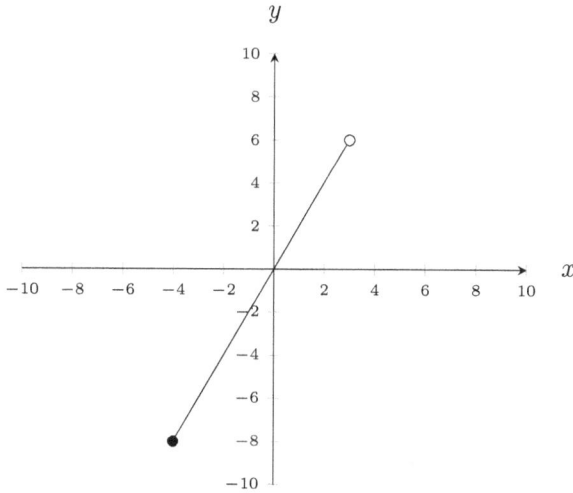

Figure O.2.11: Representation of restricted domains

3. Use the equation to find a y value for each x value in the table and record these values next to corresponding values of x in the table.
4. Find the maximum and minimum values of y in the table and choose a scale and a range for the y-axis that covers these values.
5. Generate an ordered pair from each row in the table and graph each in turn. If the point is an endpoint, use a small empty or filled circle to represent the point.
6. Connect the points in a smooth manner.
7. As a matter of style, write the equation of the graph next to the graph itself.

Example

Graph the equation

$$y = x^3 - 2x^2 - 5x + 6$$

Solution

Following Step 1 we generate a Cartesian coordinate system and choose values of x that range from -3 to 4.

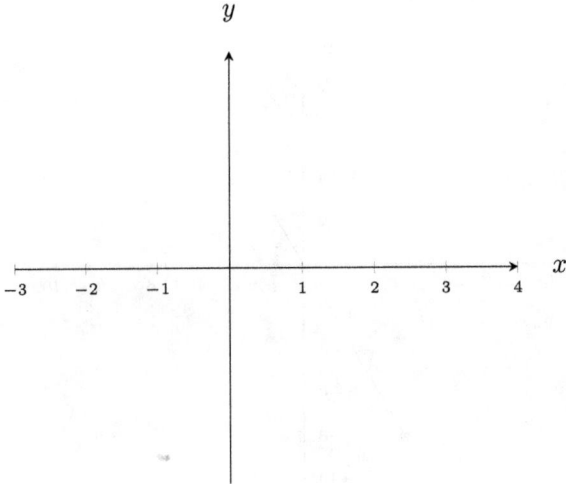

We now generate a table of x and y values and select 15 values for x distributed evenly from -3 to 4.[8]

x	y
-3	
-2.5	
-2	
-1.5	
.	
.	
.	
3.5	
4	

---

[8]To save space, we have only listed a few values of x. An actual table would have some 15 values in the x column.

We now use the equation $y = x^3 - 2x^2 - 5x + 6$ to find the values of y that correspond to the values of x that are listed in the table above. As an example, when x is -3, we have

$$
\begin{aligned}
y &= x^3 - 2x^2 - 5x + 6 \\
y &= (-3)^3 - 2(-3)^2 - 5(-3) + 6 \\
y &= -27 - 2 \times 9 + 15 + 6 \\
y &= -27 - 18 + 15 + 6 \\
y &= -24
\end{aligned}
$$

We now write the value -24 under the column headed by y next to the x value -3. This is shown below.

x	y
-3	-24
-2.5	
-2	
-1.5	
.	
.	
.	
3.5	
4	

We now repeat the calculations above with the remaining values for x, i.e., -2.5, -2, etc. This yields the following table.

x	y
-3	-24
-2.5	-9.625
-2	0
-1.5	5.625
.	
.	
.	
3.5	6.875
4	18

From the table, the maximum value of y is 18 and the minimum value is -24. We choose a scale for the y axis that runs from -30 to 20 in steps of 5. This is shown below.

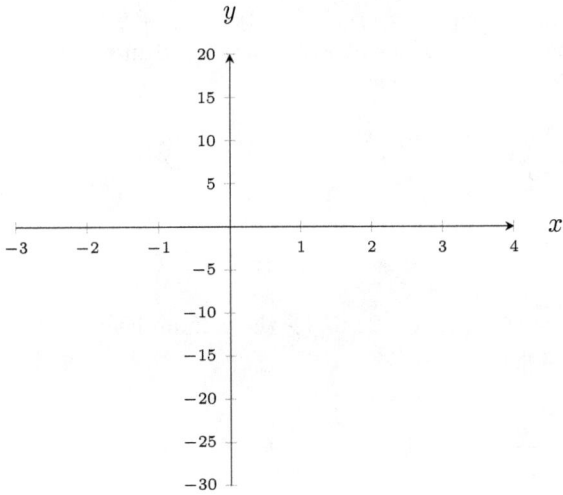

Next we plot the points that correspond to the x and y values listed in the rows of the table above. This yields the following.

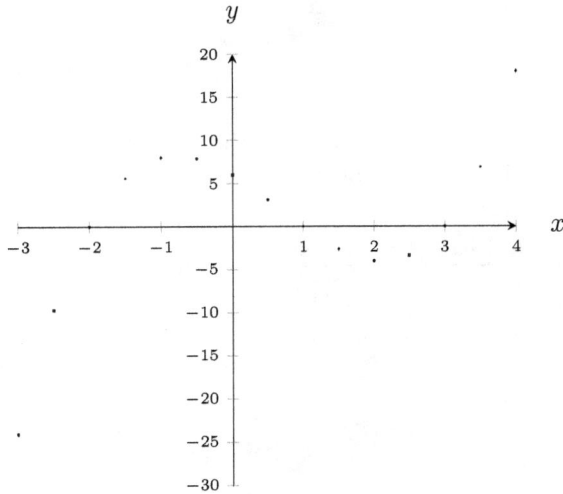

We now connect these dots in a smooth manner using a curved line.

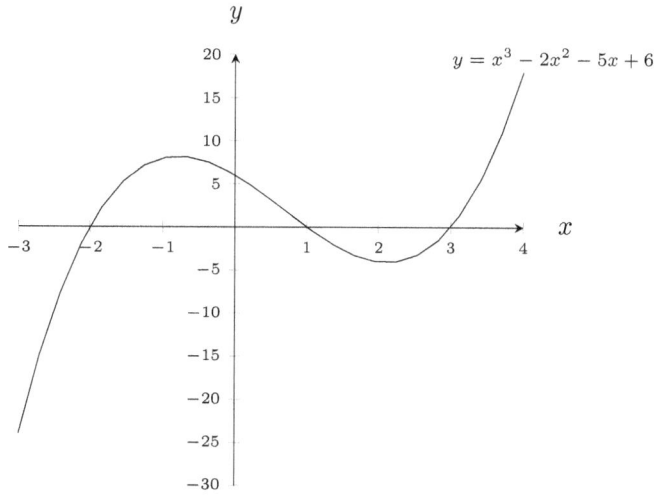

The graph shows $y = x^3 - 2x^2 - 5x + 6$

Exercise Set O.2

1. Estimate the coordinates of each point in the following diagram and write each as an ordered pair.

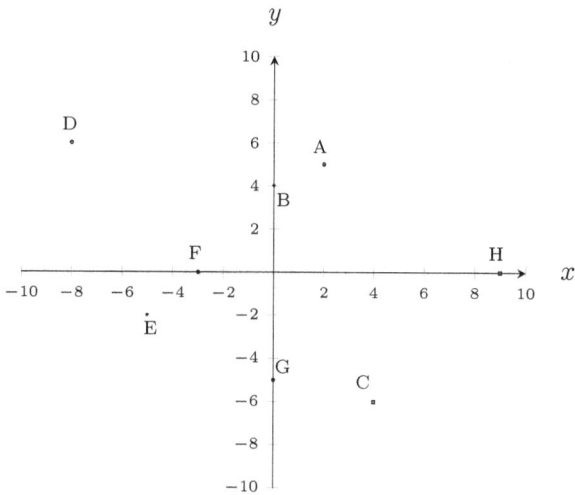

2. Estimate the coordinates of each point in the following diagram and write each as an ordered pair.

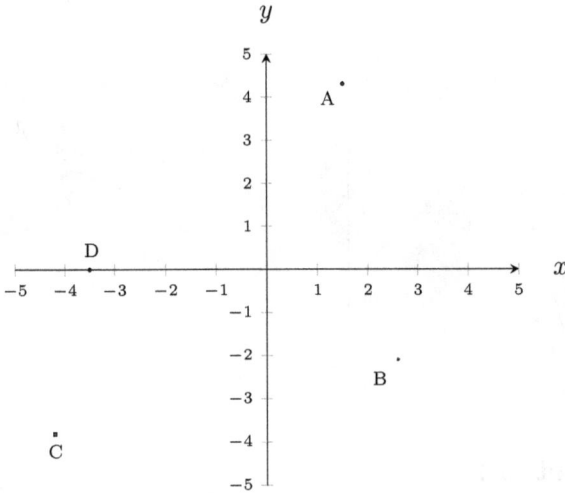

3. The relationship between x and y is such that when x is 2 y is -1.8, when x is -3 y is 2.9, when x is -4.9 y is 0, when x is 4.4 y is 3.7, and when x is 0 y is -2.7. Write these relationships as ordered pairs and graph them on the same plane. Use a Cartesian coordinate system with x values ranging from -5 to 5 with a step size of 1.

4. Graph each of the following equations by plotting points. Use a coordinate system with x values ranging from -10 to 10 with a step size of 2.

a. $y = x - 1$

b. $y = x^2$

c. $y = -x + 2$

d. $y = 4$

e. $y = \sqrt{x}$

f. $y = \log x$

g. $y = \frac{1}{x}$

h. $x = -3$

5. Graph each of the following equations by plotting points. Use a coordinate system with x values ranging from -10 to 10 with a step size of 1.

a. $y = -x, \quad -2 < x < 8$

b. $y = -\sqrt{x}, \quad x \geq 1$

c. $y = x^2 + 2, \quad x < -1$

d. $y = 4, \quad -7 \leq x \leq 5$

e. $y = -2x + 4, \quad x \in \mathbb{N}$

f. $y = -x^3, \quad x \in \mathbb{I}$

O.3   Intercepts

The intercepts are among the most basic features of a graph that one should pay close attention to as, in practical applications, there is often some significance associated with them.

The **x-intercepts** of a graph are points where the graph meets the x-axis. As an example, the graph of the equation $y = 2x$, shown below, has one x-intercept: $(0, 0)$.

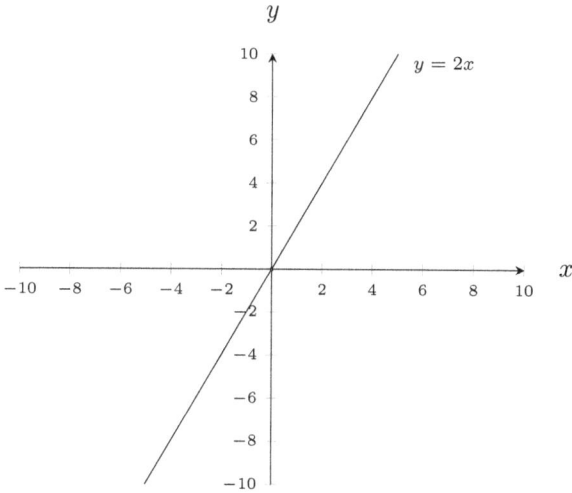

Figure O.3.12: The x-intercept of the graph of the equation $y = 2x$

The equation, $y = x^3 - 2x^2 - 5x + 6$, on the other hand, has three x-intercepts at $(-2, 0)$, $(1, 0)$ and $(3, 0)$:

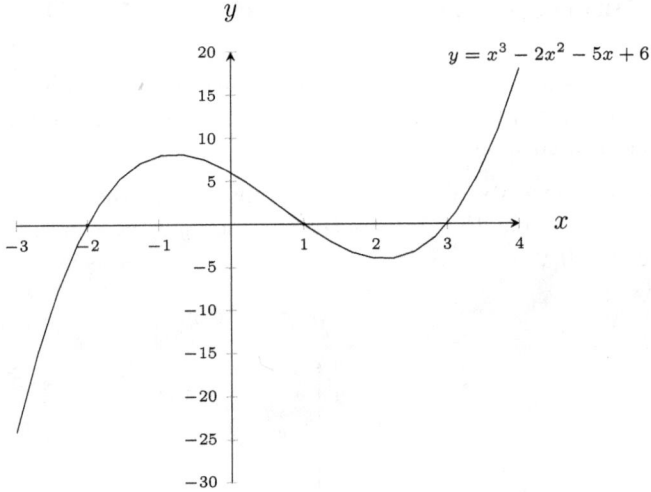

Figure O.3.13: The x-intercepts of the graph of the
equation $y = x^3 - 2x^2 - 5x + 6$

The **y-intercepts** of a graph are points where the graph meets the y-axis.
As an example, the graph of the equation $y = 2x$, shown below, has one
y-intercept: $(0, 0)$.

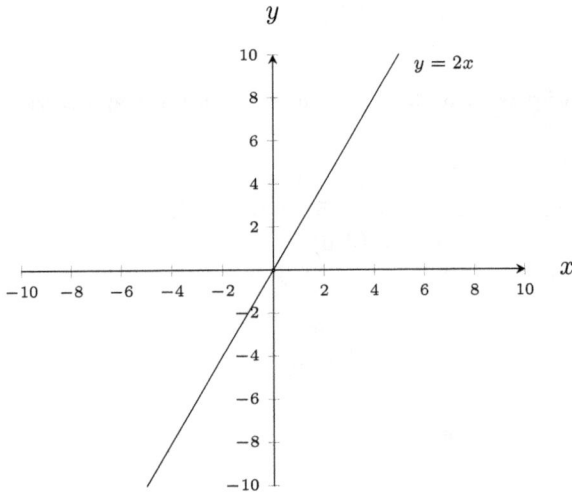

Figure O.3.14: The y-intercept of the graph of the
equation $y = 2x$

The equation, $y = x^3 - 2x^2 - 5x + 6$, also has one y-intercept at $(0, 6)$:

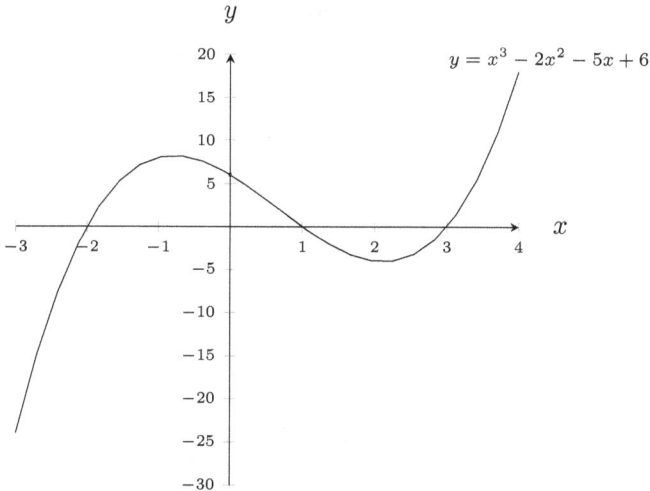

Figure O.3.15: The y-intercept of the graph of the equation $y = x^3 - 2x^2 - 5x + 6$

Since an x-intercept is on the x-axis, its y-coordinate is 0. This property of the x-intercept allows us to find the x-intercept of a model by substituting 0 for y in the associated equation and solving the resulting equation for x. Here is an example.

Example

Find the x-intercepts of

$$y = -3x + 2$$

Solution

To find the x-intercepts of the equation $y = -3x + 2$, we substitute 0 for y and solve the resulting equation for x. This yields

$$y = -3x + 2$$
$$0 = -3x + 2$$
$$-3x + 2 = 0$$
$$-3x = 0 - 2$$
$$-3x = -2$$
$$x = \frac{-2}{-3}$$
$$x = \frac{2}{3}$$

The x-intercept of the line $y = -3x + 2$ is the point $\left(\frac{2}{3}, 0\right)$.[9]

Similarly, since a y-intercept is on the y-axis, its x-coordinate is 0. This property of the y-intercept allows us to find the y-intercepts of a model by substituting 0 for x in the associated equation and solving the resulting equation for y. Here is an example.

Example

Find the y-intercepts of

$$y = -3x + 2$$

Solution

To find the y-intercepts of the equation $y = -3x + 2$, we substitute 0 for x and solve the resulting equation for y. This yields

$$y = -3x + 2$$
$$y = -3 \times 0 + 2$$
$$y = 0 + 2$$
$$y = 2$$

The y-intercept of the line $y = -3x + 2$ is the point $(0, 2)$.[10]

As we mentioned earlier, in practical applications one or both intercepts are often of significance. As an example, if the horizontal axis represents the

---

[9]Note that the x-intercept in the problem above is the point $\left(\frac{2}{3}, 0\right)$, not the number $\frac{2}{3}$.
[10]Note that the y-intercept in the problem above is the point $(0, 2)$, not the number 2.

value of temperature in degrees Celsius, t, and the vertical axis represents the value of temperature in kelvins, T, then the t-intercept would represent the value of coldest possible temperature and the T-intercept would represnt the value of the temperature at which water freezes under standard conditions.[11]

Exercise Set O.3

1. In each case find the x- and y-intercepts from the graph.

   a.

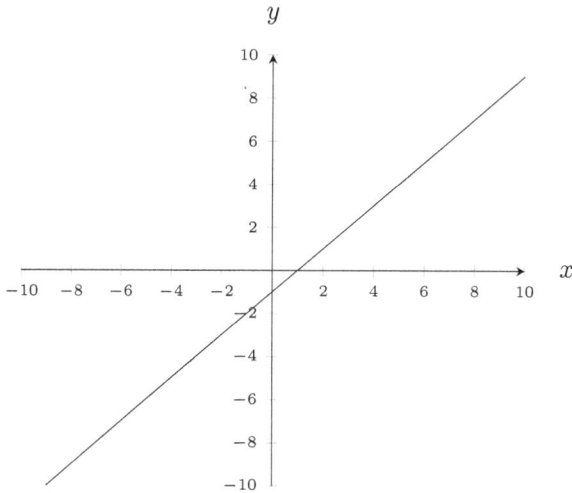

---

[11] For more on practical significance of the intercepts please see the companion textbook *Semantics and the Syntax of Algebra* by the author.

b.

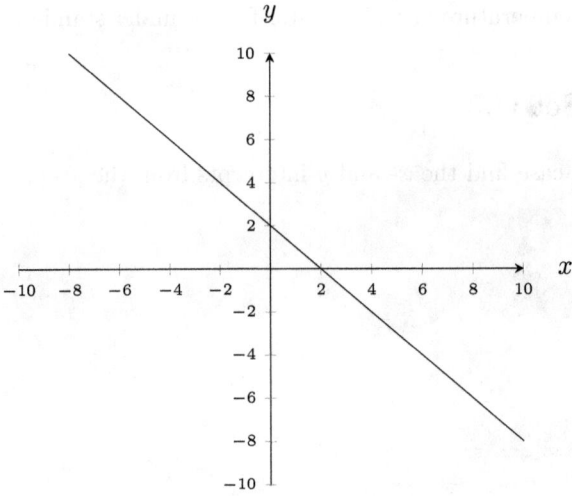

c.

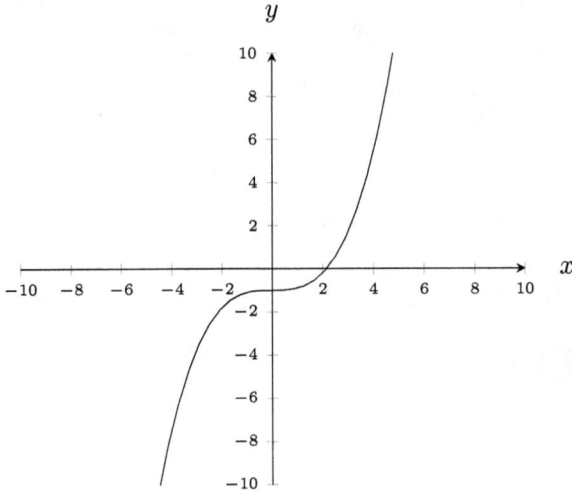

d.

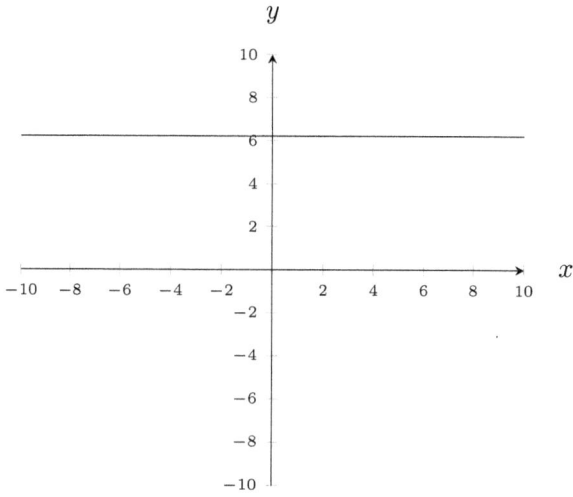

e.

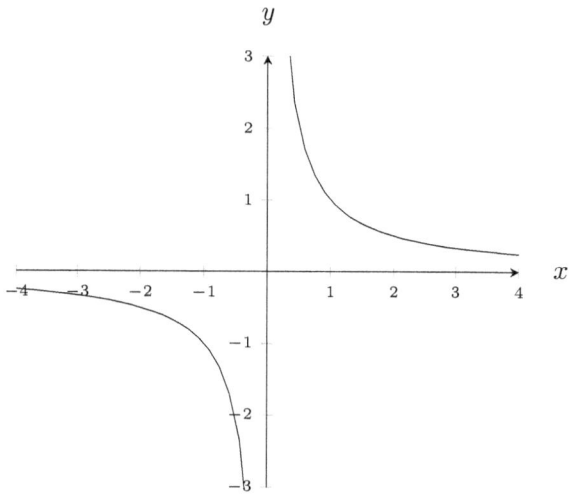

f.

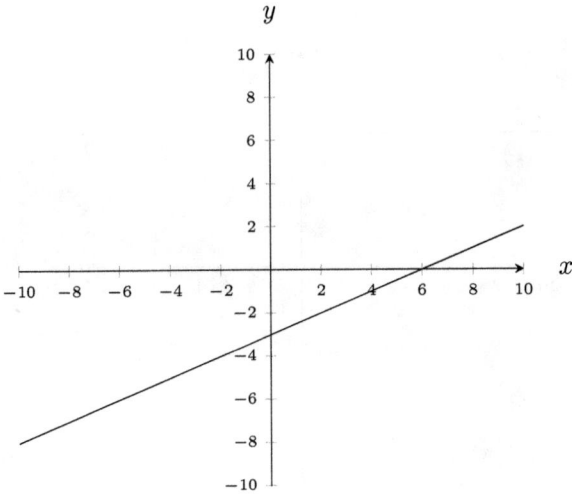

g.

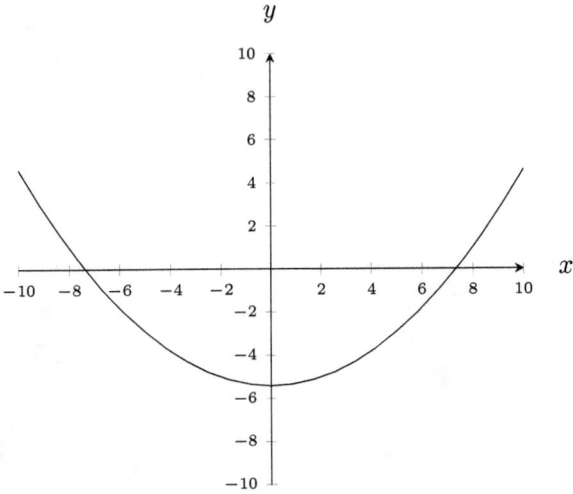

2. Find the x- and y-intercepts of the graphs of the following equations.

a. $y = 4x + 3$

b. $y = -x + 3$

c. $y - x = 4$

d. $y = -2$

e. $y = x^2 - 1$

f. $y = x^3$

g. $y = -\frac{3}{x}$

h. $x^2 + y^2 = 4$

O.4   Slope

Another concept of importance in graphs is that of the slope of a straight line. The **slope** of a nonvertical straight line is a measure of how y changes when x changes with the changes compared through division. As such, it represents a rate, referred to it as *the rate of change of y with respect to x*.

The importance of slope is in its significance in practical applications. As an example, if the horizontal axis represents the value of travel time in h and the vertical axis represents distance covered in km, then the graph of distance covered vs. travel time for motion along a straight line at constant speed would be a line with a slope that represents the speed of the car in km/h.[12]

We will use an example to explain how the slope of a nonvertical, straight line can be computed. Consider the line

$$y = 2x + 7$$

The graph of this line is shown in Figure O.4.16 below.

Consider now a case where x changes from 1 to 4, a change of 3 units. This changes in x forces a change in y from 9 to 15, a change of 6 units.[13] A change of 6 units in y for a change of 3 units in x (a ratio of 6 to 3) is the same as a change of 2 units in y for a change of 1 unit in x (a ratio of 2 to 1). So, when x changes by 1 unit, y changes by 2 units. The slope of the line is 2.

The slope of a line can, therefore, be calculated using the following equation:

$$m = \frac{y_2 - y_1}{x_2 - x_1}$$

---

[12]For more on practical significance of the slope please see the companion textbook *Semantics and the Syntax of Algebra* by the author.

[13]From the equation, $y = 2x + 7$, when x is 1, y is 9 and when x is 4, y is 15.

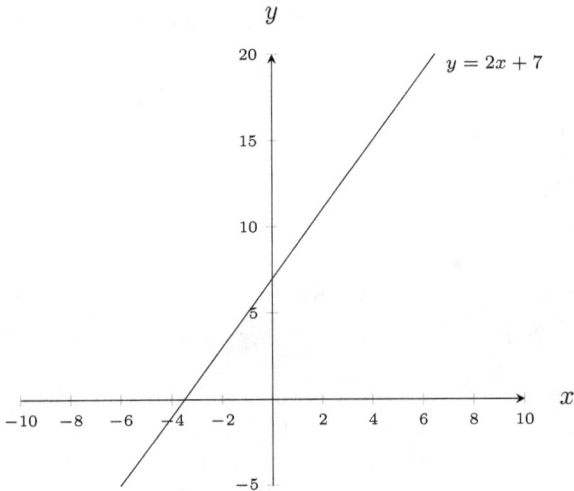

Figure O.4.16: Graph of the equation $y = 2x + 7$

In this equation m represents the slope of the line, x_1 and y_1 represent the coordinates of one point on the line, such as the point $(1, 9)$ in the example above,[14] and x_2 and y_2 represent the coordinates of another point on the line, such as the point $(4, 15)$ in the example above.[15]

In this formula, the divisor, $x_2 - x_1$, represents the change in x ($4 - 1$ in the example above) and the dividend, $y_2 - y_1$, represents the corresponding change in y. The division generates the rate of change of y with respect to x.

We can also use the delta notation to write the equation of slope. Since $\Delta y = y_2 - y_1$ and $\Delta x = x_2 - x_1$, we can rewrite the equation of slope above as

$$m = \frac{\Delta y}{\Delta x}$$

Figure O.4.17 provides a visual display of changes in x and y; values that figure in the calculation of slope.

It is interesting to note that the choice of the points, their labelling and relative positions have no effect on the value of the slope. We can choose any two (distinct) points on a line, call whichever point 1 and the other point 2 with either point to the right or left of the other and with either point above

---

[14]This means that $x_1 = 1$ and $y_1 = 9$.
[15]This means that $x_2 = 4$ and $y_2 = 15$.

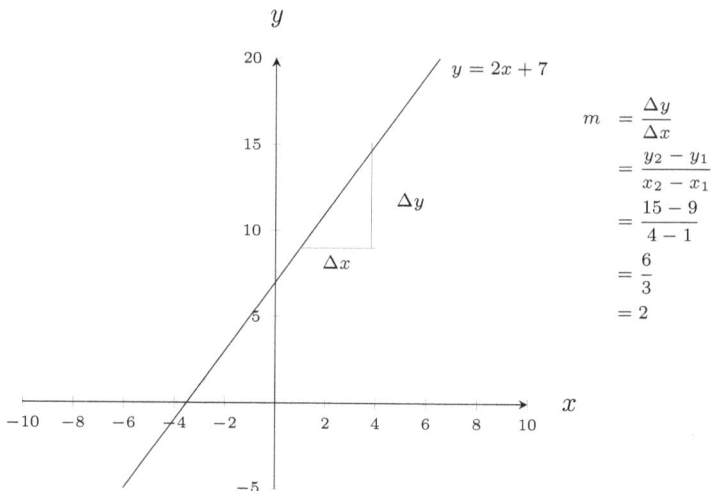

Figure O.4.17: Illustration of the concept of the slope of a straight line

or below the other and we would still arive at the same value for the slope of the line.

Note that the slope of a line may be negative, 0, or positive. In fact

- The slope of a horizontal line is 0.
- The slope of a line that rises from left to right is positive.
- The slope of a line that falls from left to right is negative.
- The slope of a vertical line is undefined.

Further to the above, provided units of equal size are used on both axes

- A line with slope between 0 and 1 makes an angle between $0°$ and $45°$ with the horizontal.
- A line with slope of 1 makes an angle of $45°$ with the horizontal.
- A line with a slope greater than 1 makes an angles with the horizontal that is greater than $45°$.
- A line with slope between 0 and -1 makes an angle between $0°$ and $-45°$ with the horizontal.
- A line with slope of -1 makes an angle of $-45°$ with the horizontal.
- A line with a slope less than -1 makes an angles with the horizontal that is less than $-45°$.

The slope increases quite rapidly in size as the angle between the line and the horizontal increases in size.

As noted above, the slope of a vertical line is undefined. That this should be the case becomes apparent once we notice that the concept of slope is related to change in y that results from change in x. However, the x value on a vertical line cannot change as all points on a vertical line have the same x-coordinate. And if one tries to use the equation of slope with the coordinates of two distinct points on a vertical line, one arrives at the undefined expression $\frac{n}{0}$, $n \neq 0$.

Exercise Set O.4

1. Calculate the slope of a straight line if

   a. y increases by 3 units when x increases by 1 unit.
   b. y increases by 6 units when x increases by 2 units.
   c. y decreases by 3 units when x increases by 1 unit.
   d. y decreases by 6 units when x increases by 2 units.
   e. y increases by 3 units when x decreases by 1 unit.
   f. y increases by 6 units when x decreases by 2 units.
   g. y decreases by 3 units when x decreases by 1 unit.
   h. y decreases by 6 units when x decreases by 2 units.

2. Calculate the slope of a straight line if

   a. y increases by 4.7 units when x increases by 2.3 units.
   b. y increases by 5.8 units when x increases by 0.1 of a unit.
   c. y decreases by 7.1 units when x increases by 20.3 units.
   d. y does not change when x increases by 3 units.

3. In each case calculate the slope of the straight line through the given points.

   a. $(3, 2)$ and $(4, 7)$
   b. $(-3, 4)$ and $(2, -1)$
   c. $(5.2, 0)$ and $(0, 0)$
   d. $(-3, 5)$ and $(-3, 2)$

   e. $(8.4, 2.1)$ and $(4.7, 9.9)$
   f. $(3, 2)$ and $(-5, 2)$
   g. $\left(\frac{1}{3}, \frac{2}{5}\right)$ and $\left(\frac{3}{4}, -\frac{1}{2}\right)$
   h. $(3.6, -2.0)$ and $(-5, -1.7)$

4. A straight line has a slope of 1. What angle does it make with the horizontal? The same scale is used on both axes.

5. A straight line has a slope of 0. What angle does it make with the horizontal?

6. A straight line has a slope of -1. What angle does it make with the horizontal? The same scale is used on both axes.

7. A straight line has a slope of -2. How does the angle that it makes with the horizontal compared to $-45°$? The same scale is used on both axes.

8. A straight line has a slope of 0.4. How does the angle that it makes with the horizontal compared to $45°$? The same scale is used on both axes.

Appendix P
On the Superiority of Natural Semantics over Standard Semantics

In the body of the textbook we presented two lines of logic for solving equations. The first, called *natural semantics*, follows the line of logic that allows one to convert the main operation on one side of the equation to its inverse operation on the other side of the equation. The second, called *standard semantics*, follows the line of logic that allows one to perform the same operation on both sides of an equation.

Of the two lines of reasoning, natural semantics is both more meaningful and more efficient.

By being *meaningful* we mean that the logic used by the scheme is in line with the manner in which we naturally reason when we solve problems. As an example, consider the following problem.

Problem

The mass of a molecule of C_2H_6 is 30.068 amu. Calculate the atomic mass of C if the atomic mass of H is 1.008 amu/atom.

A possible model for this problem is

M atomic mass of C (amu/atom)
$$M \times 2 + 1.008 \times 6 = 30.068$$

The first step in the solution to this problem is to isolate the term that contains the unknown, i.e., $M \times 2$. Following natural semantics, we isolate the term by converting the addition of the term 1.008×6 on the left side of the equation to subtraction of the same term, 1.008×6, on the right side of the equation, i.e.,

$$M \times 2 = 30.068 - 1.008 \times 6$$

The move from the model to the equation above may be expressed as follows: *If the sum of the masses of C atoms and H atoms is* 30.068 amu (which is what the equation $M \times 2 + 1.008 \times 6 = 30.068$ states), *then the mass of the C atoms is equal to the total mass minus the mass of the H atoms* (which is what the equation $M \times 2 = 30.068 - 1.008 \times 6$ states). As the reader can see, the move above maps onto the manner in which we normally reason.

The solution to the model for the problem above following standard semantics would require that we isolate the term $M \times 2$ by subtracting 1.008×6 from both sides of the equation, i.e.,

$$M \times 2 + 1.008 \times 6 - 1.008 \times 6 = 30.068 - 1.008 \times 6$$

Putting the move above by standard semantics into words, we can say that *If the sum of the masses of C atoms and H atoms is* 30.068 amu (which is what the equation $M \times 2 + 1.008 \times 6 = 30.068$ states), *to find the mass of the C atoms subtract the mass of the H atoms from both sides of the equation* (which is what the equation $M \times 2 + 1.008 \times 6 - 1.008 \times 6 = 30.068 - 1.008 \times 6$ states). This logic is hardly aligned with the manner in which we normally reason.

In addition to being more meaningful, natural semantics is also more efficient. This is evident in the example above[1] but the inefficiencies in standard semantics build up as equations become more and more complex, especially when we extend the techniques studied in solving equations to rearranging formulas. As an example of this, consider the task of solving the formula $E = \frac{1}{2}mv^2 + mgh$ for v. Following natural semantics we have

$$E = \frac{1}{2}mv^2 + mgh$$

$$\frac{1}{2}mv^2 + mgh = E$$

$$\frac{1}{2}mv^2 = E - mgh$$

$$v^2 = \frac{2}{m}(E - mgh)$$

$$v = \pm\sqrt{\frac{2}{m}(E - mgh)}$$

---

[1] Compare $M \times 2 = 30.068 - 1.008 \times 3$ to $M \times 2 + 1.008 \times 6 - 1.008 \times 6 = 30.068 - 1.008 \times 6$.

Following standard semantics we have

$$E = \frac{1}{2}mv^2 + mgh$$

$$\frac{1}{2}mv^2 + mgh = E$$

$$\frac{1}{2}mv^2 + mgh - mgh = E - mgh$$

$$\frac{1}{2}mv^2 = E - mgh$$

$$\frac{2}{m}\left(\frac{1}{2}mv^2\right) = \frac{2}{m}(E - mgh)$$

$$v^2 = \frac{2}{m}(E - mgh)$$

$$\sqrt{v^2} = \sqrt{\frac{2}{m}(E - mgh)}$$

$$|v| = \sqrt{\frac{2}{m}(E - mgh)}$$

$$v = \pm\sqrt{\frac{2}{m}(E - mgh)}$$

Index

www.ingramcontent.com/pod-product-compliance
Lightning Source LLC
Chambersburg PA
CBHW060112200326
41518CB00008B/800